Biological and Environmental Hazards, Risks, and Disasters

Hazards and Disasters Series

Biological and Environmental Hazards, Risks, and Disasters

Series Editor

John F. Shroder
Emeritus Professor of Geography and Geology
Department of Geography and Geology
University of Nebraska at Omaha
Omaha, NE 68182

Volume Editor

Ramesh Sivanpillai
Senior Research Scientist
Department of Botany | WyGISC
University of Wyoming
Laramie, WY, 82071 USA

ELSEVIER

AMSTERDAM • BOSTON • HEIDELBERG • LONDON • NEW YORK • OXFORD
PARIS • SAN DIEGO • SAN FRANCISCO • SINGAPORE • SYDNEY • TOKYO

Elsevier
Radarweg 29, PO Box 211, 1000 AE Amsterdam, Netherlands
The Boulevard, Langford Lane, Kidlington, Oxford OX5 1GB, UK
225 Wyman Street, Waltham, MA 02451, USA

Notices

Knowledge and best practice in this field are constantly changing. As new research and experience broaden our understanding, changes in research methods, professional practices, or medical treatment may become necessary.

Practitioners and researchers must always rely on their own experience and knowledge in evaluating and using any information, methods, compounds, or experiments described herein. In using such information or methods they should be mindful of their own safety and the safety of others, including parties for whom they have a professional responsibility.

To the fullest extent of the law, neither the Publisher nor the authors, contributors, or editors, assume any liability for any injury and/or damage to persons or property as a matter of products liability, negligence or otherwise, or from any use or operation of any methods, products, instructions, or ideas contained in the material herein.

British Library Cataloguing in Publication Data
A catalogue record for this book is available from the British Library

Library of Congress Cataloging-in-Publication Data
A catalog record for this book is available from the Library of Congress

ISBN: 978-0-12-394847-2

For information on all Elsevier publications visit our web site at http://store.elsevier.com/

Cover Image courtesy: NASA

In memory of my mother T.V. Padmini
who inspired me through her love,
hard work and dedication
– *Ramesh Sivanpillai*

Title and Description of the Cover Image

ALGAL BLOOM IN LAKE ERIE, USA

In October 2011, Lake Erie experienced its worst algal bloom in decades. This image captured by the Moderate Resolution Imaging Spectroradiometer (MODIS) onboard the Aqua satellite on October 9 shows this bloom. The Western basin of Lake Erie has witnessed many such blooms since 1950s due to runoff from farms, and urban and industrialized areas. However, improvements in agriculture and sewage treatment in the 1970s have reduced the number of blooms. Heavy snow in the fall of 2010 and the spring 2011, followed by high rainfall led to increased runoff from crop fields, yards, and built surfaces. This increased flow carried several pollutants including phosphorus from fertilizers into streams and rivers resulting in this bloom (Image source: NASA's Earth Observatory, Toxic algae bloom in Lake Erie, October 14, 2011, http://earthobservatory.nasa.gov/IOTD/view.php?id=76127). Also Chapter 2 (in this volume), "Algal Blooms," provides additional information about algal blooms and its impact on environment and biota.

Contents

Contributors

Chris Adriaansen, Australian Plague Locust Commission, Canberra, ACT, Australia

Kathryn J. Alftine, Department of Geographical & Sustainability Sciences, University of Iowa, Iowa City, IA, USA

Jay P. Angerer, Texas A&M AgriLife Research, Blackland Research and Extension Center, Temple, TX, USA

Kirsten M.M. Beyer, Division of Epidemiology, Institute for Health and Society, Medical College of Wisconsin, Milwaukee, WI, USA

Tim Boekhout van Solinge, Utrecht University, Utrecht, Netherlands

David R. Butler, Department of Geography, Texas State University, San Marcos, TX, USA

Norman Carreck, International Bee Research Association, Laboratory of Apiculture and Social Insects, School of Life Sciences, University of Sussex, Falmer, Brighton, UK

Rachel M. Cavin, Department of Geography, Texas State University, San Marcos, TX, USA

Ram P. Chaudhary, Research Centre for Applied Science and Technology, and Central Department of Botany, Tribhuvan University, Kirtipur, Kathmandu, Nepal

Keith Cressman, Senior Locust Forecasting Officer, Food and Agriculture Organization of the United Nations, Rome, Italy

James P. Cuda, Entomology & Nematology Department, Institute of Food & Agricultural Sciences, University of Florida, Gainesville, FL, USA

Paolo D'Odorico, Department of Environmental Sciences, University of Virginia, Charlottesville, VA, USA

René A. De Hon, Department of Geography, Texas State University, San Marcos, TX, USA

Edward Deveson, Australian Plague Locust Commission, Canberra, ACT, Australia

V. Alistair Drake, School of Physical, Environmental and Mathematical Sciences, UNSW Canberra, The University of New South Wales, Canberra, ACT, Australia; Institute for Applied Ecology, University of Canberra, Canberra, ACT, Australia

Brent Ewers, Department of Botany, University of Wyoming, Laramie, WY, USA

William E. Fox, Texas A&M AgriLife Research, Blackland Research and Extension Center, Temple, TX, USA

Benjamin A. Geaumont, North Dakota State University, Hettinger Research Extension Center, Hettinger, ND, USA

Sarah Harris, Department of Geography and Geology, Eastern Michigan University, MI, USA

John R. Hendrickson, United States Department of Agriculture, Agricultural Research Service, Mandan, ND, USA

Thomas Holmes, Southern Research Station, USDA Forest Service, Research Triangle, NC, USA

Richard A. Houghton, Woods Hole Research Center, Falmouth, MA, USA

Kevin Hyde, WY Center for Environmental Hydrology and Geophysics, University of Wyoming, Laramie, WY, USA

Jeffrey A. Lockwood, Department of Philosophy and Creative Writing Program, University of Wyoming, Laramie, WY, USA

George P. Malanson, Department of Geographical & Sustainability Sciences, University of Iowa, Iowa City, IA, USA

Robert M. May, Zoology Department, Oxford University, Oxford, UK

Suzanne McGowan, School of Geography, University Park, University of Nottingham, Nottingham, UK; School of Geography, Malaysia Campus, University of Nottingham, Semenyih, Selangor Darul Ehsan, Malaysia

John Oswald, Department of Geography and Geology, Eastern Michigan University, MI, USA

Scott Peckham, Department of Botany, University of Wyoming, Laramie, WY, USA

Sujith Ravi, Department of Earth and Environmental Sciences, Temple University, Philadelphia, PA, USA

Sagar Kumar Rimal, Ministry of Forests and Soil Conservation, Government of Nepal, Singh Durbar, Kathmandu, Nepal

Scott P. Schell, Department of Ecosystem Science and Management, University of Wyoming, Laramie, WY, USA

Kevin K. Sedivec, North Dakota State University, School of Natural Resource Sciences, Fargo, ND, USA

Ramesh Sivanpillai, Senior Research Scientist, Department of Botany | WyGISC, University of Wyoming, Laramie, WY, USA

Jake L. Snaddon, Centre for Biological Sciences, University of Southampton, Southampton, UK

Edgar C. Turner, Insect Ecology Group, Department of Zoology, University of Cambridge, Cambridge, UK

Yadav Uprety, Research Centre for Applied Science and Technology, and Central Department of Botany, Tribhuvan University, Kirtipur, Kathmandu, Nepal

Abbey F. Wick, North Dakota State University, School of Natural Resource Sciences, Fargo, ND, USA

June E. Wolfe, Texas A&M AgriLife Research, Blackland Research and Extension Center, Temple, TX, USA

James D. Woodman, Australian Plague Locust Commission, Canberra, ACT, Australia

Editorial Foreword

GENERAL HAZARDS, RISKS, AND DISASTERS

Hazards are processes that produce danger to human life and infrastructure. Risks are the potential or possibilities that something bad will happen because of the hazards. Disasters are that quite unpleasant result of the hazard occurrence that caused destruction of lives and infrastructure. Hazards, risks, and disasters have been coming under increasing strong scientific scrutiny in recent decades as a result of a combination of numerous unfortunate factors, many of which are quite out of control as a result of human actions. At the top of the list of exacerbating factors to any hazard, of course, is the tragic exponential population growth that is clearly not possible to maintain indefinitely on a finite Earth. As our planet is covered ever more with humans, any natural or human-caused (unnatural?) hazardous process is increasingly likely to adversely impact life and construction systems. The volumes on hazards, risks, and disasters that we present here are thus an attempt to increase understandings about how to best deal with these problems, even while we all recognize the inherent difficulties of even slowing down the rates of such processes as other compounding situations spiral on out of control, such as exploding population growth and rampant environmental degradation.

Some natural hazardous processes such as volcanoes and earthquakes that emanate from deep within the Earth's interior are in no way affected by human actions, but a number of others are closely related to factors affected or controlled by humanity, even if however unwitting. Chief among these, of course, are climate-controlling factors, and no small measure of these can be exacerbated by the now obvious ongoing climate change at hand (Hay, 2013). Pervasive range and forest fires caused by human-enhanced or induced droughts and fuel loadings, megaflooding into sprawling urban complexes on floodplains and coastal cities, biological threats from locust plagues, and other ecological disasters gone awry; all of these and many others are but a small part of the potentials for catastrophic risk that loom at many different scales, from the local to planet girdling.

In fact, the denial of possible planet-wide catastrophic risk (Rees, 2013) as exaggerated jeremiads in media landscapes saturated with sensational science stories and end-of-the-world Hollywood productions is perhaps quite understandable, even if simplistically shortsighted. The "end-of-days" tropes promoted by the shaggy-minded prophets of doom have been with us for centuries, mainly because of Biblical verse written in the early Iron Age during remarkably pacific times of only limited environmental change. Nowadays however, the Armageddon enthusiasts appear to want the worst to validate their death desires and prove their holy books. Unfortunately we are all entering times when just a few individuals could actually trigger societal breakdown by error or terror, if Mother Nature does not do it for us first. Thus we enter contemporaneous times of considerable peril that present needs for close attention.

These volumes we address here about hazards, risks, and disasters are not exhaustive dissertations about all the dangerous possibilities faced by the ever-burgeoning human populations, but they do address the more common natural perils that people face, even while we leave aside (for now) the thinking about higher-level existential threats from such things as bio- or cybertechnologies, artificial intelligence gone awry, ecological collapse, or runaway climate catastrophes.

In contemplating existential risk (Rossbacher, 2013), we have lately come to realize that the new existentialist philosophy is no longer the old sense of disorientation or confusion at the apparently meaninglessness or hopelessly absurd worlds of the past, but instead an increasing realization that serious changes by humans appear to be afoot that even threaten all life on the planet (Kolbert, 2014; Newitz, 2013). In the geological times of the Late Cretaceous, an asteroid collision with Earth wiped out the dinosaurs and much other life; at the present time by contrast, humanity itself appears to be the asteroid.

Misanthropic viewpoints aside, however, an increased understanding of all levels and types of the more common natural hazards would seem a useful endeavor to enhance knowledge accessibility, even while we attempt to figure out how to extract ourselves and other life from the perils produced by the strong climate change so obviously underway. Our intent in these volumes is to show the latest good thinking about the more common endogenetic and exogenetic processes and their roles as threats to everyday human existence. In this fashion, the chapter authors and volume editors have undertaken to show you overviews and more focused assessments of many of the chief obvious threats at hand that have been repeatedly shown on screen and print media in recent years. As this century develops, we may come to wish that these examples of hazards, risks, and disasters are not somehow eclipsed by truly existential threats of a more pervasive nature. The future always hangs in the balance of opposing forces; the ever-lurking, but mindless threats from an implacable nature, or heedless bureaucracies countered only sometimes in small ways by the clumsy and often feeble attempts by individual humans to improve our little lots in life. Only through improved education and understanding will any of us have a chance against such strong odds; perhaps these volumes will add some small measure of assistance in this regard.

FIGURE 1 The standard biohazard symbol is meant to be evocative of danger, and was designed to be memorable but meaningless so that people could be taught what it meant.

BIOLOGICAL ASPECTS OF HAZARDS, RISKS, AND DISASTERS

Biological hazards, also known as biohazards, refer to biological substances that pose a threat to the health of living organisms, primarily that of humans. This can include medical waste or samples of a microorganism, viruses, or toxins (from a biological source) that can affect human health. Symbolized by a striking medallion of curving, curlicue scepters (Figure 1), the sinister nature of the biohazard is evoked by the sharp and pointed nature of the otherwise round symbol.

The chapters presented in this volume are reflective not of such vector-based biohazards, but of the greater and more widespread or more generalized threats caused by the diversity of insect plagues and swarms, blooms of poisonous algae, direct animal threats, degradation of land, deforestation, desertification, ecological impacts of climate change, and even strikes upon the Earth by comets and asteroids that would so devastate life if they were large enough. The possible disruptions of the biological communities of the planet upon which humanity depends absolutely for the continuation of its own existence are most serious situations that can exert great controls on future economies. Knowing more about the nature of such generalized biohazards is an obvious need in the community of experts concerned about hazards, risks, and disasters.

Many volumes are written about the various point-source vectors of disease, contagion, and pandemics because of the insidious nature of that group of medical hazards. Less concern is generally exhibited with the diverse biologic hazards discussed in this volume, probably because of the more diffuse nature of many of those hazards discussed, and their seemingly lower impact to life, limb, or infrastructure. Nevertheless, many of these varieties of biological hazard can also do considerable damage, even to the loss of life, so greater attention needs to be paid to expositions of their many varieties.

This volume, by no means exhaustive of all the possibilities of such biohazard, still addresses numerous such problems and should be read as an introduction to a very problematic and quite diverse area of hazard occurrence.

John (Jack) Shroder
Editor-in-Chief
July 9, 2015

REFERENCES

Hay, W.W., 2013. Experimenting on a Small Planet: A Scholarly Entertainment. Springer-Verlag, Berlin, 983 p.

Kolbert, E., 2014. The Sixth Extinction: An Unnatural History. Henry Holt & Company, NY, 319 p.

Newitz, A., 2013. Scatter, Adapt, and Remember. Doubleday, NY, 305 p.

Rees, M., 2013. Denial of catastrophic risks. Science 339 (6124), 1123.

Rossbacher, L.A., October 2013. Contemplating existential risk. Earth, Geologic Column 58 (10), 64.

Acknowledgments

This book project materialized from the invaluable contributions from numerous individuals. First, I express my thanks to ***Dr David Butler*** and ***Dr George Malanson*** for the invitation to submit a chapter to this volume. Shortly after that they recommended me to serve as the editor. I thank ***Dr John Shroder*** for accepting their recommendation and entrusting this task to me. He provided incredible support while I learned the ropes as an editor. His words of wisdom helped me to move forward and bring this project to fruition. I am indebted to the authors for contributing chapters and units to this volume.

I thank ***Ms Louisa Hutchins***, associate acquisitions editor (Elsevier, UK), for the valuable support she provided since I took over the editorial responsibilities. She made herself available to answer all my questions, however, trivial they might be, contacted the authors at crucial steps, and ensured that every aspect of this project progressed smoothly. I was amazed how she could do all this despite her busy work and travel schedule. This project would not have materialized without her contribution. ***Mr Unni Kannan***, Technical Assessor (Elsevier, India) did an excellent job of scrutinizing each manuscript prior to typesetting. ***Mr Poulouse Joseph***, Production Manager (Elsevier, India) and his team did an outstanding job of taking the text, figures, and photos, and creating the impressive layout for this book. ***Ms Tharangini Sakthivel*** (Elsevier, India) worked with the authors and rest of us to keep the necessary paperwork in order. I also extend my thanks to others at Elsevier who worked on this book.

I owe a wealth of gratitude to the reviewers (table at the end of this section) who spent considerable amount of their time to review the manuscripts. All manuscripts immensely benefited through their suggestions and comments and I thank them for their valued contributions.

Identifying authors is never a trivial task and like every editor, I contacted numerous experts to contribute a chapter to this volume. While several declined my invitation, the following people took the time to provide words of encouragement and suggest names of potential authors or, at times, served as reviewers: ***Dr T. Mitchell Aide*** (University of Puerto Rico), ***Dr Dana Blumenthal*** (USDA-ARS), ***Dr Tim Collier*** (University of Wyoming), ***Dr Chris Kettle*** (ETH Zürich, Switzerland), ***Dr Anthony Fauci*** (NIH, USA), ***Dr Esther Gilman-Kehrer*** (University of Wyoming), ***Dr Ann Marie Hart*** (University of Wyoming), ***Dr Anthony Ives*** (University of Wisconsin–Madison), ***Dr William Lauenroth*** (University of Wyoming), ***Dr Jeff Pettis*** (USDA-ARS, Beltsville, MD),

Dr Ben Phalan (King's College, UK), ***Dr Lian Pin Koh*** (The University of Adelaide), ***Dr Daju Pradnja Resosudarmo*** (Center for International Forestry Research—CIFOR, Indonesia), ***Dr Tom Rudel*** (Rutgers University), ***Dr Osvaldo E. Sala*** (Arizona State University), ***Dr Scott Shaw*** (University of Wyoming), ***Dr Peter Stahl*** (University of Wyoming), and ***Dr Mark Winston*** (Simon Fraser University). I am grateful for the kind words of encouragement and assistance to identify authors and reviewers.

Mr Philip Polzer and ***Dr Kenneth L. Driese***, my colleagues at the university, deserve special mention for editing some of my text that is included in this volume. Editing someone's text is not an easy task but they did an outstanding job to add clarity. I thank them for their help.

Last but not least, I thank my family members for their patience and understanding.

This volume is by no means comprehensive or free from mistakes or omissions. If there are errors or could be further improved please send a note to me at sivanpillai.ramesh@gmail.com.

Ramesh Sivanpillai
Laramie, WY

List of Reviewers

Abinash Bhattachan, PhD, Department of Environmental Sciences, University of Virginia, Charlottesville, VA 22904, USA

Robert A. Cheke, PhD, Department of Agriculture, Health and Environment, Natural Resources Institute, University of Greenwich at Medway, Chatham Maritime, UK

Rajaraman Jayakrishnan, PhD, Dewberry, Raleigh, NC 27607, USA

William K. Lauenroth, PhD, Department of Botany, University of Wyoming, Laramie, WY 82071, USA

Jeffrey A. Lockwood, PhD, Department of Philosophy and Creative Writing Program, University of Wyoming, Laramie, WY, USA

Jennifer Lucey, PhD, Department of Biology (J2), University of York, York, YO10 5DD, UK

Rachana Giri Paudel, Department of Ecosystem Science and Management, University of Wyoming, Laramie, WY 82071, USA

Jordan Graesser, Geography Department, McGill University, Quebec H3A 0G4, Canada

Jeff Pettis, PhD, Research Entomologist, USDA-ARS Bee Research Laboratory, Bldg. 306 BARC-E, 10300 Baltimore AV., Beltsville, MD 20705, USA

Satish P. Nair, PhD, CHP, DABMP, Medical Health Physicist, F.X. Massé Associates, Inc., Health and Medical Physics Consultants, Gloucester, MA 01930, USA

Matthew Sanderson, PhD, Research Leader, USDA – Agriculture Research Service, Northern Great Plains Research Laboratory, Mandan, ND 58554, USA

Daniel Bryan Tinker, PhD, Associate Professor, Department of Botany, University of Wyoming, Laramie, WY 80271, USA

Xinyuan (Ben) Wu, PhD, Professor, Department of Ecosystem Science & Management, Texas A&M University, College Station, TX 77843, USA

Teal Wyckoff, Research Scientist, Wyoming GIS Center, University of Wyoming, Laramie, WY 82071, USA

Chapter 1

Introduction to Biological and Environmental Hazards, Risks, and Disasters

Ramesh Sivanpillai
Senior Research Scientist, Department of Botany | WyGISC, University of Wyoming, Laramie, WY, USA

The biotic components of Earth are connected by hierarchical, complex, and interconnected networks through which material and energy flow. Live cells are part of an organism, organisms are part of a population, populations are part of a community, communities are part of an ecosystem, ecosystems are part of a landscape, landscapes are part of a biome, and biomes are part of the entire biosphere. Ecologists study the components and processes at scales ranging from the physiology of small organisms to the carbon flow in the entire biosphere (Allen and Hoekstra, 1992). The structures and processes that are part of Earth's biosphere have evolved over several millions of years. When organisms are removed from their habitat or ecosystem, or introduced to a different ecosystem, alterations in the structure and processes occur, resulting in the disruption of stability of those ecosystems (Coztanza et al., 1992). Similarly, changes in abiotic components in ecosystems can alter the energy and material flows that occur within them. Any changes, minor or major, to the species composition or processes such as energy flow, pose risks and hazards to Earth's environment and its biotic components.

Accidental and intentional introduction of species to new ecosystems has resulted in adverse consequences. When modifications were made to the Welland Canal in the late 1800s and early 1900s to establish shipping connections between Lake Ontario and Lake Erie, sea lampreys (*Petromyzon marinus*) native to Atlantic Ocean entered the Great Lakes (Smith and Tibbles, 1980). This parasitic fish lacks a jaw, and sucks blood and other bodily fluids from host species (other fish) for its survival. Sea lamprey attacks do not kill their hosts in the Atlantic Ocean by virtue of millions of years of coevolution of host–parasite relationship, whereas the fish native to the Great Lakes did not have that evolutionary advantage. Sea lamprey populations exploded by the 1940s, and within the next two decades devastated native fish populations and the associated Great Lakes fishing industry (GLFC, http://www.glfc.org/sealamp/).

Biological and Environmental Hazards, Risks, and Disasters. http://dx.doi.org/10.1016/B978-0-12-394847-2.00001-2

Similarly, *Parthenium hysterophorus*, a native plant of Northeast Mexico and endemic in America has spread to Africa, Australia, Asia, and Pacific Islands in the last 100 years. Known by various names, such as whitetop weed, ragweed, congress grass, and Santa Maria feverfew, *Parthenium* is classified as one of the world's seven most devastating and hazardous weeds (Patel, 2011). This weed is one of the most troublesome and noxious weeds in India. It has caused several health problems to humans and livestock (Kohli et al., 2006). Numerous examples of such biological invasions and their impacts have been reported from almost every continent.

Certain diseases that were once considered eradicated are reappearing in some parts of the world. The U.S. Centers for Disease Control and Prevention (CDC) lists several reemerging diseases including the deadly smallpox, yellow fever, and plague (CDC, 2015), and the prevalence of drug-resistant infections is listed as a major reason. The U.S. National Institutes of Health (NIH) has published a comprehensive list of reemerging diseases (NIH, 2012). With increased and faster global travel, diseases are spreading quicker and impacting greater numbers of people in multiple continents.

It is a formidable task to capture all hazards and risks associated with the myriad processes and components in their entirety in a single volume. Topics covered in this volume represent a few of the important risks and hazards that we face today. Earlier volumes published in this book series have captured the hazards, risks, and disasters associated with water, volcanoes, landslides, earthquakes, seas and oceans, snow and ice, and wildfires. This volume addresses several hazards, risks, and disasters that could be linked to other natural phenomena or human-made activities.

Chapters included in this volume deal with several important hazards and risks. The chapter on algal blooms (*McGowan*) identifies the sources of this major problem that has increased over the past 40 years, and the risks it poses humans and the environment. An overview of recent advances in the monitoring and detection of algal blooms in addition to forecasting and treating them is included.

The next five chapters deal with risks, hazards, and disasters associated with insects or the impact of changes at their population level. Grasshoppers pose hazards to agriculture, illustrated by *Schell* using examples from western North America. Locusts, when they form swarms consisting of millions of individuals, can wipe out crops and vegetation across large geographic areas. Their impact on agriculture and vegetation in Australia, Africa, and Western Asia along with the treatment measures adopted by various national and international agencies are described under three units (*Lockwood, Adriaansen* et al., and *Cressman*) in the following chapter. This is followed by chapters on the risks and potential disasters associated with declining bee population on food production (*Carreck*), the impact of surging bark beetle populations on North American forests (*Hyde* et al.), and risks associated with the release of natural enemies to tackle invasive weeds (*Cuda*).

Human–animal interactions have been either mutually beneficial or at times hazardous to humans. *Cavin* and *Butler* provide an overview of animal hazards including zoonotic diseases and techniques used for mitigating those hazards. Species extinction and their impact on biodiversity is described in the next essay *(May)*.

Causes of environmental chronic diseases *(Beyer)* are examined next, along with responses to major disease outbreaks from different parts of the world. Insights are provided for intervening and preparing to reduce future burdens.

Land degradation and subsequent reduction in soil fertility poses a major risk to the entire human population. Following an overview (*D'Odorico* and *Ravi*), three units provide an in-depth analysis of the environmental risks associated with desertification (*Oswald and Harris*) and degradation of grassland (*Wick* et al.) and rangeland (*Jay Angerer* et al.) ecosystems.

Deforestation is a worldwide phenomenon driven by various causes in different parts. Following an overview (*Houghton*), three units highlight the causes and impact on deforestation in Southeast Asia (*Turner* and *Snaddon*), Nepal (*Chaudhary*), and Latin America (*Boekhout van Solinge*). Impacts of climate changes (*Malanson*) on species and ecosystems are described in the next chapter. Risks and threats posed by potential meteoroid and asteroid impacting the Earth are described in the final chapter (*De Hon*). Topics described in this book address several important biological and environmental risks and hazards that humanity faces today.

REFERENCES

Allen, Hoekstra, 1992. Towards a Unified Ecology. Columbia University Press, New York. NY.

CDC, 2015. Infectious Disease Information: Emerging Infectious Diseases. http://www.cdc.gov/ncidod/diseases/eid/disease_sites.htm (accessed on 08.08.15.).

Coztanza, et al. (Eds.), 1992. Ecosystem Health: New Goals for Environmental Management. Island Press, Washington, DC.

GLFC, 2015. Sea Lamprey: A Great Lakes Invader. http://www.glfc.org/sealamp/ (accessed on 08.08.2015).

Kohli, R.K., Batish, D.R., Singh, H.P., Dogra, K.S., 2006. Status, invasiveness and environmental threats of three tropical American invasive weeds (*Parthenium hysterophorus* L., *Ageratum conyzoides* L., *Lantana camara* L.) in India. Biol. Invasions 8 (7), 1501–1510.

NIH, 2012. Emerging and Re-emerging Infectious Diseases (updated in 2012). https://science.education.nih.gov/supplements/nih1/Diseases/guide/pdfs/nih_diseases.pdf (accessed on 08.08.15.).

Patel, S., 2011. Harmful and beneficial aspects of *Parthenium hysterophorus*: an update. 3 Biotech. 1 (1), 1–9.

Smith, B.R., Tibbles, J.J., 1980. Sea lamprey (*Petromyzon marinus*) in lakes Huron, Michigan, and Superior: history of invasion and control, 1936–78. Can. J. Fish. Aquatic Sci. 37 (11), 1780–1801.

Chapter 2

Algal Blooms

Suzanne McGowan
School of Geography, University Park, University of Nottingham, Nottingham, UK; School of Geography, Malaysia Campus, University of Nottingham, Semenyih, Selangor Darul Ehsan, Malaysia

ABSTRACT
Harmful algal blooms (HABs) in marine, brackish, and freshwater environments are caused by a broad range of microscopic algae and cyanobacteria. HABs are hazardous and sometimes fatal to human and animal populations, either through toxicity, or by creating ecological conditions, such as oxygen depletion, which can kill fish and other economically or ecologically important organisms. HAB hazards have increased globally over the past 40 years, because of eutrophication, translocation of exotic species via global shipping routes, climate-driven range expansions, and altered physical oceanographic conditions. Human vulnerability to HABs is greatest in communities that are nutritionally and economically reliant on fishery resources, but locally HABs also cause damage to tourist industries and have health-associated costs. Major research advances have been made in the monitoring, detection, modeling, forecast, prevention, and treatment of HABs, which have helped to mitigate health and economic risks. However, reducing HAB incidents in the future will be challenging, particularly in areas where food production and human populations (and therefore nutrient fluxes) are projected to increase. A further challenge lies in adequate communication of HAB risks and providing effective institutional structures to prepare for and respond to HAB incidents.

2.1 INTRODUCTION

Blooms are dense accumulations of microscopic algal or cyanobacterial cells within marine, brackish, and freshwater bodies, often resulting in visible discoloration of the water (Heisler et al., 2008). Most blooms are caused by planktonic algae that float in the water, but occasionally the term may describe accumulations of microscopic benthic algae or macroalgae, which grow attached to surfaces (Box 2.1). Phytoplankton blooms in coastal areas may colloquially be referred to as "red tides." Many algal species bloom as a part of their seasonal periodicity, but some algae produce toxins which are harmful to humans and other animals. The impacts of algal toxins on humans can be direct in the case

Biological and Environmental Hazards, Risks, and Disasters. http://dx.doi.org/10.1016/B978-0-12-394847-2.00002-4

BOX 2.1 Hidden Hazards of Seaweeds

Deadly concentrations of hydrogen sulphide gas emitted from thick decomposing strandlines of the seaweed *Ulva* spp. on a beach in Brittany were linked to the death of a horse, which became stuck in the algal sludge. A man accompanying the horse was left seriously ill. The incident occurred in 2009, but previously on the same beach an unexplained death of a man and the recovery of a man who lapsed into a coma, were each associated with similar bloom occurrences. The deaths of two dogs on a nearby beach were similarly associated with blankets of rotting *Ulva*. The cause of the increased blooms along the Britany coast has been linked to intensive pig farming in the area, which has increased the discharge of nitrates into the sea. Other high-profile incidents involving blooms of marine macro-algae and linked to eutrophication include the major clean-up operation to remove an *Enteromorpha* spp. bloom from the Yellow Sea in China prior to the Beijing Olympics sailing events.

of toxic exposure, resulting in death to relatively mild illness, or may arise from long-term chronic exposure, although causal links have yet to be conclusively proven (Ueno et al., 1996; Carmichael et al., 2001). Some algal blooms are also linked to the death and illness of livestock, pets, birds, and marine animals through direct toxicity or major disruption of ecological conditions. Together, blooms which cause harm to humans or other organisms are termed harmful algal blooms (HABs) (Table 2.1).

In marine ecosystems only 2% (60–80 species) of the estimated 3400–4000 phytoplankton taxa are harmful or toxic (Smayda, 1997). Most toxic taxa derive from the phylum of dinoflagellates, which are large and motile protists with flagellae (whip-like appendages that aid movement). Additionally, some species of diatoms (silica-encased and largely nonmotile algae), prymnesiophytes (flagellated "golden-brown" algae) and raphidophytes (flagellated algae) also produce potent toxins (Van Dolah, 2000). The most common route of poisoning to humans is through the ingestion of shellfish or fish, which accumulate HAB toxins, many of which are temperature-stable and so unaffected by cooking. The resulting human symptoms may be classified into around eight poisoning syndromes (Hinder et al., 2011):

1. Paralytic shellfish poisoning (PSP) is caused by saxitoxins (STXs; a group of heterocyclic guanidines), which are produced predominantly by the dinoflagellates *Alexandrium*, *Karenia*, and *Pyrodinium* spp. When ingested by humans in shellfish, they attack the peripheral nervous system leading to rapid (<1 h) onset of symptoms including tingling and numbness around the mouth and extremities, loss of motor control, drowsiness, incoherence and, at high doses, death by respiratory paralysis.
2. Neurotoxic shellfish poisoning (NSP) is caused by a suite of brevetoxins (polycyclic ethers) deriving from the dinoflagellate *Karenia* or the

TABLE 2.1 Harmful algal events collected by the HAEDAT database haedat.iode.org which is currently only operational for the ICES (International Council for the Exploration of the Sea) North Atlantic (since 1985) and from the PICES (North Pacific Marine Science Organization) North Pacific (since 2000) regions

Toxin Type	North Atlantic (1985–2013)	North Pacific (2000–2013)	% of Cases Affecting humans
Paralytic shellfish poisoning	423	406	6.5
Amnesic shellfish poisoning	175	45	0.8
Diarrhetic shellfish poisoning	654	120	1.7
Neurotoxic shellfish poisoning	49	0	53
Ciguatera fish poisoning	0	0	0
Azaspiracid shellfish poisoning	21	0	0
Aerosolized toxic effects	2	0	100
Cyanobacterial	6	0	0

raphidophyte *Chattonella* (Watkins et al., 2008). The symptoms are nausea, tingling and numbness around the mouth, loss of motor control, and severe muscular ache. As yet, no cases of human fatalities have occurred. Because *Karenia* and *Chattonella* cells are unarmored, they are susceptible to rupture (lysis), which releases toxins into waters and frequently leads to fish kills.

3. Aerosolized toxin events may arise when brevetoxins (see (2) above) released into the water become airborne through sea spray action, leading to irritation and burning of the throat and upper respiratory tract in humans.
4. Amnesic shellfish poisoning (ASP) is the only shellfish poisoning caused by a diatom. The genus *Pseudo-nitzschia* produces toxic domoic acid (a tricarboxylic amino acid) which, when ingested in shellfish, can cause gastrointestinal and neurological disturbance, disorientation, lethargy, seizures, permanent loss of short-term memory and, in rare cases, death.
5. Diarrhetic shellfish poisoning (DSP) is caused by a class of acidic polyether toxins including okadoic acid produced by dinoflagellates such as *Dinophysis fortii* or the benthic species *Prorocentrum lima.* Intoxication symptoms are mild and include gastrointestinal symptoms that usually subside within 2–3 days. Longer term effects associated with tumor growth are suspected but unconfirmed.

6. Azaspiracid shellfish poisoning (AZP) has similar symptoms to DSP (nausea, vomiting, diarrhea, and stomach cramps), leading to misdiagnosis until its polyether marine toxin azaspiracid (AZA) was identified in 1997 (Twiner et al., 2008). Recovery usually occurs within 2–3 days and no long-term symptoms have been noted. The identity of the producer organism is unconfirmed.
7. Ciguatera fish poisoning (CFP) is caused by a benthic dinoflagellate *Gambierdiscus toxicus*, which grows attached to coral reef flora, and is ingested by fish and invertebrates (Friedman et al., 2008). Humans are usually poisoned by eating piscivorous (fish-eating) fish as toxins bioaccumulate up the food chain. The symptoms are gastrointestinal upset followed by neurological problems, muscular aches, headaches, itching, tachycardia, hypertension, blurred vision, paralysis and, rarely, death.
8. Venerupin shellfish poisoning (VSP) causes gastrointestinal and nervous symptoms, delirium, and hepatic coma. The liver damage is distinctive in this syndrome and the fatality rate is quite high. The dinoflagellate species *Prorocentrum* species are thought to carry the toxin venerupin, which is usually transferred to humans via shellfish (Grzebyk et al., 1997).

New toxins, producer organisms and toxic syndromes continue to be identified. For example, yessotoxins (YTXs; produced by the dinoflagellates *Lingulodinium polyedrum, Gonygulax spinifera*, and *Protoceratium reticulatum*), previously classified with DSP toxins because they induce similar symptoms, have now been reclassified because they have a different toxicological mechanism (Tubaro et al., 2010). In contrast, reports of toxic incidents in the literature, implicating *Pfisteria* spp. in fish and human poisoning incidents (termed "estuary-associated syndrome"), are now thought to be caused by a co-occurring toxic dinoflagellate *Karlodinium veneficum*, which produces karlotoxins (Place et al., 2008; Peng et al., 2010). Other toxic algae such as the Prymnesiophytes which produce the toxin prymnesin (including the genera *Chrysochromulina* and *Prymnesium*) and the dinoflagellate *Cochlodinium polykrikoides* are more commonly associated with fish kills (Kudela et al., 2008; Manning and La Claire, 2010).

In freshwaters including lakes, ponds, reservoirs, and rivers the most common HAB organisms are cyanobacteria, forming cyano-HABs. Sometimes termed "blue-green algae," cyanobacteria are old in evolutionary terms and distinct from other eukaryotic algae because they have no cell organelles (they are prokaryotes), but they occupy a similar ecological niche. Although more common in freshwater environments, cyanobacteria also occur in marine and brackish waters. In contrast to marine HAB incidents, most freshwater cyanobacterial blooms occur on a local scale, and yet they pose a substantial threat with up to 50% of cyano-HABs being toxic (Fristachi et al., 2008). They are thus responsible for deaths and illness in many humans and animals, as well as being implicated as agents of chronic tumor promotion (Carmichael et al., 2001). Cyanobacterial toxins are widely produced by cyanobacteria from the orders of Nostocales (*Anabaena, Nodularia, Aphanizomenon, Cylindrospermopsis, Planktothrix*) and Chroococcales (*Microcystis* and *Synechococcus*), but may be separated into four major classes:

1. Hepatotoxins are cyclic peptides with differing amino acid composition that promote liver hemorrhage (Falconer, 1999). Characteristic hepatotoxins include microcystins (MCYs) (present in *Microcystis aeruginosa* and some *Anabaena, Synechococcus,* and *Planktothrix* species) and nodularin (NOD) (present in the brackish water species *Nodularia spumigena*) (Fristachi et al., 2008; O'Neil et al., 2012). Chronic exposure to lower doses of MCYs may cause progressive liver injury (Falconer, 1991).
2. Cytotoxins are usually broken down by digestion, but one cytotoxic guanidine alkaloid cylindrospermopsin (produced by *Cylindrospermopsis raciborskii* and some *Lyngbya* and *Anabaena* species) is more stable, and can withstand boiling. Exposure results in liver and kidney damage.
3. Neurotoxins are neuromuscular blocking agents that may cause progressive paralysis and death from respiratory failure. For example, the PSP-causing STX have also been found in freshwater algae such as *Anabaena circularis, C. raciborskii, Lyngbya* and *Aphanizomenon* species (Pereira et al., 2000). Other neurotoxins include anatoxin-a (a low-molecular weight alkaloid found in *Anabaena, Oscillatoria, Planktothrix,* and *Cylindrospermum* species), anatoxin-a(s) (which induces excessive salivation and is produced by *Anabaena* species), homo-anatoxin-a (produced by the benthic *Phormidium* species; Faassen et al., 2012), β-N-methylamino-L-alanine (BMAA) possibly produced by all known cyanobacterial groups (Cox et al., 2005) and palytoxin (found in *Trichodesmium* "sea sawdust" species and associated with clupeotoxism which is transmitted via fish that eat algae; Kerbrat et al., 2011).
4. Irritants include toxins such as Lyngbyatoxin-a (LTA) and debromoaplysiatoxin (DTA) that are produced by *Lyngbya* species and cause asthma-like symptoms and severe dermatitis in humans (Fristachi et al., 2008). Exposure to *Trichodesmium* species has also been anecdotally linked to dermatitis (called "pica pica" in Belize) and asthma-like symptoms in Brazil (called Tamarande fever), but the identity of the toxin remains unclear (O'Neil et al., 2012).

2.2 HISTORIC EXAMPLES OF HAB INCIDENTS

HABs and toxic incidents have been recorded for hundreds of years. Red tides were recorded in Japan during the eighth and ninth century associated with a period of economic growth (Takano, 1987 in Wyatt, 1995). Travel notes published in 1772 observed that "blood-colored" waters in the sea were more rare in Iceland than in other countries, but noted sporadic bloom occurrences along the eastern (in 1638), northwestern (in 1649), and northern (in 1712) Icelandic coasts (Olaffson and Pálsson, 1805). The deaths of many aboriginal people were described in Tierra del Fuego, Argentina in 1886 following ingestion of bivalves (Segers, 1908) and cases of suspected PSP were recorded in 1812 in Leith, Scotland (Combe, 1828). Records of freshwater cyanobacterial blooms also extend back over centuries, including twelfth century monastery records from Scotland, previously called "Monastery of the Green Loch" on account of the frequent algal blooms which still occur in its lake (Codd, 1996). Cyanobacterial blooms in Australia have been linked to a previously common outback disease "Barcoo Sprue" with similar symptoms to cyanobacterial poisoning. European explorers in 1844 described "green slime with red below" on a pond they were using as a water source, and the sickness that followed its use (Hayman, 1995). In the UK, the West Midland Meres are well known for regular algal blooms, locally termed "the breaking of the meres," described in literature in 1924, and confirmed by sedimentary analyses as being a natural feature of the lakes (McGowan et al., 1999).

2.3 HAB INCIDENTS IN RECENT DECADES

It is now understood that HAB frequency and distribution has grown in recent decades (Taylor and Trainer, 2002; Anderson, 2009; Figure 2.1). Although some of this increase may be attributed to better detection capabilities, sediment records confirm there is an increasingly serious and widespread HAB problem (Matsuoka, 1999; Barton et al., 2003; Edwards et al., 2006; McGowan et al., 2012). Currently no systematic global data collection system exists for HAB incidents, although the database HAE-DAT (http://haedat.iode.org/) (presently available for the North Atlantic and North Pacific region) is attempting to address this (Table 2.1). Estimates suggest that 60,000 marine intoxication incidents occur per year with a mortality rate of 1.5% (Van Dolah, 2000). Some 20% of all foodborne disease outbreaks in the United States result from the consumption of seafoods, with half of those resulting from naturally occurring algal toxins (Van Dolah, 2000). However, many incidents, especially mild cases go unreported. A bias in the data therefore occurs toward countries with more developed and transparent reporting procedures and toward major toxic episodes (Tables 2.2 and 2.3).

PSP is the most hazardous marine poisoning syndrome, resulting from a combination of high mortality rate (estimated at a 15% mortality rate across 2000 cases per year) (Van Dolah, 2000) and broad geographic distribution, occurring worldwide in boreal to tropical and coastal to offshore waters (Table 2.2). One of the most devastating single incidents occurred in Chile and Argentina in 1992 (300 cases, 11 dead) following an expansion in the known geographic range of *Alexandrium catanella* when some of the highest toxicity values (120,440 μg STX/100 g) were noted in mussels (Goya and Maldonado, 2014). The recurrence of blooms in 2002 (one death, 30 poisoned) led to Chile

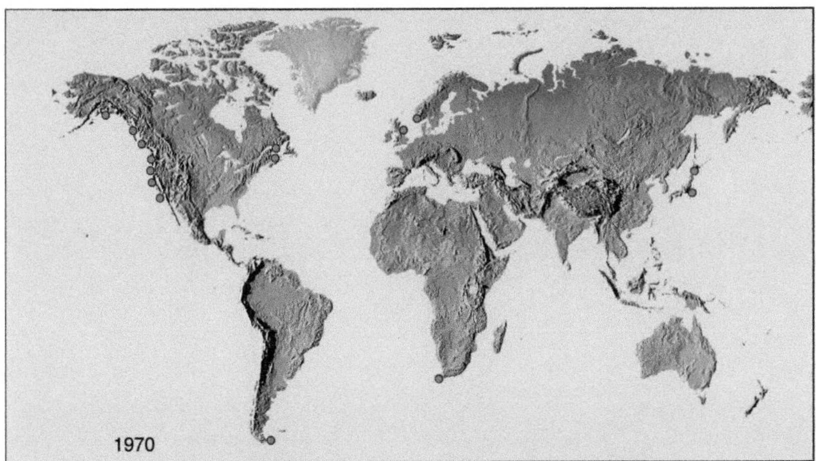

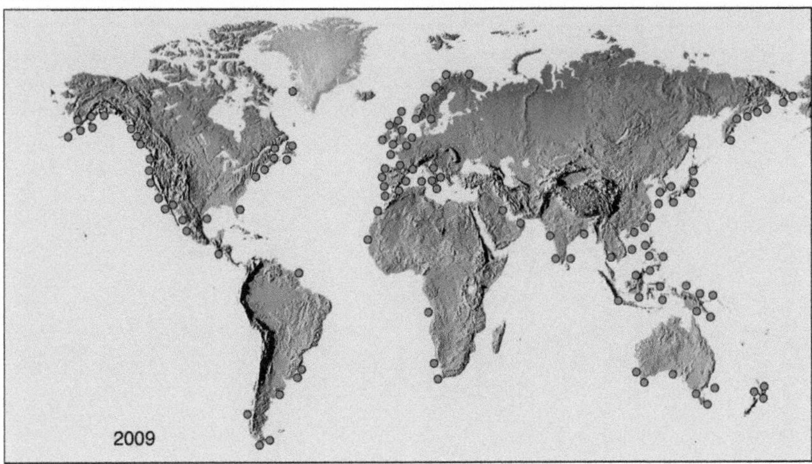

FIGURE 2.1 The documented distribution of paralytic shellfish poisoning (PSP) toxin outbreaks in 1970 and 2009, with each point representing a case when measurable levels of PSP toxins have been recorded. *Credit WHOI/US National Office for Harmful Algal Blooms.*

being declared a catastrophic area by the president because of widespread economic impacts on the shellfish industry. The Philippines has probably suffered the most recurrent and sustained losses from PSP, with an individual incident causing 21 deaths in 1983, and the cumulative death toll by 1989 being 100 people with >2000 illnesses (Hallegraeff and Maclean, 1989). In terms of mortality rate, a single VSP poisoning event in Japan claims to be the highest with 114 deaths (Grzebyk et al., 1997). The other poisoning syndromes are rarely fatal, although notable exceptions include 3 deaths in Eastern Canada in 1987 from ASP, linked to the emergence of a newly discovered poison (domoic acid), and on some occasions CFP may result in death if there is limited access to medical care.

TABLE 2.2 Summary of cases of human intoxication by marine HABs collated from Bagnis et al. (1979), European Commission (2002), Twiner et al. (2008), Watkins et al. (2008), and Hinder et al. (2011). The list is nonexhaustive and includes only widely reported incidents

Intoxication	Region	Date of Individual Incidents or Time Range	Number of Deaths and (Illnesses) or *Total Number of Incidents Per 10,000 People Per Year
Paralytic shellfish poisoning	US West Coast	1927–2011	30 (500)
	US East coast	2007, 2008–2009	ND, ND
	Mexico	1979, 1988, 1989, 1995	2, 10, 3 (99), 6 (136)
	Chile/Argentina	1991–1992, 2002	11 (300), 1 (30)
	Uruguay	2006	ND
	Canada West coast	1992	(2)
	The United Kingdom	1968	(78)
	Spain	1976, 2005–2006	(63), ND
	Portugal	2008	ND
	France	1976	(33)
	Italy	1976	(38)
	Switzerland	1976	(23)
	Germany	1976	(19)

	Morocco	1994	4 (74)
	Nicaragua	2005	1(50)
	Russia (Bering Sea)	1945, 1973	6, 2 (12)
	The Philippines	1983, 1992, 1995, 1983–1989	21 (300), 8 (141), 1 (31), 52 (843)
	Malaysia	1994	1 (13)
	India	1997	7 (500)
	New Zealand	2012	(20)
Azaspiracid shellfish poisoning	The Netherlands	1995	(8)
	Ireland	1997	(20–24)
	Italy	1998	(10)
	France	1998	(20–30)
	The United Kingdom	2000	(12–16)
Venerupin shellfish poisoning	Japan	1889, 1941	51 (81), 114 (ND)
	Norway	1979	(70)
Diarrhetic shellfish poisoning	Brazil	1990	(Several)
	Japan	1976–1982	(1300)
	France	1984–1986	(4000)
	Scandinavia	1984	(300–400)
	The United Kingdom	1997	(49)

Continued

TABLE 2.2 **Summary of cases of human intoxication by marine HABs collated from Bagnis et al. (1979), European Commission (2002), Twiner et al. (2008), Watkins et al. (2008), and Hinder et al. (2011). The list is nonexhaustive and includes only widely reported incidents—cont'd**

Intoxication	Region	Date of Individual Incidents or Time Range	Number of Deaths and (Illnesses) or *Total Number of Incidents Per 10,000 People Per Year
Amnesic shellfish poisoning (ASP)	Canada	1987	3 (153)
	Korea (ASP-like symptoms)	1992	(5)
Neurotoxic shellfish poisoning	The United States	1977, 1987, 1995–2005	ND, ND, (3009)
	New Zealand	1992–1993	(180)
Ciguatera fish poisoning	South Pacific region	1964–1977, 1973–1983	(3009),*970
	Hawaii (north Pacific)	1975–1981	*0.3
	Marshall Islands (Micronesia)	1982–1983	*2820
	French Polynesia	1979–1983	*5850
	Reunion Island (East African coast)	1986–1994	*0.78
	Queensland Australia	1965–1984	*3
	US Virgin Islands (Caribbean)	1982	*7.6
	Guadelope (Lesser Antilles)	1984	*30
	The United States (Florida)	1974–1976	*5
	Puerto Rico (Caribbean)	2005–2006	*74–170 incidents

TABLE 2.3 **Cases of human intoxication from Cyano-HABs in freshwaters collated from European Commission (2002) and Twiner et al. (2008). The list is not exhaustive**

Location and Date of Cyano-HAB	Species	Symptoms	Consequences
Drinking water			
US, 1931	*Microcystis*	Gastroenteritis	No data
US, 1976	*Schizothrix, Plectonema, Phormidium, Lyngbya*	Gastroenteritis	62% of the population supplied by the network became ill
Australia, 1979	*Microcystis*	Gastroenteritis, liver injury	149 people affected
Australia, 1981	*Microcystis*	Gastroenteritis, liver injury	No data
Brazil, 1988	*Anabaena, microcystis*	Gastroenteritis	2000 people affected, 88 deaths
Sweden, 1994	*Planktothrix agardhii*	Gastroenteritis	121 people affected
Brazil, 1996	*Aphanizomenon, Oscillatoria, Spirulina*	Hepatitis	166 people affected, 60 deaths
Recreational waters			
Canada, 1959	*Microcystis, Anabaena circinalis*	Gastroenteritis, headaches	30 people affected
UK, 1989	*Microcystis*	Gastroenteritis, vomiting	20 people affected, sore throats, 2 hospitalizations

Lethal freshwater intoxication incidents have been mostly related to contaminated drinking water supplies, including the use of water in dialysis equipment in Brazil in 1996 containing microcystins and cylindrospermopsin (Carmichael et al., 2001), which poisoned 166 and killed 60 people (Table 2.3). An earlier incident in Brazil in 1988 implicated cyanobacterial toxins in the poisoning of 2000 people and 88 deaths through drinking water from a reservoir created by the construction of the Itaparica Dam in Brazil (Teixeira et al., 1993). Although toxin testing was not conducted, *Anabaena* and *Microcystis* cells were detected in the untreated water at levels exceeding the World Health Organization (WHO) maximum acceptable level for untreated water. The "Palm Island incident" in northern Queensland, Australia in 1979 was caused by the toxic cyanobacterium *C. raciborskii* leading to the hospitalization of 149 people who had drunk water from a contaminated reservoir (Griffiths and Saker, 2003).

Other HAB syndromes are more rarely associated with human deaths, but because of their widespread nature and frequent occurrence present a significant human health risk. It is estimated that 10,000–50,000 people per year who live in or visit tropical and subtropical areas suffer from CFP, but the true incidence of CFP is unknown due to underreporting (2–10% of cases are reported) (Friedman et al., 2008). From 1960 to 1984, more than 24,000 CFP patients were reported from French Polynesia; more than six times the Pacific regional average (Friedman et al., 2008). DSP was first reported in Japan in the 1970s involving the organism *D. fortii* but since then DSP-producing organisms have been frequently detected in waters off northern Spain, Patagonia, Romania, the Philippines, and more recently in Angola and the Adriatic Sea (Ninčević-Gladan et al., 2008). Aerosolized brevetoxin incidents are frequent occurrences in the Florida (Gulf of Mexico) region where the dinoflagellate *Karenia brevis* is common. However, the emergence of a previously unknown respiratory illness in New Zealand (in 1992–1993, 1998, and 2007) from *Karenia* blooms initiated a new public health concern with 180 people affected in 1988 (Chang et al., 2001).

HABS can also cause extensive mortality to animals that live in or associated with aquatic environments (Box 2.2). Brevetoxins pose a particular

BOX 2.2 Red Tides Linked with Record Manatee Deaths

Record numbers of manatees (sea cows; *Trichechus manatus latirostris*) were killed in Southwestern Florida, during a red tide incident in March 2013. Around 276 manatees were affected with 168 cases confirmed and 108 suspected of links to brevetoxin poisoning following ELISA tests. The root cause of the dinoflagellate (*Karenia brevis*) bloom is unknown, but it is thought that the toxic algae settled from the water onto sea grasses, which were consumed by the manatees. Deaths were most prevalent in the Caloosahatchee River area where large numbers of manatees congregate.

risk to marine life because they are released through cell lysis into the water column (Brand et al., 2012). Notably, 149 endangered Florida manatees (*Trichechus manatus latirostris*) were killed in 1966 by *K. brevis* toxins, and in central New Zealand in 1988 a mass mortality of marine life including sea lions occurred during a *K. brevisulcata* bloom. Extensive fish kills can also occur through brevetoxin poisoning, as implicated during a *Chattonella* cf. *verruculosa* bloom in Delaware (Watkins et al., 2008), and in incidents from India, Japan, and Australia (Van Dolah, 2000). In some cases however, marine mortality occurs through toxin bioaccumulation up the food chain such as in the lethal domoic acid poisonings (ASP) of >100 brown pelicans and cormorants in Monterey Bay, in 1991, 100 pelicans in Baja California in 1996 and several sea lions in Monterey Bay in 1998 (Homer and Postel, 1993). Bioaccumulation of STX (PSP) was linked with the deaths of 14 humpback whales (*Megaptera novaeangliae*) in 1987 (Geraci et al., 1989). Ongoing fish kills from sustained blooms can lead to short-term declines in local populations and thus have impacts on local fisheries (Landsberg et al., 2009). Prymnesiales are regularly associated with fish poisoning episodes, including extensive kills of farmed fish along the Scandinavian coasts in 1988 linked to a 60,000 km^2 *Chrysochromulina polylepsis* bloom (Dundas et al., 1989). *Prymnesium parvum* has been linked with fish kill incidents in inland brackish ecosystems in Finland, Italy, Morocco, Greece, and Israel. Most other incidents in inland waters are associated with cyano-HABS when livestock ingest toxic water. They are common in Australia (accounting for deaths of 300 sheep, 5 cattle, and 1 horse in 1959 at Lake Bonney and for 1600 livestock deaths along the Barwon–Darling River in 1991) and South Africa (during the 1993–1996 period dozens of sheep, hundreds of cattle, and several domestic animals died associated with numerous *Nodularia* and *Microcystis* bloom incidents). In many cases however, deoxygenated conditions associated with decomposition of algal material can be hazardous to marine life, causing so-called "dead zones" in marine areas, especially semi-enclosed fjords and inland seas. Dead zones have been recorded in >400 systems covering a total area of >245,000 km^2 including the Baltic, Kattegat, Black Sea, Gulf of Mexico, and East China Sea, which are all major fishery areas (Diaz and Rosenberg, 2008). Therefore, although toxicity is an acute HAB hazard, nontoxic blooms can pose a significant environmental threat when decomposing (Box 2.1).

2.4 ECONOMIC IMPACTS OF HABs

Assessment of the global economic impacts of algal blooms is challenging because records of incidents are incomplete. Such an assessment would need to account for losses in both marine and freshwater environments and associated activities and uses within each. However, an attempt to quantify the national economic impacts of marine HABs within the United States 1987–1992 estimated that the direct costs of HABs to the US economy was 49 million USD per annum. The losses were categorized as public health impacts (45%), commercial fishery impacts including loss of revenue due to fishery closure or damage (37%), recreation and tourism impacts (13%), and monitoring and management costs (4%) (Anderson et al., 2000; Hoagland et al., 2002). This estimate did not include "economic multipliers" and was influenced by under-reporting of HAB incidents. Subsequently, the marine economy has grown and tourism has developed and so updated US estimates of HABs and hypoxia costs on the restaurant, tourism, and seafood industries were conservatively estimated as 82 million USD annually (Hoagland and Scatasta, 2006). A more recent US study focusing on health impacts estimated the cost-of-illness caused by both marine pathogens and toxins to be 900 million USD each year (Ralston et al., 2011). A national evaluation of the economic costs of eutrophication in the United States may be used to guide an estimate of cyano-HAB costs. Although not solely focused on HABs, many of the negative consequences outlined in the study were related to excessive algal growths (Dodds et al., 2008). Potential losses from eutrophication were estimated conservatively at 2.2 billion USD annually, with the greatest losses deriving from the decline in lakeside property values (0.3–2.8 billion USD per year) and recreational use (0.37–1.16 billion USD per year), with further losses arising from the recovery of threatened and endangered species (44 million USD) and drinking water (813 million USD). This study highlights the cultural (and therefore economic) importance of inland waters in the United States for recreational activities, but may not necessarily be extrapolated to other nations. For example, freshwater aquaculture in China comprises 56.4% of all world (marine and freshwater) aquaculture production, and economic valuations of Chinese inland waters would therefore likely have different priorities (FAO, 2012).

In the absence of global estimates for the economic impacts of HABs, individual case studies can provide examples of regional-scale losses. Notoriously, a *Chattonella antiqua* bloom in the Inland Seto Sea in 1972 killed over 14 million farmed fish worth 60 million USD (Imai et al., 2006). The Skagerrak area in Scandinavian coastal waters is a commercially important fishery (total catch 4–500,000 metric tons in 1988) and the fjords in this region are extensively used for aquaculture (annual production value approximately 750 million USD). The 1988 *Chrysochromulina* blooms killed 500 metric tons of caged fish with a commercial value of 5 million USD along the coast of Norway, and a further 200 sea farms (200 million USD value) were evacuated during the bloom (Dundas et al., 1989). One of the most financially devastating single incidents occurred in 1998 off the coast of Guangdong China, with the HAB spreading to Hong Kong (Yin et al., 1999). The Hong Kong press reported the deaths of hundreds of thousands of fish, especially those in mariculture cages (1000 of the 1500 mariculture cages affected), equivalent to half of the entire amount produced in Hong Kong waters the previous year, and more than all the chickens killed in Hong

Kong as a result of the Chicken Flu virus. The fish were killed by toxic dinoflagellates, including *Karenia mikimotoi* and *Karenia digitata* (Yin et al., 1999). The government estimated breeders' losses at 10.3 million USD, but fish farmers claimed the figure was at least 32 million USD.

2.5 HOW DO BLOOMS FORM?

Blooms form when there is a high concentration of algae within a particular area, caused initially by the sustained growth of algal populations, and usually accompanied by some form of physical mechanism that concentrates cells further (Glibert et al., 2005). Most bloom-forming species are predominantly photoautotrophic (they use light to fix carbon in photosynthesis) and light availability is important. However, nutrient elements such as nitrogen (N) and phosphorus (P), in addition to carbon, are also required to build cellular material, and sometimes silicon or iron can limit algal growth. Physical factors within the water influence both the availability of light and nutrients, and cell distributions within the water body. These principles govern the formation of HABS in both freshwater and marine environments.

Marine HAB species do not have inherently higher growth rates than other algae, but instead employ opportunistic or highly competitive growth strategies to achieve high densities. For example, HAB species form large, gelatinous colonies (e.g., the haptophyte *Phaeocystis*), exceptionally dense blooms (colony-forming diatoms) or allelopathic toxins (e.g., *Heterosigma*, *Chattonella*, *Prymnesium*, *Chrysochromulina*, and *K. mikimotoi*) which are inhospitable or repellant to natural-grazing organisms such as zooplankton and fish, and may ultimately lead to fish kills (Hallegraeff, 1998; Smayda and Reynolds, 2001; Granéli and Johansson, 2003). A common feature of many marine HAB species such as dinoflagellates is their ability to form cysts or resting stages that can survive dormant in sediments for long periods, and reproduce rapidly when conditions are favorable (Smayda and Reynolds, 2003). Consequently marine HAB incidents are usually highly seasonal and recur in the same areas. Physical concentrating processes are important in enhancing the intensity of blooms (Heisler et al., 2008), particularly for slow-growing taxa such as *Karenia* (Brand et al., 2012). In the marine environment, as eddies from the deep ocean cross into shallower near-shore waters, algal cells become more concentrated, and upwelling provides a nutrient source (Pitcher and Nelson, 2006). Therefore, *K. mikimotoi* blooms in the North Sea are located at tidal fronts (Holligan, 1981), and toxin production in *Pseudo-nitzschia* is much greater in nutrient-rich upwelling areas (Trainer et al., 2012). On smaller scales, vertical stratification due to salinity or temperature-induced density differences down the water column can lead to the formation of highly concentrated algal layers such as *Dinophysis* cf. *acuminata* in French coastal waters (Gentien et al., 2005), and *Dinophysis norvegica* in the Baltic Sea, which thrive in the thermocline zone between warmer surface and cooler bottom waters (Gisselson et al., 2002). Semi-enclosed estuaries and fjords with long residence times provide ideal conditions for HAB cells to thrive. Large-scale circulation systems are also important in driving HAB formation because they can influence transport of blooms over thousands of kilometers and therefore determine locations exposed to the hazard (Glibert et al., 2005).

Bloom-forming cyanobacteria form large filamentous (e.g., species from the order Nostocales) or gelatinous (e.g., species from the order Chroococcales) colonies that are non-nutritious, difficult to graze, and may produce toxic or repellant compounds (Wilson et al., 2006; Jüttner et al., 2010). Cyanobacteria compete well in eutrophic (nutrient-rich) conditions, often displacing other algal groups in terms of biomass (O'Neil et al., 2012). A distinctive feature of some cyanobacteria (e.g., the Nostocales *Anabaena, Cylindrospermopsis, Nodularia,* and *Trichodesmium*) is that they are diazotrophic, that is, they can fix N_2 for cellular growth directly from the atmosphere. Although this feature might be expected to convey an advantage when N is scarce relative to P, large-scale surveys do not bear this out (Downing et al., 2001). Instead, it has become recognized that cyano-HAB abundance often increases in response to N enrichment, and that the form of nitrogen (organic or inorganic) is key (Leavitt et al., 2006; O'Neil et al., 2012). As with marine HAB species, several cyano-HAB taxa produce resting cysts called akinetes or have dormant stages which allow them to survive in sediments. Cyanobacteria compete well in lakes with high retention times because they have slow growth rates relative to other algal groups (Paerl and Huisman, 2009). Therefore, although most cyano-HABs occur in standing waters, they may form on very slow-flowing rivers (e.g., the largest recorded cyanobacterial bloom of *Anabaena circinalis* formed along 1000 km of the Barwon–Darling River in Australia following a period of dry weather in 1991) (Bowling and Baker, 1996). Water column stability is important because cyanobacteria are able to regulate their buoyancy by creating gas vacuoles for floatation or accumulating starch granules for ballast to maintain optimal position in the water column for light harvesting (Figure 2.2; Reynolds and Walsby, 1975; Paerl, 1988). This buoyancy regulation mechanism is ineffective in windy and turbulent conditions because cells are dispersed throughout the mixed layer, but during calm conditions, thick cyanobacterial scums can accumulate on the water surface and be further concentrated if blown by the wind onto lake shores, greatly increasing the risk of toxic exposure (Codd, 1984; Reynolds et al., 1987).

2.6 VULNERABILITY

A significant proportion of the world's human population live close to coastlines (around 10% were living <10 m above sea level in the year 2000) (McGranahan et al., 2007) with the most populous areas in South, Southeast, and East Asia (Nicholls and Cazenave, 2010). Fish consumption globally provides 16% of the animal protein consumed by people, but rises to 26% of

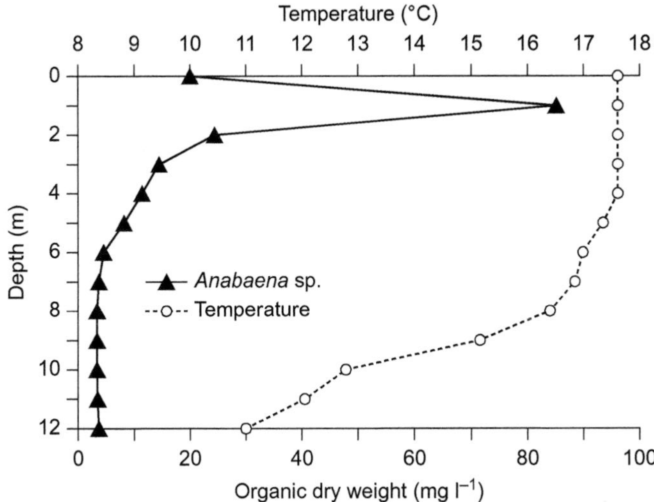

FIGURE 2.2 The vertical distribution of *Anabaena* sp. and temperature in a thermally stratified eutrophic lake during bloom conditions showing the concentration of cyaonbacterial biomass close to the water surface (Mur et al., 1999).

the diet in Asia, increasing the exposure to vector-transmitted toxins or to economic effects from fishery closure (Dewailly and Knap, 2006). In many such areas, marine resources are vital to livelihoods and national economy. For example, the Philippines has over 18,000 km of coastline with 7000 islands and over 70% of its municipalities located in coastal areas. Fisheries are the fifth most important agricultural sector economy, and account for 4% of GNP with 40–60% of the total fish catch accounted for by coastal fishing (Bajarias et al., 2006; Bureau of Fisheries and Aquatic Resources, 2010). Changes in the mode of fishing in recent decades have also changed vulnerability to HAB exposure. Global capture production of fish has remained stable at 90 million metric tons, since 2001, but aquaculture has grown strongly, increasing at 6.3% per annum to 59.9 million metric tons in 2011 (FAO, 2012). This is in part due to a growth in the popularity of shellfish farming. For example, a program in Santa Catarina State (southern Brazil), initiated in 1989 to promote mussel (*Perna perna*) production as an alternative to artisanal fishing, led to an increase in production from 190 to 5000 metric tons yr^{-1} in 6 years (Proença and Rörig, 1995). However, China is by far the biggest aquaculture producer (39% globally), with just over half of the farmed species being freshwater fish (FAO, 2012). It is predicted that "the growing harvest of nontraditional shellfish (such as moon snails, whelks, barnacles, etc.) may increase human health problems and management responsibilities" (FAO, 2004). A further change in recent years is the expansion in tourism around coastal areas, exposing new communities to hazards through ingestion of shellfish or reef fish or from inhalation of brevetoxins (Backer, 2009). Increased population has heightened demand on surface water supplies, while concurrently increasing pollution pressures (Fristachi et al., 2008). Vulnerability to freshwater toxic incidents through contaminated drinking water is greatest in countries with limited water treatment procedures (Carmichael et al., 2001). However, many livestock-poisoning incidents occur in arid regions, such as Australia and South Africa where there is limited access to water.

The expansion in HABs over the past 60 years, confirmed by sedimentary records, has exposed communities to new or more frequent hazards (Anderson, 1989; Smayda, 1990; Anderson et al., 2002; Glibert et al., 2005). Possible causes of marine HAB range expansions are the global translocation of "exotic" species within ship ballast water, transport by unusual hydrological and meteorological conditions, and the opening of new ecological niches when climate changes. Before the 1970s PSP events were only recorded in North America, Europe and Japan, but they are now common in South America, Australia, southeast Asia, and India. A well-documented range expansion is the spread of the tropical PSP-producing dinoflagellate *Pyrodinium bahamense* from Papua New Guinea to Brunei and Sabah, and then the Philippines between 1972 and 1983 in association with meteorological and hydrological changes (Hallegraeff and Maclean, 1989 in Steidinger et al., 2011). The first NSP outbreak in North Carolina during the 1987 El Niño was thought to be a range extension, but may have simply been caused when *K. brevis* was transported by the Loop Current around the tip of Florida and then north by the Gulf Stream. It is tempting, but ultimately impossible to draw conclusions about the role of climate warming in these individual incidents. However, the number of recent range expansions including the southward spread of *Gambierodiscus toxicus* along the East Coast of Australia (Hallegraeff, 2010) and the northward shift of *Ceratium* species in the North Sea (Edwards et al., 2006) add weight to this suggestion. It is clear that future warmer temperatures will influence species distributions, as is documented by paleoecological records showing a wider distribution of *Pyrodinium* during warmer conditions 50 million years ago (Hallegraeff, 2010). Evidence that this process is underway comes from the first appearance in 800,000 years of the diatom *Neodenticula seminae* in the Labrador Sea because of the reduction in Arctic Sea ice (Reid et al., 2007). Abundant evidence also shows HAB species are transported globally in ship ballast water, and although it is difficult to prove a direct role in the establishment of new HAB populations, it is thought that the establishment of *Prorocentrum minimum* in the Baltic Sea was caused by translocation in ships (Zhang and Dickman, 1999; MEECE, 2010). Other exotic species derived from shipping ballast waters, such as zebra mussels (*Dreissena polymorpha*), have antagonized *Microcystis* blooms in the North American Great Lakes' system through their avoidance of this cyanobacterium as a food source (Vanderploeg et al., 2001).

When a ship carrying a load of fertilizer sunk in a harbor in Greece, it stimulated a bloom of *Alexandrium tamarense*, and demonstrated a direct link between nutrient supply and algal growth (Glibert et al., 2005; Glibert and Burkholder, 2006). The evidence that eutrophication is a leading cause of HAB expansion is now overwhelming (Heisler et al., 2008; O'Neil et al., 2012; Figure 2.3). Classically, N is considered the limiting nutrient for marine ecosystem productivity, whereas P is limiting in freshwaters, although this paradigm is now being challenged (Schindler, 1974; Howarth and Marino, 2006). Human activities (especially agricultural intensification and fossil fuel combustion) have more than doubled the rate at which biologically available N enters the terrestrial biosphere and tripled the mean passage time of river waters, greatly increasing N export rates to coastal areas (43 Tg N per year in 2000) (Paerl et al., 2002; Galloway et al., 2008; Seitzinger et al., 2010).

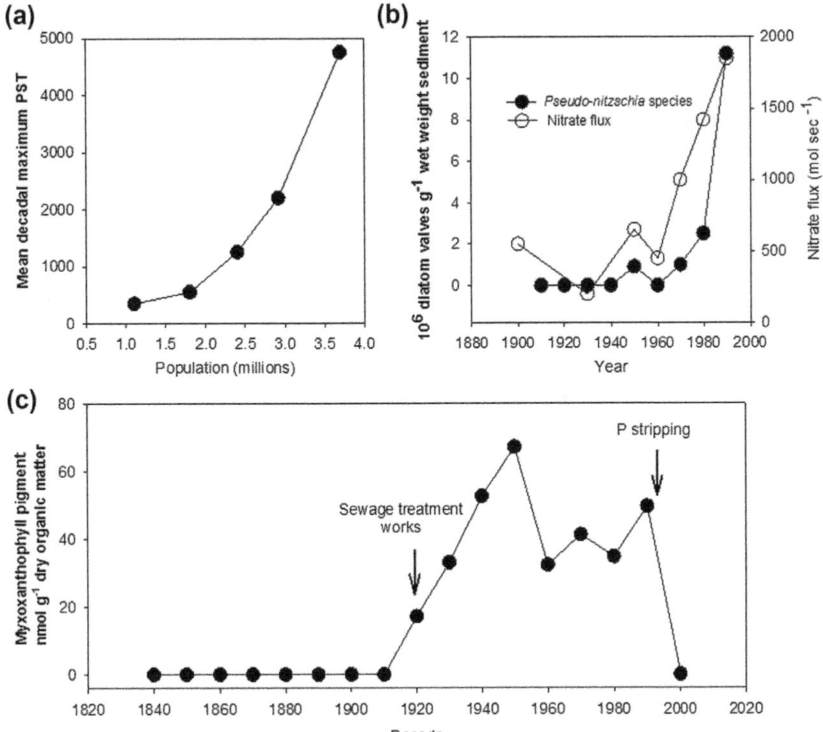

FIGURE 2.3 Examples of the relationships between human population growth, eutrophication, and Harmful algal blooms (HABs) in marine and freshwaters including (a) the correlation between paralytic shellfish toxins (PST) monitored in Puget Sound, the United States and regional population growth, redrawn from Trainer et al. (2003); (b) the abundance of *Pseudo-nitszchia* diatom valves in dated sediment cores from the Louisiana Bight, compared with the nitrate flux from Mississippi River waters, redrawn from Parsons et al. (2002); and (c) the concentrations of the cyanobacterial pigment myxoxanthophyll used to track cyano-HABs in dated sediment cores from Windermere (UK), which increased following the increased transmission of treated sewage effluent into the lake and declined following phosphorus stripping controls in the 1990s, but have returned in recent years (not recorded in this core). *Redrawn from McGowan et al. (2012).*

Fertilization of algal populations is shown by strong correlations between HAB incidents and nutrient loading in China and the US coastal waters (Paerl, 1988; Turner and Rabalais, 2003). In freshwaters, whole-lake fertilization experiments show how P fertilization may directly stimulate cyanobacterial blooms (Schindler, 1974), but there is growing evidence that both N and P influence lake cyano-HAB formation (Leavitt et al., 2006). The delivery of both nutrients through the intensification of agriculture and human population growth are correlated with the massive increases in cyano-HAB pigments in North American and European lake sediments (Hall et al., 1999; Bunting et al., 2007; McGowan et al., 2012, Taranu et al., 2015). Cyanobacterial toxins have expanded their range beyond lakes in the mid-west US, most likely linked to eutrophication (Fristachi et al., 2008). A key factor is believed to be the supply of N as urea (a commonly used nitrogen form in fertilizer), which can cause a 200–400% increase in toxin (MCY) production in freshwater cyanobacteria (Finlay et al., 2010). Similarly, urea increases domoic acid production in marine species *Pseudonitzschia australis* (Howard et al., 2007; Glibert et al., 2006). Importantly, dinoflagellates and other HAB species can utilize organic material for nutrition by ingesting dissolved or particulate material (they are mixotrophic or phagotrophic), indicating complex food webs not driven solely by photoautotrophy, which may increase HAB species resilience (Paerl, 1988; Nygaard and Tobiesen, 1993; Heisler et al., 2008).

In addition to a role in range expansions, climate change has and will influence HABs through increased warming, extreme weather events, and rising sea levels (Patz et al., 2006; Hallegraeff, 2010; IPCC, 2013). Sea-level rises of 28–98 cm predicted by 2100 (IPCC, 2013) will increase continental shelf areas, which have stable and shallow water columns suited to bloom formation. Warming temperatures may have a direct effect on HAB species by increasing growth rates, or expanding the seasonal "temperature windows" in which HAB formation is possible (Barton et al., 2003; Peperzak, 2005; Hallegraeff, 2010). Many cyanobacteria grow optimally at higher temperatures than eukaryotes, suggesting that they will be "winners" from climate warming (Paerl and Huisman, 2009). Differential effects of temperature on the timing of growth of phytoplankton and their zooplankton consumers can lead to "food mismatching," and excessive phytoplankton growth, as observed in marine and freshwater systems (Edwards and Richardson, 2004; Winder and Schindler, 2004). Warming will cause stronger thermal stratification, with projected "desertification" or productivity declines in the open ocean because of less nutrient transport from deeper layers, but potentially greater productivity where nutrients are available (e.g., high latitude and coastal areas) because of greater light exposure in the shallower upper water layer (Behrenfeld et al., 2006; Doney, 2010; MEECE, 2010). In warmer thermally stratified lakes, the water is less viscous and so buoyant cyanobacteria or motile dinoflagellates are able to stay higher in the water column (O'Neil et al., 2012). More rapid rainfall delivery during extreme events may enhance the transport of terrestrial organic material and favor mixotrophic taxa such as *Pyrodinium* (Azanza and Taylor, 2001), and hence HAB events often follow on from a period of high rainfall (Qi et al., 2004; Miller et al., 2006). Rainfall extremes may lead to excessive nutrient flux and, when followed by periods of dry weather, can provide ideal conditions for cyano-HABs (Elliott, 2010). Wind intensity can influence aeolian transport of dust to the oceans, which fertilizes with micronutrients such as iron and stimulates HABs (Lenes et al., 2001). Hurricane damage to coral reefs may, in association with coral bleaching, increase the surface area available for colonization of macroalgae and therefore increase the risk of ciguatera (Hoegh-Guldberg, 1999).

2.7 MITIGATION

Mitigation of HABs has occurred throughout history, with local knowledge being used to avoid eating ciguatoxic fish during dangerous seasons, and Queensland aboriginals digging soaks to obtain drinking water rather than using cyano-HAB contaminated standing waters (Codd, 1996). However, scientific advances in recent decades have led to the development of increasingly sophisticated mitigation options. Effective mitigation requires the accurate and swift identification and detection of both the toxic organisms and the transfer pathway to humans. A large research focus has clarified the taxonomy of many difficult HAB groups through morphometric and molecular techniques with, for example, the reorganization of some *Gymnodinium* species to the genus *Karenia* (Yang et al., 2001; Janson and Hayes, 2006; Haywood et al., 2007) and identified the chemical structure of many HAB toxins (Van Dolah, 2000). Detection of HABs using microscopy is time-consuming, and so the development of molecular probes using rDNA such as PCR (polymerase chain reaction) has improved the detection limits of the process allowing earlier detection (Penna et al., 2007; Heisler et al., 2008). Molecular detection methods may be packaged into moored and automated monitoring instruments that can be left *in situ* (Heisler et al., 2008; Anderson, 2009). Optical detection techniques for chlorophyll pigments may be used in circumstances where a single known HAB species regularly dominates the phytoplankton assemblage and deployed in sondes or automated underwater vehicles (AUVs). Physiological studies have assisted in linking producer organisms to toxins, and helped to define the conditions in which they proliferate. Rapid sensitive *in vitro* detection methods specific to each class of algal toxins have been developed over the past decade to replace ethically sensitive and imprecise mouse bioassays (Dolah and Ramsdell, 2001; Pierce and Kirkpatrick, 2001; Watkins et al., 2008). These include immunoassays such as ELISA (enzyme-linked immunosorbent assay) (Chu et al., 1989; Naar et al., 2002), enzyme inhibition assays (An and Carmichael, 1994), receptor assays (Doucette et al., 1997), and cell assays (Flanagan et al., 2001). Quantification and identification of toxins have been developed with the use of combined chromatography and mass spectrometry techniques (LC-MS) with separation methodologies developed for all major HAB toxins. Simple test kits are now available for some HAB toxins (Anderson, 2009).

The most common mitigation action to prevent HAB intoxication is the closure of areas of shellfish or fish harvesting, but this action requires useful recommendations for biotoxin limits. Recommended limits for marine and freshwater toxins in foods have been established by the Food and Agriculture Organization (FAO)/Intergovernmental Oceanographic Commission (IOC) and World Health Organization (WHO), and used to guide national regulatory bodies (FAO, 2004). The recommended limits are based on an Acute Reference Dose (ARfD) (from the human consumption of a single portion of shellfish) or when possible, a recommended daily intake. A risk assessment is carried out to derive the N(L)OAEL (No (Low) Observable Adverse Effects Level) by applying an uncertainty factor to protect more susceptible people, and may include the calculation of a safety margin (FAO, 2004; Twiner et al., 2008; Watkins et al., 2008). For example, for a 100–300 g meal of shellfish within the regulatory limit of 80 μg STX eq per 100 g mussel meat, a safety margin has been set of <1–3.8 toward mild PSP symptoms and of 1.9–7.2 of severe or fatal symptoms. Legislation such as the European Union directives in 2004 on food hygiene (853/2004/EC) is used to set regulatory limits for marine biotoxins including PSP (800 μg/kg), ASP (20 mg/kg), okadaic acid, dinophysistoxins, and pectenotoxins together (160 mg/kg), yessotoxins, (1 mg/kg), and for azaspiracids (160 μg/kg). These limits and the recommended analytical methods are periodically reviewed (Twiner et al., 2008) (853/2004/EC), updated in 2011 by the Commission Regulation no. 15/2011. Similarly many countries have implemented a regulatory framework to guide when mitigation actions need to be initiated (Watkins et al., 2008).

The ability to track the development of HABs through satellite and airborne remote sensing provides another early warning tool (Stumpf and Tomlinson, 2005). The major techniques that are employed are either optical sensing through the detection of colored phytoplankton pigments using, for example, Sea-viewing Wide Field of View (SeaWiFS), MODIS (Moderate Resolution Imaging Spectroradiometer), or the higher resolution MERIS (Medium Resolution Imaging Spectrometer) sensors or thermal sensing using, for example, Advanced Very High Resolution Radiometer (AVHRR). Optical techniques must correct for atmospheric interference, and may be complicated when there is colored dissolved organic matter and turbidity. However, successful uses have included the detection and tracking of *Trichodesmium* blooms (assisted by the highly reflective gas vacuoles in this cyanobacterium) using SeaWiFS (Westberry et al., 2005), *Alexandrium*, *Pseudo-nitzschia*, and *Heterosigma* blooms in Monterey Bay using MODIS with AVHRR to detect chlorophyll fluorescence and sea surface temperatures (SSTs) (Ryan et al., 2011) and the successful validation of MERIS algal1 and algal2 products for algal detection in Belgium (Ruddick et al., 2008). On lakes, blooms of the cyanobacterium *Microcystis* have been tracked using images from CASI-2 (Compact Airborne Spectrographic Imager) with algorithms to estimate chlorophyll and phycocyanin pigments (Hunter et al., 2008). Thermal techniques rely on detecting areas of upwelling where blooms form, and are thus often employed when the HAB organism does not dominate the algal assemblage (for example, *Alexandrium* or *Pseudo-nitzschia* species; *G. catenatum* in Portugal; *K. mikimotoi* in Ireland) (Glibert et al., 2005).

A number of engineering solutions exist that have been employed in an attempt to prevent HABs from forming. The first is a recently developed technique of sediment disturbance or resuspension, which aims to bury HAB resting cysts and prevent the recurrence of blooms (Anderson et al., 2012). The top 10–20 cm of sediment is turned over, analogous to ploughing agricultural land. Trials of this technique have suggested that it could provide a viable mitigation option, but is only useful for taxa that produce resting cysts or spores. Hydrological conditions may be manipulated in inland waters to prevent cyano-HAB development. For example, *A. circinalis* blooms only develop below threshold flow conditions in the Barwon–Darling River, and so hydrological management can prevent the repeat of major HAB episodes such as occurred in 1991 (Maier et al., 2004; Mitrovic et al., 2011). In lakes, manipulation of flushing rates and aeration to destratify the water column can also discourage the conditions, which lead to cyano-HAB formation (Visser et al., 1996; Spears et al., 2007; Hudnell et al., 2010). The application of clay-based material Phoslock® has also been shown to be effective in binding phosphorus to sediments and therefore preventing HAB formation (Meis et al., 2012). On smaller scales, barley and rice straw have been effective in discouraging phytoplankton growth through the production of allelopathic chemicals (Barrett et al., 1996; Park et al., 2006).

2.8 PREPAREDNESS

More than 50 countries (including Japan, Australia, and nations from Europe and North America) have local monitoring plans to check for the occurrence of toxic species in areas where shellfish or fish are consumed (European Commission, 2002). In Europe, EU legislation governs monitoring programs, and many other countries have monitoring embedded within a legal framework. Numerous tools are available for sophisticated monitoring of HAB events on a range of spatial and temporal scales. For example, Monterey Bay is monitored using in situ robotic "Environmental Sample Processors" containing DNA probes to track the algal species *Alexandrium*, *Pseudo-nitzschia,* and *Heterosigma*, and cELISA probes to detect algal toxins (domoic acid) (Ryan et al., 2011). Concurrently physical, chemical, and biological (chlorophyll) parameters are tracked using sondes, meteorological buoys, AUV deployments, and satellite remote sensing (MODIS and AVHRR) (Ryan et al., 2011). In Korean waters, real-time monitoring of bioluminescence properties with in situ instruments, shipboard and onshore monitoring, and a combination of satellite and airborne sensors are used to detect the location of HABs such as *C. polykrikoides* blooms (Suh et al., 2004; Kim et al., 2006). Several other countries use satellite-based imagery to assist in tracking blooms including Japan and China (the CEOHAB program) (Wang and Wu, 2009). Resolution may prove an impediment to adequate tracking when *in situ* samplers have inadequate spatial coverage, and in coastal areas where cloud cover can limit the temporal resolution of some remote-sensing techniques (Frolov et al., 2013). However, some networks are very extensive, such as the 50 biotoxin monitoring stations provided by the California Department of Public Health and can capture 60% of the regional bloom (Frolov et al., 2013).

High-quality monitoring data facilitate the ability to model bloom processes. For example, models have successfully described the relationship between nutrient loading from terrestrial areas and *P. minimum* HABs on watershed and global scales (Glibert et al., 2010). The fundamental understanding of dinoflagellate life cycles has allowed simple descriptive models to be developed, although information about excystment and encystment of other HAB groups is scarce (Steidinger and Garccés, 2006). Conceptual models have been employed to describe the ecosystem dynamics in the Baltic Sea (Lundberg, 2005) and in Florida Bay (SEACOM) (Madden and McDonald, 2009). Process-based models such as PROTECH have been used to simulate individual freshwater species including HAB organisms (Elliott, 2010) and the Harmful Algal Bloom Expert System (HABES) based on fuzzy logic models for five marine HAB species (*N. spumigena, Phaeocystis globosa, Dinophysis acuminata* and *acuta, K. mikimotoi,* and *Alexandrium minimum*) has been applied around European coasts (Blauw et al., 2006). Aside from modeling HAB organisms, physical transport modeling of oceanographic conditions is useful in determining when detected HABs are likely to reach landfall (Anderson, 2009).

The ultimate goal for HAB mitigation is the ability to forecast blooms and this lies at the interface between research and management (Glibert et al., 2010; Box 2.3). Several HAB forecasting systems are presently under development awaiting full application, and a few have been successfully applied. For example, intensive monitoring in Japan has allowed the calculation of a "red tide index" to define the severity of HABs in Ariake Sound and has identified that precipitation in the month preceding the bloom is one of the most important triggers (Tsutsumi et al., 2003 in Ishizaka et al., 2006). In Hong Kong, a combination of nutrient concentration and water mixing conditions can predict blooms with an 87% prognosis rate (Wong et al., 2009), but a combined hydrodynamic ecosystem model and remotely sensed chlorophyll data were less successful in Northwest European waters (Allen et al., 2008). The first-ever prediction of the size and extent of the "New England red tide" (*Alexandrium* bloom) was made in 2008 by combining near-time monitoring systems with a model developed through the ECOHAB and MERHABB

BOX 2.3 HAB Forecasting Helps Planning

A major *Microcystis* bloom on Lake Erie in 2014 was successfully predicted using HAB forecasting models. This knowledge enabled preparation for the bloom, which tested positive for microcystin toxins, but unfortunate weather conditions exacerbated the problems for users of the water supply from the lake. Strong winds mixed the surface blooms into deeper parts of the lake where water intake pipes were located. Therefore, almost half a million people in Toledo, Ohio were advised not to use water for drinking, cooking, or bathing in August. Cyanobacterial blooms were common the 1960s and 1990s on Lake Erie, but less so in the subsequent 20 years, and so the very recent increase in HAB incidents is causing some concern.

research initiatives (House et al., 2005). Red tide blooms of *K. brevis* off the Florida coast are "now-cast" using a combination of satellite detection and a simple transport model, which is published on the National Oceanic and Atmospheric Administration (NOAA) Harmful Algal Bloom Operational Forecast System (HAB-OFS) web site http://www.co-ops.nos.noaa.gov/hab/index.html. On Lake Erie, a combination of nutrient loading models and satellite data are used to provide an annual outlook prediction, published on the NOAA Web site (Box 2.3). Daily nowcasts of *Karlodinium micrum* in Chesapeake Bay are generated by applying a statistical habitat model to real-time estimates of temperature and salinity (derived from satellite and hydrodynamic modeling data) http://www.star.nesdis.noaa.gov/star/algalblooms.php.

Effective communication of HAB incidents and education of the public are key to disaster prevention. Some of the most damaging PSP outbreaks in the Philippines, in 1995, are thought to have been exacerbated by poor communication. When the HAB struck, the "Red Tide Task Force" failed to adequately publicize a ban on shellfish harvesting and sales, leading to threats of arrests, large numbers of poisonings and a "state of calamity" being declared, with the livelihoods of 20,000 fishermen directly affected (Damasco and Corrales, 1994). Lessons have been learned since this time, and new communication technologies now exist. Weekly HAB bulletins for Lake Erie have been published since 2008 through the National Centers for Coastal and Ocean Science Web site http://www.glerl.noaa.gov/res/Centers/HABS/lake_erie_hab/lake_erie_hab.html and since 2012 the US state and local government HABS monitoring program has produced weekly HAB bulletins in Texas and Florida to warn the tourist industry of impending brevetoxin incidents. Similar reports are issued monthly by the UK Food Standards Agency on biotoxin and phytoplankton levels on http://www.food.gov.uk/enforcement/monitoring/shellfish/ewbiotoxin. Health departments, government agencies, researchers, and grassroots organizations (e.g., the diving community) all have a role in contributing to education and outreach programs such as the campaign to educate about the dangers of CFP in Florida (Friedman et al., 2008). Other schemes have been employed to engage members of the public through increased awareness and citizen monitoring programs in the US (Heisler et al., 2008). However, it is recognized that further research into risk communication is required, to build trust among interested parties leading to more satisfactory outcomes (Bauer, 2006).

2.9 RESPONSE

Effective regulatory and institutional strategies are necessary to facilitate short-term response to HAB-contaminated water, and this varies among countries (Bauer, 2006; Box 2.4). EU legislation recommends that relaying and production areas must be monitored regularly for plankton and biotoxins (854/2004/EC) including weekly sampling of shellfish, to be increased if phytoplankton samples suggest there is an increased risk and extended to food businesses. If a risk to human health is identified, the production area must be closed, but can reopen once toxins have been below the limits for 48 h. A list of approved production and relaying areas must be kept. All EU member states must have an "official control" monitoring system. For example, in the UK this is managed by the Centre for Environment Fisheries and Aquaculture Science (CEFAS) under the Government Department for Environment, Food and Rural Affairs (DEFRA) and closures of shellfish areas are common (Hinder et al., 2011). In The Puget Sound (USA) blue mussel toxicity typically exceeds the regulatory limit for human consumption on a seasonal basis leading to annual closures (Moore et al., 2009). Health alert values for cyanobacteria have been developed in Australia (Fitzgerald et al., 1999). Water authorities are required to immediately notify health authorities if there are dangerous levels of biotoxins e.g., microcystin-LR-eq toxicity (10 μg/L), nodularin (10 μg/L), and saxitoxins (3 μg STX-eq/L)) or phytoplankton cell densities (e.g., *M. aeruginosa* (50,000 cells/mL), *N. spumigena* (50,000 cells/mL), and *A. circinalis* (20,000 cells/mL) in drinking waters, which sets a management

BOX 2.4 US Legislation and HABs

Implemented in 1998, the Harmful Algal Bloom and Hypoxia Research and Control Act (HABHCRA) have been instrumental in driving HAB research, monitoring and planning in the United States. Originally established to maintain a HAB and hypoxia research and control program in oceans, estuaries, and the Great Lakes, the act led to a great proliferation of HAB research, and expansion in monitoring, forecast, and mitigation capabilities. The act was reauthorized in 2004 and recently amended in 2014 to encompass freshwaters including the ecology and impacts of freshwater HABs and the forecasting and monitoring of lakes, rivers, estuaries, and reservoirs in recognition of the growing HAB issues within inland waters that were neglected by previous legislation.

response in motion (Fitzgerald et al., 1999). In recreational waters where, for humans, the primary route of exposure is oral from accidental or deliberate ingestion of recreational water, closures may be necessary (Fristachi et al., 2008), such as occurred when the 2010 "Great North Swim" event on Windermere in the UK Lake District was canceled due to a cyano-HAB outbreak. Few countries have a systematic cyano-HAB monitoring program to identify high risk areas, but recent amendments to US legislation should help to address this omission (Hudnell et al., 2010; Box 2.4).

Health Authorities have a key role in responding to incidents of human poisoning. In California, it is estimated that the 1−12% mortality rate from HAB poisonings is caused in part due to poor access to life support facilities (Trainer et al., 2012). This principle was demonstrated by a PSP poisoning incident in western Canada where two fishermen were airlifted to hospital, and it is thought that this rapid response was the key in saving their life (Taylor, 1992). Medical research into new treatments for toxic incidents is also an important development (Swinker et al., 2001; Friedman et al., 2008). In the US HAB poisoning incidents are reportable to the Department of Health (DOH), investigated and then recorded in the state register of notifiable diseases. This process is often managed by states. For example, in Florida, the Food and Waterborne Disease Surveillance Program coordinates with multiple DOHs, epidemiologists, the Florida Poison Information Center (FPIC), Emergency Departments in hospitals and other health care providers. The FPIC summarizes the data for the DOH, which is used to inform surveillance and monitoring. Another relevant US body is the Center for Food Safety and Nutrition (CFSAN), a part of the Food and Drug Administration (FDA), which collaborates with academia, government counterparts, and government research agencies. In other states, different groups or organizations may manage HAB incidents (Backer et al., 2001; Shoemaker and Hudnell, 2001). This complexity is something that has been recognized as a major knowledge gap that could influence HAB response (Bauer, 2006).

Various physical and chemical approaches exist for the removal of HAB material. The application of clay particles as a flocculent for the HAB cells has been effective in removing them from the water column (Imai et al., 2006). The advantage of this approach is that the removal is rapid (a few hours) and so limits the exposure of organisms to toxins, as well as removing the threat of hypoxia (Pierce et al., 2004; Sengco and Anderson, 2004; Lee et al., 2008). The method has been used quite regularly in Korea in the protection of fisheries (Anderson, 2009). Other treatments such as ozonation and ultrasonics have also been trialed in combination with the application of chemicals (e.g., copper sulfate, sterol surfactants, sodium hypochlorite, and magnesium hydroxides) as algicides (Sengco and Anderson, 2004). None of these techniques has had the success of clay application because they carry a risk of broader ecological damage beyond the original target organisms, as well as problems of toxin release that occur upon cell lysis. Proposals for the introduction of genetically engineered algae which are less likely to form HABs are contentious because of the many unknown risks (Anderson, 2009). Biomanipulation techniques may be employed, such as the introduction of filter-feeding carp, which has been successfully applied in lakes in China to remove cyano-HAB species (Xie and Liu, 2001). A straightforward mitigation technique to avoid fish kills within aquaculture cages is to relocate the caged fish beyond the bloom. This technique works well for *Chrysochromulina* blooms because the toxins operate by inducing osmotic stress on the fish, and so relocating to areas of low salinity inside fjords is an effective strategy (Dundas et al., 1989).

2.10 (4F) RECOVERY

A sufficient research base exists now to guide the measures required to reduce HAB incidents in the future, and nutrient control is a viable option. Nutrient reduction programs within individual lake watersheds are rather simpler to implement because of the smaller scale, and have been very effective in reducing algal biomass in Danish (and many other) lakes (Jeppesen et al., 2005). A recent stronger focus on integrated, river-basin management across provincial or state boundaries, has greatly improved the ability to manage effectively (Steinzor et al., 2012). For example, the Water Resources Act of 2007 in Australia and subsequent amendments legislated for interstate management of the Murray–Darling basin, and the EU Water Framework Directive (2000/60/EC) requires management by river basin district for the restoration of all surface waters to "good ecological status" by 2015. Such holistic approaches have implications for export of nutrients from terrestrial to coastal areas, linking the management of inland and coastal waters. A classic case study of the successful nutrient reduction and restoration of Lake Washington was achieved by diverting sewage effluents from the city of Seattle away from the lake and directly into the sea (Edmondson, 2005). Future management however, will require approaches that avoid shifting the problem downstream. The recent EU Marine Strategy Framework Directive (2008/56/EC) (adopted in June 2008) aims to achieve good environmental status of the EU's marine waters by 2020, and so will call for more holistic solutions for eutrophication abatement. Legislation implemented in Japan in 1973 reduced the loading of industrial pollutants to the Inland Seto Sea by half, and led to marked reductions in HABs (Imai et al., 2006; Figure 2.4). This demonstrates the potential for nutrient reduction to have direct impacts on more enclosed marine ecosystems (MEECE, 2010), but point source nutrients are easier to manage, whereas diffuse nutrient issues are more intractable (Heisler et al., 2008). The Mississippi River/Gulf Nutrient Task Force has fostered initiatives in many states to protect wetlands and manage agricultural run-off, and continues to work on reducing excessive nutrient loading. Linked watershed-HAB models can project the future likelihood of marine HABs that will occur

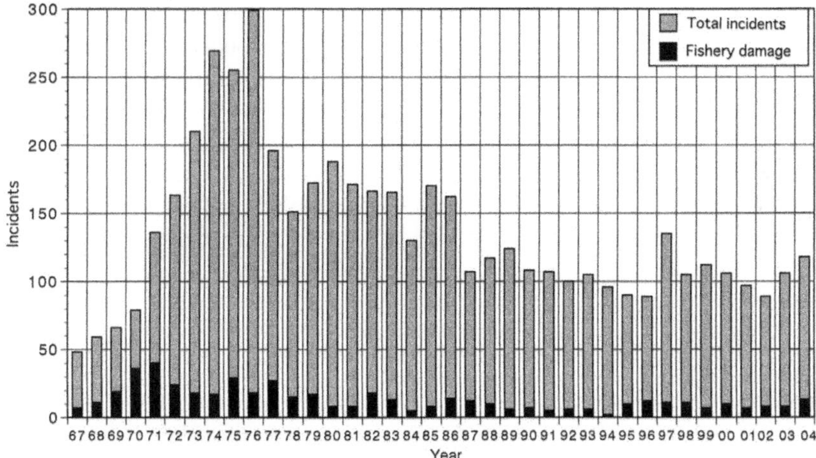

FIGURE 2.4 The occurrence of "red tides" in the Seto Inland Sea from 1967 to 2004, with closed columns indicating fishery damage incidents such as fish kills (Imai et al., 2006). The nutrient reduction legislation introduced in 1973 has led to a marked reduction in incidents. *Reproduced with permission from the Plankton Society of Japan.*

with future land-use change and population growth (Glibert et al., 2010). Future "business as usual" projections predict a global increase in DIN export from 21 Tg N yr^{-1} in 1990 to 47 Tg N yr^{-1} in 2050 with proportionally greater increases in southern and eastern Asia (Paerl et al., 2002; Howarth, 2008; Seitzinger et al., 2010). Therefore, targeted nutrient reduction efforts will be required to reduce the likelihood of HABs in coastal areas and inland waters across the world (Mayorga et al., 2010; Seitzinger et al., 2010) and patterns of human development will be key in forecasting, which areas will be most vulnerable to HABs in the future (O'Neil et al., 2012).

Any long-term recovery plans must allow for changing conditions that will occur from future climate warming. The Marine Ecosystem Evolution in a Changing Environment (MEECE) project (MEECE, 2010) www.meece.eu concluded that pollution and overfishing would have a greater impact on European marine ecosystems than climate change, but that warming would work synergistically with these other stressors. A projected decline exists already in primary productivity in temperate and southern European seas as well as an increase in higher latitude waters, corresponding to a net global decline in primary production of 9% (Behrenfeld et al., 2006; Doney, 2010; MEECE, 2010). Shallow or enclosed seas will be most sensitive to stressors such as nutrient loading, and so it is recommended that management should focus on these vulnerable areas and on variables such as nutrients that are more straightforward to manipulate. Another recommendation from the MEECE project was to enforce the Ship's Ballast Water Convention (www.imo.org) to restrict the potential for future biological invasions.

A growing recognition exists that successful environmental management is usually participatory, engaging with stakeholders at all stages of the decision-making process (Damasco and Corrales, 1994; Corrales, 1995). Such approaches have been employed for HAB management where, for example, conceptual models have been developed for the Baltic Sea to aid communication of the environmental problem with the public (Lundberg, 2005). Now, socioeconomic and cultural factors are increasingly being integrated to produce more sophisticated environmental targets and management plans (Bauer, 2006). Decision support tools have been developed to deal with these complex approaches, such as "problem structuring," which is an integrated assessment involving stakeholders, or "information synthesis," used to provide meaningful ecosystem metrics from complex monitoring data sets for communication to the public. Scenario planning can also be employed using numerical models. These interdisciplinary approaches are likely to yield more realistic and enduring solutions to the significant challenges posed by a growing human population.

REFERENCES

Allen, J., Smyth, T.J., Siddorn, J.R., Holt, M., 2008. How well can we forecast high biomass algal bloom events in a eutrophic coastal sea? Harmful Algae 8, 70–76.

An, J., Carmichael, W.W., 1994. Use of a colorimetric protein phosphatase inhibition assay and enzyme linked immunosorbent assay for the study of microcystins and nodularins. Toxicon 32, 1495–1507.

Anderson, D.M., 1989. Toxic algal blooms and red tides: a global perspective. In: Okaichi, T., et al. (Eds.), Red Tides: Biology, Environmental Science and Technology. Elsevier, pp. 11–16.

Anderson, D.M., 2009. Approaches to monitoring, control and management of harmful algal blooms (HABs). Ocean Coast. Manag. 52, 342–347.

Anderson, D.M., Cembella, A.D., Hallegraeff, G.M., 2012. Progress in understanding harmful algal blooms: paradigm shifts and new technologies for research, monitoring, and management. Ann. Rev. Mar. Sci. 4, 143–176.

Anderson, D.M., Glibert, P., Burkholder, J., 2002. Harmful algal blooms and eutrophication: nutrient sources, composition, and consequences. Estuaries 25, 704–726.

Anderson, D.M., Hoagland, P., Kaoru, Y., White, A.W., 2000. Estimated Annual Economic Impacts from Harmful Algal Blooms (HABs) in the United States. DTIC Document.

Azanza, R.V., Taylor, F.J.R., 2001. Are *Pyrodinium* blooms in the Southeast Asian region recurring and spreading? A view at the end of the millennium. Ambio J. Hum. Environ. 30, 356–364.

Backer, L.C., 2009. Impacts of Florida red tides on coastal communities. Harmful Algae 8, 618–622.

Backer, L.C., Niskar, A.S., Rubin, C., Blindauer, K., Christianson, D., Naeher, L., Rogers, H.S., 2001. Environmental public health surveillance: possible estuary-associated syndrome. Environ. Health Perspect. 109, 797.

Bagnis, R., Kuberski, T., Laugier, S., 1979. Clinical observations on 3,009 cases of ciguatera (fish poisoning) in the South Pacific. Am. J. Trop. Med. Hyg. 28, 1067–1073.

Bajarias, F.F., Relox Jr., T., Fukuyo, Y., 2006. PSP in the Philippines : three decades of monitoring a disaster. Coastal Mar. Sci. 30, 104–106.

Barrett, P., Curnow, J., Littlejohn, J., 1996. The control of diatom and cyanobacterial blooms in reservoirs using barley straw. Hydrobiologia 340, 307–311.

Barton, A., Greene, C., Monger, B., Pershing, A., 2003. The Continuous Plankton Recorder survey and the North Atlantic Oscillation: interannual-to multidecadal-scale patterns of phytoplankton variability in the North Atlantic Ocean. Prog. Oceanogr. 58, 337–358.

Bauer, M. (Ed.), 2006. Harmful Algal Research and Response: A Human Dimensions Strategy. National Office for Marine Biotoxins and Harmful Algal Bloom, Woods Hole Oceanographic Institution, Woods Hole, MA, 58 pp.

Behrenfeld, M.J., O'Malley, R.T., Siegel, D.A., McClain, C.R., Sarmiento, J.L., Feldman, G.C., Milligan, A.J., Falkowski, P.G., Letelier, R.M., Boss, E.S., 2006. Climate-driven trends in contemporary ocean productivity. Nature 444, 752–755.

Blauw, A., Anderson, P., Estrada, M., Johansen, M., Laanemets, J., Peperzak, L., Purdie, D., Raine, R., Vahtera, E., 2006. The use of fuzzy logic for data analysis and modelling of European harmful algal blooms: results of the HABES project. Afr. J. Mar. Sci. 28, 365–369.

Bowling, L., Baker, P., 1996. Major cyanobacterial bloom in the Barwon-Darling River, Australia, in 1991, and underlying limnological conditions. Mar. Freshwater Res. 47, 643–657.

Brand, L.E., Campbell, L., Bresnan, E., 2012. *Karenia*: the biology and ecology of a toxic genus. Harmful Algae 14, 156–178.

Bunting, L., Leavitt, P.R., Gibson, C.E., McGee, E.J., Hall, V.A., 2007. Degradation of water quality in Lough Neagh, Northern Ireland, by diffuse nitrogen flux from a phosphorus-rich catchment. Limnol. Oceanogr. 52, 354–369.

Bureau of Fisheries and Aquatic Resources, 2010. Fish Contribution to the Economy. http://www.bfar.da.gov.ph/pages/Programs/gma-fisheriessector.html.

Carmichael, W.W., Azevedo, S., An, J.S., Molica, R., Jochimsen, E.M., Lau, S., Rinehart, K.L., Shaw, G.R., Eaglesham, G.K., 2001. Human fatalities from cyanobacteria: chemical and biological evidence for cyanotoxins. Environ. Health Perspect. 109, 663.

Chang, F.H., Chiswell, S.M., Uddstrom, M.J., 2001. Occurrence and distribution of *Karenia brevisulcata* (Dinophyceae) during the 1998 summer toxic outbreaks on the central east coast of New Zealand. Phycologia 40, 215–221.

Chu, F., Huang, X., Wei, R., 1989. Enzyme-linked immunosorbent assay for microcystins in blue-green algal blooms. J. Assoc. Off. Anal. Chem. 73, 451–456.

Codd, G.A., 1984. Toxins of freshwater cyanobacteria. Microbiol. Sci. 1, 48–52.

Codd, G.A., 1996. Awareness of cyanobacterial or algal blooms at the Premonstratensian Monastery of the Green Loch, Soulseat Scotland, from the twelfth century, and cattle poisonings attributed to cyanobacterial hepatotoxins at this location eight hundred years later. Harmful Algal News 16, 4–5.

Combe, J.S., 1828. On the poisonous effects of the mussel (*Mytilus edulis*). Edinburgh Med. Surg. J. 29, 86–96.

Corrales, R.A., 1995. Harmful algal blooms in South-East Asia: common problems and networking. Harmful Algae News 12–13, 5–6.

Cox, P.A., Banack, S.A., Murch, S.J., Rasmussen, U., Tien, G., Bidigare, R.R., Metcalf, J.S., Morrison, L.F., Codd, G.A., Bergman, B., 2005. Diverse taxa of cyanobacteria produce β-N-methylamino-l-alanine, a neurotoxic amino acid. Proc. Natl. Acad. Sci. U.S.A. 102, 5074–5078.

Damasco, M.F., Corrales, R.A., 1994. The Philippines: social aspects of red tides. Harmful Algae News 8, 5.

Dewailly, E., Knap, A., 2006. Food from the oceans and human health. Oceanography 19, 85–93.

Diaz, R.J., Rosenberg, R., 2008. Spreading dead zones and consequences for marine ecosystems. Science 321, 926–929.

Dodds, W.K., Bouska, W.W., Eitzmann, J.L., Pilger, T.J., Pitts, K.L., Riley, A.J., Schloesser, J.T., Thornbrugh, D.J., 2008. Eutrophication of US freshwaters: analysis of potential economic damages. Environ. Sci. Technol. 43, 12–19.

Dolah, F.M., Ramsdell, J.S., 2001. Review and assessment of in vitro detection methods for algal toxins. J. AOAC Int. 84, 1617–1625.

Doney, S.C., 2010. The growing human footprint on coastal and open-ocean biogeochemistry. Science 328, 1512–1516.

Doucette, G.J., Logan, M.M., Ramsdell, J.S., Van Dolah, F.M., 1997. Development and preliminary validation of a microtiter plate-based receptor binding assay for paralytic shellfish poisoning toxins. Toxicon 35, 625–636.

Downing, J.A., Watson, S.B., McCauley, E., 2001. Predicting cyanobacteria dominance in lakes. Can. J. Fish. Aquat. Sci. 58, 1905–1908.

Dundas, I., Johannessen, O., Berge, G., Heimdal, B., 1989. Toxic algal bloom in Scandinavian waters, May–June 1988. Oceanography 2, 9–14.

Edmondson, W.T., 2005. Lake Washington. In: O'Sullivan, P.E., Reynolds, C.S. (Eds.), The Lakes Handbook: Lake Restoration and Rehabilitation. Blackwell, Oxford.

Edwards, M., Johns, D., Leterme, S., Svendsen, E., Richardson, A., 2006. Regional climate change and harmful algal blooms in the northeast Atlantic. Limnol. Oceanogr. 51, 820–829.

Edwards, M., Richardson, A.J., 2004. Impact of climate change on marine pelagic phenology and trophic mismatch. Nature 430, 881–884.

Elliott, J.A., 2010. The seasonal sensitivity of cyanobacteria and other phytoplankton to changes in flushing rate and water temperature. Glob. Change Biol. 16, 864–876.

European Commission E, 2002. Eutrophication and Health. In: Local Authorities, Health and Environment Briefing Pamphlet Series, vol. 40. Office for Official Publications of the European Communities, Luxembourg, p. 28.

Faassen, E.J., Harkema, L., Begeman, L., Lurling, M., 2012. First report of (homo)anatoxin-a and dog neurotoxicosis after ingestion of benthic cyanobacteria in The Netherlands. Toxicon 60, 378–384.

Falconer, I.R., 1991. Tumor promotion and liver injury caused by oral consumption of cyanobacteria. Environ. Toxicol. Water Qual. 6, 177–184.

Falconer, I.R., 1999. An overview of problems caused by toxic blue–green algae (cyanobacteria) in drinking and recreational water. Environ. Toxicol. 14, 5–12.

FAO, 2004. Marine Biotoxins. FAO Food and Nutrition Papers (Book 80). FAO Inter-Departmental Working Group, 80 pp.

FAO, 2012. FAO Yearbook. Fishery and Aquaculture Statistics. http://www.fao.org/fishery.

Finlay, K., Patoine, A., Donald, D.B., Bogard, M.J., Leavitt, P.R., 2010. Experimental evidence that pollution with urea can degrade water quality in phosphorus-rich lakes of the Northern Great Plains. Limnol. Oceanogr. 55, 1213.

Fitzgerald, D.J., Cunliffe, D.A., Burch, M.D., 1999. Development of health alerts for cyanobacteria and related toxins in drinking water in South Australia. Environ. Toxicol. 14, 203–209.

Flanagan, A., Callanan, K., Donlon, J., Palmer, R., Forde, A., Kane, M., 2001. A cytotoxicity assay for the detection and differentiation of two families of shellfish toxins. Toxicon 39, 1021–1027.

Friedman, M.A., Fleming, L.E., Fernandez, M., Bienfang, P., Schrank, K., Dickey, R., Bottein, M.-Y., Backer, L., Ayyar, R., Weisman, R., 2008. Ciguatera fish poisoning: treatment, prevention and management. Mar. Drugs 6, 456–479.

Fristachi, A., Sinclair, J.L., Hall, S., Berkman, J.A.H., Boyer, G., Burkholder, J., Burns, J., Carmichael, W., DuFour, A., Frazier, W., 2008. Occurrence of cyanobacterial harmful algal blooms workgroup report. In: Cyanobacterial Harmful Algal Blooms: State of the Science and Research Needs. Springer, New York, pp. 45–103.

Frolov, S., Kudela, R.M., Bellingham, J.G., 2013. Monitoring of harmful algal blooms in the era of diminishing resources: a case study of the US west coast. Harmful Algae 21, 1–12.

Galloway, J.N., Townsend, A.R., Erisman, J.W., Bekunda, M., Cai, Z., Freney, J.R., Martinelli, L.A., Seitzinger, S.P., Sutton, M.A., 2008. Transformation of the nitrogen cycle: recent trends, questions, and potential solutions. Science 320, 889–892.

Gentien, P., Donaghay, P., Yamazaki, H., Raine, R., Reguera, B., Osborn, T., 2005. Harmful algal blooms in stratified environments. Oceanography 18, 172–183.

Geraci, J.R., Anderson, D.M., Timperi, R.J., St Aubin, D.J., Early, G.A., Prescott, J.H., Mayo, C.A., 1989. Humpback whales (*Megaptera novaeangliae*) fatally poisoned by dinoflagellate toxin. Can. J. Fish. Aquat. Sci. 46, 1895–1898.

Gisselson, L.-Å., Carlsson, P., Granéli, E., Pallon, J., 2002. *Dinophysis* blooms in the deep euphotic zone of the Baltic Sea: do they grow in the dark? Harmful Algae 1, 401–418.

Glibert, P.M., Allen, J., Bouwman, A., Brown, C.W., Flynn, K.J., Lewitus, A.J., Madden, C.J., 2010. Modeling of HABs and eutrophication: status, advances, challenges. J. Mar. Syst. 83, 262–275.

Glibert, P.M., Anderson, D.A., Gentien, P., Graneli, E., Sellner, K.G., 2005. The global, complex phenomena of harmful algal blooms. Oceanography 182, 136–147.

Glibert, P.M., Burkholder, J.M., 2006. The complex relationships between increasing fertilization of the Earth, coastal eutrophication and HAB proliferation. In: Graneli, E., Turner, J. (Eds.), The Ecology of Harmful Algae. Springer-Verlag, New York, pp. 341–354.

Glibert, P.M., Harrison, J., Heil, C., Seitzinger, S., 2006. Escalating worldwide use of urea–a global change contributing to coastal eutrophication. Biogeochemistry 77, 441–463.

Goya, A., Maldonado, S., 2014. Evolution of PSP toxicity in shellfish from the Beagle Channel (Tierra del Fuego, Argentina): an overview. In: Sauvé, G. (Ed.), Molluscan Shellfish Safety. Springer, Netherlands, pp. 15–23.

Granéli, E., Johansson, N., 2003. Effects of the toxic haptophyte *Prymnesium parvum* on the survival and feeding of a ciliate: the influence of different nutrient conditions. Mar. Ecol. Prog. Ser. 254, 49–56.

Griffiths, D.J., Saker, M.L., 2003. The Palm Island mystery disease 20 years on: a review of research on the cyanotoxin cylindrospermopsin. Environ. Toxicol. 18 (2), 78–93.

Grzebyk, D., Denardou, A., Berland, B., Pouchus, Y., 1997. Evidence of a new toxin in the red-tide dinoflagellate *Prorocentrum minimum*. J. Plankton Res. 19, 1111–1124.

Hall, R.I., Leavitt, P.R., Quinlan, R., Dixit, A.S., Smol, J.P., 1999. Effects of agriculture, urbanization and climate on water quality in the Northern Great Plains. Limnol. Oceanogr. 44, 739–756.

Hallegraeff, G., 2010. Ocean climate change, phytoplankton community responses, and harmful algal blooms: a formidable predictive challenge. J. Phycol. 46, 220–235.

Hallegraeff, G.M., 1998. Transport of toxic dinoflagellates via ships' ballast water: bioeconomic risk assessment and efficacy of possible ballast water management strategies. Mar. Ecol. Prog. Ser. 168, 10–53.

Hallegraeff, G.M., Maclean, J.L., 1989. Biology, epidemiology, and management of *pyrodinium* red tides. In: Proceedings of the Management and Training Workshop, Bandar Seri Begawan, Brunei Darussalam, 23–30 May 1989. International Specialized Book Service Incorporated.

Hayman, J., 1995. Australia: cyanobacterial overgrowth and toxicity in inland waters before European settlement. Harmful Algae News 10–11, 7.

Haywood, A.J., Scholin, C.A., Marin, R., Steidinger, K.A., Heil, C., Ray, J., 2007. Molecular detection of the brevetoxin-producing dinoflagellate *Karenia brevis* and closely related species using rRNA-targeted probes and a semiautomated sandwich hybridization assay. J. Phycol. 43, 1271–1286.

Heisler, J., Glibert, P.M., Burkholder, J.M., Anderson, D.M., Cochlan, W., Dennison, W.C., Dortch, Q., Gobler, C.J., Heil, C.A., Humphries, E., 2008. Eutrophication and harmful algal blooms: a scientific consensus. Harmful Algae 8, 3–13.

Hinder, S.L., Hays, G.C., Brooks, C.J., Davies, A.P., Edwards, M., Walne, A.W., Gravenor, M.B., 2011. Toxic marine microalgae and shellfish poisoning in the British Isles: history, review of epidemiology, and future implications. Environ. Health 10, 54.

Hoagland, P., Anderson, D.M., Kaoru, Y., White, A.W., 2002. The economic effects of harmful algal blooms in the United States: estimates, assessment issues, and information needs. Estuaries 25, 819–837.

Hoagland, P., Scatasta, S., 2006. The economic effects of harmful algal blooms. In: Ecology of Harmful Algae. Springer, pp. 391–402.

Hoegh-Guldberg, O., 1999. Climate change, coral bleaching and the future of the world's coral reefs. Mar. Freshwater Res. 50, 839–866.

Holligan, P., 1981. Biological implications of fronts on the northwest European continental shelf. Philos. Trans. R. Soc. London Ser. A Math. Phys. Sci. 302, 547–562.

Homer, R., Postel, J., 1993. Toxic diatoms in western Washington waters (U.S. west coast). Hydrobiologia 269–270, 197–205.

House, C., Ogunquit, M., Braasch, E., 2005. Modeling needs related to the regional observing system in the gulf of Maine. In: Proceedings, RARGOM Report 94:1.

Howard, M.D.A., Cochlan, W.P., Ladizinsky, N., Kudela, R.M., 2007. Nitrogenous preference of toxigenic *Pseudo-nitzschia australis* (Bacillariophyceae) from field and laboratory experiments. Harmful Algae 6, 206–217.

Howarth, R.W., 2008. Coastal nitrogen pollution: a review of sources and trends globally and regionally. Harmful Algae 8, 14–20.

Howarth, R.W., Marino, R., 2006. Nitrogen as the limiting nutrient for eutrophication in coastal marine ecosystems: evolving views over three decades. Limnol. Oceanogr. 51, 364–376.

Hudnell, H.K., Jones, C., Labisi, B., Lucero, V., Hill, D.R., Eilers, J., 2010. Freshwater harmful algal bloom (FHAB) suppression with solar powered circulation (SPC). Harmful Algae 9, 208–217.

Hunter, P.D., Tyler, A.N., Willby, N.J., Gilvear, D.J., 2008. The spatial dynamics of vertical migration by *Microcystis aeruginosa* in a eutrophic shallow lake: a case study using high spatial resolution time-series airborne remote sensing. Limnol. Oceanogr. 53, 2391–2406.

Imai, I., Yamaguchi, M., Hori, Y., 2006. Eutrophication and occurrences of harmful algal blooms in the Seto Inland Sea, Japan. Plankton Benthos Res. 1, 71–84.

IPCC, 2013. Climate Change 2013: The Physical Science Basis. Working Group I Contribution to the Fifth Assessment Report of the Intergovernmental Panel on Climate Change Summary for Policymakers.

Ishizaka, J., Kitaura, Y., Touke, Y., Sasaki, H., Tanaka, A., Murakami, H., Suzuki, T., Matsuoka, K., Nakata, H., 2006. Satellite detection of red tide in Ariake Sound, 1998–2001. J. Oceanogr. 62, 37–45.

Janson, S., Hayes, P., 2006. Molecular taxonomy of harmful algae. In: Ecology of Harmful Algae. Springer, pp. 9–21.

Jeppesen, E., Sondergaard, M., Jensen, J.P., Havens, K.E., Anneville, O., Carvalho, L., Coveney, M.F., Deneke, R., Dokulil, M., Foy, B., Gerdeaux, D., Hampton, S.E., Hilt, S.,

Kangur, K., Kohler, J., Lammens, E.H.H.R., Lauridsen, T.L., Manca, M., Miracle, M.R., Moss, B., Noges, P., Persson, G., Phillips, G., Portielje, R., Romo, S., Schelske, C., Straile, D., Tatrai, I., Willen, E., Winder, M., 2005. Lake responses to reduced nutrient loading- an analysis of contemporary long-term data from 35 case studies. Freshwater Biol. 50, 1747–1771.

Jüttner, F., Watson, S.B., Von Elert, E., Köster, O., 2010. β-Cyclocitral, a grazer defence signal unique to the cyanobacterium *Microcystis*. J. Chem. Ecol. 36, 1387–1397.

Kerbrat, A.S., Amzil, Z., Pawlowiez, R., Golubic, S., Sibat, M., Darius, H.T., Chinain, M., Laurent, D., 2011. First evidence of palytoxin and 42-hydroxy-palytoxin in the marine cyanobacterium *Trichodesmium*. Mar. Drugs 9, 543–560.

Kim, G., Lee, Y.W., Joung, D.J., Kim, K.R., Kim, K., 2006. Real-time monitoring of nutrient concentrations and red-tide outbreaks in the southern sea of Korea. Geophys. Res. Lett. 33, 13.

Kudela, R.M., Ryan, J.P., Blakely, M.D., Lane, J.Q., Peterson, T.D., 2008. Linking the physiology and ecology of *Cochlodinium* to better understand harmful algal bloom events: a comparative approach. Harmful Algae 7, 278–292.

Landsberg, J.H., Flewelling, L.J., Naar, J., 2009. *Karenia brevis* red tides, brevetoxins in the food web, and impacts on natural resources: decadal advancements. Harmful Algae 8, 598–607.

Leavitt, P.R., Brock, C.S., Ebel, C., Patoine, A., 2006. Landscape-scale effects of urban nitrogen on a chain of freshwater lakes in central North America. Limnol. Oceanogr. 51, 2262–2277.

Lee, Y.-J., Choi, J.-K., Kim, E.-K., Youn, S.-H., Yang, E.-J., 2008. Field experiments on mitigation of harmful algal blooms using a Sophorolipid–Yellow clay mixture and effects on marine plankton. Harmful Algae 7, 154–162.

Lenes, J.M., Darrow, B.P., Cattrall, C., Heil, C.A., Callahan, M., Vargo, G.A., Byrne, R.H., Prospero, J.M., Bates, D.E., Fanning, K.A., 2001. Iron fertilization and the *Trichodesmium* response on the West Florida shelf. Limnol. Oceanogr. 46, 1261–1277.

Lundberg, C., 2005. Conceptualizing the Baltic Sea ecosystem: an interdisciplinary tool for environmental decision making. Ambio. J. Hum. Environ. 34, 433–439.

Madden, C.J., McDonald, A.A., 2009. Florida Bay SEACOM: Seagrass Ecological Assessment and Community Organization Model.

Maier, H.R., Kingston, G.B., Clark, T., Frazer, A., Sanderson, A., 2004. Risk-based approach for assessing the effectiveness of flow management in controlling cyanobacterial blooms in rivers. River Res. Appl. 20, 459–471.

Manning, S.R., La Claire, J.W., 2010. Prymnesins: toxic metabolites of the golden alga, *Prymnesium parvum* Carter (Haptophyta). Mar. Drugs 8, 678–704.

Matsuoka, K., 1999. Eutrophication process recorded in dinoflagellate cyst assemblages- a case of Yokohama Port, Tokyo Bay, Japan. Sci. Total Environ. 231, 17–35.

Mayorga, E., Seitzinger, S.P., Harrison, J.A., Dumont, E., Beusen, A.H., Bouwman, A., Fekete, B.M., Kroeze, C., Van Drecht, G., 2010. Global nutrient export from WaterSheds 2 (NEWS 2): model development and implementation. Environ. Modell. Software 25, 837–853.

McGowan, S., Barker, P., Haworth, E.Y., Leavitt, P.R., Maberly, S.C., Pates, J., 2012. Humans and climate as drivers of algal community change in Windermere since 1850. Freshwater Biol. 57, 260–277.

McGowan, S., Britton, G., Haworth, E.Y., Moss, B., 1999. Ancient blue-green blooms. Limnol. Oceanogr. 44, 436–439.

McGranahan, G., Balk, D., Anderson, B., 2007. The rising tide: assessing the risks of climate change and human settlements in low elevation coastal zones. Environ. Urban. 19, 17–37.

MEECE, 2010. Marine Ecosystem Evolution in a Changing Environment. Final project summary. http://www.meece.eu/.

Meis, S., Spears, B.M., Maberly, S.C., O'Malley, M.B., Perkins, R.G., 2012. Sediment amendment with Phoslock® in Clatto Reservoir (Dundee, UK): Investigating changes in sediment elemental composition and phosphorus fractionation. J. Environ. Manag. 93, 185–193.

Miller, W.D., Harding, L.W., Adolf, J.E., 2006. Hurricane Isabel generated an unusual fall bloom in Chesapeake Bay. Geophys. Res. Lett. 33, 6.

Mitrovic, S.M., Hardwick, L., Dorani, F., 2011. Use of flow management to mitigate cyanobacterial blooms in the Lower Darling River, Australia. J. Plankton Res. 33, 229–241.

Moore, S.K., Mantua, N.J., Hickey, B.M., Trainer, V.L., 2009. Recent trends in paralytic shellfish toxins in Puget Sound, relationships to climate, and capacity for prediction of toxic events. Harmful Algae 8, 463–477.

Mur, L.R., Skulberg, O.M., Utkilen, H., 1999. Cyanobacteria in the environment. In: Chorus, I., Bartram, J. (Eds.), Toxic Cyanobacteria in Water: A Guide to Their Public Health Consequences, Monitoring and Management. E & FN Spon, London and New York. An Imprint of Routledge.

Naar, J., Bourdelais, A., Tomas, C., Kubanek, J., Whitney, P.L., Flewelling, L., Steidinger, K., Lancaster, J., Baden, D.G., 2002. A competitive ELISA to detect brevetoxins from *Karenia brevis* (formerly *Gymnodinium breve*) in seawater, shellfish, and mammalian body fluid. Environ. Health Perspect. 110, 179.

Nicholls, R.J., Cazenave, A., 2010. Sea-level rise and its impact on coastal zones. Science 328, 1517–1520.

Ninčević-Gladan, Ž., Skejić, S., Bužančić, M., Marasović, I., Arapov, J., Ujević, I., Bojanić, N., Grbec, B., Kušpilić, G., Vidjak, O., 2008. Seasonal variability in *Dinophysis* spp. abundances and diarrhetic shellfish poisoning outbreaks along the eastern Adriatic coast. Bot. Mar. 51, 449–463.

Nygaard, K., Tobiesen, A., 1993. Bacterivory in algae: survival strategy during nutrient limitation. Limnol. Oceanogr. 38, 273–279.

O'Neil, J., Davis, T., Burford, M.A., Gobler, C., 2012. The rise of harmful cyanobacteria blooms: the potential roles of eutrophication and climate change. Harmful Algae 14, 313–334.

Olaffson, E., Pálsson, B., 1805. Travels in Iceland: Performed by Order of His Danish Majesty. Translated from the Danish. R. Phillips, London.

Paerl, H., Dennis, R., Whitall, D., 2002. Atmospheric deposition of nitrogen: implications for nutrient over-enrichment of coastal waters. Estuaries 25, 677–693.

Paerl, H.W., 1988. Nuisance phytoplankton blooms in coastal, estuarine and inland waters. Limnol. Oceanogr. 33, 823–847.

Paerl, H.W., Huisman, J., 2009. Climate change: a catalyst for global expansion of harmful cyanobacterial blooms. Environ. Microbiol. Rep. 1, 27–37.

Park, M.H., Han, M.S., Ahn, C.Y., Kim, H.S., Yoon, B.D., Oh, H.M., 2006. Growth inhibition of bloom-forming cyanobacterium *Microcystis aeruginosa* by rice straw extract. Lett. Appl. Microbiol. 43, 307–312.

Parsons, M.L., Dortch, Q., Turner, R.E., 2002. Sedimentological evidence of an increase in *Pseudo-nitzschia* (Bacillariophyceae) abundance in response to coastal eutrophication. Limnol. Oceanogr. 47, 551–558.

Patz, J.A., Olson, S.H., Gray, A.L., 2006. Climate change, oceans, and human health. Oceanography 19, 52.

Peng, J., Place, A.R., Yoshida, W., Anklin, C., Hamann, M.T., 2010. Structure and absolute configuration of karlotoxin-2, an ichthyotoxin from the marine dinoflagellate *Karlodinium veneficum*. J. Am. Chem. Soc. 132, 3277–3279.

Penna, A., Bertozzini, E., Battocchi, C., Galluzzi, L., Giacobbe, M.G., Vila, M., Garces, E., Lugliè, A., Magnani, M., 2007. Monitoring of HAB species in the Mediterranean Sea through molecular methods. J. Plankton Res. 29, 19–38.

Peperzak, L., 2005. Future increase in harmful algal blooms in the North Sea due to climate change. Water Sci. Technol. 51, 31–36.

Pereira, P., Onodera, H., Andrinolo, D., Franca, S., Araújo, F., Lagos, N., Oshima, Y., 2000. Paralytic shellfish toxins in the freshwater cyanobacterium *Aphanizomenon flos-aquae*, isolated from Montargil reservoir, Portugal. Toxicon 38, 1689–1702.

Pierce, R.H., Henry, M.S., Higham, C.J., Blum, P., Sengco, M.R., Anderson, D.M., 2004. Removal of harmful algal cells (*Karenia brevis*) and toxins from seawater culture by clay flocculation. Harmful Algae 3, 141–148.

Pierce, R.H., Kirkpatrick, G.J., 2001. Innovative techniques for harmful algal toxin analysis. Environ. Toxicol. Chem. 20, 107–114.

Pitcher, G.C., Nelson, G., 2006. Characteristics of the surface boundary layer important to the development of red tide on the southern Namaqua shelf of the Benguela upwelling system. Limnol. Oceanogr. 51, 2660–2674.

Place, A.R., Saito, K., Deeds, J.R., Robledo, J.A.F., Vasta, G.R., 2008. A decade of research on Pfiesteria spp. and their toxins: unresolved questions and an alternative hypothesis. In: Gessner, B.D., McLaughlin, J.B., Botana, L.M. (Eds.), Seafood and Freshwater Toxins. Pharmacology, Physiology, and Detection. CRC Press: Boca Raton, FL.

Proença, L.A., Rörig, L., 1995. Mussel production and toxic algal blooms in Santa Catarina State, southern Brazil. Harmful Algae News 12–13, 5.

Qi, Y., Chen, J., Wang, Z., Xu, N., Wang, Y., Shen, P., Lu, S., Hodgkiss, I.J., 2004. Some observations on harmful algal bloom (HAB) events along the coast of Guangdong, southern China in 1998. Hydrobiologia 512, 209–214.

Ralston, E.P., Kite-Powell, H., Beet, A., 2011. An estimate of the cost of acute health effects from food-and water-borne marine pathogens and toxins in the USA. J. Water Health 9, 4.

Reid, P.C., Johns, D.G., Edwards, M., Starr, M., Poulin, M., Snoeijs, P., 2007. A biological consequence of reducing Arctic ice cover: arrival of the Pacific diatom *Neodenticula seminae* in the North Atlantic for the first time in 800 000 years. Global Change Biol. 13, 1910–1921.

Reynolds, C.S., Oliver, R.L., Walsby, A.E., 1987. Cyanobacterial dominance: the role of buoyancy regulation in dynamic lake environments. N. Z. J. Mar. Freshwater Res. 21, 379–390.

Reynolds, C.S., Walsby, A.E., 1975. Water-blooms. Biol. Rev. 50, 437–481.

Ruddick, K., Park, Y., Astoreca, R., Borges, A., Lacroix, G., Lancelot, C., Rousseau, V., 2008. Applications of the MERIS algal pigment products in Belgian waters. In: Proceedings of the 2nd MERIS/(A) ATSR workshop ESA SP, vol. 666.

Ryan, J., Greenfield, D., Marin III, R., Preston, C., Roman, B., Jensen, S., Pargett, D., Birch, J., Mikulski, C., Doucette, G., 2011. Harmful phytoplankton ecology studies using an autonomous molecular analytical and ocean observing network. Limnol. Oceanogr. 56, 1255–1272.

Schindler, D.W., 1974. Eutrophication and recovery in experimental lakes: implications for lake management. Science 184, 897–899.

Segers, P.A., 1908. La Semana Medica. Buenos Aires.

Seitzinger, S., Mayorga, E., Bouwman, A., Kroeze, C., Beusen, A., Billen, G., Van Drecht, G., Dumont, E., Fekete, B., Garnier, J., 2010. Global river nutrient export: a scenario analysis of past and future trends. Global Biogeochem. Cycles 24, 4.

Sengco, M.R., Anderson, D.M., 2004. Controlling harmful algal blooms through clay flocculation. J. Eukaryot. Microbiol. 51, 169–172.

Shoemaker, R.C., Hudnell, H.K., 2001. Possible estuary-associated syndrome: symptoms, vision, and treatment. Environ. Health Perspect. 109, 539.

Smayda, T.J., 1990. Novel and nuisance phytoplankton blooms in the sea: evidence for a global epidemic. In: Graneli, E., et al. (Eds.), Toxic Marine Phytoplankton. Elsevier, New York, pp. 29–40.

Smayda, T.J., 1997. Harmful algal blooms: their ecophysiology and general relevance to phytoplankton blooms in the sea. Limnol. Oceanogr. 42, 1137–1153.

Smayda, T.J., Reynolds, C.S., 2001. Community assembly in marine phytoplankton: application of recent models to harmful dinoflagellate blooms. J. Plankton Res. 23, 447–461.

Smayda, T.J., Reynolds, C.S., 2003. Strategies of marine dinoflagellate survival and some rules of assembly. J. Sea Res. 49, 95–106.

Spears, B.M., Carvalho, L., Perkins, R., Kirika, A., Paterson, D.M., 2007. Sediment phosphorus cycling in a large shallow lake: spatio-temporal variation in phosphorus pools and release. Hydrobiologia 584, 37–48.

Steidinger, K., Garccés, E., 2006. Importance of life cycles in the ecology of harmful microalgae. In: Graneli, E., Turner, J.T. (Eds.), Ecology of Harmful Algae. Springer, pp. 37–49.

Steidinger, K.A., Landsberg, J., Flewelling, J., Kirkpatrick, B., 2011. Toxic Dinoflagellates. Elsevier Science Publishers, New York, NY.

Steinzor, R., Verchick, R., Vidargas, N., Huang, Y., 2012. Fairness in the Bay: Environmental Justice and Nutrient Trading. Briefing paper No. 1208. Centre for Progressive Reform.

Stumpf, R.P., Tomlinson, M.C., 2005. Remote sensing of harmful algal blooms. In: Remote Sensing of Coastal Aquatic Environments. Springer, pp. 277–296.

Suh, Y.-S., Jang, L.-H., Lee, N.-K., Ishizaka, J., 2004. Feasibility of red tide detection around Korean waters using satellite remote sensing. J. Fish. Sci. Technol. 7, 148–162.

Swinker, M., Koltai, D., Wilkins, J., Hudnell, K., Hall, C., Darcey, D., Robertson, K., Schmechel, D., Stopford, W., Music, S., 2001. Estuary-associated syndrome in North Carolina: an occupational prevalence study. Environ. Health Perspect. 109, 21.

Takano, H.A., 1987. A Guide for Studies of Red Tide Organisms. Japan Fisheries Resource Conservation Association.

Taranu, Z.E., Gregory-Eaves, I., Leavitt, P., Bunting, L., Buchaca, T., Catalan, J., Domaizon, I., Guilizzoni, P., Lami, A., McGowan, S., Moorhouse, H., Pick, F., Stevenson, M., Thompson, P., Vinebrooke, R., 2015. Acceleration of cyanobacterial dominance in north temperate-subarctic lakes during the Anthropocene. Ecol. Lett. 18 (4), 375–384.

Taylor, F.J.R., 1992. Artificial respiration saves two from fatal PSP in Canada. Harmful Algae News 3, 1.

Taylor, F.J.R., Trainer, V.L., 2002. Harmful Algal Blooms in the PICES Region of the North Pacific.

Teixeira, M., Costa, M., Carvalho, V., Pereira, M., Hage, E., 1993. Gastroenteritis epidemic in the area of the Itaparica Dam, Bahia, Brazil. Bull. PAHO 27, 244–253.

Trainer, V.L., Bates, S.S., Lundholm, N., Thessen, A.E., Cochlan, W.P., Adams, N.G., Trick, C.G., 2012. *Pseudo-nitzschia* physiological ecology, phylogeny, toxicity, monitoring and impacts on ecosystem health. Harmful Algae 14, 271–300.

Trainer, V.L., Le Eberhart, B.-T., Wekell, J.C., Adams, N.G., Hanson, L., Cox, F., Dowell, J., 2003. Paralytic shellfish poisonings toxins in Puget Sound, Washington. J. Shellfish Res. 22, 213–223.

Tsutsumi, H., Okamura, E., Ogawa, M., Takahashi, T., Yamaguchi, H., Montani, S., Kohashi, N., Adachi, T., Komatsu, T., 2003. Studies of the cross section of water in the innermost areas of Ariake Bay with the recent occurrence of hypoxic water and red tide. Oceanogr. Jpn. 12, 291–305.

Tubaro, A., Dell'Ovo, V., Sosa, S., Florio, C., 2010. Yessotoxins: a toxicological overview. Toxicon 56, 163–172.

Turner, R.E., Rabalais, N.N., 2003. Linking landscape and water quality in the Mississippi River basin for 200 years. Bioscience 53, 563–572.

Twiner, M., Rehmann, N., Hess, P., Doucette, G., 2008. Azaspiracid shellfish poisoning: a review on the chemistry, ecology, and toxicology with an emphasis on human health impacts. Mar. Drugs 6, 39–72.

Ueno, Y., Nagata, S., Tsutsumi, T., Hasegawa, A., Watanabe, M.F., Park, H.-D., Chen, G.-C., Chen, G., Yu, S.-Z., 1996. Detection of microcystins, a blue-green algal hepatotoxin, in drinking water sampled in Haimen and Fusui, endemic areas of primary liver cancer in China, by highly sensitive immunoassay. Carcinogenesis 17, 1317–1321.

Van Dolah, F.M., 2000. Marine algal toxins: origins, health effects, and their increased occurrence. Environ. Health Perspect. 108, 133.

Vanderploeg, H.A., Liebig, J.R., Carmichael, W.W., Agy, M.A., Johengen, T.H., Fahnenstiel, G.L., Nalepa, T.F., 2001. Zebra mussel (*Dreissena polymorpha*) selective filtration promoted toxic *Microcystis* blooms in Saginaw Bay (Lake Huron) and Lake Erie. Can. J. Fish. Aquat. Sci. 58, 1208–1221.

Visser, P., Ibelings, B., Van Der Veer, B., Koedood, J., Mur, R., 1996. Artificial mixing prevents nuisance blooms of the cyanobacterium *Microcystis* in Lake Nieuwe Meer, the Netherlands. Freshwater Biol. 36, 435–450.

Wang, J., Wu, J., 2009. Occurrence and potential risks of harmful algal blooms in the East China Sea. Sci. Total Environ. 407, 4012–4021.

Watkins, S.M., Reich, A., Fleming, L.E., Hammond, R., 2008. Neurotoxic shellfish poisoning. Mar. Drugs 6, 431–455.

Westberry, T., Siegel, D., Subramaniam, A., 2005. An improved bio-optical model for the remote sensing of *Trichodesmium* spp. blooms. J. Geophys. Res. Oceans (1978–2012) 110 (C6).

Wilson, A.E., Sarnelle, O., Tillmanns, A.R., 2006. Effects of cyanobacterial toxicity and morphology on the population growth of freshwater zooplankton: meta-analyses of laboratory experiments. Limnol. Oceanogr. 51, 1915–1924.

Winder, M., Schindler, D.E., 2004. Climatic effects on the phenology of lake processes. Global Change Biol. 10, 1844–1856.

Wong, K., Lee, J.H., Harrison, P.J., 2009. Forecasting of environmental risk maps of coastal algal blooms. Harmful Algae 8, 407–420.

Wyatt, T., 1995. Bibliographic notes. Harmful Algal News 10–11, 12–13.

Xie, P., Liu, J., 2001. Practical success of biomanipulation using filter-feeding fish to control cyanobacteria blooms: a synthesis of decades of research and application in a subtropical hypereutrophic lake. ScientificWorldJournal 1, 337–356.

Yang, Z., Hodgkiss, I., Hansen, G., 2001. *Karenia longicanalis* sp. nov. (Dinophyceae): a new bloom-forming species isolated from Hong Kong, May 1998. Bot. Mar. 44, 67–74.

Yin, K., Harrison, P.J., Chen, J., Huang, W., Qian, P.-Y., 1999. Red tides during spring 1998 in Hong Kong: is El Niño responsible? Mar. Ecol. Prog. Ser. 187, 289–294.

Zhang, F., Dickman, M., 1999. Mid-ocean exchange of container vessel ballast water. 1: seasonal factors affecting the transport of harmful diatoms and dinoflagellates. Mar. Ecol. Prog. Ser. 176, 243–251.

Chapter 3

Large-Scale Grasshopper Infestations on North American Rangeland and Crops

Scott P. Schell
Department of Ecosystem Science and Management, University of Wyoming, Laramie, WY, USA

ABSTRACT
Grasshopper pest species have been, and continue to be a hazard to agriculturalists in western North America. Grasshoppers are the major, above ground, insect consumer of vegetation on grasslands. They have an important role in the ecosystem as prey for other animals and in nutrient cycling. When grasshoppers damage crops or threaten to consume too much forage, insecticides are now used to control their populations. The goal of early control efforts with the first synthetic insecticides was 100% mortality of all grasshoppers. This management goal, on rangelands, has largely been replaced with Reduced Area and Agent Treatments. This IPM strategy preferentially uses a more selective insecticide applied in a manner that returns pest grasshopper populations to subeconomic densities and conserves more of their natural enemies and nonpest insect species. In the western US, the USDA-APHIS-PPQ conducts yearly surveys of pest grasshopper populations to determine the hazard of outbreak for the next year.

3.1 INTRODUCTION

When human cultures made the switch from hunting and gathering to agriculture for their sustenance, grasshoppers became an enemy instead of a resource. This is well illustrated by the history of the humans living along the shores of the Great Salt Lake in Utah, USA. Evidence of prehistoric Americans harvesting and eating grasshoppers that drowned in the salty water was found in caves near the shore (Madsen and Kirkman, 1988). The amount of highly nutritious food calories in the grasshoppers gathered by these people with little effort contrasts with the experiences of the farmers that came to the region. The Mormon pioneers first lost crops in 1848 to marching swarms of the flightless shield-backed katydid, *Anabrus simplex,* which has given the common name of

Biological and Environmental Hazards, Risks, and Disasters. http://dx.doi.org/10.1016/B978-0-12-394847-2.00003-6

Mormon cricket. Mormon crickets are relatives of grasshoppers, belonging to the family Tettigoniidae, within the insect order Orthoptera. In the years after their arrival in region, the farmers actually suffered many more crop-damaging infestations of the Rocky Mountain locust, *Melanoplus spretus*, and the migratory grasshopper, *Melanoplus sanguinipes*, than to the Mormon cricket (the difference between locusts and grasshoppers is explained under **Taxonomy**) (Bitton and Wilcox, 1978). Madsen (1989) wrote that as they initially investigated the archaeological evidence in the caves in 1984, masses of grasshoppers that he identified as the *M. sanguinipes* were washed up on the adjacent shore of the Great Salt Lake. The agriculturalists of Utah still frequently lose crop yields and forage to Mormon crickets and grasshoppers. A Utah State Cooperative Extension publication entitled *Fighting Grasshoppers and Mormon Crickets for nearly 100 Years* (undated) is available at: http://extension.usu.edu/files/publications/publication/pub__6510916.pdf (accessed August 22, 2014).

Bitton and Wilcox (1978), in their wonderfully titled article *Pestiferous Ironclads*: *The Grasshopper Problem in Pioneer Utah* includes text from a letter that Heber C. Kimball wrote to his son William, describing the serious extent and scope of the grasshopper-caused crop devastation in 1855. Nearly all of the small grain crops and 500,000 apple tree seedlings were reported lost, that year, to the depredations of the Rocky Mountain locust.

Grasshoppers, or more probably and specifically the Rocky Mountain locust, *M. spretus*, were present at high densities over large areas in 1854–56, 1867–72, and 1876–79 and caused serious damage to settler's crops in Utah and other western states and territories. In 1877, an incredibly detailed 824 page document entitled "First Annual Report to the U.S. Entomological Commission for the Year 1877" describes just about all that was then known about grasshopper and locust biology, ecology, areas damaged, and management strategies (https://archive.org/stream/repor1877unit#page/n7/mode/2up; accessed August 22, 2014); (http://books.google.com/books/about/First_and_Annual_Report_of_the_United_St.html?id=z2sqAAAAYAAJ).

A UN-FAO report on the use of insects as food states that the Ute Indian tribe provided cooked insect foodstuffs, presumably Mormon crickets or grasshoppers, to help the starving Mormon settlers in the Salt Lake region during the grasshopper outbreaks (van Huis et al., 2013). The early Mormon settlers obviously ate enough food to avoid starvation but they and their descendants never adopted orthopteran insects as a regular food source.

3.2 TAXONOMY

A short explanation of the relationships and terminology used in acridology is needed before proceeding. "Grasshopper" is the common name of members of the insect family Acrididae, belonging to the suborder Caelifera, within the order Orthoptera. The Caelifera appeared about 220 million years ago in the fossil record (Grimaldi and Engel, 2005). The suborder Ensifera is older and includes crickets, katydids, and a few other less common families (Grimaldi and Engel, 2005). The enlarged femurs and long tibias on the hind pair of legs for jumping is the common physical character of most orthopterans. In the US and Canada, an estimated 1,200 species of orthopterans occur. The Acrididae family has over 630 species (Capinera et al., 2004). The Melanoplinae subfamily contains the many of our most important pest species, including the Rocky Mountain locust. Recent work by taxonomist Otte (2012) has led to the description of an additional 80 species in the *Melanoplus* genus that cannot be determined by casual visual inspection. To clarify confusing terminology, all locusts are grasshoppers but only about a dozen species of grasshoppers fit the definition of a locust (Steedman, 1988). Locusts are grasshoppers that change in physical appearance, physiology, and behavior when they have been reared in high population densities for at least one or more generations (Pener and Simpson, 2009). This phenomenon is termed "density-dependent phase polyphenism" (Pener and Simpson, 2009). The Rocky Mountain locust, *M. spretus*, was the only grasshopper species occurring in the USA and Canada that qualified as a locust. Fortunately for mankind it went extinct in the early 1900s. The story of Rocky Mountain locust is an interesting one and is well told by J.A. Lockwood (2004) in his book "Locust: The devastating rise and mysterious disappearance of the insect that shaped the American frontier". However, several grasshopper species still extant are quite capable of causing severe and widespread damage to crops and rangeland in the western US and Canada (Pfadt, 2002).

3.3 BASIC BIOLOGY

Grasshoppers in North America are quite diverse in behavior, ecology, and physical characters. However, all orthopterans go through simple metamorphosis as the nymphs gradually, in stages called instars, develop into adults after eclosion. This means that when they hatch out of the eggs, they have the same basic body architecture and mouthparts they will have as adults. Until they become adults, they will have nonfunctional wings (there are some species of grasshoppers that never develop functional wings) and be sexually immature. Insect orders that go through complete metamorphosis, such Lepidoptera (butterflies and moths), exhibit a dramatic difference between the larvae stages and adult form and this metamorphosis is accomplished during pupation. Most of the common grasshopper species are less than 6 mm long in their first nymphal instar (Pfadt, 2002). When the grasshoppers are this size, they are very susceptible to mortality from predation and stormy weather (Pfadt, 2002). The first nymphs are so small that a driving rain can knock them into puddles were they can be caught in the water's surface tension and drown. Melanopline species of grasshopper that have been tested can swim pretty well as third or later instars and adults, and will quickly orient and swim to vertical objects to crawl out of the water (Lockwood and Schell,

1994). The number of instars required to reach adulthood can vary with species and even between males and females of the same species. Five instars for medium sized, 15–25 mm adult body length, grasshopper species are fairly common (Pfadt, 2002). As the nymphs grow they become very mobile by walking and jumping with their specialized hind legs. Most North American grasshoppers do not march in mass in a uniform direction like the Old World and Australian locust species can, and commonly do, in high-density populations. However, in western North America, large groups of the Oedipodinae species *Camnula pellucida* nymphs frequently march in a uniform direction, away from their hatching areas (Pfadt, 2002). How the direction is selected is not known but the nymphs usually stop at the nearest green vegetation (Pfadt, 2002). *Aulocara elliotti* nymphs and several species of *Melanoplus* nymphs have been recorded traveling en masse several miles in search of food after consuming all of the available plants where they hatched (Pfadt, 2002) (Figure 3.1).

The biology of the migratory grasshopper, *M. sanguinipes*, is fairly typical of the most common species. However, it can have population surges and outbreaks over large areas and be damaging to crops and rangeland in North America, north of Mexico. The migratory grasshopper occurs from the Delta River Valley in Alaska, USA, all the way to northernmost Mexico and from coast to coast. The famous Danish zoologist, Johan Christian Fabricius, first described it to science as *Gryllus sanguinipes* in 1798 (Lockwood, 2004). In subsequent years, it was given many other species names but it was eventually realized by insect taxonomists and later confirmed by molecular biology to be only one, very widespread species (Chapco and Bidochka, 1986). *Melanoplus sanguinipes* has the ability to vary its phenology to suit its particular habitat. In

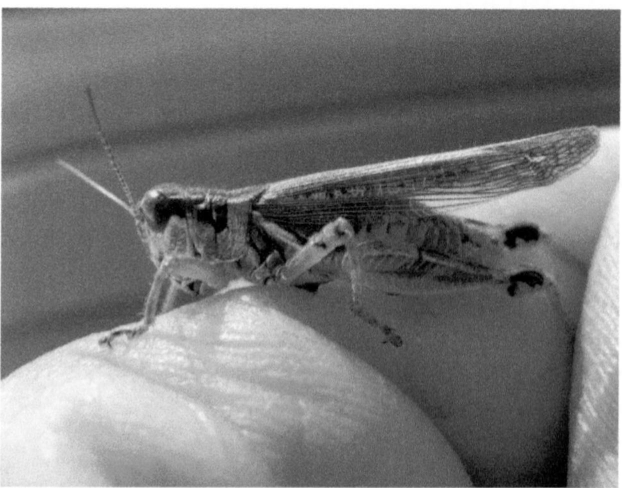

FIGURE 3.1 This adult male *Melanoplus sanguinipes* has long wings and is a capable flier. Swarms of this species have been encountered by aircraft at altitudes over 3000 m and they can easily fly 50 km in a day (Pfadt, 2002). *Scott Schell photo.*

Alaska, most eggs of this species will spend two winters in their pod in the soil before hatching. Currently, the Alaskan population of *M. sanguinipes* only reaches crop-damaging densities in even numbered years (Fielding, 2006). In contrast to 2-year life cycle in Alaska in the southern portions of its range, *M. sanguinipes* can produce nondiapausing eggs and have two generations in one year. In 2013, a control program was conducted in Arizona on the San Carlos Apache Reservation on second-generation *M. sanguinipes* nymphs damaging rangeland used by the tribe's livestock. This was reported by USDA-APHIS-PPQ-AZ officer Dewey Header, to the 2014 Annual National Grasshopper Management Board meeting that was held in Aurora, CO, USA.

The vast majority of *M. sanguinipes*, and most other grasshopper species that inhabit the temperate climate zone of North America are univoltine and spend the winter months in diapause as a partially developed embryo in an egg. Multiple eggs, the average number varies by species are inserted into a pod of hardened foamlike material. The pod material is secreted by an accessory gland in the abdomen of the female grasshopper, usually in the top 3 cm of soil although some species prefer to oviposit in the root crowns of plants (Pfadt, 2002). Most *M. sanguinipes* nymphs hatch in the spring to coincide with peak spring plant growth, usually after 527 degree days above the threshold of 10 °C (Pfadt, 2002). Other grasshopper species may hatch before or after *M. sanguinipes*, usually to coincide with growth of their preferred food plants. An example of this would be *Opeia obscura,* which prefers the warm season grasses in *Bouteloua* and *Buchloe* genera, so they have degree day and temperature thresholds that match the greening and growth of their preferred food plants (Pfadt, 2002). Several species hatch in late winter or very early spring and have evolved ways of surviving cold temperatures that would be fatal to species like *M. sanguinipes* or *O. obscura*. Even more interesting are widespread species like *Arphia conspersa*, which spend the winter as fifth instar nymphs after hatching in the late summer. The nymphs are able to survive ambient temperatures as low as −16 °C without snow cover (Pfadt, 2002).

After reaching adulthood, *M. sanguinipes* females have a preoviposition period of 2−3 weeks before they are receptive to the courtship overtures of the males. After mating, the female oviposits 5−7 days later, producing a pod containing, on average, 20 eggs (Pfadt, 2002). In laboratory conditions, a female *M. sanguinipes* can produce as many as 20 pods (Pfadt, 2002). However, in the wild there are many sources of mortality that greatly reduce a female grasshopper's chances of surviving long enough to do that (Figure 3.2).

As mentioned before, the number of eggs produced per pod and even the pod structure varies between species. For example, *Cordillacris crenulata* may only produce two to three eggs per pod (Pfadt, 2002). *Melanoplus differentialis*, a large bodied, crop-pest species, was observed to produce 194 eggs in a single pod (Pfadt, 2002). The fecundity and food plant preferences are major factors in determining how pestiferous a grasshopper species will be to agriculture (Parker, 1954).

FIGURE 3.2 The egg pod of a *Melanoplus bivittatus* torn open to reveal the numerous eggs inside. *Scott Schell photo.*

3.4 ECOLOGY

Grasshoppers are very adaptable and inhabit a wide variety of terrestrial habitats in North America. *Chorthippus curtipennis* occurs above timberline at 3350 m in Colorado, USA and at latitudes north of the Arctic Circle (Pfadt, 2002). Conversely there are species, such as *Trimerotropis pallidipennis* that occur in the hottest and driest desert region of the North America, such as Death Valley in the Mojave Desert, California, USA (Pfadt, 2002). However, regions dominated by grasslands are where grasshoppers are most diverse and abundant, with over 400 species inhabiting the 17 contiguous western states. Grasshoppers are the dominant, above-ground herbivore in grassland ecosystems (Branson et al., 2006). As such, they play an important role in the environment by being a food source for many other creatures and are involved in nutrient cycling on the grassland (Branson et al., 2006).

Grasshoppers are primarily phytophagous but are not adverse to eating other grasshoppers of their own or other species when the opportunity arises (Bomar, 1993). *Melanoplus* species will also readily eat other arthropods, their eggs and pupae, if available (Pfadt, 2002). Many grasshopper species will also readily consume dried manure, lichens, and fungi (Pfadt, 2002). I have personally seen them scavenge such things as cake doughnuts, pizza, dead rodents, and even a road-killed rattlesnake.

Melanoplus sanguinipes is a polyphagous plant feeder with a decided preference for forbs, where available (Pfadt, 2002). Crop-pest grasshoppers are generally polyphagous (Parker, 1954) (Figure 3.3).

However, some grasshopper species specialize on just grasses or just forbs and a few species even have dietary preferences restricted to a few species of

FIGURE 3.3 Five gravid *Melanoplus bivittatus* females have just brought down a sunflower (*Helianthus annuus*) seedling. This species is a threat to most commonly grown crops. *Scott Schell photo.*

plants (Pfadt, 2002). A few species of grasshoppers have even been specialized to feed on plants with defensive chemicals that discourage most herbivory. Examples of this are *Hesperotettix viridis* and the snakeweeds (*Gutierrezia* species) and *Hypochlora alba* and cudweed (*Artemisia ludoviciana*) (Pfadt, 2002).

Grasshopper species have adapted to highly variable climates and live on every continent except Antarctica, wherever enough suitable food plants grow to support them. However, other ecological factors play an important role in their distribution and propensity to have population explosions. Isely (1938) detailed a situation where a food plant, *Evax multicaulis*, for an oligophagous grasshopper species, *Acrolophitus variegatus*, grew in abundance on a soil type that either prevented survival of the grasshopper's eggs or was rejected by the ovipositing female grasshoppers. The grasshopper species was not found at that site during the course of Isely's (1938) investigation.

Large-scale grasshopper outbreaks do not occur randomly on the landscape (Schell and Lockwood, 1997). In Wyoming, USA, the areas prone to damaging grasshopper-outbreak persisting, without intervention, for multiple years is much less than the total area that can support a one year outbreak. This conclusion was drawn after analyzing many years of survey data on rangeland grasshopper populations that were collected by USDA-APHIS-PPQ (Zimmerman et al., 2004).

In intensive agriculture areas, farmers grow high densities of suitable food plants and create oviposition habitat that favors some species of grasshoppers. This "artificial" ecosystem allows for grasshoppers such as *Melanoplus bivittatus* and *Melanoplus femurrubrum* populations to persist at high densities for successive years until pesticides, epizootics of disease, or a severe weather event causes a temporary population decline (Pfadt, 2002).

3.5 GRASSHOPPER-OUTBREAK DAMAGE

Grasshopper outbreaks in the short term can cause financial losses due to crop yield declines or reduced gain on livestock being grazed in competition with the grasshoppers (Parker, 1954). Increased expenses also occur for obtaining replacement forage or conducting an insecticide treatment program (Hastings et al., 2002). Additional expenses can also occur for extending or increasing operating loans to cover the costs of treating or absorbing losses caused by the grasshopper damage. Livestock producers may have to sell animals early, experiencing a loss in potential gain and market timing (Parker, 1954). They may even have to reduce the size of their herd of mother animals (cows and ewes, etc.) due to lack of forage and reduced yields of grasshopper-damaged hay fields. The following is the text of a letter that was read to an agriculture committee of Wyoming state legislators in March of 2010 that detailed the financial impact grasshopper outbreaks can have on ranchers. It was written by a member of the Kukowski ranching family, who own the Golden Willow Ranch, of Wyarno, Wyoming. The Wyoming Stock Growers Association has awarded the Kukowski family with their 2007 Environmental Stewardship Award, which is given annually to a family or group that has exemplified good stewardship of the land. The Kukowski family also received the Leopold Conservation Award, given by the Wisconsin-based, Sand County Foundation, for their commendable environmental stewardship practices in 2006. I have inserted some text in brackets that will explain some of the specialized terms used by ranchers in the body of the letter.

"My name is Bridget Kukowski. I am a member of the Board of Agriculture and wanted to relate my family's experience with grasshopper in the last 2 years. I ranch with my family about 20 miles southeast of Sheridan. We were recipients of the Wyoming Stock Growers Association 2007 Environmental Stewardship Award. We have a commercial cow/calf and stocker yearling operation.

Our grasshopper problem started during the summer of 2008. In about mid-July of that year, we started noticing a lot of small grasshoppers. The hatch did not begin until late enough in the season that we did not really begin to see the effects until early September. In 2008, we had about 400 yearlings (calves that are weaned from their mothers but are still rapidly growing) that were supposed to be shipped in early October. By the middle of September, we had to start feeding them hay as the grasshopper had eaten all of the quality grass. The hay that had to be fed cost us about $5,000 and the yearlings were still about 40 pounds light, costing us an additional $16,000 for a calculated loss of $21,000 due to grasshoppers in the stocker enterprise (the young weaned calves are called "stocker" cattle as you would "stock" your pasture with livestock). The cow/calf side of it is harder to calculate a specific cost. We were calving in July/August, so turned the bulls out the beginning of October. Due to the grasshoppers much of the quality feed was gone, even in pastures that had not been grazed yet that year. Our breed up was poorer than usual (the conception rate of cows is affected by their nutritional condition) and more hay was needed to supplement what grazing was available.

The spring of 2009 brought many more grasshoppers. They started hatching in huge quantities in mid-May. By the end of May we shipped 230 cows off, some to leases and some sold. We chose not to run any stocker yearlings on the place. The additional lease cost plus freight was about $23,000. The cows that we did keep here, we felt a need for a protein supplement during breeding due to the grasshoppers already consuming all the high quality grass. The 30% protein blocks used for that purpose cost about $7500. In late August, we shipped another 200 pairs (a cow and her calf) off costing around $18,000 in grass lease and freight (the cost of moving the cattle by large, livestock hauling semitrucks). The calves on those cows came in 50 pounds light (a rancher makes their profit by having their animals gain a lot of body weight while they own it). I heard stories of neighbors having calves 100 pounds light. At a value of $1 per pound it does not take long for $50–$100 lost income per calf to add up. My lost revenue came to about $4000. I also know of neighbors that had to begin feeding hay in September because all their grass was gone. In early November, we shipped another 240 pairs to a wintering program because we had no winter feed left. We had a pasture we had not run a single cow in all year and it looked as bare as the rest of the pastures. We were able to utilize only ½ of our normal AUMs (animal unit month is the amount of pasture land or feed required to support a typical cow and her unweaned calf for 30 days) for the year. Even by reducing our numbers and usage here by over ½, our pastures look worse than we have ever left them, even worse than during the drought! The cost of having to send most cows off to a wintering program is $108,000. We would have to feed some hay in a normal year, so probably only half of that cost can be attributed to grasshoppers. To date, that puts our out of pocket costs, due to grasshoppers, at $127,500. This spring we are planning on spraying for grasshoppers, but are also planning on having to send about 400 cows out to a lease (pasture land rented for grazing) to make it through this coming summer. This is projected to cost about $56,000. Hopefully that will be enough of a break, for this ranch to begin to recover from the damage done by grasshoppers. We believe that grasshopper will cost us in the ballpark of $200,000 by the fall of 2010. That is the cost for just one ranch family, all of our neighbors could tell a similar story. It is imperative that the Weed and Pest Districts receive enough funding so that everyone is able to implement a spraying program. Thanks for your time and consideration.

The Wyoming State Legislature did fund the emergency insect management grant program in 2010. Those funds, along with funds from the County

Weed and Pest Districts, revenue from property taxes, along with additional funds from ranchers, resulted in 2.4 million ha being protected from grasshoppers in Wyoming in 2010. The Integrated Pest Management (IPM) methodology used is termed Reduced Area and Agent Treatments (RAATs).

3.6 LONG-TERM DAMAGE

On rangeland, good livestock producers stock their pastures to leave sufficient foliage on the good forage plants in order to maintain the physiological health of the desirable plants (Reardon and Merrill, 1976). The ungrazed portion of the plants feeds the roots on the perennial grasses preparing them for dormancy and the start of the next growing season and/or allows the production of seeds on the perennials and annuals. The desirable grasses and forbs in rangeland are called "decreasers" because their abundance and production will decrease under too intense grazing pressuring (Reardon and Merrill, 1976). Conversely undesirable forage plants and noxious weeds will increase in abundance, competing with the "decreasers" and further reducing the carrying capacity of the pastures. Grasshopper outbreaks complicate rangeland management by increasing grazing pressure tremendously. Just 14 adult *M. sanguinipes* per square meter equates to approximately 50 kg of grasshopper biomass per hectare. Grasshoppers can eat the equivalent of their body weight in forage plants daily (Hewitt, 1977; Hewitt and Onsager, 1982). In addition, to what they consume, they can make forage unavailable to livestock by clipping and dropping leaves to the ground (Hewitt, 1977; Hewitt and Onsager, 1982). A typical cow weighing 454 kg stocked at 1 per 8 ha for the Western grazing season is approximately 56.75 kg of biomass per hectare and would consume between 2.5% and 3% of her body weight daily in dry weight forage (Lyons et al., 1999). Parker (1954) estimated that forage consumption would be approximately equivalent between the cow and the grasshoppers at less than equivalent biomass population density because they consume more forage per unit of body mass than livestock. This essentially doubles the stocking rate of the rangeland. Hewitt (1977) did a detailed study of the consumption levels of several common pest grasshopper species to determine forage losses, which is necessary for calculating economic thresholds. Grasshoppers easily outcompete livestock for forage due to their small mouths that allow them to eat forage too short for larger animals to graze upon. The long-term consequences of grasshopper infestations can be severe, with the rangeland permanently degraded, losing desirable perennial plant cover (Parker, 1954; Pfadt, 2002). The increased bare ground caused by the decline in plant cover is vulnerable to wind and water erosion and the invasion of noxious weeds (Parker, 1954).

In crop-growing regions and suburban areas grasshoppers can also damage or kill windbreak tree plantings (Parker, 1954). Depending on the climate and the age of the planted trees, it can take several years plus the expense of replanting trees to get back what was lost to the grasshoppers.

3.7 PAST GRASSHOPPER—OUTBREAK MANAGEMENT

In the past, people used manual control such as smashing grasshoppers with tree branches or grain flails. Farmers set prairie fires to incinerate grasshoppers and burnt smudge pots to try and repel them from crop fields with smoke. They mechanically tilled the grasshopper egg beds to try and prevent the problem from recurring the following year. People also invented complex mechanical "hopper harvesters" to gather and kill the offending insects. They also tried trampling the grasshoppers with herds of livestock that also damaged the crops they were trying to protect. They used flocks of hungry domestic fowl, like turkeys, to feed on the grasshoppers to try and reduce their populations. These various efforts are detailed in the "First Annual Report to the U.S. Entomological Commission for the year 1877" (referenced earlier) and the "Second Report of the U.S. Entomological Commission for the years 1878 and 1879" (https://archive.org/stream/reportofunited02unit#page/n5/mode/2up; accessed August 22, 2014).

3.8 EARLY PESTICIDE CONTROL EFFORTS

With the advent of the insecticide Paris green, which is copper acetoarsenite, bait formulations were concocted to kill Mormon crickets, Rocky Mountain locusts, and grasshoppers (Cowan, 1990). The most famous grasshopper bait formulation was Norman Criddle's mixture of 100 parts horse manure to 1 part salt and 1 part Paris green with sufficient water to mix the bait. Paris green was frequently used until experiments conducted by the USDA, Bureau of Entomology in 1926 showed that dusts of sodium and calcium arsenite were superior (Shotwell and Cowan, 1928). The formulations tested found that 1 part sodium arsenite to 4 parts hydrated lime (if calcium arsenite was used then a 1:3 ratio) would kill grasshoppers and Mormon crickets and not chemically burn the foliage off the vegetation (Shotwell and Cowan, 1928). The arsenite dust formulations were used until the early 1940s when experimental trials of new toxicants for grasshoppers and Mormon cricket baits replaced the arsenites. A replacement for the arsenite dust formulations was eagerly sought due to their repellency to insects (Shotwell, 1942). Conversely, cattle were attracted to salty taste of sodium arsenite with numerous instances of them being unintentionally poisoned (Shotwell, 1942). In the 1939, the nonrepelling insect toxin sodium fluosilicate started to replace the arsenites (Cowan, 1990). Sodium fluosilicate, when formulated with rolled wheat at a 4:100 ratio, was rated as the best in a 1938 trial at a 56.5 kg of bait per hectare rate!

Post World War II, many new chlorinated hydrocarbon insecticides, also known as organochlorines, came on the market. The most infamous of this class of insecticide was DDT, however, it was not found to be very efficacious against grasshoppers (Riegert et al., 1997). Other insecticides in the organochlorines class did prove to be efficacious as bait toxicants, dusts, and sprays (Riegert et al., 1997). Parker (1954) detailed the recommended rates for the insecticides labeled for grasshoppers: toxaphene—1.13 to 1.55 kg per ha; chlordane—0.8 to 1.128 kg per ha; aldrin—105.8 to 141.2 g per ha; dieldrin—52.8 to 70.4 g per ha; heptachlor—212 to 281 g per ha. It was not detailed how much of the active ingredient is in each of those insecticide formulations.

The organochlorine insecticides do not repel insects so they also worked well as bait toxicants. The "best bait" in 1947 was "...4 ounces of aldrin in 2 quarts of diesel fuel mixed with 100 lbs of rolled wheat..." applied at a 3.4 kg per ha application rate (Cowan, 1990). The largest reported contiguous area treated was 120,481 ha in Nevada in 1953 by DC-3 airplanes spreading bait. A group of Agriculture Department officers searched for several days in the treated area "...and did not find a single live Mormon cricket..." (Cowan, 1990).

Organochlorine insecticides, such as aldrin, also became the products of choice for aerially spraying, replacing baits for logistical and economic reasons. From 1951 to 1964, 147 ml of aldrin in 9.421 of kerosene or diesel per hectare was the default formulation and rate for both Mormon crickets and grasshoppers (Pfadt and Hardy, 1987a). But by 1962, organochlorine insecticides were recognized as posing an environmental threat due to their persistence and bioaccumulation. Experiments conducted by the Entomology Research Division on the USDA starting in 1956 showed that the organochlorines insecticides accumulated in the fat and milk of livestock (Riegert et al., 1997). Reigert, Ewen, and Lockwood (1997) detailed the development of pesticide usage against grasshoppers.

Evidence points to humans causing the extinction of the Rocky Mountain locust accidentally through destruction of key source habitats during the low ebbs of their populations by the early 1900s and not by the use of early insecticides (Lockwood, 2004). It is remarkable how, with the potency, rates, and long residuals of these first synthetic insecticides, plus the actual goal of 100% grasshopper mortality that some grasshopper species were not driven to extinction. Many minor rangeland grasshoppers species, such *Dactylotum bicolor*, exist at low population densities, have low fecundity, and may not even be very mobile (Pfadt, 2002). Perhaps many grasshopper species have a few eggs every year that spend an additional 12 months safely in the soil before hatching to survive natural or manmade catastrophes. This prolonged diapause or superdiapause has been discovered in many insect species (Ushatinskaya, 1984).

During Wyoming's large-scale outbreak of rangeland grasshoppers in mid-1980s, only four insecticides were labeled for use against the grasshoppers on rangeland/pastures with a zero day grazing restrictions (Pfadt and Hardy, 1987b). Being able to treat grasshoppers without removing the livestock during the process is a practical approach to treating large blocks of western rangeland. The primary insecticide used during the large-scale outbreak in 1985–1987 was malathion, an organophosphate class pesticide (Pfadt and Hardy, 1987b). This was because malathion was economical and effective if applied within its environmental constraints. Malathion also has an ultra-low volume formulation, usually defined as equal to or less than 1 L of formulated pesticide per hectare, that allowed more area to be sprayed per plane trip (Pfadt and Hardy, 1987b). It does have the drawback of being a broad-spectrum insecticide, i.e., it is lethal to most species of insects that ingest it. Large, multiengine, war surplus airplanes modified to spray pesticides were contracted to treat the large rangeland grasshopper control projects conducted during this outbreak by the USDA-APHIS agency (Pfadt and Hardy, 1987b). The treatment blocks delineated were at a minimum of approximately 4000 ha in size. Small single engine spotter planes with USDA-APHIS employees as observers watched the treatments being applied to make sure that clogged spray nozzles or other errors in application that might cause gaps in the coverage were noted (Pfadt and Hardy, 1987b). In Wyoming alone, 1,204,800 ha were treated in 1985 out of an estimated 1,606,000 ha infested at economically damaging levels (Pfadt and Hardy, 1987b). Ground crews checked the treated areas for any areas that were missed.

A major research effort funded by the USDA for development of integrated pest management for rangeland grasshopper was given impetus after the costly grasshopper control programs of the 1980s. This occurred because of the growing regulatory and legal hurdles that come with treating grasshoppers on large areas of publically own rangeland. The result of this funding was the "Grasshopper Integrated Pest Management User Handbook" that contains information on all aspects of grasshopper control, including the results of numerous studies to document the impact grasshopper control programs and insecticides have on the environment (http://www.sidney.ars.usda.gov/grasshopper/Handbook/index.htm; accessed 6/10/2015).

3.9 RECENT OUTBREAKS

Over 2.4 million ha of primary rangeland were protected during the 2010 grasshopper outbreak in Wyoming, USA alone. Some hay fields and small grain fields gained protection from the treatments by reducing adjacent high-density grasshopper populations to levels that had no incentive to move as plenty of forage remained on the rangeland to eat. The most numerous species in most areas in the grasshopper assemblages during the outbreak was *M. sanguinipes*. The species and population density data compiled from many years of USDA-APHIS-PPQ grasshopper survey data and the geographic coordinates of the survey locations are contained in the Wyoming Grasshopper Information System but it is not currently available on the Internet.

The insecticide treatment of choice in 2010 was Dimilin® 2L, which contains the active ingredient diflubenzuron, an insect growth regulator that inhibits proper chitin formation during the arthropod growth and molting process. It is essentially nontoxic to vertebrates (Grosscurt, 2007). Diflubenzuron has been used against lice on livestock in a product called Clean-up™ and applied to koi ponds to treat infestations of anchor worm (*Lernaea* species). The rate of

Dimilin® 2L used during the outbreak was 73.4 ml per ha formulated in 589 ml water and 220 ml canola oil, with an additional 73.4 ml of commercial crop oil concentrate to act as an emulsifier (Lockwood and Schell, 1997). It was applied in alternating treated and untreated swaths, 30–40 m wide. The width of the swath depended on the contractor's equipment and contract specifics. The RAATs IPM strategy reduces the impact on nontarget insects and other wildlife and greatly reduces the cost of treatment as compared to conventional blanket treatments (Lockwood and Schell, 1997). RAATs is different from the barrier treatments often used against swarming Old World locust nymph bands. The untreated swaths are smaller in width and the treated swaths are not placed perpendicular to the direction of travel because, usually, there is no swarming and marching in single direction by the nymphs. RAATs could be described as a method to fragment a grasshopper outbreak that leads to economic control within 21 days. Only single engine, agricultural spray planes were used but frequently several planes worked each large treatment area at a time using Global Positioning System technology for precision guidance.

Diflubenzuron has to be applied while the grasshoppers are still nymphs. This necessitates intense survey, efficient organization, and coordination of aerial pesticide application to insure timely, efficacious treatment on hundreds of thousands of hectares of remote land with difficult ground access and mixed ownership. The cost to the agriculturalist of protecting the rangeland in Wyoming on a per hectare basis is about the same in 2010 than it was back during the outbreak in 1985–1987, after accounting for inflation. During the 1986 outbreak, in Wyoming, USA, landowners paid approximately $1.25 per ha for their private land to be treated as part of ≥4000 ha USDA-APHIS-PPQ grasshopper control program blocks. During that outbreak, the USDA-APHIS-PPQ agency paid the entire cost of treatment on land owned by the federal government, 50% on state-owned land, and 33% of the cost for each acre private land enrolled in the grasshopper-control program. The USDA-APHIS terminated its cost-share program in 1996 (Lockwood et al., 2000). The cost-share program, as part of the Plant Protect Act of 2000, has since been reinstituted and can be found at this URL: http://www.aphis.usda.gov/brs/pdf/PlantProtAct2000.pdf (Accessed successfully on August 28, 2014).

In addition to that, the states of Idaho and Utah, USA have been able to secure special additional funding for suppression programs against grasshopper and Mormon cricket as line items in congressional appropriations in subsequent years. However, if USDA-APHIS-PPQ contracts and conducts a grasshopper control program currently, they need to charge the landowning participants, be they private citizens, state, or federal-land management agencies some overhead to cover the additional costs to the agency to perform this task. If the majority of the infested land is owned by the federal government and leased for grazing, then the cost will be paid for by the land management agency in charge of it; most frequently the US Department of the Interior's Bureau of Land Management. In Wyoming, on private land, most landowners infested with grasshoppers now chose to have their County Weed and Pest Districts organize and contract grasshopper control programs or join with adjacent ranches and do it themselves. A 2011 grasshopper-treatment price list of Sky Aviation, Wyoming, USA agricultural aviation company, advertised a cost of $3.24 per ha protected for a 4000 ha size block of rangeland for grasshopper control using the RAATs strategy with Dimilin® 2L insecticide. The Wyoming State Legislature has, through the Wyoming Department of Agriculture, provided funds to the counties via the Emergency Insect Management Grant program. Landowners were able to participate in grasshopper control programs for as little as $2.49 per ha with the cost-share program in 2010. Adjusted for inflation, what cost $1.25 in 1986 now costs $2.72 in 2014.

At this point in time, grasshopper outbreaks cannot be prevented. They can only partially be predicted and that requires diligent yearly survey efforts. USDA-APHIS-PPQ field scouts conduct grasshopper nymph population surveys every spring. The intensity of the survey effort depends on the funds available but at a minimum long-established grasshopper survey locations, called Common Data Set points, will be sampled for population density and species present. In addition, areas of the western states, that the previous year's grasshopper hazard map predicts to be infested, are also surveyed. The adult grasshopper survey effort occurs later in the summer at the peak of the common pest grasshopper species populations. The theory behind the timing of this survey is that areas with many adult grasshoppers will produce a lot of grasshopper eggs, increasing the chances for a severe infestation the following year. The 2014 USDA-APHIS-PPQ Rangeland Grasshopper Hazard map is http://www.aphis.usda.gov/plant_health/plant_pest_info/grasshopper/downloads/hazard.pdf (accessed August 29, 2014). The map is a useful tool to help direct-limited management resources to potentially infested areas. However, our major pest grasshopper species, *M. sanguinipes*, has been documented to be able to increase from subeconomically damaging populations to outbreak densities from one year to the next (Pfadt, 2002). Receiving consistent funding to conduct comprehensive spring nymphal surveys and then always reacting quickly enough to the findings to control grasshopper outbreaks in a single season is still an elusive goal for land managers.

Hopefully, progress on effective, economical, and environmentally innocuous grasshopper IPM methods will continue to be developed as grasshopper will continue to be a biological hazard and risk to human interests in the foreseeable future.

REFERENCES

Bitton, D., Wilcox, L.P., 1978. Pestiferous ironclads: the grasshopper problem in Pioneer Utah. Utah Hist. Q. 46, 336–355.

Bomar, C.R., 1993. The Olfactory Basis for Cannibalism in Rangeland Grasshoppers (Orthoptera: Acrididae): Applications for Improved Control using Bran Baits. PhD. Dissertation, University of Wyoming. Laramie, WY, USA.

Branson, D.H., Joern, A., Sword, G.A., 2006. Sustainable management of insect herbivores in grassland ecosystems: new perspectives in grasshopper control. Bioscience 56, 743–755.

Capinera, J.L., Scott, R.D., Walker, T.J., 2004. Field Guide to Grasshoppers, Katydids, and Cricket of the United States. Comstock Publishing Associates, Cornell University Press, Ithaca, London, ISBN 0-8014-8948-2.

Chapco, W., Bidochka, M.J., 1986. Genetic variation in prairie populations of *Melanoplus sanguinipes*, the migratory grasshopper. Heredity 56, 397–408. http://dx.doi.org/10.1038/hdy.1986.62.

Cowan, F.T., 1990. The Mormon Cricket Story. Montana Agr. Exp. Special Report, 31. Montana State University.

Fielding, D., April 7, 2006. Optimal diapause strategies of a grasshopper, *Melanoplus sanguinipes*. J. Insect Sci. http://dx.doi.org/10.1673/1536-2442. Published online.

Grimaldi, D., Engel, M.S., 2005. Evolution of the Insects. xv + 755 pp. Cambridge University Press, Cambridge, New York, Melbourne, ISBN 0-521-82149-5.

Grosscurt, A.C., 2007. Dimilin, the Chitin Deposition Inhibitor Diflubenzuron for Insect Control in Forestry and Public Green, third ed. Chemtura Publication. PM250.2.

Hastings, J.D., Branting, K., Lockwood, J.A., 2002. CARMA: a Case-based rangeland management adviser. AI Mag. 23 (2), 49–62.

Hewitt, G.B., 1977. Review of Forage Losses Caused by Rangeland Grasshoppers, 1348. USDA Miscellaneous Publication, 24 pp.

Hewitt, G.B., Onsager, J.A., 1982. A method for forecasting potential losses from grasshopper feeding on northern mixed grass prairie forages. J. Range Manag. 35 (1).

van Huis, A., Van Itterbeeck, J., Klunder, H., Mertens, E., Halloran, A., Muir, G., Vantomme, P., 2013. Edible Insects: Future Prospects for Food and Feed Security. FAO Forestry Paper 171. Food and Agriculture Organization of the United Nations, Rome, ISBN 978-92-5-107596-8.

Isley, F.B., 1938. The relations of Texas acrididae to plants and soils. Ecological Monographs 8, 553–604.

Lockwood, J.A., 2004. Locust: The Devastating Rise and Mysterious Disappearance of the Insect that Shaped the American Frontier. Basic Books, New York, ISBN 0-7382-0894-9.

Lockwood, J.A., Schell, S.P., 1994. Perceptual, developmental, experiential, and physiological parameters of swimming in Melanopline grasshoppers (Orthoptera: Acrididae). J. Insect Behav. 7, 183–198.

Lockwood, J.A., Schell, S.P., November, 1997. Decreasing economic and environmental costs through reduced area and agent insecticide treatments (*RAATs*) for the control of rangeland grasshoppers: empirical results and their implications for pest management. J. Orthoptera Res. (6), 19–32.

Lockwood, J.A., Schell, S.P., Foster, R.N., Reuter, C., Rachadi, T., 2000. Reduced agent-area treatments (RAAT) for management of rangeland grasshoppers: efficacy and economics under operational conditions. Int. J. Pest Manag. 46 (1), 29–42.

Lyons, R.K., Macken, R., Forbes, D.A., 1999. Understanding Forage Intake in Range Animals. Texas Extension Agriculture Service L5152.

Madsen, D.B., 1989. A grasshopper in every pot: in the desert west, small game made big sense. Nat. Hist. 7, 22–25.

Madsen, D.B., Kirkman, J.E., 1988. Hunting hoppers. Am. Antiq. 53 (3), 593–604.

Otte, D., 2012. Eighty new *Melanoplus* species from the United States (Acrididae: Melanoplinae). Trans. Am. Entomol. Soc. 138 (1 and 2), 73–167.

Parker, J.R., 1954. Grasshoppers: A New Look at an Ancient Enemy. Farmers Bulletin No. 2064. UNT Digital Library, USDA Washington, D.C. http://digital.library.unt.edu/ark:/67531/metadc3371/.

Penner, M.P., Simpson, S.J., 2009. Locust Phase Polyphenism: An Update. Academic Press, ISBN 978-0-12-374828-7. Elsevier, Oxford.

Pfadt, R.E., 2002. Field Guide to Common Western Grasshoppers. Bulletin 912. University of Wyoming Agricultural Experiment Station, Laramie, WY, USA.

Pfadt, R.E., Hardy, D.M., 1987a. A historic look at rangeland grasshoppers and the value grasshopper control programs. In: Capinera, J.L. (Ed.), Integrated Pest Management on Rangeland: A Shortgrass Prairie Perspective. Westview Press, Boulder, CO.

Pfadt, R.E., Hardy, D.M., 1987b. Conducting a Rangeland Grasshopper Control Program: A Chronology. Bulletin B-882 September. Agriculture Experiment Station, University of Wyoming, Laramie, WY, USA.

Reigert, P.W., Ewen, Al B., Lockwood, J.A., 1997. A history of chemical control of grasshoppers and locusts 1940-1990. In: Gangwere, S.K., Muralirangan, M.C., Muralirangan, M. (Eds.), The Bionomics of Grasshoppers, Katydids and Their Kin. CABI, New York, ISBN 0-85199-141-6.

Reardon, P.O., Merrill, L.B., 1976. Vegetative response under various grazing management systems in the Edwards Plateau of Texas. J. Range Manag. 29 (3).

Schell, S.P., Lockwood, J.A., December 1997. Spatial analysis of ecological factors related to rangeland grasshopper (Orthoptera: Acrididae) outbreaks in Wyoming. Environ. Entomol. 26 (6), 1343–1353(11).

Shotwell, R.L., 1942. Evaluation of Baits and Bait Ingredients Used in Grasshopper Control. Entomology, Division of Cereal and Forage Insect Investigations Bureau of Entomology and Plant Quarantine. Technical Bulletin No. 793. United States Department of Agriculture, Washington, D.C.

Shotwell, R.L., Cowan, F.T., 1928. Some preliminary notes on the use of sodium arsenite dust and spray in the control of the Mormon cricket *Anabrus simplex* (Halde.) and the lesser migratory grasshopper *Melanoplus atlantis* Riley. J. Econ. Ento 21 (1), 222–230(9).

Steedman, A. (Ed.), 1988. Locust Handbook, second ed. Overseas Development Natural Resources Institute, London, ISBN 0-85954-232-7. 180 pp.

Ushatinskaya, R.S., 1984. A critical review of superdiapause in insects. Ann. Zool. 21, 3–30.

Zimmerman, K.M., Lockwood, J.A., Latchininsky, A.V., 2004. A spatial, markovian model of rangeland grasshopper (Orthoptera: Acrididae) population dynamics: do long-term benefits justify suppression of infestations? Environ. Entomol. 33 (2), 257–266.

Chapter 4

Locusts: An Introduction

Jeffrey A. Lockwood
Department of Philosophy and Creative Writing Program, University of Wyoming, Laramie, WY, USA

Locusts are among the most devastating pests of human agriculture. Their name is derived from the Latin *locus ustus*, meaning "burnt place," to describe the condition of the land after a swarm has passed. These insects have caused serious damage to crops and forage on every arable continent, and their depredations have become the basis for legends, myths, and—in recent times—complex political and economic programs. No pest problem spans such immense areas, with 16 million km^2 prone to outbreaks of the desert locust, *Schistocerca gregaria*, alone.

In essence, locusts are simply those species of grasshoppers that exhibit behavioral and physiological "phase changes" associated with migration under crowded conditions. This capacity to exhibit a gregarious phase is a continuous, rather than discrete, biological characteristic with some species being more locustlike than others. In this context, grasshoppers and locusts have three critical similarities.

First, both are in the family Acrididae, and thereby share a number of fundamental physiological, behavioral, and ecological similarities, including hemimetabolous development (eggs, nymphs, and adults), egg pods which are usually buried in the soil, herbivory combined with opportunistic cannibalism, susceptibility to predators and pathogens, efficient water conservation, and capacity for rapid population growth (Uvarov, 1966, 1977). Other than belonging to the same taxonomic family, the 10 species of "true" locusts (out of some 10,000 species of acridids) do not share any phylogenetic commonality. The life history strategy of locusts has evolved within rather diverse subfamilies.

Second, virtually all species of acridids are native to the habitats in which they occur. Although movements may be extensive, there are no exotic locusts (the one exception may be *Schistocerca nitens* on the Hawaiian Islands, although this species is not generally considered to be a "true" locust; Lockwood and Latchininsky, 2008). Hence, these organisms are not, by definition, targets of classical biological control programs (the use of exotic agents to control coevolved, exotic pests). Moreover, given that these species have evolved within the ecosystems where they occur, eradication is an

Biological and Environmental Hazards, Risks, and Disasters. http://dx.doi.org/10.1016/B978-0-12-394847-2.00004-8

ecologically risky and economically untenable management strategy. Considerable evidence suggests that the extirpation or chronic suppression of native species from core areas of their range, particularly those that appear to be so deeply integrated into immense, native ecosystems, is generally ill-advised (Lockwood, 1993a,b).

Third, the outbreak dynamics of acridids are the manifestation of natural processes in which the organisms exploit ephemeral resources that arise and disappear as a function of erratic weather conditions. These opportunistic resource trackers (Kemp, 1992) may have the frequency, duration, or scale of their population dynamics modified by human activities (Lockwood et al., 1988), but the essential capacity for catastrophic outbreaks is an evolved strategy to exploit unpredictably abundant resources. This life history strategy allows outbreaks to encompass immense spatial scales.

Although locusts and grasshoppers have many similarities, a number of important differences occur. First, grasshopper species are typically univoltine (one generation per year), whereas locusts are almost invariably multivoltine, at least during the development of outbreaks (Uvarov, 1966, 1977; Pedgley, 1981). Hence, locusts have an intrinsically greater capacity for population increase on an annual basis, and a lag in our response to a growing infestation has much more serious consequences.

Second, grasshopper outbreaks are often composed of many species whereas locust outbreaks are usually comprised of a single species, so the opportunity for host-specific cultural and biological control approaches is greatly enhanced (Lomer and Prior, 1992).

Third, during an outbreak, grasshoppers are more sedentary than locusts. At high densities, locust nymphs travel long distances in tightly cohesive bands. The great mobility of locusts has profound management implications. For example, large areas must be incorporated into a locust-control program, and tracking the movement of swarms becomes a critical element of a campaign. However, logistical advantages of a concentrated, mobile pest also occur. Strips of insecticides spaced at >1 km can serve as barriers during control programs for nymphal bands (Rachadi and Foucart, 1996).

Locust and grasshopper infestations often warrant intervention with chemical, biological, or cultural methods, but it is also important to understand that these insects are deeply rooted in various societies and our responses can be influenced by cultural perceptions. Perhaps most infamously, the eighth plague of Egypt in *The Book of Exodus* set the stage for both early and ongoing views of locusts as malevolent creatures. Such an adversarial framework set the stage for the use of war metaphors, which arose among acridologists in the 1930s and continue today (e.g., control programs are called "campaigns" during which we combat invading swarms).

Although locusts can be devastating to agriculture and human well-being, particularly in places of subsistence farming, some nomadic people considered the arrival of locusts to be a nutritional windfall given the protein and lipid available in the insects' bodies. And some studies suggest that the organic material left behind after a swarm passes may provide valuable fertilizer for subsequent crops—an asset that can be exploited only if the farmer survives physically or economically to enjoy the benefits.

In short, locusts are not a monolithic group of insects—nor are the ways in which we perceive and respond to them uniform. The biology and ecology of both the insects and humans interact in complex ways such that although scientists have much to gain from exchanging knowledge, we must be careful to take into consideration local contexts in terms of both problems and solutions.

As political, cultural, and communication barriers between scientists dissolve, the possibility of learning from one another's experiences (both failures and successes) promises to dramatically accelerate the rate of innovation, progress, and discovery in pest management. For example, the method of reduced agent-area treatments (in which insecticide is applied to swaths, separated by untreated buffers), which completely reshaped management of rangeland grasshoppers in the US and now serves as the standard approach (Lockwood et al., 2000, 2002), was based on the adaptation of tactics developed by African, Australian, Asian, and European scientists (Rachadi and Foucart, 1996; Musuna and Mugisha, 1997; Scherer and Célestin, 1997; Wilps and Diop, 1997; Launois and Rachadi, 1997).

The key to successful adaptation of management methods must begin with intellectual modesty and nationalistic humility so that the insights of non-scientists and experts from outside of one's country are given respect and serious consideration. It is subsequently necessary to recognize the essential similarities and differences between land-use systems and understand the political and socioeconomic contexts in which challenges and strategies have developed.

I cannot claim to understand the complexities of locust biology, ecology, and management around the world—let alone the historical, social, cultural, and political facets—as it has taken decades for me to gain insight regarding the subtleties of grasshopper management on rangeland in the US. However, I am confident that we have a great deal to gain from an open dialog, and the chapters which follow this introduction are an attempt to open such a discussion, recognizing that there are necessary simplifications in summarizing enormously complicated and sophisticated biological and human systems.

REFERENCES

Kemp, W.P., 1992. Rangeland grasshopper (Orthoptera: Acrididae) community structure: a working hypothesis. Environ. Entomol. 21, 461–470.

Launois, M., Rachadi, T., 1997. The problem of a replacement for dieldrin. In: Krall, S., Peveling, R., Diallo, D.B. (Eds.), New Strategies in Locust Control. Birkhäuser, Basel, Switzerland, p. 247.

Lockwood, J.A., 1993a. The benefits and costs of controlling rangeland grasshoppers with exotic organisms: the search for a null hypothesis and regulatory compromise. Environ. Entomol. 22, 904–914.

Lockwood, J.A., 1993b. Environmental issues involved in the biological control of rangeland grasshoppers (Orthoptera: Acrididae) with exotic agents. Environ. Entomol. 22, 503–518.

Lockwood, J.A., Latchininsky, A.V., 2008. Philosophical justifications for the extirpation of non-indigenous species: the case of the grasshopper *Schistocerca nitens* (Orthoptera) on the Island of Nihoa, Hawaii. J. Insect. Conserv. 12, 235–251.

Lockwood, J.A., Kemp, W.P., Onsager, J.A., 1988. Long-term, large-scale effects of insecticidal control on rangeland grasshopper populations. J. Econ. Entomol. 81, 1258–1264.

Lockwood, J.A., Schell, S.P., Foster, R.N., Reuter, C., Rachadi, T., 2000. Reduced agent-area treatments (RAATs) for management of rangeland grasshoppers: efficacy and economics under operational conditions. Intern. J. Pest. Manage. 46, 29–42.

Lockwood, J.A., Anderson-Sprecher, R., Schell, S.P., 2002. When less is more: Optimization of reduced agent-area treatments (RAAT) for management of rangeland grasshoppers. Crop Prot. 21, 551–562.

Lomer, C.J., Prior, C. (Eds.), 1992. Biological Control of Locusts and Grasshoppers. Redwood Press, Melksham, United Kingdom, 394 pp.

Musuna, A.C.Z., Mugisha, F.N., 1997. Evaluation of insect growth regulators for the control of the African migratory locust, Locusta migratoria migratorioides (R. & F.), in Central Africa. In: Krall, S., Peveling, R, Ba Diallo, D (Eds.), New strategies in locust control. Birkhauser, Basel, Switzerland, pp. 137–142.

Pedgley, D., 1981. Desert Locust Forecasting Manual, vol. 1. Centre for Overseas Pest Research, London, United Kingdom.

Rachadi, T., Foucart, A., 1996. L'efficacité du fipronil en traitement en barrières contre les bandes larvaires du Criquet pèlerin, *Schistocerca gregaria* (Forskål, 1775) en conditions réelles d'opérations antiacridiennes. CIRAD Document 538, Montpellier, France.

Scherer, R., Célestin, H., 1997. Persistence of benzoylphenylureas in the control of the migratory locust *Locusta migratoria capito* (Sauss.) in Madagascar. In: Krall, S., Peveling, R., Diallo, D.B. (Eds.), New Strategies in Locust Control. Birkhäuser, Basel, Switzerland, pp. 129–136.

Uvarov, B., 1966. Grasshoppers and Locusts: A Handbook of General Acridology, vol. 1. Cambridge University Press, London, 481 pp.

Uvarov, B., 1977. Grasshoppers and Locusts: A Handbook of General Acridology, vol. 2. Centre for Overseas Pest Research, London, 613 pp.

Wilps, H., Diop, B., 1997. Field investigations on *Schistocerca gregaria* (Forskål) adults, hoppers and hopper bands. In: Krall, S., Peveling, R., Diallo, D.B. (Eds.), New Strategies in Locust Control. Birkhäuser, Basel, Switzerland, pp. 117–128.

Chapter 4.1

The Australian Plague Locust—Risk and Response

Chris Adriaansen[1], James D. Woodman[1], Edward Deveson[1] and V. Alistair Drake[2,3]

[1]*Australian Plague Locust Commission, Canberra, ACT, Australia,* [2]*School of Physical, Environmental and Mathematical Sciences, UNSW Canberra, The University of New South Wales, Canberra, ACT, Australia,* [3]*Institute for Applied Ecology, University of Canberra, Canberra, ACT, Australia*

ABSTRACT

Locust plagues are natural hazards that have been historically regarded as disasters because of their impact on agricultural production. In Australia during the nineteenth and early twentieth centuries, the impacts of locusts led to significant hardships among farmers struggling to establish viable individual livelihoods. The use of pesticides for locust control and the establishment during the 1970s of coordinated response arrangements has significantly mitigated the economic and social impact of plagues.

The main risk now is of control failure, which could lead to major economic losses, but significant concern also exists about unnecessary interventions; contamination of nontarget crops, pastures, and livestock; effects on natural ecosystems; and injuries and health hazards for control staff and the general public. Establishment of a national specific-purpose locust control organization with expert staff and formal links to regional stakeholders has allowed development of appropriate and effective responses to these risks. These responses have now been formalized as defined operating procedures. Mitigating the risks of locust control is ultimately as important as mitigating the impact of the locust plagues themselves.

4.1.1 INTRODUCTION

Locust plagues are natural hazards that have historically been regarded as emergencies because of their impact on agricultural production, and particularly on subsistence farmers who have suffered famine in their wake. Migrations of swarms propagate the risks over large geographical areas and often across national boundaries, resulting in the involvement of governments and institutions worldwide in locust control as a public good. In Australia during the nineteenth and early twentieth centuries, the impacts of locusts led to significant

Biological and Environmental Hazards, Risks, and Disasters. http://dx.doi.org/10.1016/B978-0-12-394847-2.00005-X

hardships among farmers struggling to establish viable individual livelihoods and were perceived to be increasing. Since the World War II, the consolidation and diversification of agricultural enterprises, the adoption of broadscale locust control technologies and the availability of social welfare has tended to reduce the personal impacts, but farmers can still suffer significant intermittent economic losses at crucial times.

Although several pest locust species occur in Australia, damage to agriculture has been caused most frequently by swarms of the Australian plague locust *Chortoicetes terminifera* (Walker), with the south-eastern states of New South Wales, South Australia, and Victoria most severely affected. Various levels of institutional arrangement and government involvement have been established in dealing with this species, and recognition from the 1930s onward that it constitutes a national problem has led to periods of intense scientific research (Deveson, 2011). The current system of joint landholder, community, and state and federal government responsibility was established in the 1970s with the creation of the Australian Plague Locust Commission (APLC) as an intergovernment body resourced to undertake monitoring, forecasting, and control. Its responsibilities include mitigation of both the risk of plague development and the economic, human, and environmental hazards involved in managing locust populations. This chapter describes the activities undertaken to manage locust outbreaks in Australia, how these have evolved over time, and the ongoing process of review and adjustment in response to changing community expectations, legislation and technological innovation.

4.1.2 ECOLOGY OF THE AUSTRALIAN PLAGUE LOCUST

The Australian plague locust, *C. terminifera*, occurs across a range of subtropical and temperate habitat areas in all mainland Australian states, with migratory interchange between the eastern and western populations sufficient for gene flow to result in no distinguishable genetic variation between populations across its geographic range (Chapuis et al., 2011). The natural habitat of *C. terminifera* is open tussock grassland and open woodland (Hunter et al., 2001), which cover over two million square kilometers of the continent's inland. It has been suggested that overgrazing and widespread clearing of scrub and forest to increase the area available for grazing and cropping, mainly during the nineteenth and early twentieth centuries, has increased the geographical extent, frequency, and intensity of locust infestations (Clark, 1947, 1950, 1972). Rainfall, consequent soil moisture levels, and temperature are the most important abiotic factors that influence *C. terminifera* distribution and abundance, acting through their effects on vegetation growth (i.e., food supply) as well as locust development rates and activity levels (Clark, 1974; Farrow, 1979; Hunter et al., 2001; Woodman, 2010). Heavy rainfall (>40 mm) is required at intervals of no greater than approximately 6 weeks over spring, summer, and autumn to stimulate new vegetation growth in the semiarid and arid environments inhabited by this species (Hunter, 1989). Such rainfall events are inherently patchy across the landscape, with the resultant uneven vegetation distribution promoting aggregation of locust populations.

Chortoicetes terminifera generally has two to four generations between spring (September) and autumn (May) each year depending on latitude (Farrow, 1979; Hunter et al., 2001). Eggs laid in autumn may enter diapause, an anticipated dormancy in response to the decline in daylight hours experienced by the parental generation, with the proportion of diapausing eggs reaching almost 100% in mid-March at 35° S (Wardhaugh, 1980a,b; Deveson and Woodman, 2014a,b). Diapause eggs remain dormant during winter and resume development to hatch after soil temperatures increase in spring. Developing embryos also enter a quiescent state in direct response to low soil moisture levels at either of two development stages corresponding to approximately 25–30% and 40–45% of their total development time. Quiescence can occur at any time of the year and development resumes rapidly when soil moisture increases following rainfall (Wardhaugh, 1980b). *Chortoicetes terminifera* nymphs that hatch in late autumn also appear able to enter an overwintering diapause state (Wardhaugh, 1979). The interaction of development and dormancy produces a range of different potential development paths that may synchronize population timing with suitable habitat conditions, which can often be short-lived.

Of no less importance is the capacity for long-distance movement, which allows favorable habitats to be located and exploited as they arise. Daytime *C. terminifera* movements often occur with the locusts aggregated into loose, swarmlike concentrations that are highly visible; these movements generally cover only tens of kilometers per day (Casimir, 1987). Overnight migration is much less conspicuous (except with the use of radar, Drake and Wang, 2013), but regularly results in population displacements of several hundred kilometers, infesting previously clear areas, and initiating large-scale breeding events (Farrow, 1977; Symmons and Wright, 1981). A trend for movements to be northward in spring and early summer and southward in autumn has been identified and is interpreted as migration between temperate-zone winter-rainfall areas at higher latitudes and summer rainfall subtropical pastures (Deveson et al., 2005). The species' mobility and its high potential population growth rates can result in transient and translocated risks and impacts.

4.1.3 POPULATION OUTBREAKS

Population buildup of *C. terminifera* can occur rapidly over several generations, often starting from a very low base (<200 adult locusts per hectare) following an extended period of below average rainfall. The increase in locust numbers to swarm densities requires at least 10–12 months of habitat suitability (Clark, 1974; Farrow, 1979). Although population increases have occurred when favorable habitat has been confined to one or two regions, the most significant infestations or plagues (where numerous agricultural regions across two or more states are affected by swarms) arise when high rainfall and soil moisture combine with temperatures conducive to development across the majority of the eastern or western habitat range. For eastern Australia, such conditions are often associated with La Niña periods of the El Niño Southern Oscillation sea surface temperature variations over the Pacific Ocean (Allan et al., 1996; King et al., 2013).

The habitat range of *C. terminifera* in eastern Australia is very extensive, and no specific source area for large-scale outbreaks is apparent, although summer rainfall rangelands in the far southwest of Queensland are often involved in population buildup and were previously regarded as the principal plague source area (Wright, 1987). Summer populations in the subtropics can originate from immigration from the south during spring and into December (Deveson et al., 2005). Southward movements in autumn, which are intense in outbreak years, can lead to widespread swarms and subsequent oviposition in agricultural regions of South Australia, New South Wales (NSW), and Victoria. The mobility of the species greatly increases the complexity of monitoring outbreaks and of implementing early intervention management strategies, as outbreaks can occur across a variety of agricultural production areas.

Plagues of *C. terminifera* are generally short-lived compared to those of some locust species occurring in other continents. They occur roughly once each decade and commonly last between 1 and 2 years but have continued for up to 4 years when higher rainfall conditions persist (Symmons and Wright, 1981; Wright, 1986; George, 2001; Figure 4.1.1). Widespread plagues lasting more than 1 year have been recorded only four times over the past 80 years. Outbreaks are spatially variable, with high-density populations not necessarily occurring in the same regions across generations or years (Wright, 1987; Deveson, 2013).

4.1.4 HISTORY OF LOCUST OUTBREAKS AND CONTROL IN AUSTRALIA

The first historically recorded plague of *C. terminifera* occurred during 1871–1875, with swarms affecting the recently established pastoral and agricultural areas across the colonies of South Australia, Victoria, and NSW (Key, 1938). During the following half century numerous shorter-lived plagues occurred, some persisting in different regions for several years. The risk of locust plagues was regarded as increasing and, with the sudden arrival of swarms, often considered beyond control. In the absence of official knowledge or guidance the early responses by farmers included attempts to accommodate the presence of locusts by altering sowing time for crops and harvesting of hay ahead of the swarms, or by manual and mechanical control of eggs or nymphs using cultivation, trampling by livestock, or fire (Deveson, 2012).

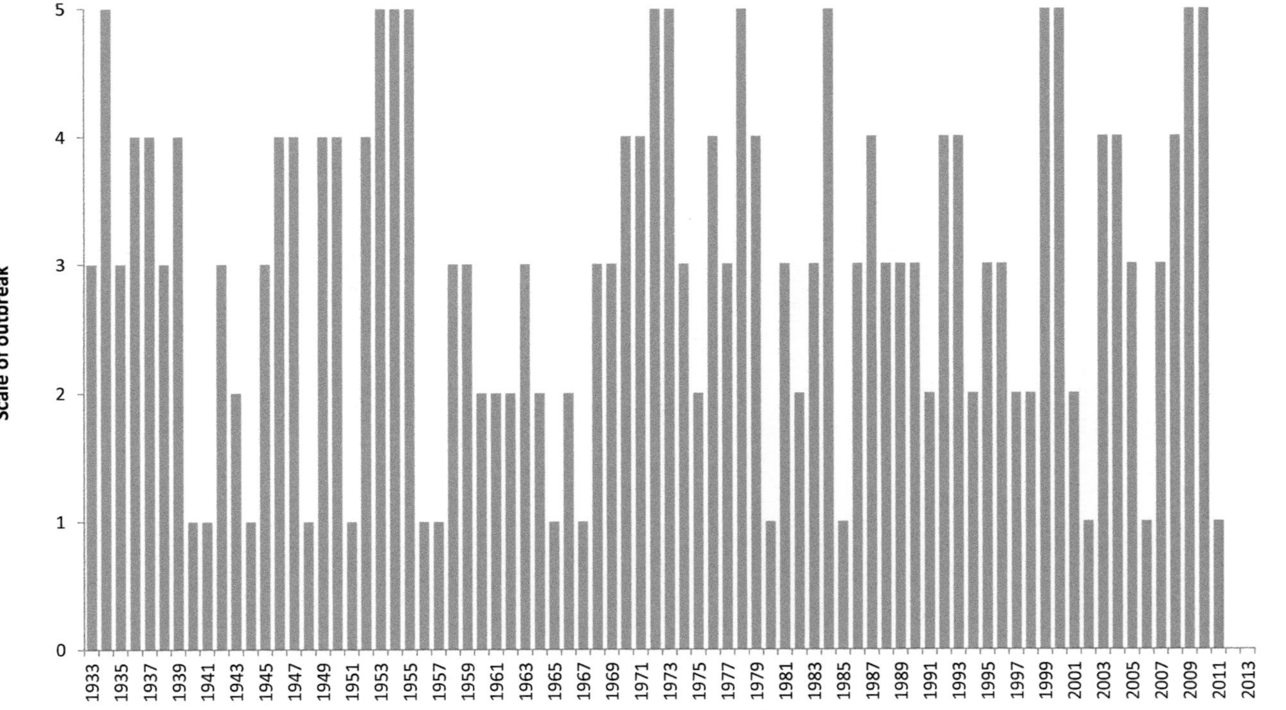

FIGURE 4.1.1 Locust infestations in eastern Australia (1933–2013). Scale: 0—no bands or swarms; 1 and 2 outbreaks—bands and swarms in one or two regions; 3 and 4 major outbreak—many bands and swarms in several regions; 5 plague—numerous regions in two or more states affected by bands and swarms. *Source: APLC, extended from Wright, 1987.*

During the early twentieth century, state governments engaged entomologists who trialled various means of locust control and converged on the use of arsenical compounds as baits and sprays (Gurney, 1934; Casimir, 1965). The hazards these posed to livestock, people, and insectivorous birds were recognized, but were considered acceptable given the public good resulting from the control (CSIR, 1938). These early entomologists reiterated the importance of coordinated communal action for control to be successful. The nature of locust population irruption and migration was realized at the time, with arguments that individual efforts were futile when the locusts originated from elsewhere.

Following the development of chlorinated hydrocarbon pesticides in the 1940s, trials in which these were broadcast from aircraft led to a new set of control technologies and more effective kill over wide areas (Hogan, 1952; CSIRO, 1954). It also brought a new set of hazards to human health and to nontarget biota, and resulted in the repeated potential exposure of operators, farmers and the environment in particular regions, especially western NSW. From the 1950s to the 1970s, landholders were supplied with organochlorine insecticides to control nymphal bands and state agriculture departments sprayed swarms using aircraft (Casimir, 1967; Hogan, 1952). Research in the 1960s by state agencies had established that bands could be seen from the air that aircraft spray technologies allowed targeted pesticide delivery over large areas, and that organophosphate insecticides provided effective control (Casimir, 1967). Since that time, the organophosphate insecticide fenitrothion has been the mainstay of aerial locust control in Australia, although other chemical pesticides such as fipronil have been employed and the biopesticide *Metarhizium acridum* (Driver and Milner), a naturally occurring pathogenic fungus effective against Orthoptera, has been used operationally since 2000.

4.1.5 ECONOMIC AND SOCIAL IMPACTS

The major quantifiable impact of *C. terminifera* outbreaks and plagues is on agricultural production, where their presence reduces both productivity and product quality. However, unlike Desert locust (*Schistocerca gregaria* Forskal) in Africa, the Middle East and western Asia and some of the Eurasian species, *C. terminifera* outbreaks do not cause humanitarian problems such as widespread famine. The direct economic impact arises from the consumption of grazing pastures and cereal grain and horticultural crops by locust nymphs and adults, often at very inopportune times (e.g., during germination or immediately prior to harvest). Swarm movements tend to disperse crop and pasture damage over much larger areas than would be expected from the size of the swarms alone.

The risk of infestation and agricultural losses is not spatially uniform even within areas of potential locust habitat, and this is reflected in the very uneven spatial pattern of APLC control of infestations (Figure 4.1.2). During the period 1978–2010, parts of the NSW Central West and Riverina regions

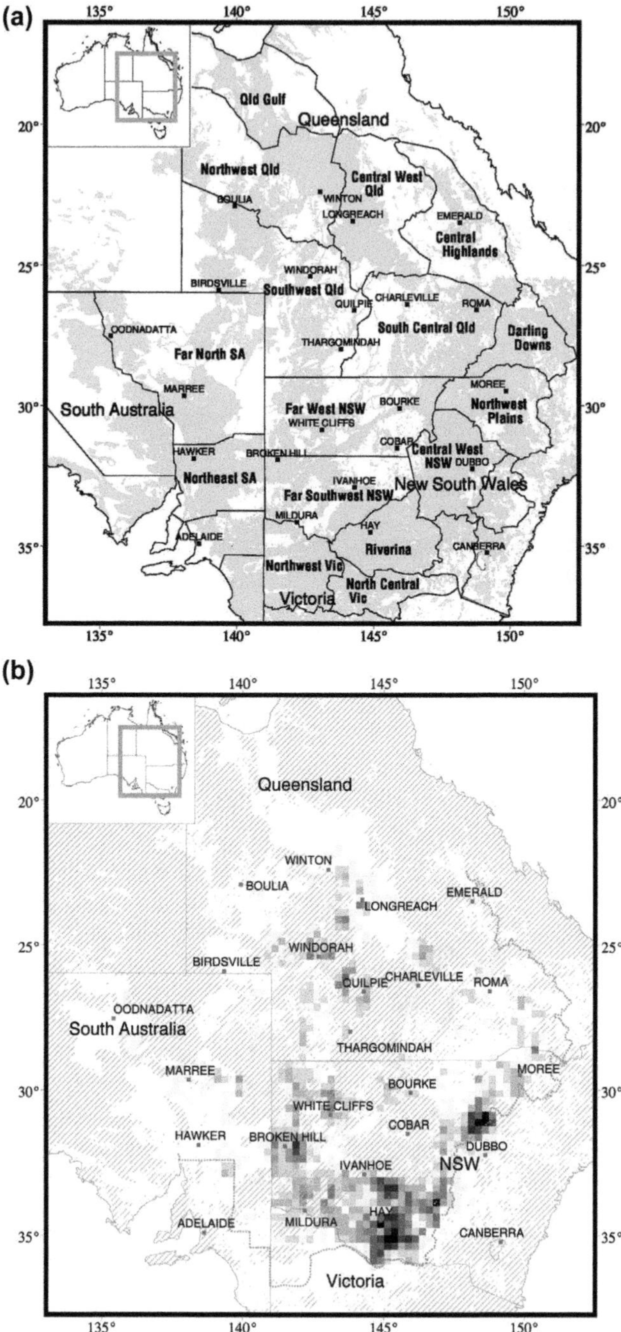

FIGURE 4.1.2 (a) Habitat regions of C. terminifera in eastern Australia. Shaded areas are potential locust habitat. (b) Frequency of Australian Plague Locust Commission (APLC) control 1978–2010. Frequency calculated as the number of seasons (either spring, summer, or autumn) in which any aerial control occurred within each 0.25 × 0.25 degree grid cell over 33 years of APLC operations. Frequency graded from white = 0 to black = 10 seasons. APLC operational boundary has been shown in red. Green hatching shows habitats mostly unsuitable for locust breeding.

experienced infestations more frequently than elsewhere, and consequently aerial control by the APLC and state agencies has also occurred more frequently in these more closely settled cropping regions. Control by landholders, using ground spray equipment, is also most frequent in these regions, and the risks associated with chemical use for locust control has therefore been geographically concentrated there. Infestations have also been frequent in pastoral areas of Southwest Queensland and Far West NSW, but APLC aerial control has been less frequent and there is little other application of insecticides in these regions.

Most major economic losses have occurred during autumn, when adult swarms damage emerging cereal and horticultural crops, often necessitating resowing. Losses can occur during spring when dense, overwintering locust egg beds (up to 1,000 egg pods per square meter each typically containing 26–60 eggs (Farrow, 1979; Woodman, 2010)) produce high-density nymphal bands near maturing grain crops. Slightly later in the season, older nymphs, fledglings, and adult locusts have been observed to eat through plant stems just below the grain head in maturing crops as they are beginning to dry off, resulting in a loss of the set grain (Love and Riwoe, 2005). In the case of native or improved pastures, locusts consume plant matter which otherwise would have contributed to liveweight gain or other production by grazing animals. A further economic impost results when fledglings and adults are present during grain crop harvesting, which often coincides with the latter stages of the first generation of locusts each season (Millist and Abdalla, 2011). The locusts can be picked up by harvesting equipment resulting in grain contaminated with insect body parts at levels exceeding the prescribed maximum set for grain to be sold for human consumption. This grain must then either be passed through expensive cleaning equipment or downgraded to feed-grain status and sold at a considerably lower price.

Arriving at a realistic figure for the economic impact of a *C. terminifera* outbreak or plague has proven very challenging, with a number of studies undertaken over the years (Wright, 1986; Love and Riwoe, 2005; Abdalla, 2007; Millist and Abdalla, 2011; Murray et al., 2013). Most of these have relied upon a series of estimates regarding locust densities across infested areas, amount and type of plant matter consumed, and the resulting loss of productivity. Notwithstanding the sensitivity of the final analysis to some of the input data and assumptions used, it has been estimated that additional losses to agriculture from the most recent plague in eastern Australia in 2010–2011 would have amounted to over AU$950 million (US$950 million at that time) had no control been undertaken. Based on this figure, the AU$50 million (US$50 million) invested by landholders, APLC, and other agencies in implementing locust control clearly provided a very positive benefit-to-cost ratio (where benefit is the avoided loss) (Millist and Abdalla, 2011). During the peak of that plague, some 40 million hectares of grazing and cropping land were infested to varying extents.

However, the risks and impacts of *C. terminifera* outbreaks and plagues extend well beyond the direct costs to agricultural productivity, and include the social effects on rural producers and communities, the risk inequality across the whole community, and the risks and hazards associated with implementing locust control programs. For example, significant outbreaks or plagues of *C. terminifera* have often occurred in the early phases of a La Niña period which has been preceded by an El Niño (or drought) period during which rural producers and their associated communities have endured several years of low rainfall, low crop yields, and limited grazing productivity. The "breaking of the drought" with the commencement of a La Niña period therefore sees a return of producer and community optimism and economic confidence, which could quickly diminish with the onset of a locust plague (Murphy, 2010).

Chortoicetes terminifera outbreaks and plagues have limited impact upon the nonfarming community, particularly those living in the major Australian capital cities which are outside preferred habitat and potential infestation areas. Road traffic traveling through swarms may be slowed and vehicles soiled, while in a few areas dwellings may require cleaning. Swarms may constitute a nuisance during outdoor activities and limited direct impact may occur, as when a local race meeting had to be canceled because flying locusts presented a safety hazard to both horses and jockeys (Ballarat Courier, 2010). These impacts are, however, very minor compared to the direct costs and losses experienced by agriculturalists and graziers. Locust plagues thus add to the perceived inequality of risk faced by rural producers compared to city dwellers.

4.1.6 THE AUSTRALIAN PLAGUE LOCUST COMMISSION AND CURRENT APPROACHES TO LOCUST MANAGEMENT IN AUSTRALIA

A series of significant widespread (level 4) outbreaks and (level 5) plagues in eastern Australia during the 20-year period from the mid-1950s (Figure 4.1.1) highlighted the difficulties of coordinating effective control across several jurisdictions, and gave rise to the decision to create the APLC in 1974. APLC is a joint venture between the Australian Government and the Member States of New South Wales, Victoria, South Australia, and Queensland. Its stated objective is to monitor and manage locust populations which pose a threat to more than one Member State, while localized outbreaks and the protection of crops and pastures remain the responsibility of regional and state agencies and the landholders themselves (APLC, 1976).

Although the initial driving rationale for the establishment of APLC was the control of *C. terminifera* outbreaks, its charter recognizes the need for ongoing population monitoring, research and development of monitoring and management techniques, and the provision of advice on impact mitigation to all stakeholders. Rather than introduce the potential for regulatory conflict by establishing specific Australian Government legislation covering APLC and its activities, it was determined that APLC would operate across the four Member States using the applicable legislation in each jurisdiction. Consequently, APLC has developed and applies locust control and other practices that are compatible with the regulatory requirements of all jurisdictions. This has proven to be a relatively minor constraint, and the clear focus in APLC's charter on providing effective national management of the risks and impacts of *C. terminifera* infestations, when combined with the propensity of this species to migrate from one state to another, has generally resulted in a high level of cooperation between jurisdictions and agencies. APLC's working strategy is to locate developing locust populations to assess whether they (or a subsequent generation) are likely to migrate to where more significant impacts might result, and if so to undertake preemptive control before migration occurs.

The lands affected by *C. terminifera* are largely held under private ownership, and the primary responsibility for locust management therefore remains with the landholder. Reporting of locusts by landholders is mandatory in some states but only encouraged in others. Requirements for landholders to undertake control also vary between states depending upon the status of locusts as a declared agricultural pest. The involvement of various levels of government is therefore based on an acknowledgment of locust control as a public good, with benefits arising from the limitation of further population development and invasive migrations (Olsen, 1965). The APLC's role is to mitigate the impacts over broad areas, often distant from where actual control is applied, rather than to provide protection to specific areas of crop and pasture.

As a species endemic to Australia (rather than an "exotic" invasive species for which eradication is the normal response objective), *C. terminifera* is an integral component of the ecosystems of the Australian inland, spanning the subtropical to cool-temperate latitudinal range. Many native Australian bird, small mammal, and reptile species utilize *C. terminifera* as an important ephemeral food source, with locust outbreaks often accompanied by significant bird assemblages (Szabo et al., 2003). In a region where rainfall, and consequently the availability of food resources, is highly variable locust outbreaks provide significant protein and lipid sources which allow for a surge of breeding activity amongst various predators (Szabo et al., 2009). Therefore, *C. terminifera* control implemented by APLC and state agencies must focus on population management during periods of outbreak rather than on pest eradication, so that an appropriate balance is achieved between the role of this species in the Australian environment and the risks which population upsurges represent to agricultural production and the broader community. The strategy is therefore one of risk mitigation and not risk elimination.

4.1.7 THE RISKS AND HAZARDS OF LOCUST CONTROL IN AUSTRALIA

The risks and hazards associated with *C. terminifera* management can be divided into two broad groups. First, there is the risk of control failure, i.e., that a locust population moves into valuable crops and causes extensive losses. Conversely, there is the risk that an intervention is made unnecessarily, i.e., effort and resources are expended controlling a locust population that would never have caused significant economic impact. Secondly, there is the risk of unintended environmental, agricultural or economic impact, or injury resulting from the control activities. Although injuries may occur to third parties, the personnel undertaking control-related activities are particularly at risk.

Australian society, along with those in other developed economies, is increasingly risk-averse and concerned about human and environmental health. Consequently, managing the risks of a locust control program has expanded to well beyond simply managing the impact of the locust population itself. Public awareness and attention helps to ensure the highest level of risk minimization and that the treatment does not end up being worse than the pest impact. A need exists for continuous improvement and development in response to changing occupational health and safety standards, greater community expectations with regard to environmental duty of care, and growing intolerance of off-target impacts. As APLC operationally represents five separate governments in Australia, very strong incentives exist for being at the forefront of best practice rather than merely compliant with legislation and regulation (Story et al., 2005).

The relative frequency with which *C. terminifera* outbreaks occur is actually beneficial for APLC's efforts to manage a number of these risks, as it necessitates regular maintenance and renewal of equipment, pesticide stocks, and infrastructure; helps to retain the experience level of APLC officers; and provides training opportunities for new officers. This development and retention of expertise is one of the most significant risk mitigation measures available to APLC.

4.1.7.1 Failure to Control

The immediate costs or losses resulting from locust plagues in eastern Australia have been estimated through various benefit—cost and other analyses. In addition, Abdalla's (2007) study explored potential longer-term costs by examining the 2004—2005 locust plague situation, where the impacts of subsequent assumed larger generations of locusts were examined in the event that early intervention to reduce initial population levels was unsuccessful. Although losses resulting from an initial locust generation might amount to over US \$50 million, it was estimated that losses in excess of US \$450 million could accumulate if subsequent locust generations were allowed to develop.

This provides some indicative measure of the potential impact from a failure to control an outbreak.

Failure to control a significant proportion of a locust population which subsequently increases or migrates can result in significant agricultural losses. Such failures may be the result of numerous factors, often constrained by the narrow time window of opportunity to plan and conduct a program during nymphal development. Logistic issues of insecticide supply and delivery, aircraft availability, access to remote locations, staff availability, and obtaining the agreement of individual landholders all set limits on the total area that can be assessed and controlled. Weather conditions frequently reduce the days available for operations during a control program. Rainfall not only curtails spraying, but can produce areas of surface water and stream flow in otherwise dry watercourses. Wind speed and direction must be within appropriate ranges for each spraying event, whereas land use and tenure limitations to insecticide use in many areas require additional planning, liaison, and buffer distances.

At the regional scale, the point of intervention, level of effective control, and reduction of overall agricultural impacts can be influenced by similar constraints. During the 2009–2010 plague, weather and other limiting factors reduced the ability to apply control measures at the earliest feasible opportunity. The plague developed after two generations of population increase in NSW and Queensland, the second following heavy inland rainfall during November 2009 to February 2010. Migrations from Southwest Queensland and further heavy rainfall associated with the development of an intense La Niña from March 2010 resulted in a third, widespread nymphal generation (Deveson, 2013). Control of the preceding summer generation in Southwest Queensland was not possible due to heavy rainfall and the extent of subsequent flooding and water hazards in potential control areas, which were widely distributed and contained few identifiable or actionable targets. A large autumn generation hatching in February–March 2010 resulted in infestations across much of NSW, Southwest Queensland, and northern South Australia. The geographic extent of the infestation was now beyond any capability for complete suppression, so APLC control was focused on the area of most intense nymphal band development in the Tibooburra area of Far West NSW. However, rainfall again intervened in the planned control program, reducing the window of opportunity for control to a 2-week period in March 2010, during which some 205,000 ha were treated (APLC, unpublished). Significant water hazards and restricted environmental areas occurred within the infested area. Although 95% of identified targets in the Tibooburra area were sprayed (APLC, unpublished), it is estimated that less than 40% of the locust population across the region was controlled (Adriaansen, 2012) and subsequent migrations into agricultural areas of South Australia, New South Wales, and Victoria resulted in extensive overwintering egg beds. These required widespread control action in the subsequent spring (September–October 2010) by all stakeholders.

4.1.7.2 Unnecessary Intervention

An inherent risk with any decision to intervene is the uncertainty involved with assessing likely future population dynamics. The complexity of population interactions, particularly the direction and scale of displacement through migration and the limited predictability of seasonal rainfall, makes inter-annual or even inter-seasonal predictions about population locations and size problematic. Additionally, decisions to intervene may be influenced by sociopolitical factors as well as a culture of aversion to risk among decision-makers.

Parasites and disease can significantly contribute to the collapse of plague populations, but they are difficult to monitor and their delayed effects make prediction of their impact difficult (Farrow, 1979; Deveson and Woodman, 2014a). Therefore, in some instances control may be carried out on populations which would have declined through natural processes. Such was possibly the case in 2010–2011, when significant late spring and summer generations of *C. terminifera* would have normally resulted in a substantial control program being undertaken by APLC, had inclement weather not intervened. Above-average rainfall during this period across most of the area infested with locusts had led to rapid and dense pasture growth, which initially favored population buildup and persistence. However, high mortality of the spring 2010 nymph generation in several locations in NSW, due to pathogen infection and subsequent egg parasitism that peaked at 90% at some sites in the NSW Riverina during summer, contributed to population collapse by autumn (Deveson and Woodman, 2014a). Such events are not unprecedented and the 1973–1974 plague is thought to have ended similarly (Bullen, 1975).

Large-scale locust control is generally publicly funded, and control actions that do not result in a significant reduction in economic losses constitute a waste of taxpayers' monies. A degree of environmental stress and a risk of off-target impacts is also always associated with a control operation. A perception that control is sometimes being implemented unnecessarily would undermine the confidence of stakeholders and the general public in the control organization and might lead to constraints or oversight requirements being imposed, with perhaps detrimental effects on operations where decisive and timely interventions are essential. Even with contemporary monitoring and forecasting capabilities (Deveson and Hunter, 2002), judgment is still required as to whether a particular developing locust population, or perhaps a subsequent generation, is likely to move into other agricultural regions and impact crops there. The possibility that the population will decline due to dry conditions or the action of natural enemies, or that it will fail to migrate or migrate in a nonthreatening direction, is routinely considered during APLC's control decision-making.

4.1.7.3 Off-target Risks and Hazards

Recognized risks are associated with the use of any pesticide. Aerial application of chemical pesticides to large areas presents hazards to human and environmental health and to trade in agricultural produce. Off-target drift onto neighboring properties and restricted or "sensitive" lands, the potential for harm to nontarget species and indirect environmental effects can be considerable (Story et al., 2005). Even when applied with best practice and only to specific high-density target areas, the hazard associated with chemical residues in produce remains (GRDC, 2010; SafeMeat, 2009). As an increasing number of graziers in inland Australia have turned to certified organic production to attract a price premium, APLC has supported the development of the biopesticide *M. acridum* and now uses this to allow some level of locust control to be applied on these properties.

Other land tenures and some legislated environmental caveats restrict the use of chemical pesticides in certain areas. Locust populations will infest environmental reserves, which can be extensive in area and contain suitable locust habitat, potentially providing a reservoir for reinfestation of adjacent treated areas. Areas of private grazing land also are the habitat of threatened species, such as the Plains-wanderer, *Pedionomus torquatus* (Gould), a ground-dwelling bird of grasslands now restricted to parts of southern NSW and northern Victoria (Story et al., 2007). In all such areas, where it is determined that control of locusts is an essential element of the broader population management objective, *M. acridum* is used, along with appropriate buffer zones if adjacent areas are treated with chemical pesticides. Although operationally less convenient than chemical pesticide due to its handling and preapplication mixing requirements and higher cost (APLC, 2012), the availability of *M. acridum* allows control to be undertaken in otherwise restricted areas. The extent to which this biopesticide is employed is highly dependent upon situational need, and has varied from below 5% of the area treated by APLC in 2009–2010 (APLC, unpublished) to as high as 46% in 2008–2009 (APLC, 2009a).

The residues resulting from the application of chemical pesticides for locust control, even when properly applied, also represent a potential unintended trade risk. APLC employs an informed consent process with landholders, to ensure that they are fully aware of the post-treatment restrictions applying to treated lands and that on-property assets at risk (such as stock water supplies and beehives) are identified and avoided (APLC, 2010). Adherence to these procedures minimizes the risk of residues in produce, including grain, meat, milk, honey, fruit, and vegetables derived from treated areas. Consequently, no export or domestic market produce from an area treated for locust control has been excluded from trade due to locust control agent residues.

Protocols have also been established to mitigate the risk of watercourses and water bodies being affected by control agents. The same downwind buffer distances (1500 m) are applied to standing water bodies as for organic production properties. When areas targeted for locust control contain dry watercourses, an assessment is made of the likelihood of significant rainfall causing run off after the proposed pesticide application. If more than 5 mm of rain is anticipated within 72 h, then application of chemical pesticides is curtailed (APLC, 2010).

Impacts on nontarget invertebrate species are being reduced through developments in pesticide application technology and procedures that allow for effective locust control while minimizing dosage rates (Hunter, 2004). Research by APLC demonstrated that reducing the ultra-low volume application rates for fenitrothion from the recommended rate of up to 508 g active ingredient (ai) per hectare to 267 g ai/ha still resulted in effective control (Hooper, 1998). This lower rate has since been adopted for all APLC aerial spraying, reducing treated area exposure rates by over 45%. A further advance involves the application of fipronil, a chemical pesticide with short-term (<6 days) residual persistence, in discrete strips up to 500 m apart. With this treatment, pesticide is applied to only 10% of the control zone, leaving large untreated areas from which nontarget species can recolonize the treated strips (Hunter, 2004). Effective control of locust nymphs is still achieved, as the aggregated mass movement of high-density nymphal infestations carries them into the treated strips. Placement of the strips varies, with the distance between them influenced by topography, rainfall and temperature as these all affect the speed at which the nymphs move.

4.1.7.4 Injury

The risk of injury, either to operational staff implementing control or to others, can be significant in the absence of appropriate mitigation measures. The inland habitat of *C. terminifera* is hot, dry, and remote, and officers undertaking population survey can be far from immediate medical attention should injury occur. Driving distances and conditions also present a significant challenge, as does the handling of pesticides during a control campaign. Pesticide exposure in and around treatment areas, to both operational staff and to members of the public, is a further concern. In all cases, appropriate policies and standards must be developed and applied to mitigate these risks by addressing the likelihood and consequences of mishandling or misapplication.

Of the various risks of injury to operators and others, aviation accidents command the highest level of attention, due to the generally catastrophic consequences of such events. Treating large areas containing the highest density portions of an outbreak population can only be efficiently achieved through the use of aircraft, both to locate high-density infestations and to apply pesticides. The risks associated with agricultural aerial operations are well documented, with an accident rate at least five times that recorded for other uses of light aircraft (ATSB, 2013). APLC has developed and rigorously applies extensive work health and safety protocols to manage the risk to its staff and contractors from aerial operations. As new safety technologies become available for light aircraft and their operators, APLC adds these to the minimum standards required for the aircraft and pilots it engages. Fatigue management of both pilots and APLC personnel is also a high priority, requiring a careful balance between responsiveness and responsibility, where the need to apply control agents to large areas in a timely manner has to be weighed against the challenges of fatigue and difficult working conditions.

APLC has adopted the "5M" structure (man, machine, materials, method, and management; FAA, 2000) for categorizing risks to its staff and others engaged in locust control operations. This approach helps to identify risks and to target controls to the source or sources of the risk. For example, addressing the machine element has resulted in APLC specifying equipment and servicing standards for the aircraft it engages, to eliminate the potential for poorly maintained aircraft being used in its operations. Similarly, the aircraft pilots are required to have extensive experience (as well as regulatory clearances) in low-level agricultural flying. To ensure adherence to these and other risk controls, APLC has developed a checklist covering the aircraft, pilot, and safety measures which must be completed before each and every APLC operation. These risk controls are included in APLC's "Instructions for Pilots" document (APLC, 2009b), which forms part of the Commission's Aviation Procedures Manual.

4.1.8 FUTURE CONSIDERATIONS

Although APLC's current strategies and procedures are generally both effective and socially appropriate for managing its locust-control task, it must be expected that developments will be required on an ongoing basis. Factors which may need to be addressed include new agricultural land uses and practices, adaptation by the pest, extension of conservation areas, and greater community expectations with regard to minimizing human injuries and health risks and avoiding ecosystem disruption. In addition, new technologies may have potential for making APLC's monitoring, forecasting, and control activities more effective, more efficient, less costly, or less risky.

APLC maintains a watching brief on potential developments relevant to its task, and reviews and revises its practices as appropriate. It undertakes or supports targeted research projects that establish whether a change of practice is needed and how any change might be implemented.

Various fundamental aspects of locust biology and ecology, examining how different environmental conditions at both fine and broad spatial scales impact upon locust development and survival, will need to be investigated to further improve the effectiveness of the Commission's monitoring and forecasting operations.

To further reflect current community expectations, the broader ecological impact of locust control measures will need to be further defined, with the aim of more explicitly understanding species and assemblage-level effects of insecticide application as well as other environmental flow-on effects.

New locust control insecticides (biological and chemical) and application methods will need to be identified and developed, with the aim of providing operational resilience in the event of deregistration of existing chemical agents following periodic safety reviews.

Also needed will be consideration and investigation of some more strategic and long-term issues which could become of increasing importance. This will include changes to agricultural land use practices and locations which could affect locust population levels and extent, and the potential effects of any changes to macro- and microclimate that could impact development and abundance of pest locust species.

4.1.9 CONCLUSIONS

Although common features occur to the problems presented by the various locust species worldwide, each has its own characteristics and requires individually tailored responses. Appropriate responses will also vary from country to country, depending on patterns of locust abundance and distribution, the biology of different locust species, the form of agriculture practiced, social and cultural norms, and perhaps most importantly the level of economic development. These differences are as evident in a consideration of risks and hazards as they are in other aspects of locust management. In the case of *C. terminifera*, two key factors have shaped the management practices employed by APLC and other government agencies. First, *C. terminifera* recessions are relatively short-lived, so that the control organization benefits from a level of continuity in maintaining the expertise and infrastructure required for implementing responses. Second, the country affected (Australia) has a developed economy in which agriculture, while important, is currently not the most significant driver of economic growth (HSBC, 2014). As a consequence, significant impacts upon agriculture such as locust plagues represent less of an overall risk to the national economy than was apparent in the first half of the twentieth century and earlier. Responses to locust plagues are therefore more measured, with an appropriate balance between risk and return. A third factor of particular importance in the context of risks and hazards is that Australian society now places an extremely high value on both human health and environmental integrity, so that locust control activities must be continuously refined to fully address the risk of injury to humans, contamination of food-production systems, and damage to ecosystems. Despite these concerns, the primary risk remains that of failure to control with its consequent potential economic losses.

REFERENCES

Abdalla, A., 2007. Benefits of Locust Control in Eastern Australia: A Supplementary Analysis of Potential Second Generation Outbreaks. Australian Bureau of Agricultural and Resource Economics Research Report 07.4.

Adriaansen, C.J., 2012. A case study on the value of early intervention. In: Strategic Review of the Australian Plague Locust Commission (APLC). R. Glanville, unpubl. Canberra. July 2012. http://www.agriculture.gov.au/__data/assets/pdf_file/0008/2364254/aplc-strategic-review-2012.pdf.

Allan, R., Lindesay, J., Parker, D., 1996. El Nino Southern Oscillation and Climate Variability. CSIRO Publishing, Melbourne.

APLC, 1976. Australian Plague Locust Commission Charter of Operations. http://www.agriculture.gov.au/animal-plant-health/locusts/role.

APLC, 2009a. Australian Plague Locust Commission Annual Activity Report 2008-09. http://www.agriculture.gov.au/SiteCollectionDocuments/animal-plant/aplc/locust/reports/activity-statement2008-09.pdf.

APLC, 2009b. Instructions for Pilots. http://www.agriculture.gov.au/SiteCollectionDocuments/animal-plant/aplc/locust/ops-documents/pilot-information-feb09.pdf.

APLC, 2010. Australian Plague Locust Commission Operations Manual Version 2.

APLC, unpublished. Australian Plague Locust Commission Locust Control Campaign Reports 2009–10.

APLC, 2012. Australian Plague Locust Commission Annual Activity Report 2011-12. http://www.agriculture.gov.au/SiteCollectionDocuments/pests-diseases-weeds/locusts/aplc/aplc-annual-activity-statement-2011-12.pdf.

ATSB, 2013. Aviation Occurrence Statistics. Australian Transport Safety Bureau Aviation Research Report AR-2103-067.

Ballarat Courier, December 6, 2010. News article "Impact of locust plague more than just economic". The Ballarat Courier newspaper.

Bullen, F.T., July 1975. Economic Effects of Locusts in Eastern Australia. Report to the Reserve Bank of Australia. Canberra.

Casimir, M., 1965. The locust problem – its history and development in New South Wales. J. Aust. Inst. Agric. Sci. 31, 267–274.

Casimir, M., 1967. The Aerial Spraying of Locusts in New South Wales from 1955 to 1970. Department of Agriculture New South Wales Science Bulletin, 86, 40.

Casimir, M., 1987. Plague locusts in New South Wales: Studies of Migration and Displacement of Populations. Department of Agriculture New South Wales Science Bulletin, 91, 21.

Chapuis, M., Popple, J.M., Berthier, K., Simpson, S.J., Deveson, E., Spurgin, P., Steinbauer, M.J., Sword, G.A., 2011. Challenges to assessing connectivity between massive populations of the Australian plague locust. Proc. R. Soc. Biol. Sci. 278, 3152–3160.

Clark, D.P., 1972. The plague dynamics of the Australian plague locust, *Chortoicetes terminifera* (Walk.). In: Hemming, C.F., Taylor, T.H.C. (Eds.), Proceedings on the International Study Conference on Current and Future Problems of Acridology, 1970. Centre for Overseas Pest Research, London, pp. 275–87.

Clark, D.P., 1974. The influence of rainfall on the densities of adult *Chortoicetes terminifera* (Walker) in central western New South Wales 1969–73. Aust. J. Zool. 23 (3), 365–386.

Clark, L.R., 1947. An Ecological Study of the Australian Plague Locust (*Chortoicetes terminifera* (Walk.) in the Bogan-Macquarie Outbreak Area, New South Wales. C.S.I.R. Bulletin No. 226. Melbourne.

Clark, L.R., 1950. On the abundance of the Australian plague locust (*Chortoicetes terminifera* (Walker) in relation to the presence of trees. Aust. J. Agric. Res. 1, 64–75.

CSIR, 1938. Council for scientific and industrial research. In: Proceedings of the First Australian Locust Conference, Melbourne, 19–22 July 1938. Resolution H (ii) 29, p. 21.

CSIRO, 1954, 27 pp.. In: Grasshopper Control Conference Held at Canberra, ACT, May 1954. CSIRO, Melbourne. Appendix B, Report by the Department of Agriculture NSW; Appendix C, Report by the Biology Branch, Department of Agriculture, Victoria 1954; Appendix E, Report by the Department of Agriculture, Western Australia – 1954.

Deveson, E.D., 2011. The search for a solution to Australian locust outbreaks: how developments in ecology and government responses influenced scientific research. Hist. Rec. Aust. Sci. 22, 1–31.

Deveson, E.D., 2012. *Naturae Amator* and the grasshopper infestations of South Australia's early years. Trans. R. Soc. South Aust. 136, 1–15.

Deveson, E.D., 2013. Satellite normalized difference vegetation index data used in managing Australian plague locusts. J. Appl. Remote Sens. 7 (1), 21, 075096.

Deveson, E.D., Drake, V.A., Hunter, D.M., Walker, P.W., Wang, H.K., 2005. Evidence from traditional and new technologies for northward migrations of Australian plague locusts (*Chortoicetes terminifera*) (Walker) (Orthoptera: Acrididae) to western Queensland. Austral Ecol. 30, 928–943.

Deveson, E.D., Hunter, D.M., 2002. The operation of a GIS-based decision support system for Australian locust management. Entomologia Sinica 9 (4), 1–12.

Deveson, E.D., Woodman, J., 2014a. Observations of *Scelio fulgidus* (Hymenoptera: Platygastridae) parasitism and development in southern NSW during the 2010 *Chortoicetes terminifera* (Orthoptera: Acrididae) locust plague. Austral Entomol. 53, 133–137.

Deveson, E.D., Woodman, J., 2014b. Embryonic diapause in the Australian plague locust relative to parental experience of cumulative photophase decline. J. Insect Physiol. 70, 1–7.

Drake, V.A., Wang, H.K., 2013. Recognition and characterization of migratory movements of Australian plague Locusts, *Chortoicetes terminifera*, with an insect monitoring radar. J. Appl. Remote Sens. 7 (1), 17, 075095.

FAA, 2000. United States Federal Aviation Administration System Safety Handbook. http://www.faa.gov/regulations_policies/handbooks_manuals/aviation/risk_management/ss_handbook.

Farrow, R.A., 1977. Origin and decline of the 1973 plague locust outbreak in central western New South Wales. Aust. J. Zool. 25, 455–489.

Farrow, R.A., 1979. Population dynamics of the Australian plague locust, *Chortoicetes terminifera* (Walker), in central western New South Wales: reproduction and migration in relation to weather. Aust. J. Zool. 27 (5), 717–745.

George, C.R.R., 2001. Use of meteorological forecasting indices and rain patterns in *Chortoicetes terminifera* (Walker) outbreak risk assessment (Bachelor of Science (Honours) thesis). University of Queensland, St Lucia.

GRDC, August 2010. Grains Research and Development Corporation Plague Locust Control Fact Sheet.

Gurney, W.B., 1934. Grasshopper swarms in the Central West. Agricultural Gazette of N.S.W., 45, 261.

Hogan, T.W., March 1952. Aerial Spraying of Locusts. J. Agric., Victoria. 50, 112–114.

Hooper, G.H.S., 1998. The changing environment of locust control in Australia. J. Orthopt. Res. 7, 113–115.

HSBC, March 2014. HSBC Agricultural Market Report.

Hunter, D.M., 1989. The response of Mitchell grass (*Astrebla* spp.) and Button grass (*Dactyloctenium radulans* (R. Br.)) to rainfall and their importance to the survival of the Australian plague locust, *Chortoicetes terminifera* (Walker), in the arid zone. Aust. J. Ecol. 14, 467–471.

Hunter, D.M., 2004. Advances in the control of locusts in eastern Australia: from crop protection to preventative control. Aust. J. Entomol. 43, 293–303.

Hunter, D.M., Walker, P.W., Elder, R.J., 2001. Adaptations of locusts and grasshoppers to the low and variable rainfall of Australia. J. Orthopt. Res. 10, 347–351.

Key, K.H.L., 1938. The Regional and Seasonal Incidence of Grasshopper Plagues in Australia. C.S.I.R. Bulletin No. 117, Melbourne.

King, A.D., Alexander, L.V., Donat, M.G., 2013. Asymmetry in the response of eastern Australia extreme rainfall to low-frequency Pacific variability. Geophys. Res. Lett. 40, 1–7.

Love, G., Riwoe, D., 2005. Economic costs and benefits of locust control in eastern Australia. Australian Bureau of Agricultural and Resource Economics Project Report 3070.

Millist, N., Abdalla, A., 2011. Benefit-cost analysis of Australian plague locust control operations for 2010–11. Australian Bureau of Agricultural and Resource Economics Project Report 43173.

Murphy, T., October 2010. Charles Sturt University Agribusiness Index.

Murray, D.A.H., Clarke, M.B., Ronning, D.A., 2013. Estimating invertebrate pest losses in six major Australian grain crops. Aust. J. Entomol. 52 (3), 227–241.

Olsen, M., 1965. The Logic of Collective Action: Public Goods and the Theory of Groups. Harvard University Press, Cambridge Massachusetts.

SafeMeat, November 2009. Locusts, Grasshoppers and Livestock Residues. Advisory brochure.

Story, P.G., Walker, P.W., McRae, H., Hamilton, J.G., 2005. A case study of the Australian Plague Locust Commission and environmental due diligence: Why mere legislative compliance is no longer sufficient for environmentally responsible locust control in Australia. Integr. Environ. Assess. Manage. 1, 245–251.

Story, P.G., Oliver, D.L., Deveson, E.D., McCulloch, L., Hamilton, J.G., Baker-Gabb, D., 2007. Estimating and reducing the amount of Plains-wanderer (*Pedionomus torquatus* Gould) habitat sprayed with pesticides for locust control in the New South Wales Riverina. Emu 107, 308–314.

Symmons, P.M., Wright, D.E., 1981. The origins and course of the 1979 plague of the Australian plague locust, *Chortoicetes terminifera* (Walker) (Orthoptera: Acrididae), including the effect of chemical control. Acrida 10, 159–190.

Szabo, J.K., Astheimer, L.B., Story, P.G., Buttemer, W.A., 2003. An ephemeral feast: birds locusts and pesticides. Wingspan 13, 10–15.

Szabo, J.K., Davy, P.J., Hooper, M.J., Astheimer, L.B., 2009. Predicting avian distributions to evaluate spatiotemporal overlap with locust control operations in eastern Australia. Ecological Applications 19, 2026–2037.

Wardhaugh, K.G., 1979. Photoperiod as a factor in the development of overwintering nymphs of the Australian plague locust, *Chortoicetes terminifera* (Walker) (Orthoptera: Acrididae). J. Aust. Entomol. Soc. 18, 387–390.

Wardhaugh, K.G., 1980a. Effects of photoperiod and temperature on the induction of diapause in eggs of the Australian plague locust, *Chortoicetes terminifera* (Walker) (Orthoptera: Acrididae). Bull. Entomol. Res. 70, 635–647.

Wardhaugh, K.G., 1980b. The effects of temperature and moisture on the inception of diapauses in eggs of the Australian plague locust, *Chortoicetes terminifera* Walker (Orthoptera: Acrididae). Aust. J. Ecol. 5, 187–191.

Woodman, J.D., 2010. High temperature survival is limited by food availability in first-instar locust nymphs. Aust. J. Zool. 58, 323–330.

Wright, D.E., 1986. Economic assessment of actual and potential damage to crops caused by the 1984 locust plague in south-eastern Australia. J. Environ. Manage. 23, 293–308.

Wright, D.E., 1987. Analysis of the development of major plagues of the Australian plague locust *Chortoicetes terminifera* (Walker) using a simulation model. Aust. J. Ecol. 12 (4), 423–438.

Desert Locust

Keith Cressman
Senior Locust Forecasting Officer, Food and Agriculture Organization of the United Nations, Rome, Italy

ABSTRACT
The desert locust is considered to be the most dangerous of all migratory pest species in the world due to its ability to reproduce rapidly, migrate long distances, and devastate crops. In order to minimize the frequency, severity, and duration of plagues, the Food and Agriculture Organization (FAO) of the United Nations operates a global early warning system based on the latest technological advances that have led to dramatic improvements in data management, analysis, and forecasting. The system can be a model for other early warning systems about migratory pests.

The desert locust (*Schistocerca gregaria*, Forskål) has the ability to change its behavior and physiology, in particular its appearance, in response to environmental conditions, and transform itself from a harmless solitarious individual to part of a collective mass of insects that form a cohesive swarm (Figure 4.2.1), which can cross continents and seas, and quickly devour a farmer's field and his entire livelihood in a single morning (Figure 4.2.2). For this reason, the desert locust is often considered as the most important and dangerous of all migratory pests in the world (Steedman, 1990).

For years the desert locust was thought to be two different insects. In the 1920s, a Russian scientist, Boris Uvarov, confirmed that it was a single species that had evolved a unique strategy for surviving in some of the harshest

FIGURE 4.2.1 A typical desert locust swarm (May 14, 2014, Addis Ababa, Ethiopia).

Biological and Environmental Hazards, Risks, and Disasters. http://dx.doi.org/10.1016/B978-0-12-394847-2.00006-1

FIGURE 4.2.2 In February 2014, locusts were present and causing damage to pearl millet (left) on the Red Sea coast in Eritrea that was due to be harvested shortly. Millet heads are extremely vulnerable and are at high risk to desert locust damage at this stage.

environments on Earth. Under normal conditions, solitarious locusts are found in low numbers scattered throughout the deserts of North Africa, the Middle East, and Southwest Asia, trying to survive in isolation by seeking shelter on sparse annual vegetation and laying eggs in moist sandy soil after intermittent rains. This arid and hyperarid area is some 16 million square kilometers in size, nearly twice as big as the United States of America, and includes about 30 countries. It is referred to as the **recession area** and the calm period without widespread and heavy infestations is called a **recession** (Figure 4.2.3).

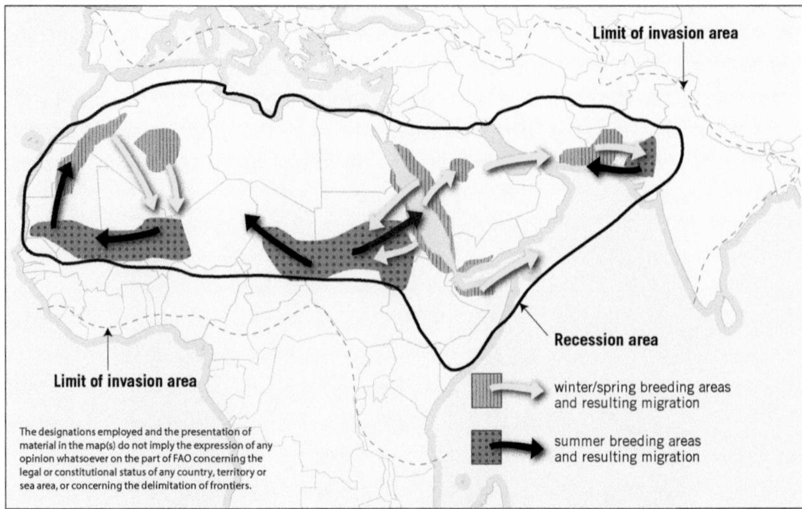

FIGURE 4.2.3 The recession and plague areas of the desert locust.

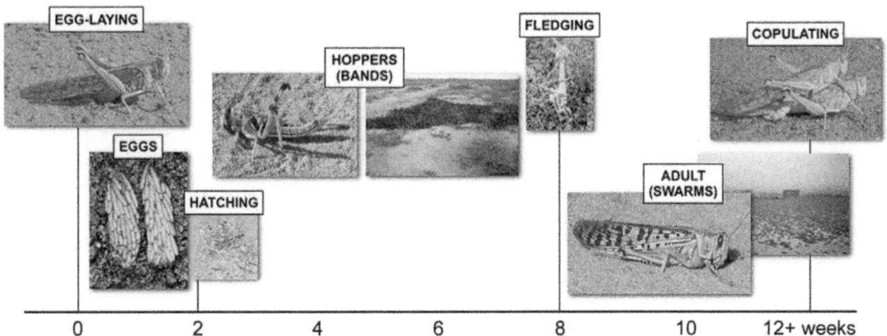

FIGURE 4.2.4 The life cycle of the desert locust.

When unusually heavy rains fall somewhere in the recession area, locusts take advantage of these rare events and multiple rapidly to increase in number. Under optimal conditions, locusts increase some 16–20 times every 3 months after a new generation of breeding (Figure 4.2.4). Once the desert habitat starts to dry out, large numbers of locusts are forced into the remaining patches of green vegetation, concentrate, come into physical contact with one another and start to behave as a single cohesive mass. They become increasingly more gregarious, initially forming small groups of hoppers (wingless nymphs) and adults that eventually fuse and form dense bands of hoppers and swarms of adults (Figure 4.2.5). This process is known as **gregarization** and the intermediate phase between **solitarious** and **gregarious**, that is, when locusts are grouping is referred to as **transiens**. Due to the sporadic nature of rainfall in the desert, fixed gregarization areas do not exist within the vast recession area. Gregarization takes place only in those parts of the recession area, where two generations of breeding can occur in rapid succession (Symmons and Cressman, 2001).

The marked increase in locust numbers on a local scale due to concentration, multiplication, and gregarization, which unless checked, can lead to the formation of hopper bands and swarms (Roffey and Popov, 1968). This is called an **outbreak**. If further rains fall, a very large increase in locust numbers and contemporaneous outbreaks can occur, followed by the production of two or more successive generations of transient-to-gregarious breeding in complimentary seasonal breeding areas. This is referred to an **upsurge**. A period of one or more years of widespread and heavy infestations, the majority of which occur as bands or swarms is called a **plague**. A major plague exists when two or more regions area affected simultaneously. During upsurges and plagues, locust swarms tend to migrate beyond the recession area, and invade an area of some 32 million square kilometers in size, equivalent to about 20% of the Earth's land surface (Figure 4.2.3). This is known as the **invasion area**.

An outbreak may develop in a relatively small area of only a few 100 square kilometers within part of a single country (Roffey et al., 1970). No fixed outbreak areas occur; instead, the location of an outbreak is a function of

FIGURE 4.2.5 The gregarization process in desert locust occurs as locusts increase in number and concentrate, consisting of: (a) solitarious hopper, (b) a small group of *transiens* hoppers, (c) a fully gregarious hopper band, (d) solitarious adult, (e) a group of *transiens* adults, and (f) a fully gregarious immature adult swarm.

the sporadic spatial and temporal nature of rainfall in the desert, subsequent vegetation development, temperature, and locust populations. An upsurge, on the other hand, can affect numerous countries or an entire region, whereas plagues usually affect a continent or more. For example, good rains fell over a widespread area of the Northern Sahel between Mauritania and Sudan during the summer of 2003. The rains also fell some 100 km further north than usual. Although locusts bred during August and September, only low densities of scattered adults were seen in the field by survey teams. Once the rains stopped and as vegetation dried out in October, tens of millions of scattered individual locusts concentrated in the few small areas, where vegetation remained green. The locusts became increasingly gregarious, and formed hopper bands and adult swarms, giving rise to four separate nonrelated outbreaks in Mauritania, Mali, Niger, and Sudan (Figure 4.2.6). The outbreaks were not controlled because they developed suddenly and occurred in remote areas so they were

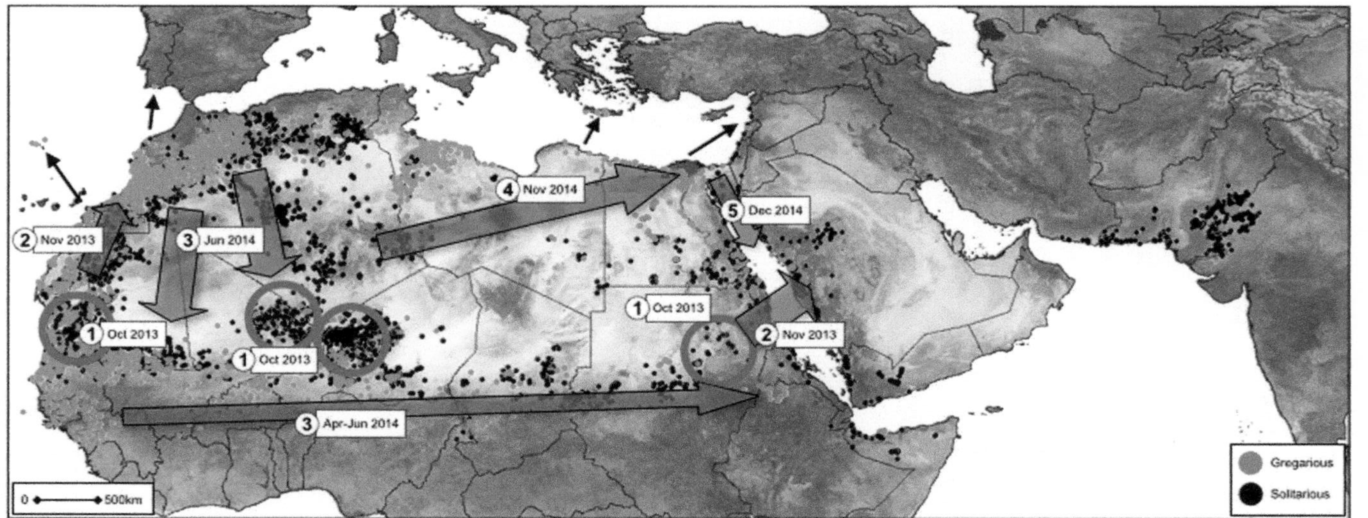

FIGURE 4.2.6 Good breeding during the summer of 2003 caused four desert locust outbreaks to occur simultaneously that were not controlled. Unusually heavy and widespread rains in late October led to an upsurge, which later spread and developed into a regional plague. By summer 2005, the plague had declined due to substantial control efforts and adverse weather.

not detected in time. Insufficient preparedness and a lack of available resources occurred in each country at that time. On 21–22 October, unusually heavy rains fell over a widespread area of West Africa extending from Dakar to the Atlas Mountains in Morocco. Some areas in Western Sahara received more rain in one day than what normally falls in an entire year. As a result, ecological conditions remained favorable for more than 6 months. Swarms that formed and were not controlled in the outbreaks in Mauritania, Mali, and Niger migrated to these areas, where 2–3 generations of breeding occurred from winter 2003 to spring 2004, giving rise to large numbers of locusts and causing an upsurge to develop. The upsurge, which in this case could be considered a regional plague, spread to 23 countries in Africa and the Middle East. It took nearly 2 years to bring it to an end, after spending more than $500 million and spraying 13 million hectares (Brader et al., 2006).

Locust plagues generally take several years to develop after a series of events in which locust numbers increase steadily (Roffey and Magor, 2001). This starts with the normally calm period of recession, followed by localized outbreaks and regional upsurges that can lead to a plague, which eventually declines, returning to a recession (Figure 4.2.7). A plague declines usually within 6 months, which is much quicker than it takes to develop. For example, it took more than 3 years for the last plague to develop but it declined within 6 months after reaching its peak. In 1985, good rains led to desert locust outbreaks in Northern Africa and around the Red Sea. Breeding continued along the Red Sea coasts, causing more swarms to form and by late 1986 an upsurge had developed in the region. Many of the swarms migrated to West Africa, where unusually heavy and widespread rain fell in late September 1987 in Northern Mauritania and Western Sahara. At least two generations of breeding occurred during the winter/spring of 1987/1988. During the following summer, swarms invaded the Sahelian countries in West Africa and

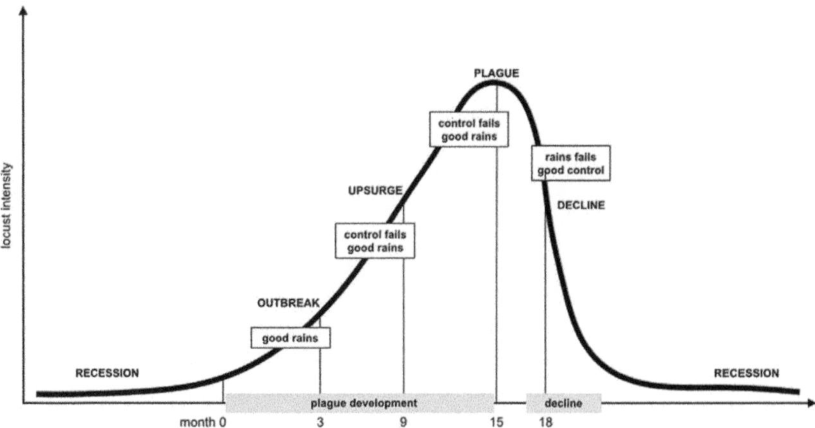

FIGURE 4.2.7 The evolution of desert locust plagues.

spread to Northeast Africa and the Arabian Peninsula, resulting in a plague that eventually declined by 1990 due to intensive control operations and poor rains.

Not all outbreaks turn into upsurges and, similarly not all upsurges become plagues. For example, an upsurge developed in 1993 after several generations of successful breeding along the Red Sea coast in the winter of 1992 (Figure 4.2.8). The resulting swarms moved to the interior of the Arabian Peninsula and bred again during spring 1993. Some of the new swarms that formed at the end of the spring moved east to India and Pakistan, while others moved west to Sudan and continued to West Africa, reaching Mauritania at the beginning of the rainy season when they bred, producing more swarms at the end of the summer that moved to Northwest Africa. Control operations brought the upsurge to an end in Southwest Asia by 1994 but it took several more years before the situation returned to normal in West Africa. Nevertheless, a plague did not develop.

Adult locusts are passive fliers and are carried by the wind. Solitarious adults fly in the early evening while swarms fly during daylight hours, starting early in the morning once the adults have warmed up and continuing until just before sunset. Swarms can fly up to 100–150 km in a single day at heights up to 2,000 m. While migrating over water, swarms can fly continuously for 20 or more hours. Locusts migrate between seasonal breeding areas. For example, summer-bred swarms often migrate from the Sahel of West Africa and Sudan to Northwest Africa, or from Sudan to the Red Sea coast; winter-bred swarms migrate from the Red Sea coastal plains to the interior of Saudi Arabia or Sudan, and spring-bred swarms can migrate from the interior of Arabia to Sudan and West Africa, or from the Horn of Africa to the Indo–Pakistan border (Pedgley, 1981). Given this terrific potential to migrate, it is not surprising that long-distance migrations have occurred in the past during upsurges and plagues, for example, from West Africa to the UK in the 1950s, and from Senegal across the Atlantic Ocean to the Caribbean in 1988.

The first records of desert locust plagues date from Pharaonic Egypt and have been documented throughout history. During the first 60 years in the twentieth century, 5 major plagues occurred, lasting up to 14 years. Plagues were present in nearly 4 out of every 5 years (Figure 4.2.9). Since 1963 a dramatic decline has occurred in the frequency and duration of plagues, and now plagues occur perhaps only once in every 10–15 years and rarely last more than 3 years. Consequently, the control strategy adopted and implemented by countries has shifted from curative to preventive. The cost of preventive control, which can be considered as an investment in food security, is substantially less than controlling a plague. For example, $500 million was spent to stop the 2003–2005 upsurge or regional plague, which is equivalent to 170 years of preventive control in West Africa.

The decline in desert locust plagues in the past 50 years can be attributed to a number of factors such as the introduction of chemical pesticides, improved

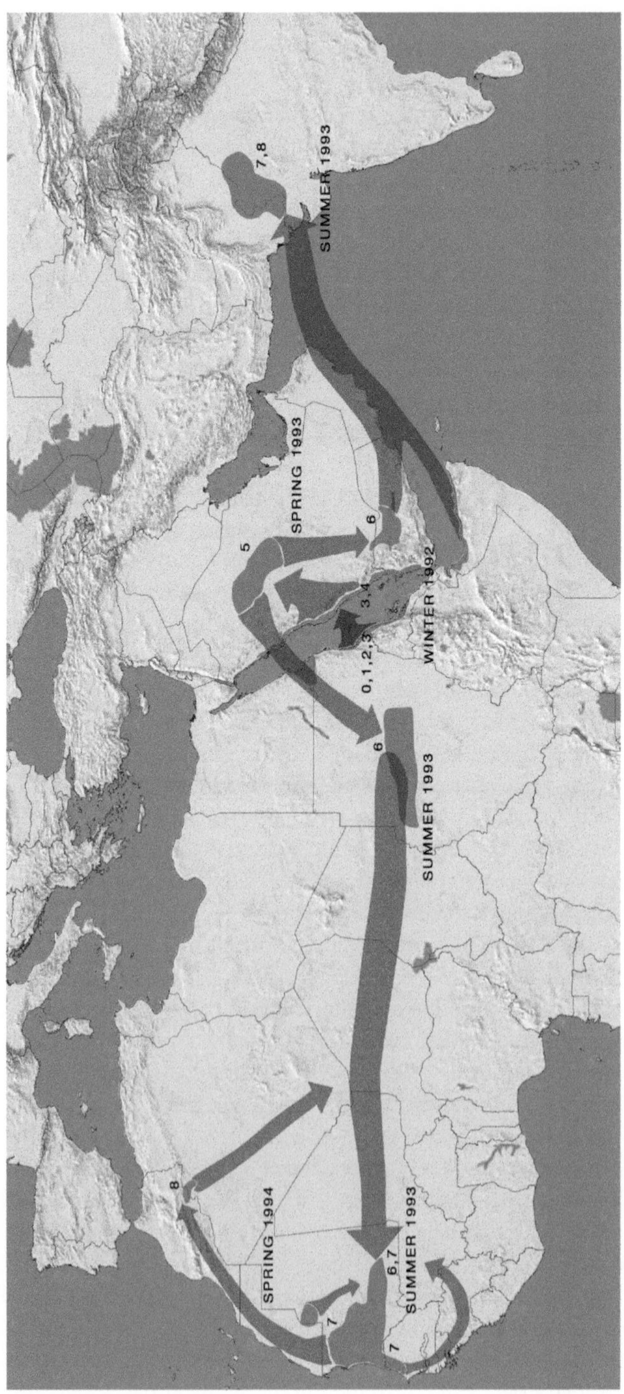

FIGURE 4.2.8 The development and spread of the 1992–1994 upsurge in which eight generations of breeding occurred, affecting Northern Africa, the Middle East, and Southwest Asia.

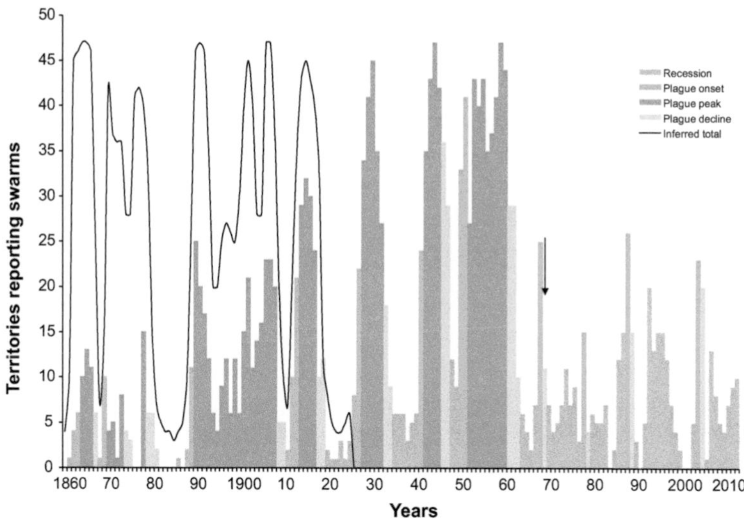

FIGURE 4.2.9 Plague and recession periods of the desert locust, 1860–2014.

transportation and infrastructure, and advances in technologies related to precision spraying, communications, geopositioning, spatial analysis, remote sensing, and early warning (Magor et al., 2007).

4.2.1 MONITORING AND FORECASTING

The early warning system for desert locust is based on more than 75 years of collaboration. It dates back to the early twentieth century and is perhaps one of the oldest systematic, pest-monitoring systems in the world. At that time, field teams and local scouts scoured the desert on camels, looking for locusts and recording their observations in notebooks. This information eventually found its way to the capital cities in the affected countries. In 1930, the Anti-Locust Research Centre (ALRC, London, UK) started to collect, map, analyze, and archive locust data from affected countries. A centralized information unit, the Desert Locust Information Service (DLIS), was established for the regular collection, exchange, and analysis of locust, weather, and ecological data. The systematic collection and mapping of the data showed that breeding of the desert locust coincided with rainfall, both seasonal and sporadic, and migration was associated with downwind movements. The importance of high-quality data on locust infestations, ecological conditions, and weather for predicting the scale, timing, and location of breeding and migration emerged. Countries agreed on a standard set of data to be collected on locust infestations, habitats, and weather to be transmitted to DLIS. This collaboration between locust-affected countries and DLIS formed the basis of the early warning system and continues to this day.

In 1943, DLIS began to issue alerts, monthly bulletins, and forecasts used for planning and undertaking control campaigns based on the analysis of information received from the field. In August 1978, the largest specialized agency of the United Nations, the Food and Agriculture Organization (FAO), assumed responsibility for global monitoring, data analysis, and monthly bulletins, and DLIS operations shifted to FAO's headquarters in Rome, Italy.

4.2.2 TECHNOLOGICAL ADVANCES

From the late 1980s onward, substantial improvements to the locust early warning system were achieved with the introduction and adoption of new technologies in a number of different fields (Figure 4.2.10).

Telecommunications. The first observations made by teams in the field were written down in a narrative style and hand carried or sent through the postal system as letters, often arriving weeks or months later. Telegrams and telex were used to transmit information from capital cities in the affected countries. Although there was a flow of information from affected countries, it was irregular, usually arriving too late and the data were often incomplete or vague. During the last major plague of 1987–1989, FAO installed facsimile machines in the key desert locust countries that were used for transmitting information and reports to DLIS. In turn, DLIS transmitted its monthly bulletins via fax rather than telex, reducing the time spent in preparation and distribution. Additional information could also be faxed relatively easily, such as daily synoptic charts, rainfall graphs, and maps of survey itineraries, locust

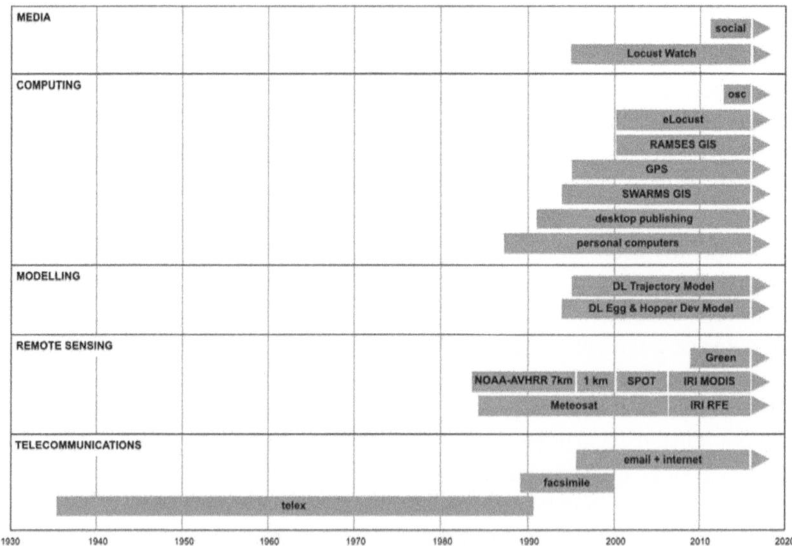

FIGURE 4.2.10 Technological developments related to locust early warning, 1935 to present.

infestations, and forecasts that previously were not possible by telex. This information helped to strengthen analysis of the locust situation and habitat conditions, and improve early warning.

Computing. Further improvements in telecommunications coincided with the widespread introduction and use of the personal computer in the late 1980s. Prior to this, typewriters and telex machines were used. The personal computer opened up nearly endless possibilities for managing, analyzing, and summarizing data, and preparing reports and bulletins. Initially, this was accomplished by using relatively simple database, word processing, and spreadsheet programs. In the early 1990s, desktop publishing was introduced, suddenly making it possible to prepare professional-level bulletins that seamlessly integrated images, maps, graphics, and text.

Geospatial data. Similar improvements occurred regarding the collection and analysis of geospatial data. The first handheld global positioning systems (GPS) appeared on the market in the late 1980s. Initial models were large and bulky, expensive, slow, and not very accurate because the system was intentionally degraded for civilian use. As handheld units became increasingly smaller and more affordable, it allowed for their gradual introduction and eventual adoption by national locust programs, so that by the late 1990s teams were using GPS units to determine the latitude and longitude coordinates of survey and control operations. In this way, precise locations of surveys, locust infestations, and control could be pinpointed on a map. In 2000, the selective availability was turned off, and GPS accuracy improved to within about 10 m or less.

In the early 1990s, with the prevalence of personal computing, geographic information systems (GIS) were introduced in response to an increasing interest in maps and mapping. A typical GIS consists of two components, a database and a mapping application. In order for the data to be displayed accurately on maps, it must be georeferenced, containing a latitude and longitude reference. In 1994, DLIS commissioned the Natural Resources Institute (NRI) and the Geography Department at the University of Edinburgh, both in the UK, to develop a GIS for operational locust monitoring and early warning. By 1996, *Schistocerca* **War**ning and **M**anagement **S**ystem (SWARMS) was being used by DLIS on a daily basis to manage and analyze environmental and locust data. It was one of the first GIS used for operational monitoring rather than map production purposes (Healey et al., 1996). SWARMS is a custom GIS consisting of an Oracle database that hosts all survey and control data received from locust-affected countries since the early 1990s, historical records dating from the 1930s, meteorological data and remote sensing imagery, and ESRI's ArcGIS software for querying, display and spatial analysis. SWARMS is a server-based system that supports several PC-based workstations, allowing users to access the same data set and work simultaneously. The system allows the forecaster to rapidly access large volumes of data in different formats and display them together in order to analyze the current weather, ecological conditions, and locust situation, and estimate future developments. Prior to GPS and GIS technology, location coordinates had to be determined from a paper map and infestations were plotted by hand on large transparent overlays using colored pencils. This was extremely labor intensive and time-consuming, especially during periods of increased locust activity when a large team of plotters was required to manage the sizable volume of data.

Difficulties in managing large volumes of data were not only a problem in DLIS but also in locust-affected countries. FAO addressed this issue by developing a smaller, less-complicated GIS that could be used by nationally designated locust information officers in the key frontline countries. **R**econnaissance **A**nd **M**anagement **S**ystem of the **E**nvironment of ***S**chistocerca* (RAMSES) was introduced in 2000, operating on a personal computer using Microsoft Access database and ESRI's ArcView software. In 2014, an open-source platform-independent version was developed using OpenJump GIS and Postgres spatial database that takes advantage of the last advances in spatial analysis and open-source software.

Internet. In 1996, e-mail and Internet services were introduced at FAO, and DLIS was one of the first users. Within a very short period of time, e-mail replaced facsimile as the mode of sending and receiving data and bulletins between countries and DLIS. This greatly enhanced the ability to easily and widely disseminate information in a timely manner. By 2000, all locust-affected countries had Internet services and were using e-mail every day to share data and information. The Internet also provided sudden access to a great wealth of information and knowledge that could be applied to locust early warning.

eLocust. Despite the numerous technological advances and associated advantages to locust early warning, one major obstacle persisted in the timely flow of high-quality data. Standardized forms had been developed and adopted for use by all countries in order to improve data quality. Locust officers were well trained in completing the forms in the field. National locust information officers were trained on how to enter the data from the forms into RAMSES. Completed forms and RAMSES data export files were transmitted to DLIS by e-mail. A strong, reliable information network had become established and was operating reliably on a daily basis between locust-affected countries and DLIS. But the weak link in the early warning chain remained the transmission of data in near-real time from the field to the national locust center in each affected country.

In 2000, a prototype data logger was developed for national field officers as a proof of concept to demonstrate that survey and control data could be enter digitally into a database in the field by the locust officer. The system, eLocust, consisted of a handheld Psion 5mx palmtop computer and a custom database linked to a mapping application. It was connected to a handheld GPS. Alkaline batteries powered both devices. But the system lacked transmission capabilities; data were only saved to the internal memory of the Psion. Five years later, an all-in-one system, eLocust2, was developed in collaboration with Novacom Services (France) that consisted of a rugged, data logger that was touch screen and handheld, with custom software in English and French, and an antenna that connected to the GPS network of satellites for determining the coordinates of the position of the survey or control position in the field and to the Inmarsat satellite for date transmission in near-real time. More than 400 units were distributed to frontline countries in 2006 for use by survey and control teams. The national locust information officer receives eLocust2 data via e-mail and the Internet, downloads it to the PC, and imports it into the RAMSES GIS. The data, as well as the position of the field teams, can also be accessed through a secure Webpage on the Internet. For the first time in more than 75 years of monitoring locusts, field observations and survey and control results became instantly available to decision-makers and forecasters (Figure 4.2.11). This revolutionized early warning and preventive control of the desert locust.

In 2014, eLocust2 was replaced with an updated version, eLocust3, that takes advantage of the latest technological developments. It consists of an Android-based 10.1-inch rugged Panasonic FZ-A1 ToughPad tablet with custom applications for data entry and viewing Landsat imagery and the latest dynamic greenness map and rainfall estimates in the field without requiring Internet access, a camera and video, a low-profile wireless (Bluetooth) antenna for data transmission, and a digital library of references that includes technical guidelines, standard operating procedures, field guides, and user manuals for survey and control equipment. The system checks the integrity of the data prior to transmission to ensure that mandatory data have been collected and entered correctly. This helps to ensure that the data are complete and of high quality. The updated maps on eLocust3 can be used to help guide the survey teams to places, where green vegetation and locusts may be present, thus reducing the large areas of empty desert that must be checked.

Remote sensing. Although satellites available for civilian use cannot detect locust infestations, remote sensing is used to help estimate where it has rained and where ecological conditions may be favorable for breeding. Since the early 1980s, DLIS analyzed visible and infrared Meteosat imagery to determine clouds that might produce sufficient rainfall for locust survival and breeding in Africa. It remained difficult to know where it had rained in the Middle East and Southwest Asia as equivalent imagery for those regions was not available. The FAO Remote Sensing Centre produced decadal maps of cold-cloud duration that estimated rainfall from cold clouds. This technique was acceptable during the summer over the Sahel of Northern Africa but did not detect rainfall reliably from low-level warmer clouds in winter breeding areas along the Red Sea coasts. Satellite sensors, meteorological numerical models, and rainfall algorithms have improved in the past 15 years, and new products have been developed to estimate rainfall on a

FIGURE 4.2.11 The flow of data used for desert locust early warning in which data are collected during field surveys, immediately entered into eLocust3 and transmitted by satellite in real time to national locust centers and the FAO Desert Locust Information Service in Rome where analyses are undertaken using a custom geographic information system (GIS) in order to provide assessments, forecasts, and early warning.

local, regional, or global level. Satellite-based, rather than model-based, products are used, because the former are better at estimating the spatial distribution of rainfall. DLIS has been using rainfall estimates since 2006 as an enhanced means of estimating rainfall in breeding areas of the desert locust, rather than relying on data from the relatively few national meteorological stations.

A significant evolution also occurred during the past three decades in remote sensing imagery for detecting green vegetation, shifting from 7-km resolution NOAA-AVHRR normalized difference vegetation index imagery in the mid-1980s, to 1-km resolution imagery 10 years later. In 2000, 1-km resolution SPOT imagery replaced NOAA-AVHRR, taking advantage of the SPOT sensor that was specifically designed for vegetation monitoring. In 2006, SPOT imagery was superseded by higher resolution 250-m MODIS imagery. Despite the dramatic improvements in spatial resolution, remotely sensed vegetation imagery continues to suffer from two limitations: accuracy and dissemination. Although resolution has increased nearly 800-fold, it is still not sufficient to detect the thin green vegetation that hosts desert locust (Dinku et al., 2010). In other words, imagery commonly indicates that an area is dry when in reality it is green, the so-called false negatives. With increased resolution, comes increased file size for each image and difficulties in data management. This makes it challenging to distribute high-resolution imagery such as MODIS to affected countries by Internet, e-mail, or FTP because many countries have very slow and erratic connections.

New, higher-resolution products from the latest generation of satellites, such as the European Space Agency's PROBA-V and Sentinel 3 with resolutions up to 100 and 30 m respectively, offer significant improvements in the detection of green vegetation in desert locust habitats within the next few years.

Modeling. In the mid-1990s, DLIS began using two, custom-developed models to estimate the development rates of eggs and hoppers and to estimate the migration routes of swarms. The Desert Locust Egg and Hopper Development Model relies on the well-known relationship of air and soil temperature on the development of eggs and hoppers, using long-term temperature means from the nearest meteorological station (Reus and Symmons, 1992). The Desert Locust Trajectory Model estimates the displacement of locust swarms forward and backward in time using 6-h meteorological and forecast data for up to 10 days from the European Centre for Medium-range Weather Forecasts (ECMWF), consisting of temperature, pressure, wind direction, and speed at several atmospheric levels between the surface and 500 hPa with a resolution of 0.25–1.0 degree square.

4.2.3 EARLY WARNING

New advances in technologies have led to a paradigm shift in locust early warning from that of collecting information for interpreting and forecasting breeding and migration to predicting habitat development and the development of outbreaks, upsurges, and plagues. In the past three decades, the system has shifted from camels to four-wheel drive vehicles, from telex to e-mail, from map reading to GPS, from narratives to handheld data loggers, from manual plotting to GIS, and from weather station reports to satellite-based rainfall estimates and greenness maps. GPS, RAMSES and SWARMS GIS, and Internet and eLocust3 have replaced the traditional tools of paper, colored pencils, maps, and telephone.

The current early warning system consists of a variety of integrated elements that all must function smoothly and reliably in order to provide accurate and timely information and alerts on a regular basis to a large international audience. The first step is the collection, recording, and transmission of survey and control data from the field by national teams using eLocust3. The data are received by e-mail at the National Locust Control Centre (NLCC) in each country. The National Locust Information Officer (NLIO) processes the data by importing it into RAMSESv4 where it is checked for completeness and accuracy before it is inserted into a standard database. The data are then exported and summarized, and both items are sent by e-mail to DLIS in Rome within a few days of the survey or control operations. In Rome, the data are checked and corrected prior to importing into the SWARMS GIS.

The NLIO at the NLCC and the Senior Locust Forecasting Officer in DLIS use RAMSESv4 and SWARMS GIS, respectively, to analyze the data in conjunction with rainfall estimate imagery, greenness maps, previous survey, and control results, and historical locust, ecology, and weather data. The objective of the analysis is to understand the current situation and estimate potential developments. The results of the NLCC analysis are used for planning field operations such as the timing, location, and extent of survey and control operations; whereas, the DLIS analysis focuses on the timing, location and scale of current, and future breeding and migration in order to provide advice and early warning to countries. As DLIS maintains locust, ecology, and weather data that cover the entire recession area, its analyses are global and regional in nature, whereas the NLCC analyses are restricted to the national level.

DLIS also prepares detailed case studies of particular situations, especially outbreaks and upsurges, in order to better understand locust population and plague dynamics. DLIS distributes a variety of products by e-mail, the Internet, and social media such as situation updates, bulletins, warnings, alerts, case studies, reference, and training material.[1]

4.2.4 CHALLENGES

In the past decade, a new challenge is facing the locust early warning system. Political unrest and instability, national border disputes and sensitivities, kidnappings, mines, and conflict have led to insecurity in many parts of the recession area. It is becoming increasingly difficult for survey and control teams to access many important areas, where desert locust may be present and breeding. These areas can only be accessed when accompanied by a military escort or are simply closed because they are deemed too unsafe. For example, an outbreak occurred along the Algerian/Libyan border in early 2012 that

1. Locust Watch (www.fao.org/ag/locusts), Facebook (www.facebook.com/faolocust), Twitter (www.twitter.com/faolocust), Slideshare (www.slideshare.net/faolocust), and YouTube.

normally would be controlled without any problem since both countries have strong, well-resourced national locust units. However, this was not possible in 2012 due to political upheaval in Libya that affected the national locust program and made it unsafe to undertake the necessary control operations in the outbreak area. As a result, swarms formed and invaded Mali and Niger during the summer where at least one generation of breeding occurred. Survey and control operations were not possible in northern Mali due to civil unrest so breeding occurred unchecked within a large area and locusts increased and spread to other countries. In Niger, large military escorts had to accompany field teams to ensure their safety in the north but this slowed down and hampered survey and control operations.

Nearly a dozen important locust habitats and breeding areas straddle both sides of a common international boundary such as Mauritania/Western Sahara, Algeria/Libya, Mali/Niger, Sudan/Egypt, Sudan/Eritrea, Eritrea/Ethiopia, Ethiopia/Somalia, Yemen/Saudi Arabia, India/Pakistan, and Iran/Pakistan. Border areas are by nature sensitive places, but during periods of intense or prolonged conflict, such areas can be dangerous or simply off-limits. For example, a joint Iran/Pakistan team conducts an annual survey in spring breeding areas on both sides of their common border, south of Afghanistan, to confirm habitat conditions and check for locust infestations and breeding. The results of the joint spring survey are used for planning the summer campaign along the Indo–Pakistan border. The month-long joint survey has been carried out every year since 1995 but recent insecurity in Baluchistan, Pakistan has prevented the Iranian team from participating in the past few years.

As a consequence, the NLCC's ability to monitor ecological conditions and locust infestations as well as to undertake the necessary control operations is gradually being compromised. This is resulting in an increasing number of spatial gaps within the early warning system, where no information from ground observations is made by field teams. Remote sensing can help address such gaps but satellite-based estimates are not a substitute for *in situ* verification on the ground. Thus, in many cases, DLIS must forecast the current situation as well as future developments.

Sustaining an effective early warning system, whose foundation is based on national surveillance relies on a number of important elements in each locust-affected country. First and foremost, the NLCC must be a fully funded unit that is autonomous and centralized, with dedicated resources and well-trained staff that can be shifted easily from one side of the country to other at a moment's notice in order to monitor field conditions and respond quickly to infestations and invasions. The individuals that make up the unit, such as locust survey and control officers, and NLIOs need to be energetic, motivated, and curious. NLCC master trainers must provide regular training to national staff on a continual basis in survey methodologies, data collection and transmission, and the use of standard equipment. Field teams need to be properly equipped with GPS, maps, eLocust3, radios, 4WD vehicles, and camping equipment.

Financial support required for survey, control, reporting, and training should be a standard item in the annual national budget of the Ministry of Agriculture in all locust-affected countries. Field officers should receive incentives to encourage their participation in field operations, especially when such operations entail being away from families or in remote areas for several weeks or months. The NLCC should manage field teams effectively and provide feedback regarding their performance. Field officers should be reminded of their critical role within the early warning system as the primary source of information and data. In this way, they are the most important stakeholders within the global system. Lastly, the regular monitoring of desert locust habitats should become a routine activity that is done on a regular basis every year, especially after good rainfall. All of these elements must be fully integrated if early warning is to be effective in reducing the duration, severity, and frequency of desert locust plagues.

4.2.5 CONCLUSION

The success of the early warning system for desert locust depends on a well-organized and funded NLCC in every locust-affected country that can monitor field conditions and respond to locust infestations by: (1) conducting ground surveys and control operations; (2) collecting and transmitting accurate geospatial data rapidly; (3) using a GIS to analyze the data; (4) keeping all stakeholders informed on a regular and timely basis through simple well-targeted outputs; (5) sharing reports within a robust and reliable information network; and (6) maintaining a cadre of well-trained and dedicated individuals. Each national component should be fused together and feed into a centralized DLIS that has a global overview of the situation. The overall strength of the system will be only as strong as its weakest link.

No doubt exists that technological advances have led to dramatic improvements in locust early warning, resulting in plagues that occur less frequently and of shorter duration. Yet, despite the decline of plagues, the desert locust remains a very serious and important threat. More than $500 million, 13 million liters of pesticide, and 2 years were required to bring the last regional desert locust plague under control in Northern Africa. Even with this effort, locust spread to the Middle East, control operations were conducted in 23 countries, up to 100% cereal loss occurred, 3 out 5 household heads went into debt in Mauritania, and $100 million was spent on food aid. The annual cost of preventive control for the 10 frontline countries in West and Northwest Africa is $3.3 million. The cost of the 2003–2005 regional plague was equivalent to 170 years of preventive control. It seems obvious that undisputed benefits exist in continuing efforts to adopt new technologies for improving and sustaining monitoring, early warning, and preventive control in order to prevent desert locust plagues, protect food security, and reduce hunger throughout the world. The success and lessons learned from the desert locust early warning system can be adopted and modified for use in early warning systems of other migratory pests.

REFERENCES

Brader, L., Djibo, H., Faye, F.G., Ghaout, S., Lazar, M., Luzietoso, P.N., Ould Babah, M.A., 2006. Towards a More Effective Response to Desert Locusts and Their Impacts on Food Security, Livelihoods and Poverty. Multilateral Evaluation of the 2003-05 Desert Locust Campaign. FAO, Rome, 114 pp.

Dinku, T., Ceccato, P., Cressman, K., Connor, S.J., 2010. Evaluating detection skills of satellite rainfall estimates over desert locust recession regions. J. Appl. Meteorol. Climatol. 49 (6), 1322–1332.

Healey, R.G., Roberston, S.G., Magor, J.I., Pender, J., Cressman, K., 1996. A GIS for desert locust forecasting and monitoring. Int. J. Geogr. Inf. Syst. 10 (1), 117–136.

Magor, J.I., Ceccato, P., Dobson, H.M., Pender, J., Ritchie, L., 2007. Preparedness to prevent desert locust plagues in the central region, an historical review. In: FAO Desert Locust Technical Series. AGP/DL/TS/35. Rome.

Pedgley, D.E., 1981. Desert Locust Forecasting Manual, vol. I. Centre for Overseas Pest Research, London viii + 268 pp.

Roffey, J., Magor, J.I., 2001. Desert locust population parameters. In: FAO Desert Locust Technical Series. AGP/DL/TS/30. Rome.

Roffey, J., Popov, G.R., 1968. Environmental and behavioural processes in a desert locust outbreak. Nature (London) 219, 446–450.

Roffey, J., Popov, G., Hemming, C.F., 1970. Outbreaks and recession populations of the desert locust, *Schistocerca gregaria* (Forsk.). Bull. Ent. Res. 59, 675–680.

Reus, J.A., Symmons, P.M., 1992. A model to predict the incubation and nymphal development periods of the desert locust, *Schistocerca gregaria* (Orthoptera: Acrididae). Bull. Ent. Res. 82, 517–520.

Steedman, A. (Ed.), 1990. Locust Handbook. Natural Resources Institute, Chatham, 204 pp.

Symmons, P.M., Cressman, K., 2001. Desert Locust Guidelines 1. Biology and Behaviour. Food and Agriculture Organization, Rome, 43 pp.

Chapter 4.3

Other Locusts

Ramesh Sivanpillai
Senior Research Scientist, Department of Botany | WyGISC, University of Wyoming, Laramie, WY, USA

Locusts are a major threat to global food security and could adversely impact the livelihoods of farmers in numerous countries (Simpson and Sword, 2008;

	Common Name	Scientific Name	Habitat	Source
1.	Asian migratory locust	*Locusta migratoria migratoria*	Asia and Europe	COPR (1982) and FAO
2.	Brown locust	*Locustana pardalina*	South Africa and Southern Namibia	COPR (1982) and Crooks and Cheke (2014)
3.	Central American locust	*Schistocerca piceifrons piceifrons*	Central America	COPR (1982)
4.	Italian locust	*Calliptamus italicus*	Asia and Europe	COPR (1982) and FAO
5.	Madagascar migratory locust	*Locusta migratoria capito*	Madagascar and numerous Islands in the Indian Ocean	COPR (1982) and Latchininsky (2013)
6.	Moroccan locust	*Dociostaurus maroccanus*	Asia and Europe	COPR (1982), FAO, and Latchininsky (2013)
7.	Oriental migratory locust	*Locusta migratoria manilensis*	Asia	COPR (1982)
8.	Red locust	*Nomadacris septemfasciata*	Southern Africa including Madagascar	COPR (1982)
9.	South American locust	*Schistocerca cancellata*	Central South America	COPR (1982)

Biological and Environmental Hazards, Risks, and Disasters. http://dx.doi.org/10.1016/B978-0-12-394847-2.00007-3

Latchininsky, 2010). The United Nations Food and Agriculture Organization along with numerous national agencies monitor, forecast, and manage locust populations in order to minimize the damages. Chapters 14 and 6 focused on Australian plague locusts and desert locusts respectively, this chapter contains pertinent information about other locusts, their habitats, and additional sources of information about them. A few more species are considered as borderline between locusts and grasshoppers (Latchininsky, 2013) and are not included in this listing.

REFERENCES

COPR, 1982. The Locust and Grasshopper Agricultural Manual. Centre for Overseas Pest Research, London, UK.

Crooks, W.T.S., Cheke, R.A., 2014. Soil moisture assessments for brown locust *Locustana pardalina* breeding potential using synthetic aperture radar. J. Appl. Remote Sens. 8 (1), 084898. http://dx.doi.org/10.1117/1.JRS.8.084898.

Food and Agriculture Organization of the United Nations (FAO). Locust Watch: Locusts in Caucasus and Central Asia. http://www.fao.org/ag/locusts-CCA/en/ (In English and Russian) (accessed on 07.07.15.).

Latchininsky, A.V., 2010. Locusts. In: Breed, M.D., Moore, J. (Eds.), Encyclopedia of Animal Behavior, vol. 2. Academic Press, Oxford, UK, pp. 288–297.

Latchininsky, A.V., 2013. Locusts and remote sensing: a review. J. Appl. Remote Sens. 7 (1), 075099. http://dx.doi.org/10.1117/1.JRS.7.075099.

Simpson, S.J., Sword, G.A., 2008. Locusts. Curr. Biol. 18 (9), R364–R366. http://dx.doi.org/10.1016/j.cub.2008.02.029.

Chapter 5

Decline of Bees and Other Pollinators

Norman Carreck
International Bee Research Association, Laboratory of Apiculture and Social Insects, School of Life Sciences, University of Sussex, Falmer, Brighton, UK

ABSTRACT
Declines of bees and other pollinating insects have been documented for a long time, but attracted the attention of the world's press with the phenomenon of "colony collapse disorder" of honey bees in the USA in 2006. This has led to much research into the causes. The scientific consensus is that there is no single cause of pollinator declines, but that the primary long-term driver is land-use changes, resulting in reduced food and fewer nest sites. In addition, weather causes short-term fluctuations, and many other interacting environmental causal factors occur, including pests and diseases, and farming practices including the use of pesticides, and climate change.

5.1 INTRODUCTION

In 2006, the world's newspaper headlines were full of reports that honey bees (*Apis mellifera*) in the USA were dying from a mysterious new affliction dubbed "colony collapse disorder" (CCD). The media attention was driven mainly by the fact that a shortage of honey bees threatened pollination of the multi-billion dollar almond crop in California (Ratnieks and Carreck, 2010). CCD itself was a very precisely defined set of symptoms (vanEnglesdorp et al., 2009), resembling damage seen in Europe caused by viruses associated with the parasitic mite *Varroa destructor* (Carreck, 2009b), and seems likely in hindsight to have been caused by another virus, Israeli acute paralysis virus (Cox-Foster et al., 2007).

But the honey bee is just one of perhaps 25,000 species of bee worldwide, and dramatic losses of honey bees had been seen many times before (Underwood and vanEngelsdorp, 2007). For example, more than 100 years ago, many colonies of honey bees in the UK were lost due to a condition dubbed "the Isle of Wight disease" by the media (Neumann and Carreck, 2010).

Biological and Environmental Hazards, Risks, and Disasters. http://dx.doi.org/10.1016/B978-0-12-394847-2.00008-5

Actually a concern has existed about long-term declines of bees and other pollinators for many years. In a pioneering study, Williams (1982, 1986) showed by comparing records before and after 1960 that many species of bumble bees (*Bombus* spp.) in Britain had become very rare, and a key international conference "The conservation of bees" was held in London, UK in 1995 (Matheson et al., 1996). Similarly, butterfly declines in Britain have been well documented (Warren et al., 2001) and more recently, comparative studies in Britain and the Netherlands have demonstrated parallel declines of many bee and hoverfly species and the plants that they support (Biesmeijer et al., 2006). A Europe-wide assessment concluded that 9% of all European bee species were at risk of extinction, although worryingly, insufficient information was available for half the species to make a proper assessment (Nieto et al., 2014). Conversely, some common bumble bee species remain very common, and do not appear to have suffered declines.

Pollination is the transfer of pollen from the male part of a flowering plant to the female part, allowing seed production. Insect pollinators, of which bees are generally considered the most important, also include hoverflies, butterflies, and wasps. Bees include the fully social honey bees and stingless bees, the semisocial bumble bees, and numerous species of solitary bees. It has been estimated that the production of 84% of crop species cultivated in Europe depends directly on insect pollination, especially by bees (Williams, 1994), and 70% of the 124 staple food crops in the world require pollinators (Klein et al., 2007). It has been suggested that some crop yields could be limited by pollinator shortage (Garibaldi et al., 2009; Potts et al., 2010). The total economic value of insect pollination worldwide has been estimated as over $162 billion (€153 billion; Gallai et al., 2009).

The surge in concern about pollinator declines has led to a number of international initiatives to study such declines. Because of their economic importance, honey bees have been well studied, such as by the COLOSS association (Williams et al., 2012), but much less information is available about many other species (Nieto et al., 2014). Strong scientific consensus seems to exist that no single cause of pollinator decline exists, but instead a series of interacting factors occur. The main long-term driver is undoubtedly changes in land use, whereas weather causes short-term fluctuations, but overlying these are a range of other factors including pests and diseases, climate change, and pesticide use.

5.2 LAND-USE CHANGES

Since the beginning of agriculture some 12,000 years ago, humans have altered the landscape from its natural state. However, the pace of this change has increased in recent years, especially in the second half of the twentieth century. Modern machinery and the use of agrochemicals, new crop varieties, and more intensive farming systems have all reduced the availability of nectar and pollen as food for pollinators, together with a reduction in suitable nest sites for nonmanaged pollinators (Williams and Carreck, 1994; Westrich, 1998). Likewise, urbanization has reduced seminatural vegetation, yet paradoxically suburban gardens can commonly provide both good forage and suitable nest sites for bees (Osborne et al., 2008), better than intensive farmland.

5.3 WEATHER

Studies in the USA and Europe (e.g., Steinhauer et al., 2014; van der Zee et al., 2014) of losses of winter colonies of honey bees have shown that no clear regional pattern exists to colony loss. Heavy losses occur in one region one year, but in that location the following year losses may be low, and heavy losses may occur elsewhere. Clear relationships exist, however, between adverse weather conditions and heavy winter losses. For example, a number of European countries such as Britain and Ireland experienced heavy losses (greater than 30%) in the winter of 2012–2013, which coincided with poor weather conditions (van der Zee et al., 2014). This poor weather may manifest itself in several ways. In these countries in 2012, the summer was cold and wet, leading to small honey bee colonies that failed to develop well. Poor weather during queen mating led to a high incidence of failing queens, and the subsequent cold winter was followed by a long cool damp spring, preventing colonies from building up normally. In the spring, a trade-off always occurs between the old winter bees dying, and new bees being produced. Poor spring conditions can lead to colonies starving or dwindling to nothing.

5.4 PEST AND DISEASES

Pests and diseases of the honey bee have been well studied. Initial attention focused on bacterial brood diseases such as American and European foulbrood, infestation by the tracheal mite *Acarapis woodi*, the microsporidian gut parasites *Nosema* spp., and 30 or so viruses (Bailey and Ball, 1991; Morse and Flottum, 1997). The parasitic mite *V. destructor* has spread across the globe over the last 60 years, and probably remains the greatest threat to beekeeping with *A. mellifera* worldwide (Neumann and Carreck, 2010; Figure 5.1). This mite can feed on both honey bee larvae and pupae, reproducing in the bee brood cells, and during feeding vectors a number of normally harmless viruses, principally deformed wing virus, causing crippled bees, shortened bee life, and the loss of the colony. Resistance of the mite to the most commonly used chemical acaricides has led to a limited choice of control measures, and progress on breeding varroa-resistant bees has been slow (Carreck, 2011).

Other new threats to honey bees emerge from time to time. The Asian hornet *Vespa velutina*, which attacks honey bees, is currently spreading across

FIGURE 5.1 The *Varroa destructor* equator of global honey bee colony losses. So far, elevated colony losses have recently been reported from Europe, the USA, the Middle East, and Japan, but not from South America, Africa, or Australia. Colonies of Africanized honey bees in South America, and African ones survive without *V. destructor* treatment and the mite has not yet been introduced into Australia. This global picture indicates a central role of this particular ectoparasitic mite for honey bee colony losses. *From Neumann and Carreck (2010). © International Bee Research Association. Reproduced with permission of the editors of the Journal of Apicultural Research.*

Western Europe, and the small hive beetle *Aethina tumida*, native to sub-Saharan Africa, and which has caused extensive damage in parts of the USA, was found in Italy for the first time in September 2014, leading to costly but unsuccessful attempts to eradicate it (Mutinelli et al., 2014).

Other species of bees and other pollinators also suffer from pests and diseases (Alford, 1975; Meeus et al., 2011), although these have been less well studied. It has recently been discovered that a number of diseases thought to be unique to honey bees also commonly occur in bumble bees. The extent to which this reflects their natural distribution, or whether diseases have passed from managed honey bees to wild bumble bees, or from managed bumble bees to wild populations, is not fully understood. Mounting evidence occurs, however, that global trade in honey bees and managed bumble bees has exacerbated pest and disease problems (Fürst et al., 2014; McMahon et al., 2015).

5.5 CLIMATE CHANGE

Climate change will have differing effects on different bee species, and both direct and indirect effects may occur. The western honey bee *A. mellifera*, although having its origins in tropical regions, is a true generalist, and is now managed on all continents apart from Antarctica, over a very wide climatic range, and is thus unlikely to be greatly affected by climate change per se. On the other hand, bumble bees (*Bombus* spp.) are temperate creatures, well suited to foraging under cool conditions (Heinrich, 1979), and may have their range restricted by rising temperatures. Climate change may have both winners and losers. For example the tree bumble bee *Bombus hypnorum*, although common on mainland Europe, was only found for the first time in Southern England in 2001 (Goulson and Williams, 2001), but has since spread northwards and is seemingly ideally suited to much of Britain.

Indirect effects can include changes in the forage plants available to bees, as climate change makes areas more or less suited to certain plant species that form key diets of bees. This is most likely to affect specialist pollinator species that have a restricted range of food plants. Changes in phenology may also affect bees, as the flowering of key food plants may no longer coincide with the active period of specialist bees. Similarly, this could have an adverse effect on the pollination of rare and threatened plant species, if asynchrony occurs between the flowering of the plant and its specialist pollinator.

5.6 PESTICIDES

Problems with pesticides and bees have been known since the middle of the twentieth century (Carreck, 2008). Early problems occurred with the spraying of heavy metal compounds such as arsenical compounds in fruit orchards, but became especially severe with the use of the first generation of synthetic insecticides, the organochlorine and organophosphate compounds, which were highly toxic to bees when used as sprays. Newer generations of insecticides, first the carbamates, and later the synthetic pyrethroids, led to substantial reductions in incidents of pesticide poisoning (Figure 5.2).

The introduction of a new class of systemic insecticides, the neonicotinoids, intended to be used as seed dressings rather than sprays, should in theory have made the world even safer for pollinators. From their first use in the 1990s, however, concern about their effects on pollinators arose (Carreck, 2008; Ratnieks and Carreck, 2010). Concern about losses of honey bees after foraging on crops of sunflowers and maize (Figure 5.3) treated with one neonicotinoid, imidacloprid, led to restrictions on their use in countries such as France, although it is unclear whether the damage observed at that time was indeed due to the chemical, or to inadequate control of varroa, as resistant mites were first found in France at the same time. More recently, various laboratory-based studies (e.g., Gill et al., 2012; Henry et al., 2012; Whitehorn et al., 2012) have suggested that neonicotinoids have subtle, sublethal adverse effects on both bumble bees and honey bees, and these studies were influential in the imposition of a Europe-wide moratorium on their use of three neonicotinoid compounds on "bee friendly" crops in 2013 (Carreck and Ratnieks, 2013).

So far, however, such adverse effects of neonicotinoid compounds on bees have not actually been demonstrated in the field, and it remains unclear

FIGURE 5.2 Incidents involving the poisoning of honey bees investigated by the UK Wildlife Incident Investigation Scheme, and those confirmed to have been due to the approved use of a compound. No confirmed incident involving honey bees and the approved use of an agricultural pesticide has occurred since 2003. The single incident in 2010 involved a beekeeper who had treated his hives with wood preservative. The majority of poisoning incidents in recent years have involved misuse of compounds such as bendiocarb for destroying wild honey bee colonies. *Data from WIIS:* http://www.pesticides.gov.uk/guidance/industries/pesticides/topics/reducing-environmental-impact/wildlife. *From Carreck and Ratnieks (2014). © International Bee Research Association. Reproduced with permission of the editors of the Journal of Apicultural Research.*

FIGURE 5.3 Honey bee, *Apis mellifera* foraging on maize, *Zea mays*. *Photo © N. Carreck.*

whether bees in the field actually experience the levels used in these laboratory-based studies (Carreck and Ratnieks, 2014). A further complication is that toxicity testing has been almost exclusively carried out on honey bees (with some testing on bumble bees in recent years), but this may not necessarily be the best model system, as other bee species may be more susceptible. In a meta-analysis of many studies, Arena and Sgolastra (2014) found that in a comparison of LD_{50} values (the dose causing 50% death of a test population—a way of comparing toxicity of different compounds to different species), although much variability occurred, stingless bee species appear to be more sensitive to pesticides than the honey bee *A. mellifera*, whereas bumble bees and other solitary bees seem to be less sensitive. In addition, however, the large colonies of honey bees and stingless bees may give them more ability to survive the loss of individuals than species such as bumble bees with smaller colonies, and solitary species. Clearly more studies are needed. Despite this controversy, the overwhelming evidence is, however, that modern pesticides are much less harmful to bees than were the compounds extensively used between the 1950s and the 1980s.

5.7 OTHER CAUSES

Many other explanations for pollinator decline have been put forward, some more plausible than others, and these include genetically modified crops, various forms of electromagnetic radiation, especially the use of mobile phones, nanotechnology, and alien abduction (Carreck, 2009a, 2014), but there is little scientific evidence to support these.

5.8 WHAT CAN BE DONE?

Despite the pessimistic headlines, some evidence exists that declines of some bee species have at least slowed in some areas in recent years. Because no single cause exists for pollinator decline, likewise no single solution exists. For honey bees, understanding pests and diseases are the keys to overcoming them. For all species, providing food such as pollen and nectar throughout the season is necessary, and for nonmanaged bees, providing suitable nests sites is essential (Jones and Munn, 1998). Mixtures of nectar and pollen plants (Carreck and Williams, 1997, 2002; Carreck et al., 1999) can be planted as field margins, commonly as part of agri-environment schemes (Heard et al., 2007; Matheson and Carreck, 2014). In the UK, the government has introduced a National Pollinator Strategy (Defra, 2014) which aims to encourage all who manage land to incorporate provision for bees in land management plans and activities. Much that individuals can also do, exists in terms of planting suitable plants for bees (Kirk and Howes, 2012) in their gardens, and using less intensive management of land to provide suitable nest sites for bees and other pollinating insects.

REFERENCES

Alford, D.V., 1975. Bumble Bees. Harper Collins, London, UK, ISBN 978-0706701388, 352 pp.

Arena, M., Sgolastra, F., 2014. A meta-analysis comparing the sensitivity of bees to pesticides. Ecotoxicology 23, 324–334. http://dx.doi.org/10.1007/s10646-014-1190-1.

Biesmeijer, J.C., Roberts, S.P.M., Reemer, M., Ohlemüller, R., Edwards, M., Peeters, T., Schaffers, A.P., Potts, S.G., Kleukers, R., Thomas, C.D., Settele, J., Kunin, W.E., 2006. Parallel declines in pollinators and insect-pollinated plants in Britain and the Netherlands. Science 313 (351). http://dx.doi.org/10.1126/science.1127863.

Bailey, L., Ball, B.V., 1991. Honey Bee Pathology. Academic Press, London, UK, ISBN 0-12-073481-8, 193 pp.

Carreck, N.L., 2008. Pesticides and honey bees. Br. Beekeep. Assoc. News 173, 3–4.

Carreck, N.L., 2009a. Why do we avoid the obvious explanations for colony losses? Br. Beekeep. Assoc. News 178, 3–4.

Carreck, N.L., 2009b. Can studies of Kashmir bee virus and *Varroa destructor* aid our understanding of "Colony Collapse Disorder"?. In: Proceedings of XXXXIst International Apicultural Congress, Montpellier, France, 15th–20th September 2009, p. 146.

Carreck, N.L. (Ed.), 2011. Varroa - Still a Problem in the 21st Century? International Bee Research Association, Cardiff, UK, ISBN 978-0-86098-269-2, 78 pp.

Carreck, N.L., 2014. Electromagnetic radiation and bees again. Bee World 91 (4), 101–102. http://dx.doi.org/10.1080/0005772X.2014.11417624.

Carreck, N.L., Ratnieks, F.L.W., 2013. Will neonicotinoid moratorium save the bees? Res. Fortnight 415, 20–22.

Carreck, N.L., Ratnieks, F.L.W., 2014. The dose makes the poison: have "field realistic" rates of exposure of bees to neonicotinoid insecticides been over estimated in laboratory studies? J. Apic. Res. 53 (5), 607–614. http://dx.doi.org/10.3896/IBRA.1.53.5.08.

Carreck, N.L., Williams, I.H., 1997. Observations on two commercial flower mixtures as food sources for beneficial insects in the UK. J. Agric. Sci, Cambridge, 128, 397–405. http://dx.doi.org/10.1017/S0021859697004279.

Carreck, N.L., Williams, I.H., 2002. Food for insect pollinators on farmland: insect visits to the flowers of annual seed mixtures. J. Insect Conserv. 6, 13–23. http://dx.doi.org/10.1023/A:1015764925536.

Carreck, N.L., Williams, I.H., Oakley, J.N., 1999. Enhancing farmland for insect pollinators using flower mixtures. Aspects of Applied Biology 54: Field Margins and Buffer Zones – Ecology, Management and Policy 101–108.

Cox-Foster, D.L., Conlan, S., Holmes, E.C., Palacios, G., Evans, J.D., Moran, N.A., Quan, P.-L., Briese, T., Hornig, M., Geiser, D.M., Martinson, V., VanEngelsdorp, D., Kalkstein, A.L., Drysdale, A., Hui, J., Zhai, J., Cui, L., Hutchison, S.K., Simons, J.F., Egholm, M., Pettis, J.S., Lipkin, W.I., 2007. A metagenomic survey of microbes in honey bee colony collapse disorder. Science 318 (5848), 283–287. http://dx.doi.org/10.1126/science.1146498.

Department for Environment, Food and Rural Affairs (Defra), 2014. The National Pollinator Strategy: For Bees and Other Pollinators in England. UK Department for Environment, Food and Rural Affairs, London UK, 36 pp.

vanEngelsdorp, D., Evans, J.D., Saegerman, C., Mullin, C., Haubruge, E., Nguyen, B.K., Frazier, M., Frazier, J., Cox-Foster, D., Chen, Y.-P., Underwood, R., Tarpy, D.R., Pettis, J.S., 2009. Colony collapse disorder: a descriptive study. PLoS One 4 (8), e6481. http://dx.doi.org/10.1371/journal.pone.0006481.

Fürst, M.A., McMahon, D.P., Osborne, J.L., Paxton, R.J., Brown, M.J.F., 2014. Disease associations between honey bees and bumble bees as a threat to wild pollinators. Nature 506, 364–366. http://dx.doi.org/10.1038/nature12977.

Garibaldi, L.A., Aizen, M.A., Cunningham, S.A., Klein, A.M., 2009. Pollinator shortage and global crop yield. Commun. Integr. Biol. 2 (1), 37–39.

Gill, R.J., Ramos-Rodriguez, O., Raine, N.E., 2012. Combined pesticide exposure severely affects individual and colony level traits in bees. Nature 491, 105–119. http://dx.doi.org/10.1038/nature11585.

Gallai, A., Salles, J.M., Settele, J., Vaissière, B.E., 2009. Economic valuation of the vulnerability of world agriculture confronted with pollinator decline. Ecol. Econ. 68, 810–821.

Goulson, D., Williams, P.H., 2001. *Bombus hypnorum* (Hymenoptera: Apidae), a new British bumble bee? Br. J. Nat. Hist. 14, 129–131.

Heard, M.S., Carvell, C., Carreck, N.L., Rothery, P., Osborne, J.L., Bourke, A.F.G., 2007. Landscape context not patch size determines bumble bee density on flower mixtures sown for agri-environment schemes. Biol. Lett. 3, 638–641. http://dx.doi.org/10.1098/rsbl.2007.0425.

Heinrich, B., 1979. Bumble Bee Economics. Harvard University Press, Cambridge, MA, USA, ISBN 0-674-08580-9, 245 pp.

Henry, M., Béguin, M., Requier, F., Rollin, O., Odoux, J.-F., Aupinel, P., Aptel, J., Tchamitchian, S., Decourtye, A., 2012. A common pesticide decreases foraging success and survival in honey bees. Science 336, 348–351. http://dx.doi.org/10.1126/science.1215039.

Jones, H.R., Munn, P.A. (Eds.), 1998. Habitat Management for Wild Bees and Wasps. International Bee Research Association, Cardiff, UK, ISBN 0 86098 235 1, 38 pp.

Kirk, W.D.J., Howes, F.N., 2012. Plants for Bees. International Bee Research Association, Cardiff, UK, ISBN 978-0-86098-271-5, 311 pp.

Klein, A.-M., Vaissiere, B.E., Cane, J.H., Steffan-Dewenter, I., Cunningham, S.A., Kremen, C., Tscharntke, T., 2007. Importance of pollinators in changing landscapes for world crops. Proc. R. Soc. B: Biol. Sci. 274 (1608), 303–313. http://dx.doi.org/10.1098/rspb.2006.3721.

Matheson, A., Buchmann, S.L., O'Toole, C., Westrich, P., Williams, I.H. (Eds.), 1996. The Conservation of Bees. Linnean Society of London/International Bee Research Association/Academic Press, London, UK, ISBN 0-12-479740-7, 254 pp.

Matheson, A., Carreck, N.L. (Eds.), 2014. Forage for Pollinators in an Agricultural Landscape. International Bee Research Association, Cardiff, UK, ISBN 978-0-86098-277-7, 75 pp.

McMahon, D.P., Fürst, M.A., Caspar, J., Theodorou, P., Brown, M.J.F., Paxton, R.J., 2015. A sting in the spit: widespread cross-infection of multiple RNA viruses across wild and managed bees. J. Animal Ecol. http://dx.doi.org/11.1111/1365-2656.12345.

Meeus, I., Brown, M.J.F., De Graaf, D.C., Smagghe, G., 2011. Effects of invasive parasites on bumble bee declines. Conserv. Biol. 25 (4), 662–671. http://dx.doi.org/10.1111/j.1523-1739.2011.01707.x.

Morse, R.A., Flottum, K. (Eds.), 1997. Honey Bee Pests, Predators and Diseases, third ed. A.I. Root Co., Medina, OH, USA, ISBN 0-936028-10-6, p. 718.

Mutinelli, F., Montarsi, F., Federico, G., Granato, A., Ponti, A.M., Grandinetti, G., Ferrè, N., Franco, S., Duquesne, V., Rivière, M.-P., Thiéry, R., Hendrix, P., Ribière-Chabert, M., Chauzat, M.-P., 2014. First report on the detection of *Aethina tumida* Murray (*Coleoptera: Nitidulidae.*) in Italy. J. Apic. Res. 53 (5), 569–575. http://dx.doi.org/10.3896/IBRA.1.53.5.08.

Neumann, P., Carreck, N.L., 2010. Honey bee colony losses. J. Apic. Res. 49, 1–6. http://dx.doi.org/10.3896/IBRA.1.49.1.01.

Nieto, A., Roberts, S.P.M., Kemp, J., Rasmont, P., Kuhlmann, M., García Criado, M., Biesmeijer, J.C., Bogusch, P., Dathe, H.H., De la Rúa, P., De Meulemeester, T., Dehon, M., Dewulf, A., Ortiz-Sánchez, F.J., Lhomme, P., Pauly, A., Potts, S.G., Praz, C., Quaranta, M., Radchenko, V.G., Scheuchl, E., Smit, J., Straka, J., Terzo, M., Tomozii, B., Window, J., Michez, D., 2014. European Red List of Bees. Publication Office of the European Union, Luxembourg, 98 pp.

Osborne, J.L., Martin, A.P., Shortall, C.R., Todd, A.D., Goulson, D., Knight, M.E., Hale, R.J., Sanderson, R.A., 2008. Quantifying and comparing bumble bee nest densities in gardens and countryside habitats. J. Appl. Ecol. 45, 784–792. http://dx.doi.org/10.1111/j.1365-2664.2007.01359.x.

Potts, S.G., Biesmeijer, J.C., Kremen, C., Neumann, P., Schweiger, O., Kunin, W.E., 2010. Global pollinator declines: trends, impacts and drivers. Trends Ecol. Evol. 25 (6), 345–353. http://dx.doi.org/10.1016/j.tree.2010.01.007.

Ratnieks, F.L.W., Carreck, N.L., 2010. Clarity on honey bee collapse? Science 327, 152–153. http://dx.doi.org/10.1126/science.1185563.

Steinhauer, N.A., Rennich, K., Wilson, M.E., Caron, D.M., Lengerich, E.J., Pettis, J.S., Rose, R., Skinner, J.A., Tarpy, D.R., Wilkes, J.T., vanEngelsdorp, D., 2014. A national survey of managed honey bee 2012–2013 annual colony losses in the USA: results from the Bee Informed Partnership. J. Apic. Res. 53 (1). http://dx.doi.org/10.3896/IBRA.1.53.1.01.

Underwood, R., vanEngelsdorp, D., 2007. Colony collapse disorder: have we seen this before? Bee Cult. 35, 13–18.

Warren, M.S., Hill, J.K., Thomas, J.A., Asher, J., Fox, R., Huntley, B., Roy, D.B., Telfer, M.G., Jeffcoate, S., Harding, P., 2001. Rapid responses of British butterflies to opposing forces of habitat and climate change. Nature 414 (6859), 65–69.

Westrich, P., 1998. Habitat requirements of British bees and wasps. In: Jones, H.R., Munn, P.A. (Eds.), Habitat Management for Wild Bees and Wasps. International Bee Research Association, Cardiff, UK, ISBN 0-86098-235-1, pp. 4–12.

Whitehorn, P.R., O'Connor, S., Wackers, F.L., Goulson, D., 2012. Neonicotinoid pesticide reduces bumble bee colony growth and queen production. Science 336, 351–352. http://dx.doi.org/10.1126/science.1215025.

Williams, I.H., 1994. The dependence of crop production within the European Union on pollination by honey bees. Agric. Zool. Rev. 6, 229–257.

Williams, I.H., Carreck, N.L., 1994. Land use changes and honey forage plants. In: Matheson, A. (Ed.), Forage for Bees in an Agricultural Landscape. International Bee Research Association, Cardiff, ISBN 0 86098 217 3, pp. 7–20.

Williams, G.R., Dietemann, V., Ellis, J.D., Neumann, P., 2012. An update on the COLOSS network and the "BEEBOOK": standard methodologies for *Apis mellifera* research. J. Apic. Res. 51 (2), 151–153. http://dx.doi.org/10.3896/IBRA.1.51.2.01.

Williams, P.H., 1982. The distribution and decline of British bumble bees (*Bombus* Latr.). J. Apic. Res. 21 (4), 236–245. http://dx.doi.org/10.1080/00218839.1982.11100549.

Williams, P.H., 1986. Environmental change and the distributions of British bumble bees (*Bombus* Latr.). Bee World 67 (2), 50–61. http://dx.doi.org/10.1080/0005772X.1986.11098871.

van der Zee, R., Brodschneider, R., Brusbardis, V., Charrière, J.-D., Chlebo, R., Coffey, M.F., Dahle, B., Drazic, M.M., Kauko, L., Kretavicius, J., Kristiansen, P., Mutinelli, F., Otten, C., Peterson, M., Raudmets, A., Santrac, V., Seppälä, A., Soroker, V., Topolska, G., Vejsnæs, F., Gray, A., 2014. Results of international standardised beekeeper surveys of colony losses for winter 2012–2013: analysis of winter loss rates and mixed effects modelling of risk factors for winter loss. J. Apic. Res. 53 (1). http://dx.doi.org/10.3896/IBRA.1.53.1.02.

Chapter 6

Bark Beetle-Induced Forest Mortality in the North American Rocky Mountains

Kevin Hyde [1], Scott Peckham [2], Thomas Holmes [3] and Brent Ewers [2]

[1] *WY Center for Environmental Hydrology and Geophysics, University of Wyoming, Laramie, WY, USA,* [2] *Department of Botany, University of Wyoming, Laramie, WY, USA,* [3] *Southern Research Station, USDA Forest Service, Research Triangle, NC, USA*

ABSTRACT

The epidemic of mortality by insects and disease throughout the Northern American Rocky Mountains exceeds previous records both in severity and spatial extent. Beetle attacks weaken trees and introduce blue-stain fungi that induce hydraulic failure leading to mortality. The magnitude of this outbreak spurs predictions of major changes to biogeochemical cycling and hydrologic response, changes in species assemblages, and increased wildfire risk. Review of emerging empirical studies reveals conflicting evidence of changes and limited environmental threats. However, widespread forest mortality generates net economic costs and losses by reducing or eliminating market and nonmarket value. Potential deadfall may threaten human life and infrastructure and add costs of programs for hazard-tree reduction. Although forest regeneration following insect epidemics indicates resilient ecological systems, synergistic interactions of beetle kill with other disturbance processes, exacerbated by warming temperatures and drought may stimulate longer-term environmental concerns.

6.1 INTRODUCTION

6.1.1 The Nature and Extent of Mortality by Insects and Disease

Commencing in the late twentieth century, epidemic infestations of bark-boring beetles escalated throughout the Rocky Mountains of North America. Estimates of forest mortality by insects and disease (commonly referred to as "beetle kill") range from 6 to 11 million hectares (Meddens et al., 2012). The extent, duration, and severity of the outbreak are unprecedented in recent history (Raffa et al., 2008). Endemic beetle populations transitioned to epidemic levels as environmental conditions exceeded thresholds of natural

Biological and Environmental Hazards, Risks, and Disasters. http://dx.doi.org/10.1016/B978-0-12-394847-2.00009-7

population controls (Raffa et al., 2008; Six et al., 2014). Population explosions are attributed to a synergistic combination of climate change (Bentz et al., 2010; Kurz et al., 2008; Logan and Powell, 2009), drought (Adams et al., 2012; Greenwood and Weisberg, 2008), dense forest structure resulting from fire suppression, and increased forest homogeneity from forest management practices (Raffa et al., 2008). The magnitude of the forest mortality and prediction of increased frequency and severity of future outbreaks (Hicke et al., 2006) raise concerns about multiple impacts addressed in this essay: altered biogeochemical cycling and hydrologic response, reductions in overall forest condition and economic value, future wildfire behavior, vegetation response and changes to forest composition, and interacting disturbance processes.

Nine species are identified as aggressive bark beetles, each associated with a host tree species (see Bentz et al., 2009 for an overview of beetle biology and infestation). Two beetle species are primarily responsible for forest mortality. The mountain pine beetle (*Dendroctonus ponderosae*) prefers lodgepole (*Pinus contorta*), ponderosa (*Pinus ponderosa*), limber (*Pinus flexilis*), and whitebark pine (*Pinus albicaulis*) as host tree species. At higher elevations, the spruce bark beetle (*Dendroctonus rufipennis*) utilizes Engelmann spruce (*Picea engelmannii*) to complete its life cycle. Although both species of beetles bore through the bark of their host, feed on phloem, and lay eggs, neither boring nor phloem feeding kills the trees. Both beetle species carry strains of the blue-stain fungi (*Grosmannia clavigera*) (Six and Bentz, 2003; Six and Wingfield, 2011), and through their action infect the host tree xylem with fungi, which spreads rapidly blocking the tree's water transport system, causing tree death due to hydraulic failure (Hubbard et al., 2013; Knight et al., 1991; Yamaoka et al., 1995).

Progression of beetle kill is described in five phases, green, red, gray, tree fall, and regeneration, and impacts vary by phase (Mikkelson et al., 2013; Pugh and Gordon, 2013). In the green phase needles appear green, even after beetle attack has killed the tree (Wulder et al., 2006). During the red phase, needles change color to red and brown and fall to the ground (Pugh and Gordon, 2013). Only bare stems and branches of standing trees remain in the gray phase (Wulder et al., 2006) and dead trees fall from wind throw or simply gravity as roots decay. Regeneration of vegetation on the forest floor has been observed throughout the progression of beetle kill as the canopy opens and completion for water use by mature trees declines (Figure 6.1). Although several studies report on local regeneration patterns (Collins et al., 2011; Hadley and Veblen, 1993; Seidl et al., 2008), data on regeneration trends in the North American Rocky Mountains are limited. Forest mortality due to beetle infestations starts as scattered pockets and spread with nonlinear increase (Figure 6.2). Patterns of mortality vary within stands (Figure 6.3) due, in part, to mixed species composition and tree density. Mortality also varies between stands in close proximity to each other (Figure 6.4).

FIGURE 6.1 Regeneration following beetle kill of lodgepole pine (*Pinus contorta*) stand (Figure 6.1) with abundant seedlings and saplings. *Photo: Hyde, K., August 25, 2014.*

6.2 EFFECTS OF BARK BEETLE IMPACTS

6.2.1 Nitrogen

It has been hypothesized that the beetle epidemic would alter the forest nitrogen (N) cycle (Edburg et al., 2012). Although disturbances such as logging or severe storms can greatly increase stream N (Likens et al., 1970; Rhoades et al., 2013), and some model simulations showed a strong increase in available N locally due to beetle-caused mortality (Edburg et al., 2011), it was unclear how this excess N would move through the larger ecosystem. Although both litter N concentration and N-mineralization rates may increase in the early stages of beetle infestation (Griffin et al., 2011; Morehouse et al., 2008; Norton et al., 2015), stream N concentrations in a Colorado forest showed no change in a watershed heavily impacted by bark beetles (Rhoades et al., 2013) and stream-water nitrate was not significantly related to the percent of the basin impacted by beetles (Clow et al., 2011). The findings suggest that residual vegetation in the watershed was utilizing the excess N available from increased litter inputs and decreased uptake by the dominant forest canopy trees (Rhoades et al., 2013) rather than being transported out of the system in streamflow.

6.2.2 Carbon

Although beetle-caused forest mortality does not immediately change the total amount of ecosystem carbon like logging or wildfire, beetle impacts to carbon cycling in conifer systems have the potential to greatly alter the forest carbon balance (i.e., the rate of C exchange between ecosystem and atmosphere). Some modeling studies projected a large emission of carbon to the atmosphere in the years following beetle mortality (Edburg et al., 2011; Kurz et al., 2008).

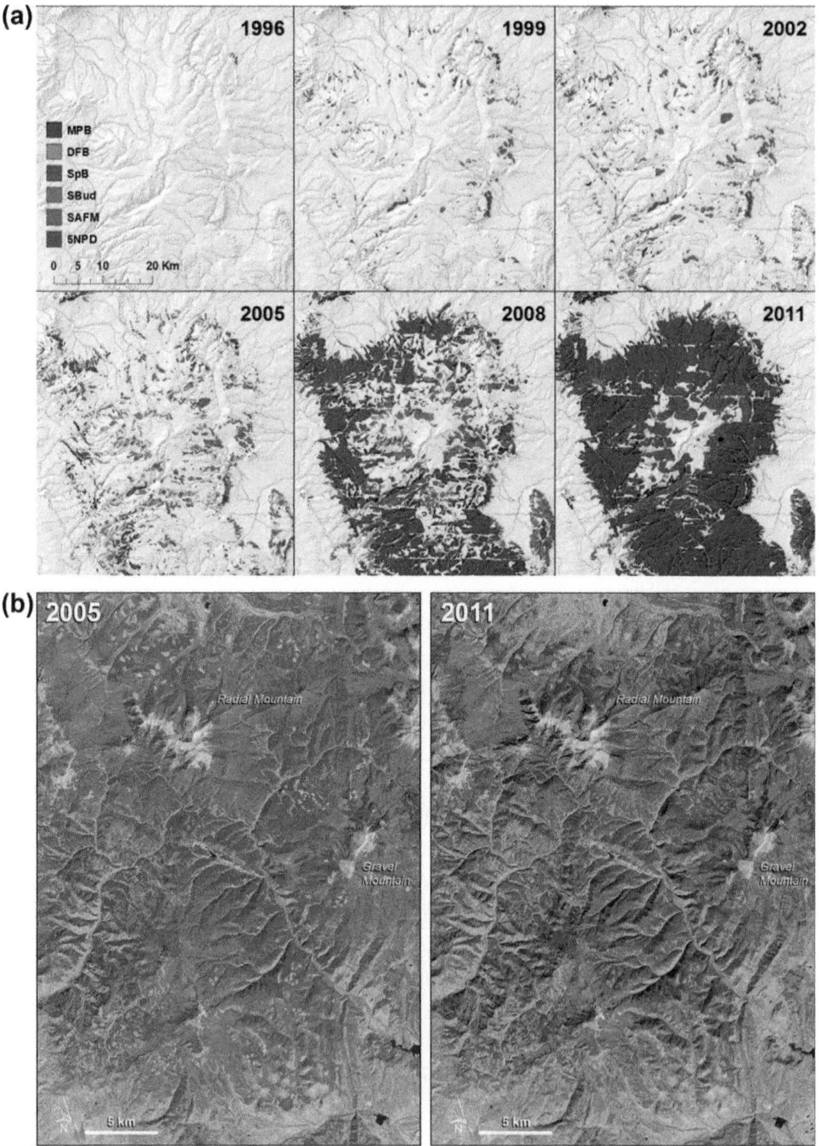

FIGURE 6.2 (a) Progression in 3 years increments of tree mortality due to insects and disease from 1996 to 2011, Snowy Range of the Medicine Bow Mountains, WY, USA. Six agents are mapped: MPB—mountain pine beetle (*Dendroctonus ponderosae*), DFB—Douglas fir beetle (*Dendroctonus pseudotsugae*), SPB—spruce beetle (*Dendroctonus rufipennis*), SBud—spruce budworm (*Choristoneura fumiferana*), SAFM—subalpine fir mortality complex, 5NPD—five needle pine decline. Mortality by SAFM and 5NPD involves multiple insect and disease pathogens *(Maps: Hyde, K. Source data: Aerial detection survey, Region 2, USDA Forest Service. Available at: http://www.fs.usda.gov/detail/r2/forest-grasslandhealth/?cid=fsbdev3_041629. (accessed 27.11.14))*, (b) Die-off in the lodgepole pine (*Pinus contorta*) forests near Grand Lake, Colorado, USA due to beetle infestation illustrated by Landsat-5 Thematic Mapper images acquired in Sept 2005 (left) and Sept 2011 (right). Forest die-off is most noticeable in the middle of this image pair. Brown areas in the 2011 image are mostly lodgepole pine stands, while the areas that remained green are either Engelmann spruce (*Picea engelmannii*) or aspen (*Populus tremuloides*) dominated forests that escaped infestation *(Image-pair: Sivanpillai, R. Source data: Simmon, R., Enhanced Landsat images, NASA Earth Observatory. Available at: http://earthobservatory.nasa.gov/IOTD/view.php?id=78677.).*

FIGURE 6.3 Heterogeneous patterns of tree mortality reflecting, in part, mixed species composition—Engelmann spruce (*Picea engelmannii*) and subalpine fir (*Abies lasiocarpa*), Snowy Range of the Medicine Bow Mountains, WY, USA. *Photos: Hyde, K., August 12, 2014.*

However, measurements at instrumented field sites suggest alternative trajectories. In Canada, researchers reported that stands with high levels of mortality were net carbon sinks within 5 years of infestation (Brown et al., 2012), whereas in the United States, Reed et al. (2014) reported no change in maximum carbon uptake as mortality in the stand increased from 30% to 78%. Despite canopy mortality levels exceeding 50%, drastic reduction in carbon uptake was not observed, and suggests that the remaining live trees and understory vegetation play a crucial role in the carbon cycle during the mortality event (Bowler et al., 2012; Brown et al., 2012; Collins et al., 2012b; Reed et al., 2014). However, in a high-elevation ecosystem of Engelmann spruce−subalpine fir, Frank et al. (2014) observed a 51% decrease in growing season C uptake where the majority of the canopy trees had died. These studies suggest that beetle-caused impacts to the carbon cycle may be dependent on the specific forest ecosystem affected.

6.2.3 Water

The primary impacts of beetle kill relative to water resources is loss of canopy leading to direct and indirect changes to hydrologic processes. A conceptual model based on a simple water balance approach follows that widespread forest mortality and canopy loss permit more rain and snow to reach the forest floor and significantly reduce transpiration. Therefore, more water moves into the soil and transfers to groundwater and streamflow; ideas confirmed by predictive modeling exercises (Mikkelson et al., 2011). Several hypotheses extend from expected increases in streamflow. Water yields may increase, accompanied by higher peak flows, more frequent and extreme flooding, and elevated erosion of hillslope and stream channel (Adams et al., 2012; Redding et al., 2008). Sediments produced through erosion may impair water quality

FIGURE 6.4 Progressive stages of beetle outbreak in stands of lodgepole pine (*Pinus contorta*): (a). no evidence of infestation, (b). 20–30% of trees killed, (c). mortality exceeded 80–90%. These stands are located within 100 m of each other illustrating the spatial variability of beetle-induced mortality even at fine scales. Note regeneration in stand (b) and (c) with more detail in Figure 6.2. *Photos: Hyde, K., August 25, 2014.*

and threaten aquatic habitat and human uses. Baseflow during low-flow periods may also increase, with higher flows extending later into the warmer summer months. A contrary hypothesis suggests that water yields may decrease in drier forests under warming climate conditions (Adams et al., 2012).

The expectation of increased stream discharge following forest disturbance has been driven in part by observations of large increases in runoff following wildfire (see Moody et al., 2013) and forest harvest (e.g., Stednick, 1996; Troendle and King, 1985), however, the analogies are limited (Adams et al., 2012; Pugh and Gordon, 2013). Harvest and wildfire cause rapid and commonly spatially continuous change to forest structure, where severe wildfire especially disrupts the entire biophysical system. Extreme fire events consume all biomass and destroy soil structure. By comparison, the gradual processes of mortality by insects and disease are generally not spatially continuous (Figure 6.3) and do not impair the potential of remaining vegetation to use water otherwise not consumed by impacted trees.

Results from empirical studies of observed hydrologic changes following forest mortality by insects and disease are mixed and contradictory (Table 6.1). Studies from 40 and 50 years ago reported stream discharge following beetle epidemics were observed to modestly increase within a few years immediately following an outbreak (Potts, 1984) or were most pronounced 15 years later (Bethlahmy, 1974). A more recent study found that groundwater contributions to streamflow increased due to reduced summer transpiration as trees succumbed to beetle attacks (Bearup et al., 2014). However, comparative analysis of discharge from multiple catchments found no detectable changes in streamflow (Hyde et al., 2014; Somor, 2010) and in one catchment, streamflow significantly declined (Somor, 2010). No changes or declines in streamflow may be due to observed compensations where increases in evaporation due to open canopy outpaces decreased transpiration (Biederman et al., 2014a), or where increased transpiration from accelerated growth of recovering vegetation (Ewers et al., 2013) outpaces decreased transpiration from tree mortality (Brown et al., 2014). Response of surviving trees adjacent to gaps opened by dead trees may provide an additional compensatory mechanism. Living roots rapidly fill the root gaps (Parsons et al., 1994b) and may explain muted impacts both with nutrient cycling (Parsons et al., 1994a) and hydrologic response as living roots exploit freed resources.

Snowmelt infiltration drives the overall hydrology and therefore streamflow of many of the areas impacted by beetle infestations. Complex process interactions controlling snow accumulation and melt are therefore expected to strongly influence hydrologic response to widespread tree mortality (Mikkelson et al., 2011; Pugh and Gordon, 2013; Pugh and Small, 2012, 2013). Impacted processes include evapotranspiration, soil infiltration, solar transmission and energy storage, needle fall, wind speed, canopy interception, sublimation, snowmelt timing, and water yield and the relative process role and interactions vary by the stages of infestation and mortality (see Pugh and Gordon, 2013 for a thorough discussion of processes and interactions). Research into interactions between snow processes and beetle infestation are beginning to emerge (Table 6.1). Low ablation rates (loss of snow from evaporation, sublimation, and wind redistribution) and increased ground-snow accumulation were found to lengthen the snowmelt period

TABLE 6.1 Summary of empirical studies of streamflow change and snow hydrology related to impacts of beetle infestations itemizing study location and dominant conifer species

Location	Dominant Species	Summary	Sources
Streamflow Change Studies			
Colorado, USA	Lodgepole pine	30 ± 15% increase in late-season contribution of groundwater to streamflow, probably from transpiration losses—hydrograph separation and stable-isotope analysis	Bearup et al. (2014)
Colorado, USA	Engelmann spruce	Modest increases in streamflow, most pronounced 15 years following epidemic outbreak	Bethlahmy (1974)
British Columbia, Canada	Lodgepole pine	Increased transpiration from spurred growth of residual vegetation compensated for reduced water use	Brown et al. (2014)
Wyoming, USA	Mixed subalpine conifer	No detectable changes in discharge after accounting for variable snow inputs	Hyde et al. (2014)
Montana, USA	Lodgepole pine/mixed conifer	15% Increase in average annual water yield with earlier peak flow and 10% increase in low flows	Potts (1984)
Colorado and Montana, USA	Lodgepole pine/mixed subalpine conifer	No detectable changes in discharge observed in 7 catchments—significantly reduced streamflow in one catchment	Somor (2010)
Snow Hydrology Studies			
Colorado, USA	Lodgepole pine	Snowpack depth and density increased in dead stands	Pugh and Gordon (2013)
British Columbia, Canada	Lodgepole pine/mixed subalpine conifer	Litter from dead and dying trees accumulates on snowpack, decreases albedo and increase snowpack ablation	Winkler et al. (2010)
Colorado and Wyoming, USA	Lodgepole pine	No change in streamflow—vapor loss from evaporation and snowpack sublimation counteracts decreased transpiration and canopy interception losses	Biederman et al. (2014a)
Colorado and Wyoming, USA	Lodgepole pine	Increased sublimation losses from snow-surface compensates for decreased loss from canopy interception and sublimation	Biederman et al. (2014b)
British Columbia, Canada	Lodgepole pine	Snow accumulation while ablation rates remained similar leading to longer snowmelt period	Boon (2007)

(Boon, 2007). Needles dropped from dead and dying trees interact with other factors to alter snow-surface albedo and ablation rates (Winkler et al., 2010). Snow accumulated to greater depths in dead stands due to reduced interception and sublimation, compared to nearby live stands (Pugh and Small, 2012, 2013). Yet another study observed no difference in peak snow—water equivalent between infested and uninfected forest stands, concluding that increased snowpack sublimation compensated for reduced canopy sublimation (Biederman et al., 2014b).

6.2.4 Vegetation Response

Concerns about long-term vegetation response to beetle kill include possible changes to species assemblages, widespread die-off, and changes to wildfire risk. A combination of field and modeling studies in Colorado and Wyoming, USA, indicated that in lodgepole pine stands where subalpine fir seedlings established prior to an outbreak, fir will become the dominant species as vegetation recovers (Collins et al., 2011). Increases in fir density are expected to vary by lodgepole forest type and moisture conditions (Kayes and Tinker, 2012). Impacts of beetle infestations may be amplified by other forest disturbance processes, especially wildfire, and drought (Adams et al., 2012), contributing to current global trends in widespread forest die-off under warming climates (Allen et al., 2010).

Evidence is mixed on changes to wildfire risk due to insect-related mortality. The time since the onset of infestation strongly influences changes in fire behavior and confidence varies in the likelihood of the range of possible changes to fire behavior (see Hicke et al., 2012 and Jenkins et al., 2014 for thorough reviews). In the Intermountain West, USA, insect epidemics in any coniferous stands may result in increased fire spread rates, fire intensities, and potential for passive crown fires (Jenkins et al., 2008), and downed trees in impacted lodgepole pine stands are expected to increase coarse fuel loads (Collins et al., 2012a). Yet in contrast to these predictions of increased fire severity, field observations found fire severity to be unrelated to the severity of prefire beetle outbreaks throughout coniferous stands in the Northern Rocky Mountains, USA (Harvey et al., 2014).

6.2.5 Economic Costs and Losses

Concerns about the direct threat to life and property from falling trees killed by the bark beetle outbreaks have prompted a widespread campaign to identify and remove hazardous trees throughout public lands in the United States (see http://www.fs.usda.gov/main/barkbeetle/safety). The cost of hazardous tree removal is substantial. For example, on the Laramie Ranger District of the Medicine Bow—Routt National Forest hazardous trees are identified and removed adjacent to highways (Figure 6.5), power lines, and forest roads and

FIGURE 6.5 Hazardous trees marked for removal (blue blaze) along Hwy 130 through the Snowy Range of the Medicine Bow Mountains, WY, USA. *Photo: Hyde, K., December 12, 2014.*

within and adjacent to campgrounds.[1] Cost for physical removal of hazardous trees ranges from $1.5K to $2K per acre with costs along highways at the upper end of the range or higher. This estimate does not include personnel time to cruise and mark trees or time to manage contracting and sales administration. Since 2010 approximately 100 miles of highways and forest roads have been treated at an average of 15 acres per mile for an estimated cost of $2.25M. To put this in perspective, six Ranger Districts occur in this National Forest and as many as 97 National Forests occur in the Western US where hazardous tree removal may be required.

In addition to the costs of removal of hazardous trees, bark beetle epidemics cause substantial economic losses to timber markets and nonmarket economic values. Where large volumes of timber are killed, salvage operations create a short-run pulse of timber and drive prices down. This results in a transfer of wealth from timber producers (growers) to timber consumers (mills); producers lose and consumers gain and the net effect is an unambiguous loss of economic value (Holmes, 1991). For example, analysis of the timber market, based on three decades of data concluded that southern pine beetle (SPB) (*Dendroctonus frontalis* Zimmermann) epidemics resulted in an average annual loss of $13M in the U.S. South (Pye et al., 2011). Although timber mortality from bark beetle epidemics can increase timber prices over the long run, due to relative timber scarcity, the average annual long-run

1. This information provided via personal communication with S. Alberts on January 7, 2015.

market impacts of SPB have historically been small relative to the short-run market impacts (Pye et al., 2011).

The potential for economic benefit from salvage logging in areas impacted by bark beetles depends upon the scale of the outbreak relative to the processing capacity of local timber markets. The current scale of processing capacity in Idaho, Montana, Washington, Oregon, California, and South Dakota appears adequate to absorb the supply of timber salvaged from the current MPB epidemic while maintaining positive economic returns for producers and consumers (Prestemon et al., 2013). Although the relatively small scale of timber processing capacity in Colorado and Wyoming, where substantial volumes of salvageable timber occur, is too small to economically salvage all of the timber killed by MPB, the reopening of a sawmill in Wyoming to salvage timber from the Medicine Bow National Forest shows that new economic opportunities exist. In another example, timber salvage operations in British Columbia resulting from the recent MPB outbreak created a short-term boom for the regional timber economy as well as the service and retail sectors (Patriquin et al., 2007). However, in the longer run, the magnitude of the MPB epidemic suggested that the scarcity of future timber supplies will not allow the regional timber economies to return to preepidemic levels once all merchantable timber has been salvaged.

A second category of economic damage resulting from bark beetle outbreaks is the impact on nonmarket economic values such as recreation and landscape aesthetics. An early study of the recreational impacts of MPB, using the travel cost method, was conducted on the Targhee National Forest by comparing use of campgrounds with different levels of MPB infestations (Michalson, 1975). The authors found a substantial loss of economic value to campers, due to less frequent visits and shorter stays, in the areas with heavier MPB infestations, and the annual losses on this Forest were estimated to exceed $500K. A similar study was conducted in East Texas reservoir campgrounds and found that economic losses from SPB outbreaks exceeded $1M for some campgrounds (Leuschner and Young, 1978). However, the authors correctly recognized that although some campers may be discouraged from using specific campgrounds where bark beetle activity is severe, they may choose to utilize other, unattacked sites. Substitution of unattacked campgrounds, where available, was found to reduce economic impacts by 85–90%.

In contrast to the studies of bark beetle impacts on actual recreational use, as described above, another approach is to interview recreationists to determine how their recreational choices might change due to infestations by bark beetle, and how much they are willing to pay for specific levels of healthy tree cover (Walsh and Olienyk, 1981). Using this approach in the Colorado Rocky Mountains Front Range, the authors found that a 1% decrease in the number of live trees found on recreation sites results in a 0.28% decrease in recreation demand. Despite the magnitude of the recent MPB outbreak in the Western

US, we are unaware of recent studies documenting the impact of bark beetles on recreation demand.

However, economists have conducted studies of the impact of the recent MPB outbreak on the value of residential properties using the hedonic valuation method. A loss in the value of homes located in residential forests impacted by bark beetle outbreaks results from the loss of aesthetic quality of surrounding landscapes and the costs of removing hazardous trees. Econometric analysis was used to analyze transactions of properties sold in Grand County, Colorado, which had experienced heavy infestations of MPB (Price et al., 2010), and it was found that the loss to homeowners was approximately $648/tree for trees located within 0.1 km of their home. A second hedonic valuation study of the impact of MPB on residential property values was conducted in Larimer and Boulder Counties, Colorado (Cohen et al., 2014). The authors estimated that, for homes with host trees within 0.1 km, the average loss per home was roughly $66K. Losses to residents in these counties who sold their home during the study period were estimated to be about $36M and, when these estimates were extrapolated to all homes in the study area, the total losses were about $137M.

6.3 SUMMARY

The epidemic of mortality by insects and disease throughout the North American Rocky Mountains exceeds previous records both in severity and spatial extent. Despite the magnitude of this outbreak, predictions of major changes to biogeochemical cycling and hydrologic response, changes in species assemblages, and increased wildfire risk have not been observed. In fact, evidence supports the conclusion that ecosystems recover quickly from impacts from beetle outbreak and local effects are muted at broader scales (Ewers et al., 2013). Forest regeneration following insect epidemics indicates highly resilient ecological systems. On the other hand, direct threats to life and infrastructure from falling trees have prompted costly programs to remove hazardous trees. Further, direct economic losses accrue as forest mortality reduces or eliminates market and nonmarket value.

However, as suggested by Adams et al. (2012), long-term environmental threats to impacted forests may emerge as warming temperatures exacerbate widespread drought. Beetle kill in water-stressed forests may synergistically interact with other anthropogenic and natural disturbance processes (Bigler et al., 2005; Kaiser et al., 2013; Raffa et al., 2008; Temperli et al., 2013) resulting in transformative landscape change (Adams, 2013; Allen et al., 2010; Breshears et al., 2005; Wilcox et al., 2012). In spite of the resiliency so far observed in the impacted forests, the possibility of wide-scale vegetation die-off indicates need to diligent monitor for long-term effects of mortality by insects and disease.

REFERENCES

Adams, H.D., Luce, C.H., Breshears, D.D., Allen, C.D., Weiler, M., Hale, V.C., Smith, A., Huxman, T.E., 2012. Ecohydrological consequences of drought-and infestation-triggered tree die-off: insights and hypotheses. Ecohydrology 5 (2), 145–159.

Adams, M.A., 2013. Mega-fires, tipping points and ecosystem services: managing forests and woodlands in an uncertain future. For. Ecol. Manag. 294, 250–261.

Allen, C.D., Macalady, A.K., Chenchouni, H., Bachelet, D., McDowell, N., Vennetier, M., Kitzberger, T., Rigling, A., Breshears, D.D., Hogg, E.H., Gonzalez, P., Fensham, R., Zhang, Z., Castro, J., Demidova, N., Lim, J.-H., Allard, G., Running, S.W., Semerci, A., Cobb, N., 2010. A global overview of drought and heat-induced tree mortality reveals emerging climate change risks for forests. For. Ecol. Manag. 259 (4), 660–684.

Bearup, L.A., Maxwell, R.M., Clow, D.W., McCray, J.E., 2014. Hydrological effects of forest transpiration loss in bark beetle-impacted watersheds. Nat. Clim. Change 4, 481–486.

Bentz, B., Logan, J., MacMahon, J., Allen, C.D., Ayres, M., Berg, E., Carroll, A., Hansen, M., Hicke, J., Joyce, L., 2009. Bark Beetle Outbreaks in Western North America: Causes and Consequences. University of Utah Press, Snowbird, UT.

Bentz, B.J., Régnière, J., Fettig, C.J., Hansen, E.M., Hayes, J.L., Hicke, J.A., Kelsey, R.G., Negrón, J.F., Seybold, S.J., 2010. Climate change and bark beetles of the western United States and Canada: direct and indirect effects. BioScience 60 (8), 602–613.

Bethlahmy, N., 1974. More streamflow after a bark beetle epidemic. J. Hydrol. 23 (3–4), 185–189.

Biederman, J., Harpold, A., Gochis, D., Ewers, B., Reed, D., Papuga, S., Brooks, P., 2014a. Increased evaporation following widespread tree mortality limits streamflow response. Water Resour. Res. 50 (7), 5395–5409.

Biederman, J.A., Brooks, P., Harpold, A., Gochis, D., Gutmann, E., Reed, D., Pendall, E., Ewers, B., 2014b. Multiscale observations of snow accumulation and peak snowpack following widespread, insect-induced lodgepole pine mortality. Ecohydrology 7 (1), 150–162.

Bigler, C., Kulakowski, D., Veblen, T.T., 2005. Multiple disturbance interactions and drought influence fire severity in Rocky Mountain subalpine forests. Ecology 86 (11), 3018–3029.

Boon, S., 2007. Snow accumulation and ablation in a beetle-killed pine stand in Northern Interior British Columbia. J. Ecosyst. Manag. 8 (3).

Bowler, R., Fredeen, A.L., Brown, M., Black, T.A., 2012. Residual vegetation importance to net CO_2 uptake in pine-dominated stands following mountain pine beetle attack in British Columbia, Canada. For. Ecol. Manag. 269, 82–91.

Breshears, D.D., Cobb, N.S., Rich, P.M., Price, K.P., Allen, C.D., Balice, R.G., Romme, W.H., Kastens, J.H., Floyd, M.L., Belnap, J., 2005. Regional vegetation die-off in response to global-change-type drought. Proc. Natl. Acad. Sci. U. S. A. 102 (42), 15144–15148.

Brown, M.G., Black, T.A., Nesic, Z., Foord, V.N., Spittlehouse, D.L., Fredeen, A.L., Bowler, R., Grant, N.J., Burton, P.J., Trofymow, J., 2014. Evapotranspiration and canopy characteristics of two lodgepole pine stands following mountain pine beetle attack. Hydrol. Process. 28 (8), 3326–3340.

Brown, M.G., Black, T.A., Nesic, Z., Fredeen, A.L., Foord, V.N., Spittlehouse, D.L., Bowler, R., Burton, P.J., Trofymow, J.A., Grant, N.J., Lessard, D., 2012. The carbon balance of two lodgepole pine stands recovering from mountain pine beetle attack in British Columbia. Agric. For. Meteorology 153, 82–93.

Clow, D.W., Rhoades, C., Briggs, J., Caldwell, M., Lewis Jr., W.M., 2011. Responses of soil and water chemistry to mountain pine beetle induced tree mortality in Grand County, Colorado, USA. Appl. Geochem. 26, S174–S178.

Cohen, J., Blinn, C.E., Boyle, K.J., Holmes, T.P., Moeltner, K., 2014. Hedonic valuation with translating amenities: mountain Pine beetles and host trees in the Colorado front range. Environ. Resour. Econ. 61, 1–30.

Collins, B., Rhoades, C., Battaglia, M., Hubbard, R., 2012a. The effects of bark beetle outbreaks on forest development, fuel loads and potential fire behavior in salvage logged and untreated lodgepole pine forests. For. Ecol. Manag. 284, 260–268.

Collins, B.J., Rhoades, C.C., Battaglia, M.A., Hubbard, R.M., 2012b. The effects of bark beetle outbreaks on forest development, fuel loads and potential fire behavior in salvage logged and untreated lodgepole pine forests. For. Ecol. Manage. 284, 260–268.

Collins, B.J., Rhoades, C.C., Hubbard, R.M., Battaglia, M.A., 2011. Tree regeneration and future stand development after bark beetle infestation and harvesting in Colorado lodgepole pine stands. For. Ecol. Manag. 261 (11), 2168–2175.

Edburg, S.L., Hicke, J.A., Brooks, P.D., Pendall, E.G., Ewers, B.E., Norton, U., Gochis, D., Gutmann, E.D., Meddens, A.J.H., 2012. Cascading impacts of bark beetle-caused tree mortality on coupled biogeophysical and biogeochemical processes. Front. Ecol. Environ. 10 (8), 416–424.

Edburg, S.L., Hicke, J.A., Lawrence, D.M., Thornton, P.E., 2011. Simulating coupled carbon and nitrogen dynamics following mountain pine beetle outbreaks in the western United States. J. Geophys. Res.-Biogeosci. 116.

Ewers, B., Norton, U., Borkhuu, B., Reed, D., Peckham, S., Biederman, J., King, A., Gochis, D., Brooks, P., Harpold, A., 2013. Bark Beetle Impacts on Ecosystem Processes Are over Quickly and Muted Spatially, Annual Meeting of the American Geophysical Union. American Geophysical Union, San Francisco pp. H11D-1189.

Frank, J.M., Massman, W.J., Ewers, B.E., Huckaby, L.S., Negron, J.F., 2014. Ecosystem CO_2/H_2O fluxes are explained by hydraulically limited gas exchange during tree mortality from spruce bark beetles. J. Geophys. Res.-Biogeosci. 119 (6), 1195–1215.

Greenwood, D.L., Weisberg, P.J., 2008. Density-dependent tree mortality in pinyon-juniper woodlands. For. Ecol. Manag. 255 (7), 2129–2137.

Griffin, J.M., Turner, M.G., Simard, M., 2011. Nitrogen cycling following mountain pine beetle disturbance in lodgepole pine forests of Greater Yellowstone. For. Ecol. Manage. 261 (6), 1077–1089.

Hadley, K.S., Veblen, T.T., 1993. Stand response to western spruce budworm and Douglas-fir bark beetle outbreaks, Colorado Front Range. Can. J. For. Res. 23 (3), 479–491.

Harvey, B.J., Donato, D.C., Turner, M.G., 2014. Recent mountain pine beetle outbreaks, wildfire severity, and postfire tree regeneration in the US Northern Rockies. Proc. Natl. Acad. Sci. 111 (42), 15120–15125.

Hicke, J.A., Johnson, M.C., Hayes, J.L., Preisler, H.K., 2012. Effects of bark beetle-caused tree mortality on wildfire. For. Ecol. Manag. 271, 81–90.

Hicke, J.A., Logan, J.A., Powell, J., Ojima, D.S., 2006. Changing temperatures influence suitability for modeled mountain pine beetle (Dendroctonus ponderosae) outbreaks in the western United States. J. Geophys. Res. Biogeosci. 111 (G2) (2005–2012).

Holmes, T.P., 1991. Price and welfare effects of catastrophic forest damage from southern pine beetle epidemics. For. Sci. 37 (2), 500–516.

Hubbard, R.M., Rhoades, C.C., Elder, K., Negron, J., 2013. Changes in transpiration and foliage growth in lodgepole pine trees following mountain pine beetle attack and mechanical girdling. For. Ecol. Manag. 289, 312–317.

Hyde, K., Miller, S., Anderson-Sprecher, R., Ewers, B., Sweatman, H., 2014. "Excess Water" Following Deforestation by Beetle Kill? Annual Meeting of the American Geophysical Union. American Geophysical Union, San Francisco.

Jenkins, M.J., Hebertson, E., Page, W., Jorgensen, C.A., 2008. Bark beetles, fuels, fires and implications for forest management in the Intermountain West. For. Ecol. Manag. 254 (1), 16–34.

Jenkins, M.J., Runyon, J.B., Fettig, C.J., Page, W.G., Bentz, B.J., 2014. Interactions among the mountain pine beetle, fires, and fuels. For. Sci. 60 (4), 489–501.

Kaiser, K.E., McGlynn, B.L., Emanuel, R.E., 2013. Ecohydrology of an outbreak: mountain pine beetle impacts trees in drier landscape positions first. Ecohydrology 6 (3), 444–454.

Kayes, L.J., Tinker, D.B., 2012. Forest structure and regeneration following a mountain pine beetle epidemic in southeastern Wyoming. For. Ecol. Manag. 263, 57–66.

Knight, D.H., Yavitt, J.B., Joyce, G.D., 1991. Water and nitrogen outflow from lodgepole pine forest after two levels of tree mortality. For. Ecol. Manage. 46 (3–4), 215–225.

Kurz, W.A., Dymond, C.C., Stinson, G., Rampley, G.J., Neilson, E.T., Carroll, A.L., Ebata, T., Safranyik, L., 2008. Mountain pine beetle and forest carbon feedback to climate change. Nature 452 (7190), 987–990.

Leuschner, W.A., Young, R.L., 1978. Estimating the southern pine beetle's impact on reservoir campsites. For. Sci. 24 (4), 527–537.

Likens, G.E., Bormann, F.H., Johnson, N.M., Fisher, D.W., Pierce, R.S., 1970. Effects of forest cutting and herbicide treatment on nutrient budgets in Hubbard Brook watershed-ecosystem. Ecol. Monogr. 40 (1), 23–47.

Logan, J.A., Powell, J.A., 2009. Ecological Consequences of Climate Change Altered Forest Insect Disturbance Regimes.

Meddens, A.J., Hicke, J.A., Ferguson, C.A., 2012. Spatiotemporal patterns of observed bark beetle-caused tree mortality in British Columbia and the western United States. Ecol. Appl. 22 (7), 1876–1891.

Michalson, E., 1975. Economic impact of mountain pine beetle on outdoor recreation. South. J. Agric. Econ. 7 (2), 43–50.

Mikkelson, K., Maxwell, R., Ferguson, I., Stednick, J., McCray, J., Sharp, J., 2011. Mountain pine beetle infestation impacts: modeling water and energy budgets at the hill-slope scale. Ecohydrology 6 (1), 64–72.

Mikkelson, K.M., Bearup, L.A., Maxwell, R.M., Stednick, J.D., McCray, J.E., Sharp, J.O., 2013. Bark beetle infestation impacts on nutrient cycling, water quality and interdependent hydrological effects. Biogeochemistry 115, 1–21.

Moody, J.A., Shakesby, R.A., Robichaud, P.R., Cannon, S.H., Martin, D.A., 2013. Current research issues related to post-wildfire runoff and erosion processes. Earth Sci. Rev. 122, 10–37.

Morehouse, K., Johns, T., Kaye, J., Kaye, M., 2008. Carbon and nitrogen cycling immediately following bark beetle outbreaks in southwestern ponderosa pine forests. For. Ecol. Manage. 255 (7), 2698–2708.

Norton, U., Ewers, B.E., Borkhuu, B., Brown, N.R., Pendall, E., 2015. Soil nitrogen five years after bark beetle infestation in Lodgepole Pine forests. Soil Sci. Soc. Am. J. 79 (1), 282–293.

Parsons, W.F., Knight, D.H., Miller, S.L., 1994a. Root gap dynamics in lodgepole pine forest: nitrogen transformations in gaps of different size. Ecol. Appl. 4, 354–362.

Parsons, W.F., Miller, S.L., Knight, D.H., 1994b. Root-gap dynamics in a lodgepole pine forest: ectomycorrhizal and nonmycorrhizal fine root activity after experimental gap formation. Can. J. For. Res. 24 (8), 1531–1538.

Patriquin, M.N., Wellstead, A.M., White, W.A., 2007. Beetles, trees, and people: regional economic impact sensitivity and policy considerations related to the mountain pine beetle infestation in British Columbia, Canada. For. Policy Econ. 9 (8), 938–946.

Potts, D.F., 1984. Hydrologic impacts of a large-scale mountain pine beetle (*Dendroctonus ponderosae Hopkins*) epidemic. JAWRA J. Am. Water Resour. Assoc. 20 (3), 373–377.

Prestemon, J.P., Abt, K.L., Potter, K.M., Koch, F.H., 2013. An economic assessment of mountain pine beetle timber salvage in the west. Western Journal of Applied Forestry 28 (4), 143–153.

Price, J.I., McCollum, D.W., Berrens, R.P., 2010. Insect infestation and residential property values: A hedonic analysis of the mountain pine beetle epidemic. Forest Policy and Economics 12 (6), 415–422.

Pugh, E., Gordon, E., 2013. A conceptual model of water yield effects from beetle-induced tree death in snow-dominated lodgepole pine forests. Hydrol. Process. 27, 2048–2060.

Pugh, E., Small, E., 2012. The impact of pine beetle infestation on snow accumulation and melt in the headwaters of the Colorado River. Ecohydrology 5 (4), 467–477.

Pugh, E.T., Small, E.E., 2013. The impact of beetle-induced conifer death on stand-scale canopy snow interception. Hydrol. Res. 44 (4).

Pye, J.M., Holmes, T.P., Prestemon, J.P., Wear, D.N., 2011. Economic Impacts of the Southern Pine Beetle.

Raffa, K.F., Aukema, B.H., Bentz, B.J., Carroll, A.L., Hicke, J.A., Turner, M.G., Romme, W.H., 2008. Cross-scale drivers of natural disturbances prone to anthropogenic amplification: the dynamics of bark beetle eruptions. Bioscience 58 (6), 501–517.

Redding, T., Winkler, R., Teti, P., Spittlehouse, D., Boon, S., Rex, J., Dubé, S., Moore, R., Wei, A., Carver, M., 2008. Mountain Pine Beetle and Watershed Hydrology, Mountain Pine Beetle: From Lessons Learned to Community-based Solutions.

Reed, D.E., Ewers, B.E., Pendall, E., 2014. Impact of mountain pine beetle induced mortality on forest carbon and water fluxes. Environ. Res. Lett. 9 (10), 105004.

Rhoades, C.C., McCutchan, J.H., Cooper, L.A., Clow, D., Detmer, T.M., Briggs, J.S., Stednick, J.D., Veblen, T.T., Ertz, R.M., Likens, G.E., Lewis, W.M., 2013. Biogeochemistry of beetle-killed forests: explaining a weak nitrate response. Proc. Natl. Acad. Sci. 110 (5), 1756–1760.

Seidl, R., Rammer, W., Jäger, D., Lexer, M.J., 2008. Impact of bark beetle (*Ips typographus* L.) disturbance on timber production and carbon sequestration in different management strategies under climate change. For. Ecol. Manag. 256 (3), 209–220.

Six, D.L., Bentz, B.J., 2003. Fungi associated with the North American spruce beetle, *Dendroctonus rufipennis*. Can. J. For. Res. 33 (9), 1815–1820.

Six, D.L., Biber, E., Long, E., 2014. Management for mountain pine beetle outbreak suppression: does relevant science support current policy? Forests 5 (1), 103–133.

Six, D.L., Wingfield, M.J., 2011. The role of phytopathogenicity in bark beetle-fungus symbioses: a challenge to the classic paradigm, Annual Review of Entomology, vol. 56. Annual Review of Entomology, Annual Reviews, Palo Alto, pp. 255–272.

Somor, A., 2010. Quantifying streamflow change following bark beetle outbreak in multiple central Colorado catchments. In: Graduate College, University of Arizona, pp: 87.

Stednick, J.D., 1996. Monitoring the effects of timber harvest on annual water yield. J. Hydrol. 176 (1), 79–95.

Temperli, C., Bugmann, H., Elkin, C., 2013. Cross-scale interactions among bark beetles, climate change, and wind disturbances: a landscape modeling approach. Ecol. Monogr. 83 (3), 383–402.

Troendle, C.A., King, R.M., 1985. The effect of timber harvest on the Fool Creek watershed, 30 years later. Water Resour. Res. 21 (12), 1915–1922.

Walsh, R.G., Olienyk, J.P., 1981. Recreation Demand Effects of Mountain Pine Beetle Damage to the Quality of Forest Recreation Resources in the Colorado Front Range. Department of Economics, Colorado State University.

Wilcox, B.P., Seyfried, M.S., Breshears, D.D., McDonnell, J.J., 2012. Ecohydrologic connections and complexities in drylands: new perspectives for understanding transformative landscape change. Ecohydrology 5 (2), 143–144.

Winkler, R., Boon, S., Zimonick, B., Baleshta, K., 2010. Assessing the effects of post-pine beetle forest litter on snow albedo. Hydrol. Process. 24 (6), 803–812.

Wulder, M.A., Dymond, C.C., White, J.C., Leckie, D.G., Carroll, A.L., 2006. Surveying mountain pine beetle damage of forests: a review of remote sensing opportunities. For. Ecol. Manag. 221 (1), 27–41.

Yamaoka, Y., Hiratsuka, Y., Maruyama, P.J., 1995. The ability of *Ophiostoma clavigerum* to kill mature lodgepole pine trees. Eur. J. For. Pathol. 25 (6–7), 401–404.

Chapter 7

Novel Approaches for Reversible Field Releases of Candidate Weed Biological Control Agents: Putting the Genie Back into the Bottle

James P. Cuda

Entomology & Nematology Department, Institute of Food & Agricultural Sciences, University of Florida, Gainesville, FL, USA

ABSTRACT

Biological control programs for weeds are coming under increased scrutiny by environmental groups in the USA. Legal challenges not only have stopped the release and redistribution of one group of biological control agents for weeds, but also have apparently delayed the issuance of permits for other biological control agents that have been recommended for field release. In this paper, three experimental approaches are described for predicting the field host specificity of some biological control candidates for weeds in the proposed country of introduction without any risk of permanent establishment of the insects. All three approaches are based on scientifically proven concepts and will facilitate biological control releases on an experimental basis before full-scale implementation.

7.1 INTRODUCTION

The invasion of a new habitat by a nonnative plant is due in part to the absence of natural enemies that normally limit the reproduction and spread of the plant in its native range (Williams, 1954; Keane and Crawley, 2002). In classical biological control, specialist natural enemies are imported from the native range of an adventive pest and released in small numbers in attempt to establish a permanent reproducing population (Frank, 2000). In this context, classical biological control of weeds involves the planned introduction of natural enemies to permanently suppress populations of a nonnative invasive

Biological and Environmental Hazards, Risks, and Disasters. http://dx.doi.org/10.1016/B978-0-12-394847-2.00010-3

plant (Cuda et al., 2008a; Cuda, 2014). Selective control of the target weed occurs when one or more of these coevolved natural enemies are reunited with the invasive plant in its adventive range. Typically, these natural enemies are nonnative, target-specific organisms, e.g., arthropods, nematodes, or plant pathogens. Biological weed control has been studied and used for managing invasive plants worldwide for over a century (Winston et al., 2014), and has developed into a complicated and highly regulated science.

One of the strengths of classical weed biocontrol, i.e., its sustainability or permanency, however, may contribute to unintended and negative consequences or "revenge effects" (Tenner, 1996). For instance, once a biocontrol agent is released and establishes reproducing populations, it cannot be recalled if nontarget species are negatively impacted. For this reason, the candidate agent must undergo rigorous testing in approved containment laboratories before field release. Typically, the organism is exposed to a series of carefully chosen nontarget plants in replicated no-choice tests to determine the "fundamental host range" and then multiple-choice trials to demonstrate "preference" for a particular host (Sheppard et al., 2005). Taken together, the results of these tests are used to predict the organism's "field host specificity" (Sheppard et al., 2005). In the USA, the U.S. Department of Agriculture's Animal and Plant Health Inspection Service, Plant Protection Quarantine unit (APHIS PPQ) controls the release approval process (Buckingham, 1994; Horner, 2004). A voluntary multiagency Technical Advisory Group administered by APHIS PPQ reviews information ("petitions") provided by the requesting scientist prior to making a recommendation to APHIS PPQ concerning the release of an agent (Buckingham, 1994; Cofrancesco and Shearer, 2004).

Currently, a perception exists among biological control researchers of weeds in the USA that it is becoming increasingly more difficult to obtain permits for releasing natural enemies of invasive weeds under a regulatory system where release decisions are based almost entirely on fundamental host range data (Hinz et al., 2014) or minor damage to individual plants in laboratory tests (Blossey, 2014). As of 2009, a total of 27 petitions were submitted by researchers to the TAG requesting field release of candidate biological control agents of weeds (TAG, 2015). Biological control agents in 10 of these petitions reviewed by the TAG, or 37%, were recommended for field release. Yet, no release permits have been issued by APHIS PPQ since 2009 (TAG, 2015). This may be a coincidence but 2009 is the same year that the Center for Biological Diversity (CBD) filed the first of two lawsuits against the USDA APHIS and the U.S. Fish & Wildlife Service for violating provisions of the Endangered Species Act and National Environmental Policy Act in connection with the biological control program for saltcedar (*Tamarix* spp.) (CBD, 2009, 2013; DeLoach et al., 2014). On June 15, 2010, USDA APHIS subsequently issued a moratorium against further use of the saltcedar leaf beetle *Diorhabda* spp. outside of a containment facility (Gruver, 2010).

To be sure, the risks associated with classical biological weed control are high. Host and habitat specificity often are difficult to predict from laboratory studies alone, and natural enemy releases are permanent and irreversible. If the biological control agent is found not to be entirely host-specific post-release or impacts a federally listed endangered species like the southwestern willow flycatcher, *Empidonax traillii extimus* Phillips (Passeriformes: Tyrannidae), in the example of saltcedar biological control (DeLoach et al., 2014), legal challenges by the CBD or other environmental groups will continue to delay or perhaps even stop the issuance of release permits by APHIS PPQ.

Limited or controlled release studies to demonstrate field host specificity in the proposed area of introduction are not currently an option as part of the current permitting process. Consequently, new approaches for risk assessment must be developed and adopted by the US regulatory community in order to correctly arrive at a *finding of no significant impact,* or FONSI (Horner, 2004). Otherwise, potentially beneficial and host-specific biological control agents may continue to be rejected based solely on the results of overly conservative laboratory testing (Hinz et al., 2014).

One approach for predicting field host specificity is open-field testing in the native range (Fowler et al., 2012). Although this has been done successfully with some shared crop or ornamental plants (Medal et al., 2002; Gandolfo et al., 2007), quarantine restrictions in other countries prohibit exporting test plants from the USA into a weed's native range because they could become invasive (Fowler et al., 2012). A more logical approach would involve *temporary and reversible* experimental releases of some candidate biological control agents in the intended release area. For instance, data on potential host range expansion, as well as direct and indirect food web effects, can be obtained by taking advantage of the biological attributes of certain organisms (e.g., arrhenotoky, whereby unmated females produce only male offspring by parthenogenesis), using radiation-induced, inherited F1 sterility for lepidopteran candidate biological agents, or perhaps even genetically modified organisms. Advantages of these novel approaches include the exposure of the biological control agent to the actual environmental conditions it would experience if approved for release, accurate prediction of field host specificity, and more importantly, the ability to stop releases of the biological control agent after one or more generations if nontarget damage or unintended ecological effects are detected.

This paper examines several innovative methods for field risk assessment of candidate weed biological control agents using the invasive Brazilian peppertree, *Schinus terebinthifolia* Raddi (Anacardiaceae) as a case study. Two of the aforementioned methods already have been tested with Brazilian peppertree natural enemies, and a third could be adopted based on recent advances in biotechnology.

7.2 BRAZILIAN PEPPERTREE CASE STUDY

In the USA, Brazilian peppertree is an aggressive, rapidly colonizing woody shrub of disturbed habitats, natural communities, and conservation areas in southern California, Hawaii, Texas, and Florida (USDA NRCS, 2015). In Hawaii, Brazilian peppertree is recognized as one of the most significant nonindigenous threats to federally listed endangered and threatened native plants (USFWS, 1998). In California, several flocks of cedar waxwings (*Bombycilla cedrorum* Vieillot, Passeriformes: Bombycillidae) recently died from trauma resulting from collision with hard objects after engorging on overripe berries of Brazilian peppertree (Kinde et al., 2012).

Introduced into Florida from South America as a landscape ornamental in the late nineteenth century (Barkley, 1944), Brazilian peppertree escaped cultivation and presently dominates entire ecosystems in central and south Florida (Morton, 1978). This invasive shrub grows rapidly, tolerates a wide range of environmental conditions, and is a prolific seed producer (Ewel et al., 1982; Spector and Putz, 2006). Brazilian peppertree also contributes to other invasive species problems. For example, it is an important alternate host for the naturalized Caribbean root weevil *Diaprepes abbreviatus* L., a major pest of commercial citrus, ornamental plants, and some agronomic crops in Florida and California (McCoy et al., 2003; Jetter and Godfrey, 2009). It was reported recently that the invasive black spiny-tailed iguana (*Ctenosaura similis* (Gray), Squamata: Iguanidae), introduced from Latin America, is highly dependent upon Brazilian peppertree for its survival in southwest Florida during the winter months (Jackson and Jackson, 2007).

The rapid growth and spread of Brazilian peppertree in Florida may be due in part to hybrid vigor (Williams et al., 2007; Geiger et al., 2011), allelopathy (Overholt et al., 2012), the absence of competitive plant taxa, and/or natural enemies (Hoshovsky and Randall, 2000; Cuda et al., 2006). A lack of natural enemies was the rationale for initiating a classical biological control program against Brazilian peppertree in Hawaii during the 1950s (Krauss, 1963; Yoshioka and Markin, 1991).

Conventional methods for controlling Brazilian peppertree (burning, various forms of physical extraction, and spraying of herbicides) are expensive and not always effective. For example, during the financial year 2010–2011, the Florida Fish and Wildlife Conservation Commission reported that nearly $7 million were spent by governmental agencies in Florida to control terrestrial invasive plants, including Brazilian peppertree (FWC, 2011). Rodgers et al. (2012) estimate that ~283,000 ha of south and central Florida are invaded by Brazilian peppertree, and expenditures to control the tree by the South Florida Water Management District alone were approximately $1.7 million in 2011. Doren et al. (1990) calculated that it would cost $20 million ($37,000 per ha) and take 20 years to restore a 2000 ha parcel in the Florida Everglades by bulldozing and burning the existing Brazilian peppertrees and completely removing the plowed rock substrate to prevent reinvasion of the site.

During the mid-1980s, a classical biological control program was initiated in Florida (Bennett et al., 1990). The long-term goal of this research project was to introduce a complex of specialist natural enemies into Florida that are capable of selectively attacking and reducing the invasiveness of Brazilian peppertree. Several arthropod natural enemies that occur in South America were identified that are capable of restricting seed production and reducing the vigor and growth rate of seedlings and young plants (Habeck et al., 1994; Cuda et al., 2006; McKay et al., 2009). Biological and host range studies were initiated for some of the more promising natural enemies to determine their suitability for release in Florida (Medal et al., 1999; Hight et al., 2003; Martin et al., 2004; Cuda et al., 2005, 2013b; Burckhardt et al., 2011; Oleiro et al., 2011; Wheeler et al., 2011; Manrique et al., 2012, 2014; McKay et al., 2012; Rendon et al., 2012; Diaz et al., 2014a,b).

7.2.1 Biological Approach

One of the first insects considered as a potential biological control agent of Brazilian peppertree in Florida, USA was the defoliating sawfly *Heteroperreyia hubrichi* Malaise (Hymenoptera: Pergidae) (Habeck et al., 1994). Laboratory testing showed the insect is host specific to Brazilian peppertree (Medal et al., 1999; Hight et al., 2003; Vitorino et al., 2000; Cuda et al., 2005; Pedrosa et al., 2006). Simulated herbivory studies also suggested that defoliation by the sawfly could reduce the competitiveness of Brazilian peppertree (Treadwell and Cuda, 2007). The insect was recommended for release in Florida by the TAG in 1997 (Medal et al., 1999; Hight et al., 2002; Cuda et al., 2013a). However, it was later discovered that the larvae contained compounds that could be toxic to livestock and small mammals (Dittrich et al., 2004). A larval toxicity study completed in Brazil showed that cows ingesting large quantities of *H. hubrichi* larvae (up to 300 mature larvae over 4 days) did not exhibit clinical signs of hepatotoxicity normally associated with these toxins (Dittrich et al., 2004). In addition, blood and liver chemistry profiles from samples taken from the test animals were found to be within normal reference ranges, even at the highest dose of sawfly larvae ingested by the cattle (Dittrich et al., 2004).

Another concern was the potential impact on insectivorous native and migratory birds in Florida that may attempt to consume the larvae. To address this issue, replicated laboratory feeding trials with red-winged blackbirds, *Agelaius phoeniceus* L. (Passeriformes: Icteridae), were conducted by the USDA APHIS Wildlife Services Laboratory, Gainesville, FL. The results of the short-term feeding trials showed the larvae were distasteful and would not be consumed by adult birds in the wild. The U.S. Fish and Wildlife Service Office, Vero Beach, Florida, agreed that adult birds probably would learn to

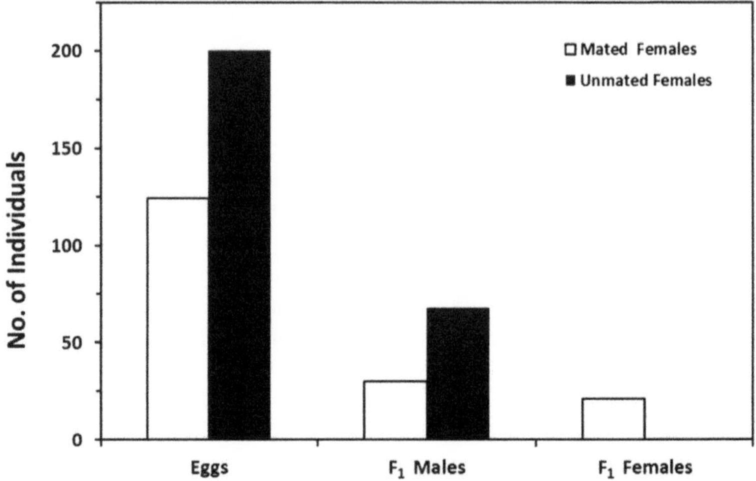

FIGURE 7.1 Fecundity (number of eggs) and F_1 adults produced from mated and unmated females of the pergid sawfly *Heteroperreyia hubrichi*. Unmated females deposited 200 eggs that produced only male progeny (68 adults) whereas mated females laid 124 eggs that produced both sexes (30 males and 21 females).

avoid the larvae. However, the U.S. Fish and Wildlife Service expressed concern about the effects of the larvae on nestlings that may not have developed a regurgitation reflex to reject the larvae. Consequently, a release permit was not issued by APHIS PPQ (Cuda et al., 2013a).

The impact on nesting birds could have been tested with a pilot release of unmated female sawflies. The sawfly *H. hubrichi* exhibits uniparental reproduction, or arrhenotoky. Arrhenotoky is a form of parthenogenesis in which unmated females produce larvae that develop only into males (Figure 7.1).

Except for the Lepidoptera, this form of reproduction occurs in most insect orders used in weed biological control programs, e.g., Thysanoptera, Hymenoptera, Hemiptera, and Coleoptera (Normark, 2003). By conducting single sex releases (unmated females only), the negative effects of *H. hubrichi* larvae on nesting birds as well as other nontarget organisms could be monitored during the temporary field evaluations of the biocontrol agent. However, if unanticipated damage to nontarget flora or fauna was discovered during or soon after the single-sex sawfly releases, no further releases would be made. Stopping the sawfly releases would effectively prevent the insect from establishing permanent, reproducing populations (Cuda et al., 2008b).

7.2.2 Autocidal Approach

7.2.2.1 Radiation Induced

The F_1 sterile insect technique (F_1SIT) is another method that has been proposed for field testing some candidate biological control agents for weeds

(Delfosse, 2005). The insect order Lepidoptera, which is second only to the Coleoptera in terms of its importance in biological control programs for weeds (Bellows and Headrick, 1999), is ideally suited for F_1SIT field testing of candidate biological control agents (Greany and Carpenter, 2000). Production of F_1 sterile progeny that are produced by exposing the parental generation to low doses of gamma radiation would facilitate observations on survival and impact of the F_1 larvae on nontarget plants under actual field conditions, with no risk of permanent establishment because the F_1 adults are completely sterile.

Episimus unguiculus Clarke (Lepidoptera: Tortricidae), formerly known as *Episimus utilis* Zimmerman, was introduced into Hawaii for classical biological control of Brazilian peppertree in the 1950s (Krauss, 1963). Like the sawfly, the larval stage or caterpillar of *E. unguiculus* attacks the foliage of Brazilian peppertree in its native range (McKay et al., 2009). Under the confined and artificial laboratory conditions imposed during the no-choice fundamental host range testing process in Florida, *E. unguiculus* accepted the economically important *Pistacia* spp. and several native *Rhus* spp. for oviposition and development (Cuda et al., 2013a). One of the reasons is that the terpenoid chemical compounds in the leaves and bark of Brazilian peppertree are more similar to *Pistacia* species than those isolated from other species of the genus *Schinus* (Campello and Marsaioli, 1975). This finding and the laboratory-induced oligophagy exhibited by *E. unguiculus* suggest that this candidate biocontrol agent could threaten production of cultivated pistachio, *Pistacia vera* L., in the USA. However, *E. unguiculus* exhibited a clear preference for Brazilian peppertree over cultivated pistachio in the multiple-choice tests, despite their similar chemistries (Cuda et al., 2010). More importantly, pistachio is cultivated in the western USA under very different climatic conditions (hot, dry summers, and cool winters) than Florida (tropical conditions with high humidity). Because of "geographical incompatibility" (Sheppard et al., 2005), the environmental conditions where cultivated pistachio production occurs would not be conducive to the establishment of *E. unguiculus*.

In the laboratory, Moeri et al. (2009) treated virgin adult males and females with increasing doses of gamma radiation, and were either inbred or out-crossed to nontreated *E. unguiculus* adults. The radiation dose at which treated females were 100% sterile was 200 Gy (Figure 7.2). The dose at which F_1 females and males were 100% sterile was 225 Gy (Figure 7.3).

For field testing, adult male moths would be treated with 225 Gy and upon mating with untreated females, the F_1 generation adults would be completely sterile (Moeri et al., 2009).

A petition was submitted in September 2009 but the TAG did not recommend field release based solely on the results of the no-choice tests (Cuda et al., 2013a), even though F_1SIT approach was tested and proposed as a possible mitigation method for risk assessment (LRN, 2007; Cuda et al., 2008b).

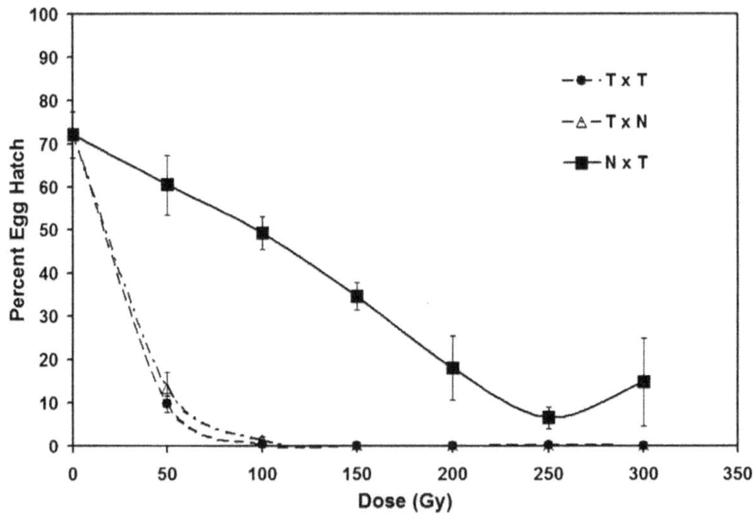

FIGURE 7.2 Fertility (mean percentage of hatched eggs) of *Episimus unguiculus* adults for three crosses (T♀ × T♂, T♀ × N♂, and N♀ × T♂) treated with increasing dose of gamma radiation. T♀ × T♂; $y = 63.6209 - 0.7619x + 0.0020x^2$; $R^2 = 0.782$; T♀ × N♂; $y = 60.0056 - 0.6843x + 0.0017x^2$; $R^2 = 0.790$; N♀ × T♂; $y = 64.4286 - 0.1999x$; $R^2 = 0.649$. *Moeri et al. (2009); reprinted with permission.*

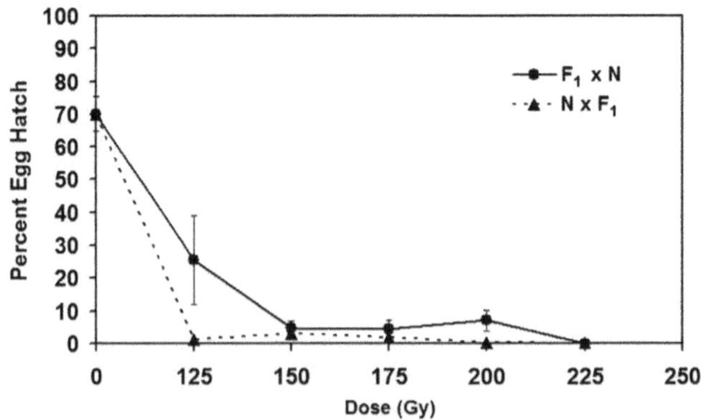

FIGURE 7.3 Fertility (mean percentage of hatched eggs) of F_1 crosses (F_1♀ × N♂, N♀ × F_1♂) of *Episimus unguiculus* adults as a result of radiation administered to parental males. F_1♀ × N♂; $y = 63.8547 - 0.3176x$; $R^2 = 0.575$; N♀ × F_1♂; $y = 55.1634 - 0.6017x + 0.0016x^2$; $R^2 = 0.608$. *Moeri et al. (2009); reprinted with permission.*

7.2.2.2 Transgenic

A new genetic-based sterility method was recently developed and tested for controlling mosquitoes and pest Lepidoptera (Harris et al., 2011; Morrison et al., 2012). This transgenic biotechnology, referred to as Release of Insects carrying a Dominant Lethal gene (RIDL), essentially is a genetic enhancement of the radiation-mediated F_1SIT approach (Harris et al., 2011; Morrison et al., 2012). The advantages of RIDL are that genetically engineered male insects can compete successfully with wild males, and 100% mortality occurs in engineered progeny carrying the lethal transgene (Harris et al., 2011; Morrison et al., 2012). Furthermore, the lethal transgenic system can be repressed in the presence of dietary antidotes (Harris et al., 2011; Morrision et al., 2012).

To date, this proven technology has been tested only on pest insects and also the silkworm, *Bombyx mori* L. (Tan et al., 2013). Perhaps the RIDL approach could be adopted for prerelease field testing of candidate biological control agents for weeds in the future.

7.3 CONCLUSION

In a special issue of the journal *Biological Control* that addresses the benefits and risks of classical biological control of weeds, Delfosse (2005) challenged biological control researchers on weeds to continue developing and improving host specificity, testing protocols in order to more accurately predict field host specificity, including the adoption of F_1SIT. More recently, Fowler et al. (2012) argued that researchers need to improve risk-assessment procedures "for assessing whether a fundamental host is likely to become a realized host without having to rely on native-range open-field specificity testing." Other researchers are now questioning whether classical biological in the USA will continue to be a "viable tool" for invasive plant management under the current regulatory climate, and make the case for a fundamental transformation of host specificity, testing procedures, and risk-assessment protocols (Blossey, 2014; Hinz et al., 2014).

In this paper, three different temporary, reversible approaches were presented for accurately predicting the field host specificity of a candidate biological control agent for weeds. These scientifically proven methods could be incorporated into the risk assessment process, if necessary, to precisely identify the plant species that will support agent populations in the proposed release area or to reveal unintended ecological impacts before reproducing individuals of a permitted biocontrol agent are released. The onus now is on the regulatory community to incorporate these temporary field release methods into the current decision-making and permitting process.

Classical biological control of weeds is not without some level of risk (Delfosse, 2000). A guiding principle for reducing risks when controlling established marine invaders is "…proceeding with preferred alternatives on an experimental basis before full-scale implementation and monitoring results to determine the effects of the approach. A primary means of reducing risk and proceeding experimentally is to make a biocontrol action, in effect, reversible." (Bax et al., 2001, p. 1241). Perhaps the time has come for the regulatory community to seriously consider adopting one or more of the aforementioned procedures for reversible releases of candidate biological control agents of weeds for risk-assessment purposes.

ACKNOWLEDGMENTS

The author thanks Veronica Manrique for the *H. hubrichi* sawfly rearing data, and is grateful to Howard Frank for critically reviewing an earlier draft of this manuscript and Jane Medley for assistance with the figures.

REFERENCES

Barkley, F.A., 1944. *Schinus* L. Brittonia 5, 160–198.

Bax, N., Carlton, J.T., Mathews-Amos, A., Headrich, R.L., Howarth, F.G., Purcell, J.E., Rieser, A., Gray, A., 2001. The control of biological invasions in the world's oceans. Conserv. Biol. 15, 1234–1246.

Bellows, T.S., Headrick, D.H., 1999. Arthropods and vertebrates in biological control of weeds. In: Bellows, T.S., Fisher, T.W. (Eds.), Handbook of Biological Control, Principles and Applications of Biological Control. Academic Press, San Diego, pp. 505–516.

Bennett, F.D., Crestana, L., Habeck, D.H., Berti-Filho, E., 1990. Brazilian peppertree – prospects for biological control. In: Delfosse, E.S. (Ed.), Proceedings of the VII International Symposium on Biological Control of Weeds, 6–11 March 1988, Rome, Italy. Ist. Sper. Patol. Veg. (MAF), Rome, Italy, pp. 293–297.

Blossey, B., 2014. Understanding and managing ecological impacts: can biocontrol remain a viable tool for managers in the USA? In: Impson, F.A., Kleinjan, C.A., Hoffmann, J.H. (Eds.), Proceedings of the XIV International Symposium on Biological Control of Weeds, Kruger National Park, South Africa, 2–7 March 2014. University of Capetown, South Africa, p. 187.

Buckingham, G.R., 1994. Biological control of aquatic weeds. In: Rosen, D., Bennett, F.D., Capinera, J.L. (Eds.), Pest Management in the Subtropics: Biological Control – a Florida Perspective. Intercept, Andover, UK, pp. 413–480.

Burckhardt, D., Cuda, J.P., Manrique, V., Diaz, R., Overholt, W.A., Williams, D.A., Christ, L.R., Vitorino, M.D., 2011. *Calophya latiforceps*, a new species of jumping plant lice (Hemiptera: Calophyidae) associated with *Schinus terebinthifolius* (Anacardiaceae) in Brazil. Fla. Entomol. 94, 489–499.

Campello, J.P., Marsaioli, A.J., 1975. Terebenthifolic acid and bauerenone. Phytochemistry 14, 2300–2302.

Center for Biological Diversity (CBD), 2009. Complaint for Declaratory and Injunctive Relief. http://www.biologicaldiversity.org/species/birds/southwestern_willow_flycatcher/pdfs/tamarisk_complaint.pdf.

Center for Biological Diversity (CBD), 2013. Complaint for Declaratory and Injunctive Relief. http://www.biologicaldiversity.org/species/birds/southwestern_willow_flycatcher/pdfs/lawsuit_2_20130930_COMPLAINT.pdf.

Cofrancesco Jr., A.F., Shearer, J.F., 2004. Technical advisory group for biological control agents of weeds. In: Coombs, E.M., Clark, J.K., Piper, G.L., Cofrancesco Jr., A.F. (Eds.), Biological Control of Invasive Plants in the United States. Oregon State University Press, Corvallis, Oregon, pp. 38–41.

Cuda, J.P., 2014. Chapter 8: Introduction to biological control of aquatic weeds. In: Gettys, L.A., Haller, W.T., Petty, D.G. (Eds.), Biology and Control of Aquatic Plants: A Best Management Practices Handbook, third ed. Aquatic Ecosystem Restoration Foundation, Marietta, GA, pp. 51–58.

Cuda, J.P., Ferriter, A.P., Manrique, V., Medal, J.C. (Eds.), 2006. Interagency Brazilian peppertree *(Schinus terebinthifolius)* management plan for Florida. Recommendations from the Brazilian Peppertree Task Force, second ed., Florida Exotic Pest Plant Council, South Florida Water Management District, West Palm Beach, FL.

Cuda, J.P., Medal, J.C., Vitorino, M.D., Habeck, D.H., 2005. Supplementary host specificity testing of the sawfly *Heteroperreyia hubrichi*, a candidate for classical biological control of Brazilian peppertree, *Schinus terebinthifolius*, in the USA. BioControl 50, 195–201.

Cuda, J.P., Medal, J.C., Overholt, W.A., Vitorino, M.D., Habeck, D.H., 1999, revised 2013a. Classical Biological Control of Brazilian Peppertree (*Schinus Terebinthifolia*) in Florida. ENY-820, UF/IFAS Extension. University of Florida, Gainesville. http://edis.ifas.ufl.edu/pdffiles/IN/IN11400.pdf.

Cuda, J.P., Medal, J., Pedrosa-Macedo, J., Manrique, V., Overholt, W.A., April 2010. Biology and fundamental host range of *Episimus unguiculus* (Lepidoptera: Tortricidae), a new candidate for biological control of Brazilian peppertree, *Schinus terebinthifolius* (Anacardiaceae) in Florida. In: Program and Abstracts, Florida Exotic Pest Plant Council 25th Annual Symposium. Changes in Latitude, 5–8. Crystal River, Florida, pp. 14–15. http://www.fleppc.org/Symposium/2010/FLEPPC.Final.Program.2010.pdf.

Cuda, J.P., Gillmore, J.L., Medal, J.C., Garcete-Barrett, B., Overholt, W.A., 2013b. Biology and fundamental host range of the stem boring weevil *Apocnemidophorus pipitzi* (Coleoptera: Curculionidae), a candidate biological control agent for Brazilian peppertree. In: Wu, Y., Johnson, T., Sing, S., Raghu, S., Wheeler, G., Pratt, P., Warner, K., Center, T., Goolsby, J., Reardon, R. (Eds.), Proceedings of the XIII International Symposium on Biological Control of Weeds, 11–16 September 2011, Waikoloa, Hawaii, USA. FHTET-2012-07, Morgantown, West Virginia, p. 47.

Cuda, J.P., Charudattan, R., Grodowitz, M.J., Newman, R.M., Shearer, J.F., Tamayo, M.L., Villegas, B., 2008a. Recent advances in biological control of submersed aquatic weeds. J. Aquat. Plant Manag. 46, 15–32.

Cuda, J.P., Moeri, O.E., Overholt, W.A., Manrique, V., Bloem, S., Carpenter, J.E., Medal, J.C., Pedrosa-Macedo, J.H., 2008b. Novel approaches for risk assessment: feasibility studies on temporary reversible releases of biocontrol agents. In: Julien, M.H., Sforza, R., Bon, M.C., Evans, H.C., Hatcher, P.E., Hinz, H.L., Rector, B.G. (Eds.), Proceedings, XII International Symposium on Biological Control of Weeds. CAB International, Warlingham, UK, p. 102.

Delfosse, E.S., 2000. Biological control: important tool for managing invasive species. Agricultural Research. USDA ARS 48, 2.

Delfosse, E.S., 2005. Risk and ethics in biological control. Biol. Control 35, 319–329.

DeLoach, C.J., Carruthers, R.I., Knutson, A.E., Bean, D., Tracy, J.L., Moran, P.J., Herr, J.C., Gaskin, J.F., Dudley, T.L., Ritzi, C.M., Michaels, G.J., 2014. Biological control of saltcedar (*Tamarix*: Tamaricales) in the Western USA: control and environmental effects progress since 2010. In: Impson, F.A., Kleinjan, C.A., Hoffmann, J.H. (Eds.), Proceedings of the XIV International Symposium on Biological Control of Weeds, Kruger National Park, South Africa, 2–7 March 2014. University of Capetown, South Africa, pp. 191–192.

Diaz, R., Moscoso, D., Manrique, V., Williams, D., Overholt, W.A., 2014a. Native range density, host utilization and life history of *Calophya latiforceps* (Hemiptera: Calophyidae): an herbivore of Brazilian peppertree (*Schinus terebinthifolia*). Biocontrol Sci. Technol. 24, 536–553.

Diaz, R., Manrique, V., Munyaneza, J.E., Sengoda, V.G., Adkins, S., Hendricks, K., Roberts, P.D., Overholt, W.A., 2014b. Host specificity testing and examination for plant pathogens reveal that the gall-inducing psyllid *Calophya latiforceps* is safe to release for biological control of Brazilian peppertree. Entomol. Exp. Appl. 1–14. http://dx.doi.org/10.1111/eea.12249.

Dittrich, R.L., Macedo, J.H.P., Cuda, J.P., Biondo, A.W., 2004. Brazilian peppertree sawfly larvae toxicity in bovines, Abstract 29. In: Proceedings of the Joint Annual Meeting of the American College of Veterinary Pathologists (55th) and American Society for Veterinary Clinical Pathology (39th), Orlando, FL, 13–17 November, vol. 33, No. 3, p. 191. http://dx.doi.org/10.1111/j.1939-165X.2004.tb00373.x.

Doren, R.F., Whiteaker, L.D., Molnar, G., Sylvia, D., 1990. Restoration of former wetlands within the hole-in-the-doughnut in Everglades National Park. In: Webb Jr., F.J. (Ed.), Proceedings of the 7th Annual Conference on Wetlands Restoration and Creation. Hillsborough Community College, Institute of Florida Studies, Tampa, Florida, pp. 33–50.

Ewel, J., Ojima, D., Karl, D., Debusk, W., 1982. *Schinus* in Successional Ecosystems of Everglades National Park. South Florida Research Center Report T-676. Everglades National Park, National Park Service, Homestead, Florida.

Frank, J.H., 2000. Glossary of Expressions in Biological Control. Univ. Florida online. http://ipm.ifas.ufl.edu/Education_Extension/gloss.shtml. viewed on March 11, 2015.

Fowler, S.V., Paynter, Q., Dodd, S., Groenteman, R., 2012. How can ecologists help practitioners minimize non-target effects in weed biocontrol? J. Appl. Ecol. 49, 307–310.

Florida Fish and Wildlife Conservation Commission (FWC), 2011. Upland Invasive Exotic Plant Control Projects Fiscal Year 2010–2011, 5pp. http://myfwc.com/media/2072931/UplandPlantManage_10-11.pdf.

Gandolfo, D., McKay, F., Medal, J.C., Cuda, J.P., 2007. Open–field host specificity testing of *Gratiana boliviana* (Coleoptera: Chrysomelidae), a biological control agent of tropical soda apple (Solanaceae) in the United States. Fla. Entomol. 90, 223–228.

Geiger, J.H., Pratt, P.D., Wheeler, G.S., Williams, D.A., 2011. Hybrid vigor for the invasive exotic Brazilian peppertree (*Schinus terebinthifolius* Raddi., Anacardiaceae) in Florida. Int. J. Plant Sci. 172, 655–663.

Greany, P.D., Carpenter, J.E., 2000. Use of nuclear techniques in biological control. In: Tan, K.H. (Ed.), Area-Wide Control of Fruit Flies and Other Insect Pests. Penerbit Universiti Sains Malaysia, Penang, pp. 221–227.

Gruver, M., 2010. USDA Stops Using Beetles vs. Invasive Saltcedar. The Washington Examiner. June 21, 2010. http://www.biologicaldiversity.org/news/center/articles/2010/washington-examiner-06-21-2010.html.

Habeck, D.H., Bennett, F.D., Balciunas, J.K., 1994. Biological control of terrestrial and wetland weeds. In: Rosen, D., Bennett, F.D., Capinera, J.L. (Eds.), Pest Management in the Subtropics: Biological Control—a Florida Perspective. Intercept, Andover, United Kingdom, pp. 523–547.

Harris, A.F., Nimmo, D., McKemey, A.R., Kelly, N., Scaife, S., Donnelly, C.A., Beech, C., Petrie, W.D., Alphey, L., 2011. Field performance of engineered male mosquitoes. Nat. Biotechnol. 29, 1034–1037.

Hight, S.D., Horiuchi, I., Vitorino, M.D., Wikler, C., Pedrosa-Macedo, J.H., 2003. Biology, host specificity tests and risk assessment of the sawfly *Heteroperreyia hubrichi*, a potential biological control agent of *Schinus terebinthifolius* in Hawaii. BioControl 48, 461–476.

Hight, S.D., Cuda, J.P., Medal, J.C., 2002. Chapter 24, Brazilian peppertree. In: Van Driesche, R., Blossey, B., Hoddle, M., Lyon, S., Reardon, R. (Eds.), Biological Control of Invasive Plants in the Eastern United States. USDA Forest Service Publication FHTET-2002-04, Morgantown, WV, pp. 311–321.

Hinz, H.L., Schwarzländer, M., Gassmann, A., Bourchier, R.S., 2014. Successes we may not have had: a retrospective analysis of selected weed biological control agents in the United States. Invasive Plant Sci. Manag. 7, 565–579.

Horner, T., 2004. Permitting. In: Coombs, E.M., Clark, J.K., Piper, G.L., Cofrancesco Jr., A.F. (Eds.), Biological Control of Invasive Plants in the United States. Oregon State University Press, Corvallis, Oregon, pp. 42–46.

Hoshovsky, M.C., Randall, J.M., 2000. Management of invasive plant species. In: Bossard, C.C., Randall, J.M., Hoshovsky, M.C. (Eds.), Invasive Plants of California's Wildlands. University of California Press, Berkeley, Calif, pp. 19–28.

Jackson, J.A., Jackson, B.J.S., 2007. An apparent mutualistic association between invasive exotics: Brazilian pepper (*Schinus terebinthifolius*) and black spiny-tailed iguanas (*Ctenosaura similis*). Nat. Areas J. 27, 254–257.

Jetter, K., Godfrey, K., 2009. Diaprepes root weevil, a new California pest, will raise costs for pest control and trigger quarantines. Calif. Agric. 63, 121–126.

Keane, R.M., Crawley, M.J., 2002. Exotic plant invasions and the enemy release hypothesis. Trends Evol. Ecol. 17, 164–170.

Kinde, H., Foate, E., Beeler, E., Uzal, F., Moore, J., Poppenga, R., 2012. Strong circumstantial evidence for ethanol toxicosis in cedar waxwings (*Bombycilla cedrorum*). J. Ornithol. 153, 995–998. http://dx.doi.org/10.1007/s10336-012-0858-7.

Krauss, N.L.H., 1963. Biological control investigations on Christmas berry (*Schinus terebinthifolius*) and emex (*Emex* spp.). Proc. Hawaii. Entomol. Soc. 18, 281–287.

Landcare Research Newsletter (LRN), August 2007. Sterilizing insects for a good cause. In: What's New in Biological Control of Weeds? Issue 41, p. 10.

Manrique, V., Diaz, R., Pogue, M.G., Vitorino, M.D., Overholt, W.A., 2012. Description and biology of *Paectes longiformis* (Lepidoptera: Euteliidae), a new species from Brazil and potential biological control agent of Brazilian peppertree in Florida. Biocontrol Sci. Technol. 22, 163–185.

Manrique, V., Diaz, R., Condon, T., Overholt, W.A., 2014. Host range tests reveal *Paectes longiformis* is not a suitable biological control agent for the invasive plant *Schinus terebinthifolia*. BioControl 59, 761–770.

Martin, C.G., Cuda, J.P., Awadzi, K.D., Medal, J.C., Habeck, D.H., Pedrosa-Macedo, J.H., 2004. Biology and laboratory rearing of *Episimus utilis* (Lepidoptera: Tortricidae), a candidate for classical biological control of Brazilian peppertree, *Schinus terebinthifolius* (Anacardiaceae), in Florida. Environ. Entomol. 33, 1351–1361.

McCoy, C.W., Stuart, R.J., Nigg, H.N., 2003. Seasonal life stage abundance of *Diaprepes abbreviatus* in irrigated and non-irrigated citrus plantings in central Florida. Fla. Entomol. 86, 34–42.

McKay, F., Oleiro, M., Walsh, G.C., Gandolfo, D., Cuda, J.P., Wheeler, G.S., 2009. Natural enemies of Brazilian peppertree (*Schinus terebinthifolius*: Anacardiaceae) from Argentina: their possible use for biological control in the USA. Fla. Entomol. 92, 292–303.

McKay, F., Oleiro, M., Vitorino, M.D., Wheeler, G., 2012. The leafmining *Leurocephala schinusae* (Lepidoptera: Gracillariidae): not suitable for the biological control of *Schinus terebinthifolius* (Sapindales: Anacardiaceae) in continental USA. Biocontrol Sci. Technol. 22, 477–489.

Medal, J.C., Sudbrink, D., Gandolfo, D., Ohashi, D., Cuda, J.P., 2002. *Gratiana boliviana*, a potential biocontrol agent of *Solanum viarum*: quarantine host-specificity testing in Florida and field surveys in South America. BioControl 47, 445–461.

Medal, J.C., Vitorino, M.D., Habeck, D.H., Gillmore, J.L., Pedrosa, J.H., De Sousa, L.D., 1999. Host specificity of *Heteroperreyia hubrichi* Malaise (Hymenoptera: Pergidae), a potential biological control agent of Brazilian peppertree (*Schinus terebinthifolius* Raddi). Biol. Control 14, 60–65.

Moeri, O.E., Cuda, J.P., Overholt, W.A., Bloem, S., Carpenter, J.E., 2009. F1 sterile insect technique: a novel approach for risk assessment of *Episimus unguiculus* (Lepidoptera: tortricidae), a candidate biological control agent of *Schinus terebinthifolius* in the Continental USA. BioControl Sci. Technol. 19 (Suppl. 1), 303–315.

Morrison, N.I., Simmons, G.S., Fu, G., O'Connell, S., Walker, A.S., Dafa'alla, T., Walters, M., Claus, J., Tang, G., Jin, L., Marubbi, T., Epton, M.J., Harris, C.L., Staten, R.T., Miller, E., Miller, T.A., Alphey, L., 2012. Engineered repressible lethality for controlling the pink bollworm, a lepidopteran pest of cotton. PLoS One 7, e50922, 1–10.

Morton, J.F., 1978. Brazilian pepper – its impact on people, animals and the environment. Econ. Bot. 32 (4), 353–359.

Normark, B.B., 2003. The evolution of alternative genetic systems in insects. Annu. Rev. Entomol. 48, 397–423.

Oleiro, M., Mc Kay, F., Wheeler, G.S., 2011. Biology and host range of *Tecmessa elegans* (Lepidoptera: Notodontidae), a leaf-feeding moth evaluated as a potential biological control agent for *Schinus terebinthifolius* (Sapindales: Anacardiaceae) in the United States. Environ. Entomol. 40, 605–613.

Overholt, W.A., Cuda, J.P., Markle, L., 2012. Can novel weapons favor native plants? Allelopathic interactions between *Morella cerifera* (L.) and *Schinus terebinthifolius* Raddi. J. Torrey Bot. Soc. 139, 356–366.

Pedrosa-Macedo, J.H., Poulmann, W., Stolle, L., Ukan, D., Cuda, J.P., Medal, J.C., 2006. Greenhouse mass rearing of a defoliating sawfly for biological control of Brazilian peppertree. Floresta 36, 371–378.

Rendon, J., Chawner, M., Dyer, K., Wheeler, G.S., 2012. Life history and host range of the leaf blotcher *Eucosomophora schinusivora*: a candidate for biological control of *Schinus terebinthifolius* in the USA. Biocontrol Sci. Technol. 22, 711–722.

Rodgers, L., Bodle, M., Black, D., Laroche, F., 2012. Status of nonindigenous species. In: South Florida Environmental Report, vol. I. The South Florida Environment. South FloridaWater Management District, West Palm Beach, Florida.

Sheppard, A.W., van Klinken, R.D., Heard, T.A., 2005. Scientific advances in the analysis of direct risks of weed biological control agents to nontarget plants. Biol. Control 35, 214–226.

Spector, T., Putz, F.E., 2006. Biomechanical plasticity facilitates invasion of maritime forests in the southern USA by Brazilian pepper (*Schinus terebinthifolius*). Biol. Invasions 8, 255–260.

Technical Advisory Group for Biological Control Agents of Weeds (TAG), 2015. TAG Petitions-APHIS Actions updated January 13, 2015. http://www.aphis.usda.gov/plant_health/permits/tag/downloads/TAGPetitionAction.pdf.

Tan, A., Fu, G., Jin, L., Guo, Q., Li, Z., Niu, B., Meng, Z., Morrison, N.I., Alphey, L., Huang, Y., 2013. Transgene-based, female-specific lethality system for genetic sexing of the silkworm, *Bombyx mori*. Proc. Natl. Acad. Sci. 110, 6766–6770.

Tenner, E., 1996. Why Things Bite Back: Technology and the Revenge Effect. Alfred A. Knopf, Inc, New York, 346 pp.

Treadwell, L.W., Cuda, J.P., 2007. Effects of defoliation on growth and reproduction of Brazilian peppertree (*Schinus terebinthifolius*). Weed Sci. 55, 137–142.

USDA, NRCS, 2015. The Plants Database. http://plants.usda.gov (March 11, 2015). National Plant Data Team, Greensboro, NC 27401–4901 USA.

U.S. Fish and Wildlife Service (USFWS), 1998. Draft Recovery Plan for Multi-Island Plants, 63. U.S. Fish and Wildlife Service, Portland, OR. Federal Register, pp. 51946–51947.

Vitorino, M.D., Pedrosa-Macedo, J.H., Cuda, J.P., 2000. Biology and specificity tests of the sawfly, *Heteroperreyia hubrichi* Malaise, 1955 (Hymenoptera: Pergidae) a potential biological control agent for Brazilian peppertree – *Schinus terebinthifolius* Raddi (Anacardiaceae). In: Spencer, N.R. (Ed.), Proceedings of the X International Symposium on Biological Control of Weeds, 4–14 July 1999. Montana State University, Montana, USA, pp. 645–650.

Wheeler, G.S., Geiger, J., McKay, F., Rendonc, J., Chawnerc, M., Pratt, P.D., 2011. Defoliating broad-nosed weevil, *Plectrophoroides lutra*; not suitable for biological control of Brazilian pepper (*Schinus terebinthifolius*). Biocontrol Sci. Technol. 21, 89–91.

Williams, D.A., Muchugu, E., Overholt, W.A., Cuda, J.P., 2007. Colonization patterns of the invasive Brazilian peppertree, *Schinus terebinthifolius*, in Florida. Heredity 98, 284–293.

Williams, J.R., 1954. The biological control of weeds. In: Report of the Sixth Commonwealth Entomological Congress, London, 7–16 July 1954, pp. 95–98.

Winston, R.L., Schwarzländer, M., Hinz, H.L., Day, M.D., Cock, M.J.W., Julien, M.H., 2014. In: Biological Control of Weeds: A World Catalogue of Agents and Their Target Weeds, fifth ed. USDA Forest Service, Forest Health Technology Enterprise Team, Morgantown, West Virginia. FHTET-2014-04. 838 pp.

Yoshioka, E.R., Markin, G.P., 1991. Efforts of biological control of Christmas berry (*Schinus terebinthifolius*) in Hawaii. In: Center, T.D., Doren, R.F., Hofstetter, R.L., Myers, R.L., Whiteaker, L.D. (Eds.), Proceedings of the Symposium of Exotic Pest Plants, 2–4 November 1988, Miami, FL. U.S. Department Interior, National Park Service, Washington, DC, pp. 377–387.

Chapter 8

Animal Hazards—Their Nature and Distribution

Rachel M. Cavin and David R. Butler
Department of Geography, Texas State University, San Marcos, TX, USA

ABSTRACT
Humans interact with animals in a variety of ways, whether the animal is domesticated or wild, or the interaction is intentional or accidental. Some of these interactions with animals can be hazardous to the human involved, by causing injury, disease, or damage to property. The purpose of this chapter is to provide an overview of animal hazards, and an introduction to some techniques used for management and mitigation of hazards related to animal populations. The chapter is subdivided into animal attacks, which involve direct and aggressive harm from an animal, animal accidents characterized by indirect, accidental harm by an animal, diseases from animals, known as zoonotic diseases, which make up the majority of all infectious pathogens, and property damage from animals covering loss of crops, livestock, and infrastructure to animal activity.

8.1 INTRODUCTION

Popular English-language, natural hazards textbooks in use in colleges and universities in North America portray a diversity of natural hazards that are typically subdivided into geophysical or meteorological/climatological hazards, with some recognition of astronomical/extraterrestrial hazards such as asteroid and comet impacts (e.g., Bryant, 1991; Coch, 1995; Hyndman and Hyndman, 2013). Nowhere to be seen are the diverse and geographically widespread hazards presented to humans directly and indirectly by animals. In this chapter, we address the question, and illustrate the importance, of an appreciation for the breadth of animal hazards that people face in a variety of places around the earth. We subdivide animal hazards into animal attacks, animal accidents, diseases contracted from animals, and property damage and losses caused by animals.

8.2 ANIMAL ATTACKS

Animal attacks occur in a variety of settings around the world, and involve both wild and domesticated animals. Domesticated animal attacks are the

Biological and Environmental Hazards, Risks, and Disasters. http://dx.doi.org/10.1016/B978-0-12-394847-2.00011-5

subject of a variety of studies (Clark et al., 1991; Weese et al., 2002; Chowdhury et al., 2013) that illustrate the widespread nature of the hazard. Most problems resulting from domesticated animal bites and scratches are not because of the inherent trauma, but because of secondary infections (Weese et al., 2002). These authors also reported that a recent study states that 3–18% of dog, and 28–80% of cat bite wounds in humans become infected. Bites from domesticated animals are also linked to other natural hazards, such as in the work by Warner (2010) that illustrated how displaced domesticated dogs and cats were traumatized by the passage of a major hurricane and reacted with an increased incidence of animal bites of humans, largely within 72 h of passage of the hurricane.

Wild animal attacks are enhanced in nature preserves such as national parks, and at seashores and adjacent ocean waters, by the reduced persecution of animals in such locations and the increased density of human use of natural environments (Tiefenbacher et al., 2003). Visitors to national parks and nature preserves also originate from many different places and arrive with a diverse set of attitudes toward wild animals. Encounters have the potential for negative outcomes for both people and the animal attacking, because wild animals attacking humans are typically located and exterminated.

Three kinds of wild animal attacks can be envisioned and categorized—a *charge* occurs when an animal runs toward a person or persons in a threatening fashion but does not make contact with the person(s) involved; when an animal comes in contact with a person and induces injuries such as bites, lacerations, scratches, or stings, and the person survives the attack, the encounter is defined as an *injury*; and if a victim dies as a result of injuries received from an animal, a *fatality* has occurred (DeChano and Butler, 2002). Wild animals such as the North American grizzly bear (*Ursus arctos*, Figure 8.1), or African and Asian megafauna such as elephants and rhinos (Figure 8.2) have been implicated in all three forms of attacks.

Wild animal attacks around the world include a broad diversity of animals on land, in the water, and in the air. We cannot describe in detail here the nature of all these animal attacks. Instead, we offer a representative cross section of animals involved in attacks and their general distribution, as listed in Table 8.1. We have not listed insect/arachnid stings or bites here, but instead address those hazards in Section 8.4 of this chapter.

8.3 ANIMAL ACCIDENTS

Animal attacks are certainly not the only fashion by which animals pose threats to humans. Human/animal accidents are fairly common occurrences. Here we briefly examine motor vehicle accidents with animals, bird strikes by planes, and a variety of miscellaneous accidents.

Motor vehicle collisions with animals occur around the world and pose serious hazards to both humans and animals (Bashir and Abu-Zidan, 2006).

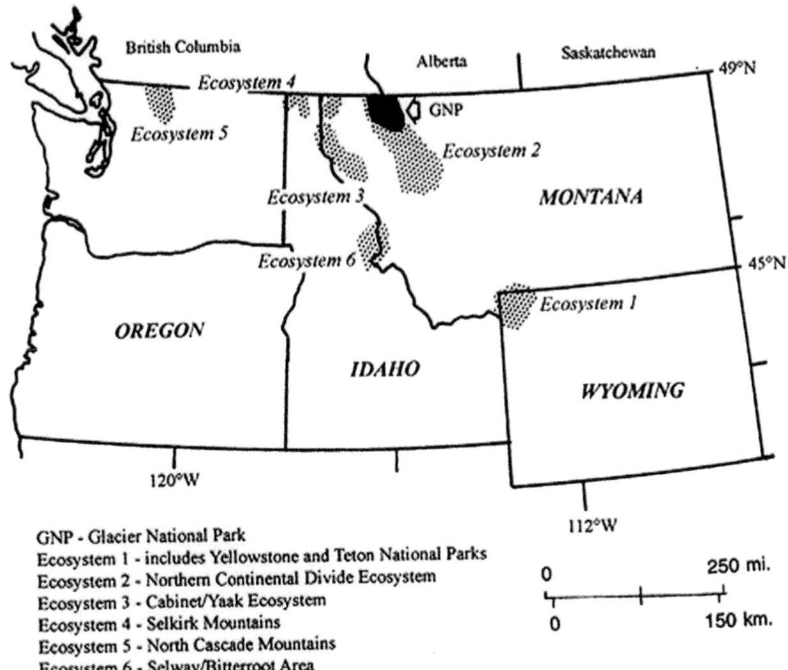

FIGURE 8.1 Grizzly bear habits in the conterminous US (above), with Glacier National Park illustrated. Glacier National Park, where the photograph of the grizzly bear was taken, has seen numerous grizzly bear attacks during its more than 100-year history (DeChano and Butler, 2002).

FIGURE 8.2 The African white rhino (a) and African elephant (b) have been known to charge vehicles visiting locations such as Kruger National Park, South Africa, seen here in both photos. Injuries and fatalities have also resulted from attacks from both species (Table 8.1).

A variety of large animals including moose, camels, deer, and kangaroos are involved in human/animal vehicle collisions (Seiler, 2005), but smaller wild animals or domesticated ones such as cattle, sheep, dogs, or cats can cause a driver to swerve and cause an accident (Langley and Hunter, 2001). The number of such events is quite high, e.g., in Sweden alone over 4,500 moose–vehicle incidents per year are reported, with several resulting fatalities (Seiler, 2005). Recommendations for preventing animal/vehicle collisions include warning signs for drivers, underpasses or overpasses for animals in areas of high animal population density, and reflectors that frighten animals (Bashir and Abu-Zidan, 2006). When implemented carefully, with mindfulness of the needs of the targeted species, highway over- and underpasses (shown in Figure 8.3) have been successful in reducing the number of traffic incidents involving animals by giving animals an alternative to crossing the road (Butler, 1993; Bashir and Abu-Zidan, 2006). The effectiveness of wildlife road reflectors, which reflect light from car headlights to the sides of roads and encourage animals to wait until the lights pass before proceeding, has been debated, with studies finding incidents where the reflectors are both effective and ineffective in reducing traffic accidents involving animals (Schafer and

TABLE 8.1 Representative List of Wild Animals Involved in Attacks on Humans, from Literature Sources

Animal	Location	Citations
Cougar/mountain lion	US and Canada	Beier (1991, 1992) and Tiefenbacher et al. (2003)
Grizzly bear	US and Canada	Floyd (1999) and DeChano and Butler (2002)
Black bear	US and Canada	Floyd (1999) and Herrero et al. (2011)
Polar bear	US and Canada far north	Floyd (1999)
Sloth bear	Indian subcontinent	Bargali et al. (2005)
Coyote	North America	Timm et al. (2004), White and Gehrt (2009) and Poessel et al. (2013)
Elephant	Africa, India	Durrheim and Leggat (1999)
Rhinoceros	Africa	Durrheim and Leggat (1999)
Wolf	North America	Jenness (1965)
Tiger	Asia	Nyhus and Tilson (2004a,b), and Neumann-Denzau and Denzau (2010)
Leopard	Africa	Durrheim and Leggat (1999)
Lion	Africa	Begg et al. (2007) and Durrheim and Leggat (1999)
Hippopotamus	Africa	Durrheim and Leggat (1999)
African buffalo	Africa	Durrheim and Leggat (1999)
Zebra	Africa	Durrheim and Leggat (1999) and Toovey et al. (2004)
Spotted hyena	Africa	Begg et al. (2007)
Crocodile	Africa	Begg et al. (2007)
Snake (variety of species)	Worldwide	Gold et al. (2002) and Cruz et al. (2009)
Jellyfish	Worldwide oceans	Fenner (2005) and Fenner et al. (2010)
Shark	Worldwide oceans	Hazin et al. (2008) and Rtshiladze et al. (2011)

FIGURE 8.3 An underpass installed along the US Highway 2 in Glacier National Park for use by mountain goats that frequently crossed the road to access a salt deposit nearby.

Penland, 1985; Reeve and Anderson, 1993; Gulan et al., 2006). The recognition that collisions may be inevitable in some areas has led to the installation of "roo bars" (from kangaroo) on the front of vehicles (Bashir and Abu-Zidan, 2006). Departments of Transportation in the United States strive to increase public safety on roadways through better management and mitigation of hazards. The frequency of collisions with wildlife on roadways make it a major public safety problem, in 2002 it was estimated that 1.1 billion dollars in damages and 150 human lives were lost in the United States from traffic accidents involving wildlife (Gibby and Clewell, 2006).

The primary accident hazard from birds is bird–aircraft collisions. Referred to as bird strikes in the literature, these collisions have the potential to cause severe damage to aircraft, especially if the bird is large or if the aircraft collides with a flock of birds (Blackwell and Wright, 2006). Examples of the potential severity of airborne collisions with birds are outlined by Thorpe (2003). One such example involved a Boeing 737 that collided with a flock of speckled pigeons, resulting in the ingestion of multiple birds into each engine, causing double engine failure followed by a failed crash landing. Thirty-five passengers were killed, with 21 more injured as a result of this particular accident (Thorpe, 2003). Several researchers (e.g., Richardson and West, 2000; Thorpe, 2003; Mode et al., 2005) have focused attention on the widespread nature of this hazard, for both civilian and military aircrafts around the world. Little success has occurred in preventing such accidents on a global scale.

Miscellaneous animal accidents cover a variety of events, including falling from camels (Abu-Zidan et al., 2012), falling from horses (both amateur equestrians and professional racing jockeys) (McCrory et al., 2006; Ball et al., 2007), and train wrecks and fatalities produced by the outburst of beaver dams and resultant flooding (Butler and Malanson, 2005). The property damage associated with beaver-dam outburst flooding is described in Section 8.5 of this chapter, below.

8.4 DISEASES CONTRACTED FROM ANIMALS

Animals can carry and transmit pathogens, fungi, and parasites that are harmful to human health. Zoonotic diseases refer specifically to diseases that can spread to human from animal populations (CDC, 2014a). This class of disease poses a tremendous global health threat; many of the deadliest epidemics are classified as zoonotic. The relevance of zoonotic diseases is increasing, as over one half of all infectious diseases are zoonotic in origin, and three quarters of all emerging diseases are zoonotic (Taylor et al., 2001). According to the World Health Organization (2014a), emerging diseases are those that have infected humans for the first time, or those that have begun to infect a larger number of people across a greater geographic range. A trend exists of more people becoming infected with an ever increasingly diverse set of infectious zoonotic diseases around the globe.

Most zoonotic diseases originate in wildlife populations, when domesticated animals or humans interact with carrier species in the wild. Wolfe et al. (2005) state that three risk factors are related to the emergence of new zoonotic pathogens from wild animals: the diversity of microbes present in the wildlife population in that region, known as the "zoonotic pool," how environmental change impacts the diversity and number of pathogens present in the region, and the frequency of contact with those wild populations by humans or domesticated animals (Wolfe et al., 2005).

In addition to infectious disease, animals can cause other health problems when interacting with people. Animals can cause allergic reactions in individuals with an allergy to some aspect of the animal, such as a bee sting or dog dander. Allergies can be a mild nuisance or as serious as anaphylaxis, a severe allergic reaction that can result in death (Klotz et al., 2009). Animals can also increase exposure of humans to parasites, some of which, like ticks, carry zoonotic diseases themselves (Robertson et al., 2000; Jongejan and Uilenberg, 2004). Injuries from animals, such as bite wounds or scratches, have a high chance of becoming infected by bacteria, the animal carried (Weese et al., 2002).

In the sections below, diseases and other health hazards related to human–animal interaction are categorized by the interaction responsible for transmission. This section is intended to demonstrate the risk factors, severity, and nature of human health hazards resulting from interaction with animals, rather than to serve as a comprehensive review of zoonotic disease.

8.4.1 Contracted from Animal Bites and Stings

Some zoonotic diseases are present in the saliva of carrier species, and can be introduced into the human body through broken skin from bites (Dantas-Torres et al., 2012). Additionally, both biting and stinging animals can cause the body to have dangerous allergic reactions in individuals with allergies to that animal (Klotz et al., 2009). Some of the diseases responsible for the most human deaths are transmitted through bites from animals, particularly mosquitos. Table 8.2 shows examples of diseases transmitted through animal bites, with carrier species, global death tolls (when available), and geographic distribution information provided as well.

TABLE 8.2 Examples of Diseases Commonly Spread to Humans Through Animal Bites and Stings

Disease Name	Carriers	Deaths (Region, Year)	Geographic Distribution
Lyme disease	Ticks	0 (United States, 2006)	North America, Europe, Asia
West Nile virus	Mosquitos	119 (United States, 2013)	Africa, Middle East, Europe, North America, West Asia
Dengue fever	Mosquitos	22,000 (Global, annually)	Tropical and subtropical regions
Trypanosomiasis (sleeping sickness)	Tsetse flies	100,000 (Global, annually)	Sub-Saharan Africa
Malaria	Mosquitos	627,000 (Global, 2012)	Tropical and subtropical regions

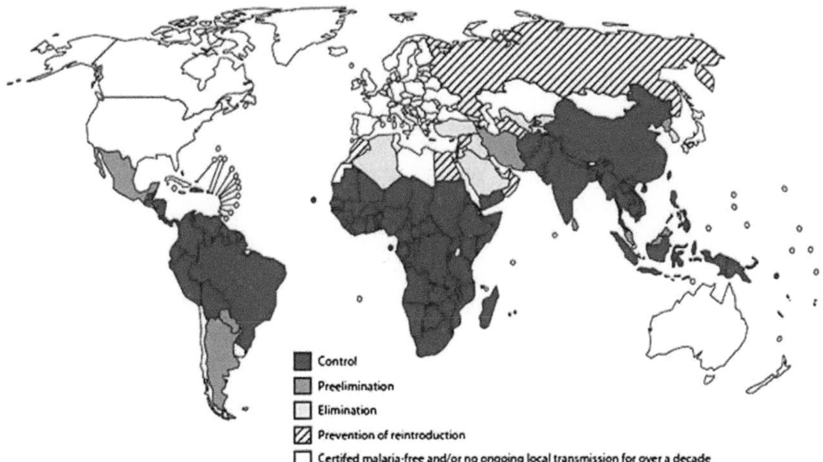

FIGURE 8.4 The status of malaria control measures in affected countries. Control indicates a reduction in malaria mortality and elimination indicates complete prevention of malaria transmission. *All maps are used with permission from the CDC. Photographs are property of co-author David R. Butler.*

Mosquitos transmit malaria, a parasitic disease that was responsible for approximately 627,000 deaths in 2012 (WHO, 2013). The risk for malaria is greatest in developing countries in or near the tropics (Figure 8.4). Despite being preventable and curable, malaria continues to have a high death toll in many developing countries (WHO, 2013). Mosquitos also transmit dengue fever, a viral infection that has infected populations in over 100 countries within Africa, the Americas, Asia, the Pacific Islands, and the Mediterranean (WHO, 2014b). The geographic range of dengue fever continues to grow, presently 40% of the global population is at risk of contracting dengue fever (WHO, 2014b). Additional diseases transmitted by mosquito bites include, but are not limited to, West Nile virus, yellow fever, and multiple types of encephalitis (Reiter, 2001; CDC, 2014b,c; WHO, 2014c,h).

Ticks are another important carrier of zoonotic diseases. In the United States, Lyme disease is the most common vector-borne disease reported to the Centers for Disease Control and Prevention (CDC), and tick bites are the source of infection (CDC, 2008a). Between 1992 and 2006, the CDC received 248,074 reports of Lyme disease cases in the United States (CDC, 2008a). Additionally, cases of Lyme disease occur in Europe and Asia, though the numbers reported here are limited to the Unites States cases. Although seldom fatal, Lyme disease can cause abnormalities in multiple body systems, including dermatologic, musculoskeletal, neurologic, and cardiac complications and symptoms. These diverse symptoms can make it difficult to diagnose, particularly if cases occur outside of the northeast and north central states where the highest frequency of cases occur in the United States (CDC, 2008a). The tsetse fly in Africa transmits trypanosomiasis or human African sleeping sickness, through a bite (WHO, 2014d). Although the geographic range of the tsetse fly is confined to sub-Saharan Africa, the disease is believed to kill 100,000 people a year, and is the leading cause of death in some African countries (Picozzi et al., 2005; WHO, 2014d).

Rocky Mountain spotted fever (RMSF) is an additional tick-borne disease caused by the bacterium *Rickettsia rickettsii.* This organism is a cause of potentially fatal human illness in North and South Americas, and is transmitted to humans by the bite of infected tick species. In the United States, these include the American dog tick (*Dermacentor variabilis*), Rocky Mountain wood tick (*Dermacentor andersoni*), and brown dog tick (*Rhipicephalus sanguineus*). Symptoms of RMSF include fever, abdominal pain, vomiting, headache, and muscle pain (CDC, 2015). A rash, which gives the disease its name along with the characteristic fever, may also develop but is often absent in the first few days of the disease. In spite of the name of the disease, the greatest incidences in the United States occur in the contiguous states of North Carolina, Oklahoma, Arkansas, Tennessee, and Missouri (CDC, 2015). Peak number of cases occur in the months of June and July. The US Centers for Disease Control note that the frequency of reported cases of RMSF is highest among males, American Indians, and people at least 40 years old. Risk of infection increases for individuals who experience frequent exposure to dogs and who live near areas with high grass or wooded areas where ticks may reside in greater number (CDC, 2015).

Rabies is a serious zoonotic disease that is commonly transmitted to humans through contact with the saliva of an infected animal, likely from an animal bite or through mucus membranes (CDC, 2014d). Out of all of the rabies fatalities that occur each year, 99% take place in developing nations (WHO, 2004). Rabies is considered a neglected zoonosis, because despite the availability of accessible and affordable solutions to rabies prevalence in developing countries, the disease continues to be uncontrolled in many places (WHO, 2004). Dogs are the source of 54% of the rabies exposures, whereas 42% were caused by exposure from infected wildlife, and 4% of exposures were related to bats in particular (WHO, 1998). Bats are particularly significant hosts of zoonotic diseases that can be spread through a variety of interactions with humans and other mammals; they have been found to host more viruses per species than rodents. This is thought to be because of flight and associated immunological adaptations related to the fast metabolism of bat species, although it is not known for certain (O'Shea et al., 2014).

An important consideration in health risks associated with animals is allergies, which can potentially lead to anaphylactic shock from bites and stings. Although this does not involve infection with pathogens from an animal the way zoonotic diseases do, it is related in that exposure to a particular animal causes bodily harm. Many animals can cause allergic reactions in humans, however allergies to the stings of wasps, bees, and hornets are most common (Klotz et al., 2009). Additionally, various rodents, reptiles, arachnids, insects, and other taxon groups can cause allergic reactions from bites and stings (Klotz et al., 2009).

8.4.2 Contracted from Consumption of Infected Animal Products

Humans are often exposed to zoonotic pathogens through the consumption of improperly prepared foods (EFSA, 2014). These foods are commonly animal products, or other products that have been contaminated by the zoonotic pathogen in some way (EFSA, 2014; WHO, 2014g). Foodborne zoonotic diseases can be deadly, particularly if proper treatment for associated health problems such as severe dehydration is not administered (WHO, 2014g). Examples of foodborne zoonotic diseases are listed in Table 8.3 with additional information regarding disease fatalities and geographic distribution.

One of the most common foodborne zoonotic diseases is due to *Salmonella*. *Salmonella* is present in the digestive system of a wide variety of natural and domesticated animals, and transmission often occurs through the consumption of contaminated food (EFSA, 2014; WHO, 2014g). Additionally, *Salmonella* can be contracted as a result of direct contact with infected animals; this mode of transmission will be discussed in the following subsection. Elder and young people are most vulnerable to severe cases of *Salmonella*, though less than 1% of all cases are fatal (EFSA, 2014; WHO, 2014e). Despite

TABLE 8.3 Examples of Diseases That Can Be Spread to Humans via the Consumption of Contaminated or Infected Animal Products

Disease Name	Carriers/Reservoirs	Annual Deaths (Region, Year)	Geographic Distribution
Trichinellosis	Many species, including horses, dogs, and pigs	0 (European Union, 2012)	Global
Campylobacteriosis	Common in wild and domesticated birds and mammals	31 (European Union, 2012)	Global
Salmonella	Many domestic and wild animals	61 (European Union, 2012)	Global
Escherichia coli	Many animals, particularly ruminants	91 (United States, annually)	Global
Listeriosis	Many domestic and wild animals	198 (European Union, 2012)	Global

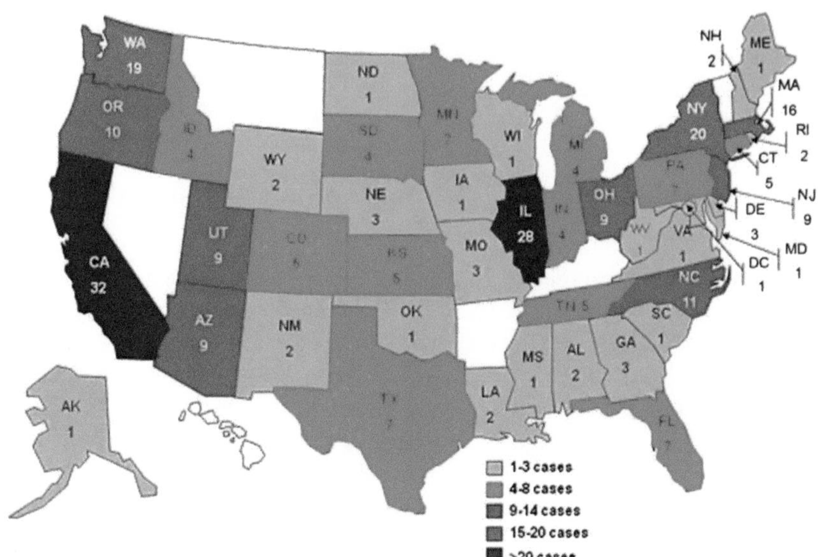

FIGURE 8.5 The number of infections of *Salmonella* Montevideo in the United States as of April 28, 2010. *All maps are used with permission from the CDC. Photographs are property of co-author David R. Butler.*

the low mortality rate, it is estimated that *Salmonella* causes over 100,000 fatalities globally each year (WHO, 2014e). The number of cases in each state of *Salmonella* Montevideo during an outbreak from July 2009 to April 2010 is shown in Figure 8.5.

Escherichia coli is another prevalent zoonotic foodborne pathogen. In the United States, 350 *E. coli* outbreaks were reported between 1982 and 2002 in 49 states (Rangel et al., 2005). It is estimated that 74,000 cases of *E. coli* occur each year in the United States (Mead et al., 1999). Raw beef is often the suspected or confirmed source of *E. coli* outbreaks, though other foods can harbor the disease and transmit it to humans when consumed (EFSA, 2014). Previously, outbreaks from fast food beef were a major concern because of events like the 1993 multistate outbreak that was linked to undercooked beef from a restaurant chain (Rangel et al., 2005). However, revisions of regulations on beef preparation has proven effective, and an outbreak stemming from fast food has not been reported since 1995 (Rangel et al., 2005). *Escherichia coli* continues to infect people all over the world, and direct transmission from infected animals continues to grow more common in addition to foodborne transmission (Rangel et al., 2005).

Campylobacter has been the most frequently reported foodborne zoonotic pathogen in the European Union since 2005 (EFSA, 2014). In 2012, there were 501 outbreaks of *Campylobacter* in Europe alone with the number of reported cases at 214,268 (EFSA, 2014). The main reservoirs of *Campylobacter* are the digestive system of wild and domesticated birds and mammals, the bacteria is present in many livestock animals and pets (EFSA, 2014). Major transmission sources include eating contaminated poultry, contact with live poultry or other infected animals such as pets. Other significant sources of infection include cross contamination and drinking raw milk or contaminated water, though these cause fewer infections.

Listeria is another bacteria that is transmitted through consumption of contaminated food. *Listeria* occurs throughout the environment, in water, soil, and forage, but also in infected animals (EFSA, 2014). Consumption from contaminated food is the most common route of transmission, but direct transmission from infected animals occurs occasionally (EFSA, 2014). The bacteria pose the biggest threat to the young and elderly, and when pregnant women are infected the bacteria can spread to the fetus, causing spontaneous abortion or stillbirth (EFSA, 2014). There were 1642 confirmed cases of listeriosis in the European Union in 2012, 198 of which resulting in fatalities (EFSA, 2014). Outbreaks of *Listeria* occur frequently is the United States as well, a serious outbreak occurred in 2011 in cantaloupes grown on a farm in Colorado. A total of 147 cases were reported to the CDC in 28 states, with 33 deaths, all of which were traced back to contaminated cantaloupes from one particular farm (CDC, 2012b).

Additional examples of zoonotic foodborne pathogens are trichinellosis and bovine tuberculosis. Trichinellosis is cause by a parasitic nematode species that is mostly hosted in mammal species and transmitted through eating improperly prepared foods that had been infested by larvae of the parasite (EFSA, 2014). Bovine tuberculosis is a zoonotic disease that infects a large variety of mammals, including humans. It is similarly transmitted through consumption of contaminated food, particularly raw dairy products (EFSA, 2014).

8.4.3 Contracted from the Handling of Animals or Animal Products

Many diseases can be transmitted to humans directly from contact with the infected animal or products, such as a butcher or hunter coming in contact with animal meat. These diseases enter the body through vulnerable paths such as inhalation, mucus membranes, or cuts and scratches on the skin (WHO, 2008, 2014f). Many of the disease examples in this section are serious illnesses that are currently, or feared to become global pandemics. Some also have potential for biological weapons, in addition to transmission from animal populations (Webster and Hulse, 2005; WHO, 2008). Table 8.4 gives regional annual deaths, species known to transmit the disease to humans, and the current geographic range of the illnesses.

One disease that poses a threat to individuals with close contact to herbivores is anthrax (WHO, 2008). Humans contract anthrax when the bacterial

TABLE 8.4 Examples of Diseases Spread to Humans through the Handling of Animals or Animal Products

Disease Name	Carriers	Annual Deaths (Region, Year)	Geographic Distribution
Salmonella	Common in reptiles and amphibians	1[a] (United States, 2007)	Global
Anthrax	Warm blooded animals, particularly herbivores	10 (Gambia, 1970–1974)	Global
Avian flu	Birds	25 (Global, 2013)	Africa, Asia, Eastern Europe
Ebola	Fruit bats, infected wildlife	50 (D.R.C., Uganda, 2012)	Central and West Africa
Swine flu	Pigs	284,200 (Global, 2009)	Global
HIV	Emerged from SIV in simians	2,800,000 (Global, 2006)	Global

SIV, simian immunodeficiency virus.
[a]*Only for Salmonella associated with handling of reptiles and amphibians.*

spores enter the body, such as through skin lesions or inhalation, but also through consumption of contaminated meat or contact with infected humans in some instances (Heyworth et al., 1975; WHO, 2008). Anthrax transmission is generally an occupational hazard that disproportionately affects butchers, farmers, veterinarians, and other individuals often in contact with large herbivores (WHO, 2008). The species that often transmit anthrax inhale spores from the soil while grazing (WHO, 2008). Anthrax spores have been used in bioterrorism on multiple occasions, and have been involved in many more threats (WHO, 2008).

Salmonella, mentioned in the previous section on transmission of disease from consumption, is also frequently transmitted through direct contact with infected animals. In particular, reptiles and amphibians are major sources of *Salmonella* contracted through direct contact with animals (CDC, 2007, 2014e). It can also be contracted from animal waste, for example from cleaning the water in a pet turtle aquarium (CDC, 2007, 2014e). In the United States, selling turtles with a carapace smaller than four inches wide is prohibited, because small children may put turtles this size in their mouth which makes transmission of *Salmonella* likely (CDC, 2007, 2008b). Despite these restrictions, outbreaks of *Salmonella* occur, such as a multistate outbreak in 2006 with 103 cases, many of which were linked to contact with infected reptiles and amphibians (CDC, 2008b).

Two zoonotic diseases that have received a lot of attention recently are the influenza viruses known as avian flu and swine flu. Avian flu occurs in birds, including poultry and waterfowl, and has recently begun transmitting to humans exposed to infected bird populations (Webster and Hulse, 2005). The virus currently primarily occurs in southeast Asia, but the recent discovery of infected migratory waterfowl has raised concern that the geographic range of the virus may be extended by the movement of infected birds (Chen et al., 2005; WHO, 2014i). When transmission to humans has occurred, China and Thailand have suppressed the potential for additional human infections through mass culling of live birds (Webster and Hulse, 2005). Although the virus currently cannot spread between humans, it is feared this capability could develop in the avian flu strains (Webster and Hulse, 2005). Swine flu circulates in pig populations, and unlike avian flu once the virus enters humans it has the capability to pass from human to human (Wang and Palese, 2009). In 2009, a swine flu outbreak occurred that resulted in 18,500 laboratory confirmed deaths, with researchers estimating that the actual death toll was closer to 284,200 global deaths from the pandemic (Dawood et al., 2012). Influenza viruses vary in pandemic potential, but have the potential to be extremely deadly diseases (Wang and Palese, 2009). Concern regarding the potential of these influenza viruses to emerge as major pandemics continues to grow.

In the late 1970s a new zoonotic disease emerged in Africa, Ebola hemorrhagic fever (WHO, 2014f). This virus is transmitted to humans through contact with the bodily fluids of infected animals (WHO, 2014f). Species known to transmit the disease include chimpanzees, gorillas, fruit bats, monkeys, forest antelopes, and porcupines, with fruit bats believed to be the natural host of the disease (WHO, 2014f). No known treatment, drug, or vaccine exists for Ebola virus, and the disease has a very high mortality rate (WHO, 2014f). The distribution of outbreaks of Ebola virus is shown in Figure 8.6, with symbology for the species and size of each outbreak. Sporadic outbreaks have occurred since the emergence of the virus in the 1970s, and the worst outbreak in history occurred in 2014. As of the time this chapter was written, it was estimated that there had been 3069 cases with 1552 deaths in the affected African nations, with prevention of new infections not yet reached (CDC, 2014f).

Another recent zoonotic emergence with great significance is the human immunodeficiency virus (HIV). HIV, although now primarily spread through human to human transmission, originated through cross-species transmission from primates infected with simian immunodeficiency virus (SIV) (Sharp and Hahn, 2011). The virus was first discovered in the 1980s, in patients who had developed acquired immunodeficiency syndrome (AIDS), from initial HIV infections (Greene, 2007). Since the earliest known cases, it is estimated that HIV/AIDS has led to at least 25,000,000 deaths around the world (Greene, 2007).

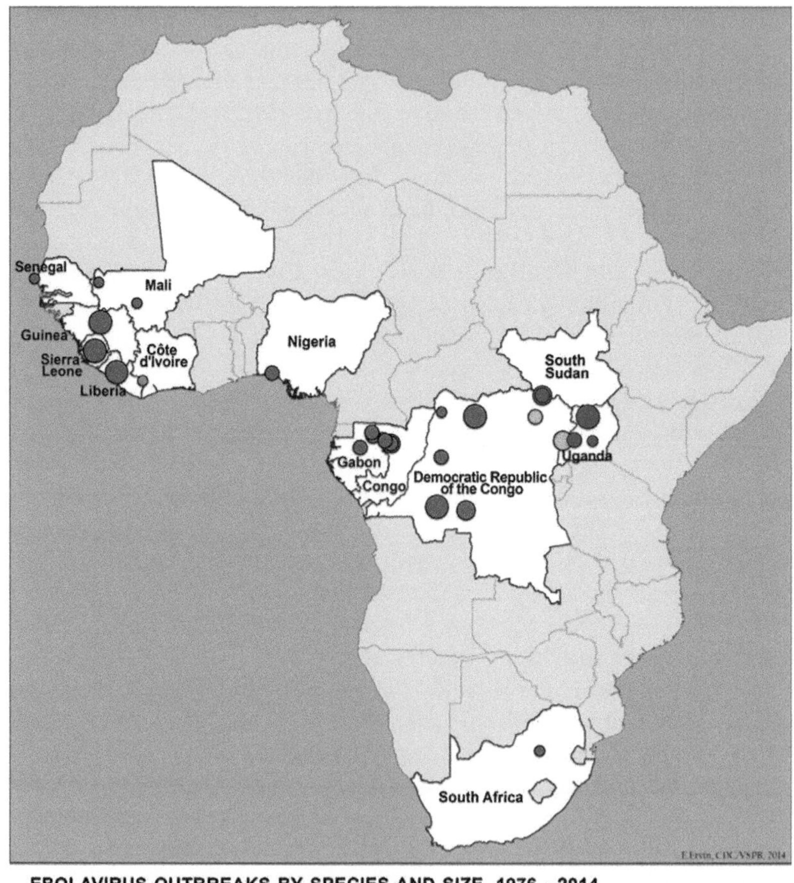

FIGURE 8.6 Ebola virus outbreaks by species and number of cases. *All maps are used with permission from the CDC. Photographs are property of co-author David R. Butler.*

8.4.4 Contracted from Contact with, and Inhalation of Animal Waste

Animal waste is another common way in which zoonotic diseases are spread. Pathogens in animal waste can contaminate food or water, or enter the body directly through inhalation, skin lesions, and other routes vulnerable to pathogen entry (Izurieta et al., 2008; Petrovay and Balla, 2008; CDC, 2012a).

TABLE 8.5 Examples of Diseases Spread to Humans Through Contact With, or Inhalation of Animal Waste

Disease Name	Carriers	Annual Deaths (Region, Year)	Geographic Distribution
Giardiasis[a]	All animals	0 (United States, 2009)	Global
Psittacosis/ chlamydiosis	Birds	2 (Hungary, 2005)	Global
Hantavirus	Mice and rats	27 (United States, 1993)	The Americas, Europe, Asia
Leptospirosis	Rodents and other animals	115 (Kerala, India, 2003)	Global

[a]*Giardiasis rarely kills people in developed countries, however, is believed to cause many deaths in developing countries. Surveillance and reporting of deaths from giardiasis in these nations is poor, so the numbers are unknown.*

Table 8.5 gives the species known to transmit diseases, the annual regional fatalities, and the geographic distributions of example zoonotic diseases.

Leptospirosis is considered to be the most widespread zoonotic disease on earth (Izurieta et al., 2008). The disease is caused by bacteria called leptospires that can live both in the environment and in animal hosts (Izurieta et al., 2008). Leptospires need moisture, and can contaminate freshwater sources and infect a large number of animals, however, rats are an important reservoir and source of human infections (Izurieta et al., 2008). Humans are often exposed to leptospires through contact with urine of an infected animal (Izurieta et al., 2008). The disease can be difficult to diagnose because it manifests similarly to other illnesses, with symptoms including fever, aches, headache, jaundice, and chills (Izurieta et al., 2008). The illness occurs most frequently in tropical and subtropical locations, however it can occur in temperate regions as well (Izurieta et al., 2008).

Giardiasis is an illness caused by a protozoa that is generally contracted from contaminated water (CDC, 2012a). *Giardiasis* is the most common intestinal parasite in the United States, and is common throughout the rest of the world, especially where sanitation and water quality is poor (CDC, 2012a; Torgerson et al., 2014). The protozoa are transmitted through contact with the feces of infected animals, often in water sources (CDC, 2012a). In the United States alone there were 19,888 cases in 2010 (CDC, 2012a). Figure 8.7 shows the incidence of *Giardiasis* per 100,000 people in the United States in 2010.

Psittacosis is another zoonotic disease spread to humans through contact with infected animal waste. This illness is generally associated with birds, and the bacteria that cause the illness is spread when an individual inhales feces or

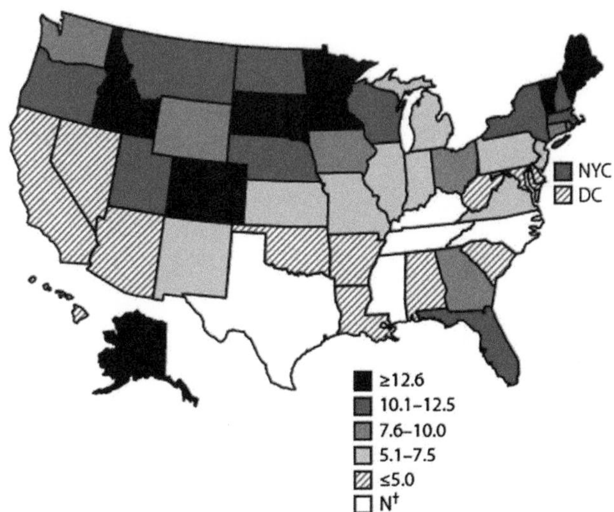

FIGURE 8.7 The incidence of giardiasis per 100,000 people in the United States in 2010. *All maps are used with permission from the CDC. Photographs are property of co-author David R. Butler.*

secretions from birds that were infected with the bacteria (Petrovay and Balla, 2008). In 2005, there was an outbreak of psittacosis in Hungary that was linked to two separate poultry processing plants (Petrovay and Balla, 2008). A total of 140 cases of psittacosis were reported, and two severe cases resulted in death (Petrovay and Balla, 2008). Psittacosis is primarily a hazard to those who interact closely with birds, either as pets or in an occupational setting, such as a poultry processing plant (Petrovay and Balla, 2008).

A group of infections commonly spread through contact with rat and mouse feces are known as hantavirus, with two different resulting diseases: the European and Asian hemorrhagic fever with renal syndrome (HFRS) and the American hantavirus pulmonary syndrome (HPS) (CDC, 2014g). Hantavirus infections primarily spread from rodent to human, with human to human transmission of HFRS being extremely rare and having never been observed in cases of HPS (CDC, 2014h,i). Home infestations with rodents are the main risk associated with contracting hantavirus, and transmission most often occurs from contact with or inhalation of rodent feces, urine, or debris from nesting materials (CDC, 1994, 2014g,h,i). The most severe outbreak of HPS occurred in 1993, with 27 confirmed deaths and 48 total cases, however, HPS continues to have similar, though slightly lower case numbers; in 2013 there were 21 cases and with 9 deaths (CDC, 2013).

8.4.5 Summary

Zoonotic diseases pose a major hazard for human health all over the globe. The majority of all diseases, and of emerging diseases in particular, have a zoonotic origin and transmission capability (Taylor et al., 2001). Recent emergence of diseases like Ebola and HIV, changes in the geographic range of diseases like *malaria*, and first time transmissions to humans of diseases like avian flu are all cause for attempting to understand and control these illnesses to minimize the public health burden they cause. Zoonotic illnesses can be transmitted as a result of many different interactions with infected animals, including animal bites, consumption of animals, direct contact with animals and animal products, and contact with animal waste. It is important to exercise caution in all of these interactions with animals, and to understand the inherent risks that zoonotic pathogens pose.

8.5 PROPERTY DAMAGE AND LOSSES CAUSED BY ANIMALS

Humans rely on infrastructure and production for food, transportation, shelter, communications, and all other aspects of modern society. However, in addition to the harm to human health and wellness that the previous three subsections of this chapter emphasized, animals can damage and destroy critical material components of human life. Buildings, transportation, electricity, water, agricultural, and many other types of essential infrastructure are vulnerable to various animal species all over the world. In the 2005 fiscal year, the United States Department of Agriculture spent $12,500,000 to protect property from damage from wildlife as part of the Wildlife Services Program (USDA, 2006). The global financial burden associated with animal damages is presumably tremendous. This section will provide examples of species responsible for particular damages to infrastructure, including a separate section on agricultural damages to crops and livestock.

8.5.1 Infrastructure Damage and Loss

Wildlife can cause damages to numerous forms of infrastructure that are essential to the functioning of human society. Buildings are vulnerable when wildlife attempts to gain entry, causing foundation or structural damage, as well as damaging electrical components of buildings (USDA, 2006). Beavers, ground hogs, armadillos, and other aquatic or burrowing animals are often responsible for damage to foundations or structures through borrowing or damming activities, whereas birds and rodents are most often responsible for electrical damages (USDA, 2006). In some areas, snakes can cause major problems. In Guam, brown tree snakes cause regular power outages (USDA, 2006). Some animal waste is corrosive, such as that of bats and roosting birds, and can damage machinery, paint, and other materials when animals roost nearby (USDA, 2006). Many believe beavers are responsible for more property damage than any other species (USDA, 2006). Despite their tremendous ecological benefits, beavers that colonize the wrong area can be economically detrimental. As a result of flooding associated with beaver damming activities, and the failure of beaver dams, roads, bridges, dams and dikes, sewers, and other water infrastructure are all vulnerable to damages attributable to beavers (Butler and Malanson, 2005; USDA, 2006). Beavers also have a taste for ornamental trees, and can cause damage to landscaping (USDA, 2006).

Vehicles are also vulnerable to animal associated damage and destruction. Car and aircraft collisions are expensive, and occasionally fatal hazards associated with wildlife. In the United States, deer are most often involved in car collisions, but other animals posing threats include elk, antelope, bear, hogs, and moose (USDA, 2006). The animals that collide with vehicles globally vary tremendously, but size and prevalence of the animal are good indicators of severity and frequency of crashes. There are over a million collisions with deer in the United States each year, and in Sweden 4,500 collisions are reported each year involving moose (Seiler, 2005; USDA, 2006). Aviation is commonly affected by collisions with birds, referred to as bird strikes, which can damage or destroy aircraft, sometimes injuring and killing passengers (Richardson and West, 2000; Thorpe, 2003). Collisions with birds have led to the destruction of at least 80 civil aircrafts, and an additional 283 military aircraft from 32 countries (Richardson and West, 2000; Thorpe, 2003). The United States Air Force spends approximately $33,000,000 for losses and repairs associated with bird strikes alone, and commercial airlines are estimated to spend somewhere between $78,200,000 and $391,400,000 for damages and delays associated with collisions with birds (Allan, 2000; Clearly et al., 2000).

8.5.2 Crop and Livestock Damage and Loss

Pests are a major concern in agriculture, as insects alone consume an estimated 14% of the total global agricultural yield despite prevention strategies in place (Oerke et al., 1994). Additionally, predators often choose livestock as prey, causing economic losses for ranchers all over the world. The species that act as pests vary tremendously around the world, this chapter will provide several examples of pests from various parts of the world, with associated losses provided where available.

Despite the widespread use of pesticides in most parts of the world, insects continue to pose a problem for farmers. In addition to direct consumption of crops, insect pests spread pathogens that infect plants, further reducing viable crop yields (Hilder and Boulter, 1999). Insect-related losses are a major motivation for the development of genetically modified crops that are engineered for enhanced robustness and resistance to insect pests (Hilder and Boulter, 1999). In addition to the global struggle with insects, many other animals can be agricultural pests. In 13 African countries, the animals that were most frequently ranked as the worst offenders in crop damage and loss were elephants, monkeys and baboons, bush pigs, cane rats, and buffalos

(Naughton-Treves and Treves, 2005). Elephants were consistently ranked as a worst offender whether the area being surveyed was savannah or forest (Naughton-Treves and Treves, 2005). Another study in the Peppara Wildlife Sanctuary area of India ranked wild boars, elephants, porcupine, blacknaped hare, bonnet macaques, mouse and barking deer, palm civet, and bandicoot rats are the most destructive animals in terms of crop damage and loss (Jayson and Christopher, 2008).

Livestock are another area inherently vulnerable to damage and loss from wildlife interaction. Carnivores are species that are often in conflict with nearby humans, and globally their populations are falling. Predation on livestock is a major source of this conflict, and a primary motivation for killing carnivores to protect animal herds and flocks (Nyhus and Tilson, 2004a; Patterson et al., 2004; Miquelle et al., 2005). Tigers are one such carnivore in conflict because of predation on wildlife. In Sumatra, Indonesia, tigers killed at least 870 livestock between 1978 and 1997, and additional 100 livestock per year in Eastern Russia (Nyhus and Tilson, 2004a; Miquelle et al., 2005). On ranches in Kenya that are located adjacent to the Maasai Mara National Reserve, hyenas killed 74 livestock, leopards killed 55, and lions killed an additional 18 livestock between March 2003 and April 2004 (Kolowski and Helekamp, 2006). Lions alone killed 433 livestock between 1996 and 1999 on two ranches in Southeastern Kenya (Patterson et al., 2004). In the United States, the recovery of wolf populations has not been without conflict. Between 1987 and 2001, 148 cattle, 356 sheep, and 37 dogs were killed by wolves in Montana, Idaho, and the Yellowstone area (Bangs and Shivik, 2001). In 2011, Rocky Mountain gray wolves were delisted from the federal endangered species list, and authority was given to states to manage populations and handle conflicts with ranchers (Montana Fish, Wildlife, and Parks, 2014; Utah Division of Wildlife Resources, 2012). However, in certain regions they are still considered federally protected, giving less power for management of wolves and the conflicts they can cause (Utah Division of Wildlife Resources, 2012). Coyotes are another predator that poses a threat to livestock in North America, with a range extending across all of North America, far larger than that of the gray wolf. Coyotes prey upon small mammals most often, but frequently attack livestock when the opportunity presents itself, with sheep and goats as common victims (Lehner, 1976; Treves et al., 2011). Increasingly, coyotes are moving into suburban and urban areas, and predation on pets as well as direct conflict with humans and urban property are becoming more frequent over time (Gehrt, 2007).

Preventative measures are often taken in protecting livestock and pets from predation. Fencing and guard animals are commonly utilized to make predation more difficult (Gehring et al., 2011). In Africa, fencing is increasingly used to form a barrier between livestock and protected wildlife from contracting diseases from wildlife outside the fence (Lindsey et al., 2012). When preventative measures are not enough, predators are often culled through various means, including hunting, trapping, and poisoning (Lehner, 1976; Nyhus and Tilson, 2004a; Patterson et al., 2004; Miquelle et al., 2005). All of these measures have ecological implications, either directly to the predator and their place in the ecosystem, or a cascading, indirect influence on ecosystem functioning.

8.6 SUMMARY

Animals pose hazards to humans in many ways, including damaging infrastructure and agricultural yields. Animals cause millions of dollars in damages to property per year in the United States alone, globally the impact of animals on property and agriculture is tremendous (USDA, 2006). Crops and livestock are preyed upon by a variety of animal species, lowering yields that would otherwise be used for sustenance or profits. Management of the animals is difficult, and unfortunately around the world the solution currently in place is often to kill the animals responsible for damages. As human settlements sprawl, increasingly encroaching on animal habitats, hazards associated with animal interactions with human property will likely continue and increase in frequency. The same may be said for animal attacks upon humans, and human/animal accidents. Disease from animal vectors will likely remain the most geographically widespread and potentially deadly of all forms of animal hazards and human population continues to grow.

REFERENCES

Abu-Zidan, F.M., Hefny, A.F., Eid, H.O., Bashir, M.O., Branicki, F.J., 2012. Camel-related injuries: prospective study of 212 patients. World J. Surg. 36, 2384–2389.

Allan, J.R., 2000. The costs of bird strikes and bird strike prevention. In: Clark, L. (Ed.), Proceedings of the National Wildlife Research Center Symposium, Human Conflicts with Wildlife: Economic Considerations. U.S. Department of Agriculture, Natural Wildlife Research Center, Fort Collins, Colorado, USA, p. 8.

Ball, C.G., Ball, J.E., Kirkpatrick, A.W., Mulloy, R.H., 2007. Equestrian injuries: incidence, injury patterns, and risk factor for 10 years of major traumatic injuries. Am. J. Surg. 193, 636–640.

Bangs, E., Shivik, J., 2001. Managing Wolf Conflict with Livestock in the Northwestern United States, vol. 3. Carnivore Damage Prevention News, pp. 2–5.

Bargali, H.S., Akhtar, N., Chauhan, N.P.S., 2005. Characteristics of sloth bear attacks and human casualties in North Bilaspur Forest Division, Chhattisgarh, India. Ursus 16, 263–267.

Bashir, M.O., Abu-Zidan, F.M., 2006. Motor vehicle collisions with large animals. Saudi Med. J. 27, 1116–1120.

Begg, C., Begg, K., Muemedi, O., 2007. Preliminary Data on Human – Carnivore Conflict in Niassa National Reserve, Mozambique, Particularly Fatalities Due to Lion, Spotted Hyaena and Crocodile. Sociedade para a Gestão e Desenvolvimento da Reserva do Niassa, Mozambique.

Beier, P., 1991. Cougar attacks on humans in the United States and Canada. Wildl. Soc. Bull. 19, 403–412.

Beier, P., 1992. Cougar attacks on humans: an update and some further reflections. In: Proceedings of the Fifteenth Vertebrate Pest Conference. University of California—Davis, Davis California, pp. 365–367.

Blackwell, B.F., Wright, S.E., 2006. Collisions of Red-tailed Hawks (*Buteo jamaicensis*), Turkey Vultures (*Cathartes aura*), and Black Vultures (*Coragyps atratus*) with Aircraft: Implications for Bird Strike Reduction. USDA National Wildlife Research Center. Paper 410.

Bryant, E.A., 1991. Natural Hazards. Cambridge University Press, Cambridge, UK, 294 pp.

Butler, D.R., 1993. The impact of mountain goat migration on unconsolidated slopes in Glacier National Park, Montana. Geogr. Bull. 35 (2), 98–106.

Butler, D.R., Malanson, G.P., 2005. The geomorphic influences of beaver dams and failures of beaver dams. Geomorphology 71, 48–60.

Centers for Disease Control, 2015. Rocky Mountain Spotted Fever (RMSF). Retrieved from: http://www.cdc.gov/rmsf/stats/index.html.

Centers for Disease Control and Prevention, 2014a. Zoonotic Disease: When Humans and Animals Intersect. Retrieved from: http://www.cdc.gov/24-7/cdcfastfacts/zoonotic.html.

Centers for Disease Control and Prevention, 2014b. Mosquito-Borne Diseases. Retrieved from: http://www.cdc.gov/ncidod/diseases/list_mosquitoborne.htm.

Centers for Disease Control and Prevention, 2014c. West Nile Virus and Other Arboviral Diseases – United States, 2013. Retrieved from: http://www.cdc.gov/mmwr/preview/mmwrhtml/mm6324a1.htm.

Centers for Disease Control and Prevention, 2014d. Rabies. Retrieved from: http://www.cdc.gov/rabies/index.html.

Centers for Disease Control and Prevention, 2014e. Reptiles, Amphibians, and Salmonella. Retrieved from: http://www.cdc.gov/Features/SalmonellaFrogTurtle/.

Centers for Disease Control and Prevention, 2014f. 2014 Ebola Outbreak in West Africa. Retrieved from: http://www.cdc.gov/vhf/ebola/outbreaks/guinea/index.html.

Centers for Disease Control and Prevention, 2014g. Hantavirus. Retrieved from: http://www.cdc.gov/hantavirus/index.html.

Centers for Disease Control and Prevention, 2014h. Hantavirus Pulmonary Syndrome. Retrieved from: http://www.cdc.gov/hantavirus/hps/index.html.

Centers for Disease Control and Prevention, 2014i. Hemorrhagic Fever with Renal Syndrome (HFRS). Retrieved from: http://www.cdc.gov/hantavirus/hfrs/index.html.

Centers for Disease Control and Prevention, 2013. Annual U.S. HPS Cases and Case-fatality, 1993–2013. Retrieved from: http://www.cdc.gov/hantavirus/surveillance/annual-cases.html.

Centers for Disease Control and Prevention, 2012a. Giardiasis Surveillance – United States, 2009–2010. Retrieved from: http://www.cdc.gov/mmwr/preview/mmwrhtml/ss6105a2.htm.

Centers for Disease Control and Prevention, 2012b. Multistate Outbreak of Listeriosis Linked to Whole Cantaloupes from Jensen Farms, Colorado. Retrieved from: http://www.cdc.gov/listeria/outbreaks/cantaloupes-jensen-farms/082712/index.html.

Centers for Disease Control and Prevention, 2008a. Surveillance for Lyme Disease – United States, 1992–2006. Retrieved from: http://www.cdc.gov/MMWR/PREVIEW/MMWRHTML/ss5710a1.htm.

Centers for Disease Control and Prevention, 2008b. Multistate Outbreak of Human Salmonella Infections Associated with Exposure to Turtles – United States, 2007–2008. Retrieved from: http://www.cdc.gov/mmwr/preview/mmwrhtml/mm5703a3.htm.

Centers for Disease Control and Prevention, 2007. Turtle-associated Salmonellosis in Humans – United States, 2006–2007. Retrieved from: http://www.cdc.gov/mmwr/preview/mmwrhtml/mm5626a1.htm.

Centers for Disease Control and Prevention, 1994. Hantavirus Pulmonary Syndrome – United States, 1993. Retrieved from: http://www.cdc.gov/mmwr/preview/mmwrhtml/00025007.htm.

Chen, H., Smith, G.J.D., Zhang, Z.Y., Qin, K., Wang, J., Li, K.S., Webster, R.G., Peiris, J.S.M., Guan, Y., 2005. H5N1 virus outbreak in migratory waterfowl. Nature 436 (14), 191–192.

Chowdhury, R., Mukherjee, A., Naskar, S., Lahiri, S.K., 2013. A study on knowledge of animal bite management and rabies immunization among interns of a government medical college in Kolkata. Int. J. Med. Public Health 3, 17–20.

Clark, M.A., Sandusky, G.E., Hawley, D.A., Pless, J.E., Fardal, P.M., Tare, L.R., 1991. Fatal and near-fatal animal bite injuries. J. Forensic Sci. 36, 1256–1261.

Clearly, E.C., Wright, S.E., Dolbeer, R.E., 2000. Wildlife Strikes to Civilian Aircraft in the United States 1990–1999. Federal Aviation Administration, Washington, DC, USA, p. 76.

Coch, N.K., 1995. Geohazards – Natural and Human. Prentice Hall, Englewood Cliffs, NJ, 481 pp.

Cruz, L.S., Vargas, R., Lopes, A.A., 2009. Snakebite envenomation and death in the developing world. Ethn. Dis. 19, 42–46.

Dantas-Torres, F., Chomel, B.B., Otranto, D., 2012. Ticks and tick-borne diseases: a one health perspective. Trends Parasitol. 29 (10), 437–446.

Dawood, F.S., Iyliano, A.D., Reed, C., Meltzer, M.I., Shay, D.K., Cheng, P., Bandaranayake, D., Breiman, R.F., Brooks, W.A., Buchy, P., Feikin, D.R., Fowler, K.B., Gordon, A., Hien, N.T., Horby, P., Huang, Q.S., Katz, M.A., Krishnan, A., Lal, R., Montgomery, J.M., Mølbak, K., Pebody, R., Presanis, A.M., Razuri, H., Steens, A., Tinoco, Y.O., Wallinga, J., Yu, H., Vong, S., Bresee, J., Widdowson, M., 2012. Estimated global mortality associated with the first 12 Months of 2009 pandemic influenza A H1N1 virus circulation: a modelling study. Lancet Infect. Dis. 12, 687–695.

DeChano, L.M., Butler, D.R., 2002. An analysis of attacks by grizzly bears (*Ursus arctos horribilis*) in Glacier National Park, Montana. Geogr. Bull. 44, 30–41.

Durrheim, D.N., Leggat, P.A., 1999. Risk to tourists posed by wild mammals in South Africa. J. Travel Med. 6, 172–179.

European Food Safety Authority and European Centre for Disease Prevention and Control, 2014. The European Union summary report on trends and sources of zoonoses, zoonotic agents and food-borne outbreaks in 2012. EFSA J. 12 (2), 3547–3858.

Fenner, P.J., 2005. Venomous jellyfish of the world. South Pac. Underwater Med. Soc. J. 35, 131–138.

Fenner, P.J., Lippmann, J., Gershwin, L.-A., 2010. Fatal and nonfatal severe jellyfish stings in Thai waters. J. Travel Med. 17, 133–138.

Floyd, T., 1999. Bear-inflicted human injury and fatality. Wilderness Environ. Med. 10, 75–87.

Gehring, T.M., VerCauteren, K.C., Cellar, A.C., 2011. Good fences make good neighbors: implementation of electric fencing for establishing effective livestock-protection dogs. Hum. Wildl. Interact. 5 (1), 106–111.

Gehrt, S.D., 2007. Ecology of coyotes in urban landscapes. In: Nolte, D.L., Arjo, W.M., Stalman, D.H. (Eds.), Proceedings of the 12th Wildlife Damage Management Conference, pp. 303–311.

Gibby, R.A., Clewell, R., 2006. Evaluation of Wildlife Warning Systems and Other Countermeasures. Nevada Department of Transportation Technical Document NV-RDT-06-010.

Greene, W.C., 2007. A history of AIDS: looking back to see ahead. Eur. J. Immunol. 37, S94–S102.

Gold, B.S., Dart, R.C., Barish, R.A., 2002. Bites of venomous snakes. N. Engl. J. Med. 347, 1–10.

Gulan, S., McCabe, G., Rosenthal, I., Wolfe, S.E., Anderson, V.L., 2006. Evaluation of wildlife reflectors in reducing vehicle-deer collisions on Indiana interstate 80/90. Indiana Department of Transportation Technical Report FHWA/IN/JTRP-2006/18.

Hazin, F.H.V., Burgess, G.H., Carvalho, F.C., 2008. A shark attack outbreak off recife, Pernambuco, Brazil: 1992–2006. Bull. Mar. Sci. 82, 199–212.

Herrero, S., Higgins, A., Cardoza, J.E., Hajduk, L.I., Smith, T.S., 2011. Fatal attacks by American black bear on people: 1900–2009. J. Wildl. Manage. 75, 596–603.

Heyworth, B., Ropp, M.E., Voos, U.G., Meinel, H.I., Darlow, H.M., 1975. Anthrax in Gambia: an epidemiological study. Br. Med. J. 4, 79–82.

Hilder, V.A., Boulter, D., 1999. Genetic engineering of crop plants for insect resistance – a critical review. Crop Prot. 18, 177–191.

Hyndman, D., Hyndman, D., 2013. Natural Hazards and Disasters. Brooks/Cole, Belmont, CA, 555 pp.

Izurieta, R., Galwankar, S., Clem, A., 2008. Leptospirosis: the "mysterious" mimic. J. Emerg. Trauma Shock 1 (1), 21–33.

Jayson, E.A., Christopher, G., 2008. Human–Elephant Conflict in the Southern Western Ghats: a case study from the Peppara Wildlife Sanctuary, Kerala, India. Indian For. 134, 1309–1325.

Jenness, S.E., 1965. Arctic wolf attacks scientist – a unique Canadian incident. Arctic 38, 129–132.

Jongejan, F., Uilenberg, G., 2004. The global importance of Ticks. Parasitology 129, S3–S14.

Klotz, J.H., Klotz, S.A., Pinnas, J.L., 2009. Animal bites and stings with anaphylactic potential. J. Emerg. Med. 36, 148–156.

Kolowski, J.M., Holekamp, K.E., 2006. Spatial, temporal, and physical characteristics of livestock depredations by large carnivores along a Kenya reserve border. Biol. Conserv. 128, 529–541.

Langley, R.L., Hunter, J.L., 2001. Occupational fatalities due to animal-related events. Wilderness Environ. Med. 12, 168–174.

Lehner, P.N., 1976. Coyote behavior: implications for management. Wildl. Soc. Bull. 4 (3), 120–126.

Lindsey, P.A., Masterson, C.L., Beck, A.L., Romañach, S., 2012. Ecological, social, and financial issues related to fencing as a conservation tool in Africa. In: Somers, M.J., Hayward, M.W. (Eds.), Fencing for Conservation: Restriction of Evolutionary Potential or Riposte to Threatening Processes? Springer, Berlin, pp. 215–234.

McCrory, P., Turner, M., LeMasson, B., Bodere, C., Allemandou, A., 2006. An analysis of injuries resulting from professional horse racing in France during 1991–2001: a comparison with injuries resulting from professional horse racing in Great Britain during 1992–2001. Br. J. Sports Med. 40, 614–618.

Mead, P.S., Slutsker, L., Dietz, V., McCaig, L.F., Brescee, J.S., Shapiro, C., Griffin, P.M., Tauxe, R.V., 1999. Food-related illness and death in the United States. Emerg. Infect. Dis. 5 (5), 607–625.

Miquelle, D., Nikolaev, I., Goodrich, J., Litvinov, B., Smirnov, E., Suvorov, E., 2005. Searching for the Coexistence Recipe: a case study of conflicts between people and tigers in the Russian Far East. In: Woodroffe, R., Thirgood, S., Rabinowitz, A. (Eds.), People and Wildlife: Conflict or Coexistance? Cambridge University Press, Cambridge, pp. 305–322.

Mode, N.A., Hackett, E.J., Conway, G.A., 2005. Unique occupational hazards of Alaska: animal-related injuries. Wilderness Environ. Med. 16, 185–191.

Montana Fish, Wildlife, and Parks, 2014. Wolf Program. Retrieved from: http://fwp.mt.gov/fishAndWildlife/management/wolf/.

Naughton-Treves, L., Treves, A., 2005. Socio-ecological factors shaping local support for wildlife: crop-raiding by elephants and other wildlife and Africa. In: Woodroffe, R., Thirgood, S., Rabinowitz, A. (Eds.), People and Wildlife: Conflict or Coexistance? Cambridge University Press, Cambridge, pp. 252–277.

Neumann-Denzau, G., Denzau, H., 2010. Examining certain aspects of human-tiger conflict in the Sundarbans Forest, Bangladesh. Tiger Pap. 37, 1–11.

Nyhus, P., Tilson, R., 2004a. Agroforestry, elephants, and tigers: balancing conservation theory and practice in human-dominated landscapes of Southeast Asia. Agric. Ecosyst. Environ. 104, 87–97.

Nyhus, P., Tilson, R., 2004b. Characterizing human-tiger conflict in Sumatra, Indonesia: Implications for conservation. Oryx 38 (1), 68–74.

Oerke, E.C., Dehne, H.W., Schönbeck, F., Weber, A., 1994. Crop Production and Crop Protection: Estimated Losses in Major Food and Cash Crops. Elsevier, New York, p. 808.

O'Shea, T.J., Cryan, P.M., Cunningham, A.A., Fooks, A.R., Hayman, D.T.S., Luis, A.D., Peel, A.J., Plowright, R.K., Wood, J.L.N., 2014. Bat flight and zoonotic viruses. Emerg. Infect. Dis. 20 (5), 741–745.

Patterson, B.D., Kasiki, S.M., Selempo, E., Kays, R.W., 2004. Livestock predation by lions (*Panthera leo*) and other carnivores on ranches neighboring Tsavo National Parks, Kenya. Biol. Conserv. 119, 507–516.

Petrovay, F., Balla, E., 2008. Two fatal cases of psittacosis caused by Chlamydophila psittaci. J. Med. Microbiol. 57 (10), 1296–1298.

Poessel, S.A., Breck, S.W., Teel, T.L., Shwiff, S., Crooks, K.R., Angeloni, L., 2013. Patterns of human–oyote conflicts in the Denver metropolitan area. J. Wildl. Manage. 77, 297–305.

Picozzi, K., Fèvre, E.M., Welburn, S.C., 2005. Sleeping sickness in Uganda: a thin line between two fatal diseases. Br. Med. J. 331 (7527), 1238–1241.

Rangel, J.M., Sparling, P.H., Crowe, C., Griffin, P.M., Swerdlow, D.L., 2005. Epidemiology of Escherichia coli O157:H7 outbreaks, United States, 1982–2002. Emerg. Infect. Dis. 11 (4), 603–609.

Reeve, A.F., Anderson, S.H., 1993. Ineffectiveness of Swareflex reflectors at reducing deer-vehicle collisions. Wildl. Soc. Bull. 21, 127–132.

Reiter, P., 2001. Climate change and mosquito-borne disease. Environ. Health Perspect. 109, 141–161.

Richardson, J.W., West, T., 2000. Serious birdstrike accidents to military aircraft: updated list and summary. In: International Bird Strike Committee, 25th Meeting. Amsterdam, Netherlands, p. 31.

Robertson, I.D., Irwin, P.J., Lymbery, A.J., 2000. The role of companion animals in the emergence of parasitic zoonosis. Int. J. Parasitol. 30, 1369–1377.

Rtshiladze, M.A., Andersen, S.P., Nguyen, D.Q.A., Grabs, A., Ho, K., 2011. The 2009 Sydney shark attacks: case series and literature review. Aust. N.Z. J. Surg. 81, 345–351.

Schafer, J.A., Penland, S.T., 1985. Effectiveness of Swareflex reflectors in reducing deer-vehicle accidents. J. Wildl. Manage. 49 (3), 774–776.

Seiler, A., 2005. Predicting locations of moose-vehicle collisions in Sweden. J. Appl. Ecol. 42, 371–382.

Sharp, P.M., Hahn, B.H., 2011. Origins of HIV and the AIDS Pandemic. Cold Spring Harb. Perspect. Med. 1 (1), 1–22.

Taylor, L.H., Latham, S.M., Woolhouse, M.E., 2001. Risk factors for human disease emergence. Philos. Trans. R. Soc. Biol. Sci. 356 (1411), 983–989.

Thorpe, J., 2003. Fatalities and destroyed civil aircraft due to bird strikes, 1912–2002. In: International Bird Strike Committee, 26th Meeting. Warsaw, Poland, p. 28.

Tiefenbacher, J.P., Shuey, M.L., Butler, D.R., 2003. A spatial evaluation of cougar-human encounters in U.S. National Parks: the cases of Glacier and Big Bend National Parks. In: Harveson, L.A., Harveson, P.M., Adams, R.W. (Eds.), Proceedings of the 6th Mountain Lion Workshop. Texas Parks and Wildlife Department, Austin, TX, pp. 43–50.

Timm, R.M., Baker, R.O., Bennett, J.R., Coolahan, C.C., 2004. Coyote Attacks: An Increasing Suburban Problem. In: Proceedings of the 21st Vertebrate Pest Conference. University of California – Davis, Davis, CA, pp. 47–57.

Torgerson, P.R., de Silva, N.R., Fèvre, E.M., Kasuga, F., Rokni, M.B., Zhou, X., Sripa, B., Gargouri, N., Willingham, A.L., Stein, C., 2014. The global burden of foodborne parasitic diseases: an update. Trends Parasitol. 30 (1), 20–26.

Toovey, S., Annandale, Z., Jamieson, A., Schoeman, J., 2004. Zebra bite to a South African tourist. J. Travel Med. 11, 122–124.

Treves, A., Martin, K.A., Wydeven, A.P., Wiedenhoeft, J.E., 2011. Forecasting environmental hazards and the application of risk maps to predator attacks on livestock. BioScience 61 (6), 451–458.

United States Department of Agriculture, 2006. Protecting Property, Infrastructure, and Transportation in Rural and Urban Settings. USDA Animal and Plant Health Inspection Service, Austin, Texas, USA, p. 3.

Utah Division of Wildlife Resources, 2012. Wolf Management in Utah. Retrieved from. https://wildlife.utah.gov/pdf/fact_sheets/wolves.pdf.

Wang, T.T., Palese, P., 2009. Unraveling the mystery of swine influenza virus. Cell 137, 983–985.

Warner, G.S., 2010. Increased incidence of domestic animal bites following a disaster due to natural hazards. Prehosp. Disaster Med. 25, 188–190.

Webster, R., Hulse, D., 2005. Controlling avian flu at the source. Nature 435 (26), 415–416.

Weese, J.S., Peregrine, A.S., Armstrong, J., 2002. Occupational health and safety in small veterinary practice: part 1 – nonparasitic zoonotic diseases. Can. Vet. J. 43, 631–636.

White, L.A., Gehrt, S.D., 2009. Coyote attacks on humans in the United States and Canada. Hum. Dimens. Wildl. 14, 419–432.

Wolfe, N.D., Daszak, P., Kilpatrick, A.M., Burke, D.S., 2005. Bushmeat hunting, deforestation, and prediction of zoonotic disease. Emerg. Infect. Dis. 11 (12), 1822–1827.

World Health Organization, 2014a. Emerging Diseases. Retrieved from: http://www.who.int/topics/emerging_diseases/en/.

World Health Organization, 2014b. Dengue and Severe Dengue. Retrieved from: http://www.who.int/mediacentre/factsheets/fs117/en/.

World Health Organization, 2014c. West Nile Virus. Retrieved from: http://www.who.int/mediacentre/factsheets/fs354/en/.

World Health Organization, 2014d. Trypanosomiasis, Human African (Sleeping Sickness). Retrieved from: http://www.who.int/mediacentre/factsheets/fs259/en/.

World Health Organization, 2014e. Salmonella (Non-Typhoidal). Retrieved from: http://www.who.int/mediacentre/factsheets/fs139/en/.

World Health Organization, 2014f. Ebola Virus Disease. Retrieved from: http://www.who.int/mediacentre/factsheets/fs103/en/.

World Health Organization, 2014g. Cumulative Number of Confirmed Human Cases for Avian Influenza A(H5N1) Reported to WHO, 2003–2014. Retrieved from: http://www.who.int/influenza/human_animal_interface/EN_GIP_20140124CumulativeNumberH5N1cases.pdf?ua=1.

World Health Organization, 2014h. Impact of Dengue. Retrieved from: http://www.who.int/csr/disease/dengue/impact/en/.

World Health Organization, 2014i. Cumulative Number of Confirmed Human Cases for Avian Influenza A(H5N1) Reported to WHO, 2003–2014. Retrieved from: http://www.who.int/influenza/human_animal_interface/EN_GIP_20140124CumulativeNumberH5N1cases.pdf?ua=1.

World Health Organization, 2013. World Malaria Report 2013. World Health Organization, Geneva, Switzerland, 284 pp.

World Health Organization, 2008. Anthrax in Humans and Animals, fourth ed. WHO, Geneva, Switzerland. 208 pp.

World Health Organization, 2004. WHO Expert Consultation on Rabies. World Health Organization, Geneva, Switzerland, 121 pp.

World Health Organization, Division of Emerging and other Communicable Diseases Surveillance and Control, 1998. World Survey of Rabies for the Year 1996. World Health Organization, Geneva, Switzerland, 29 pp.

Chapter 9

Loss of Biodiversity: Concerns and Threats

Robert M. May
Zoology Department, Oxford University, Oxford, UK

ABSTRACT

Over the past century, the number of documented extinctions in well-studied taxonomic groups of plants and animals has, on average, increased at rates 100–1000 times faster than those seen over the half-billion-year sweep of the fossil record. It must, however, be emphasized that for most groups, particularly invertebrates, we are very uncertain how many species occur on Earth today, much less the rates at which they are being extinguished.

9.1 INTRODUCTION

In what follows, this chapter first outlines how little we know about the number of distinct eukaryotic species alive on Earth today. It next discusses what we know—and do not know—about current rates of species extinction. In conclusion, the chapter outlines some of the reasons why we should be concerned about such losses, at rate orders of magnitude above the average seen over the fossil record; these reasons range from the ethical to the very practical.

9.2 HOW MANY SPECIES?

Unfortunately, analysis of the causes and consequences of accelerating extinction rates is impeded by the rudimentary state of our knowledge. This in turn—as discussed further below—derives more from past intellectual fashions than any dispassionate assessment of scientific priorities.

The first systematic study of how many distinct life forms—species—exist on our planet today began late in our intellectual history. It lagged the foundation of the French and British Academies of Science by roughly a century.

Biological and Environmental Hazards, Risks, and Disasters. http://dx.doi.org/10.1016/B978-0-12-394847-2.00012-7

Linnaeus' binary codification in *De Rerem Naturae*, which recognized a global total of around 9,000 species of plants and animals, is dated 1758. In many ways, the legacy of this century-long lag still lingers.

Currently around 1.5−1.6 million distinct species of plants and eukaryotes animals have been named and recorded (May, 1999). Even this number—analogous to the number of books in the British Library, which is precisely known—is uncertain to within around 10%, because the majority of species are invertebrate animals of one kind or another, for most of which the records are still on filing cards in separate museums and other institutions. Lacking a synoptic database it is hard to sort out problems with synonyms (the same species being separately identified and differently named in two or more places).

Currently, new species are being identified at the rate of roughly 13,000−15,000 a year, while at the same time earlier synonyms are being resolved at around 3,000 each year, for a net addition of roughly new 10,000 species per year.

So much for what is known. But how many species may there be in total on Earth today? Recent estimates lie in the range 5−15 million (Erwin, 1982; Hamilton et al., 2010). Lower numbers, and also much higher ones, also have their advocates. Even if we take a low estimate of 3 million still to be identified, at the current rates just noted, the job would take 300 years. Organizing better databases, and using molecular information about newly discovered species' genomes ("bar coding life"), promises to speed up this distressingly slow task. Even so, the craft of collecting material in the field will remain a seriously rate-limiting step.

These lamentable uncertainties result partly from what a management consultant would call inefficiencies in the distribution of the relevant workforce. Although the taxonomy of taxonomists is in itself poorly documented, rough estimates suggest that it is approximately evenly divided among vertebrates, plants, and invertebrates (Gaston and May, 1992). But there are around 10 plant species and at least 100 invertebrate species (possibly more than 1,000) for each vertebrate species. The labor force is even more inefficiently divided if one considers the research literature on conservation biology. Here, an analysis of the 2,700 papers published in *Conservation Biology* and *Biological Conservation* between 1979 and 1998 showed 69% devoted to vertebrates, 20% to plants, and 11% to all invertebrates (with half of these being Lepidoptera, which appear to have the status of honorary birds) (Clark and May, 2002). And conservation action, as indicated, for example, by WWF's Annual Report, is almost wholly devoted to charismatic megavertebrates. This is understandable in view of public attitudes, but arguably unfortunate in terms of preserving ecosystem functioning. The argument that protecting vertebrate biodiversity will more-or-less automatically also preserve invertebrate biodiversity does not survive close examination (Prendagast et al., 1993).

9.3 EXTINCTION RATES

The pressures currently being inflicted on natural communities of plants and animals are huge and increasing. Since Darwin published the *Origin of Species* roughly 150 years ago, human numbers have increased sevenfold, and the energy use per person has increased by a similar factor, resulting in a 50-fold increase in our overall impact on our planet's ecosystems. Vitousek et al.'s (1986) estimate that humanity takes to itself, directly or indirectly, roughly 40% of terrestrial net primary productivity has recently been validated by satellite images of the land area modified by us. Even more extraordinary, of all the atmospheric nitrogen fixed in 2008, 55% came from the Harber–Bosch chemical process rather than the natural biogeochemical processes which created, and which struggle to maintain, the biosphere (Sachs, 2008).

If we do not know how many species have been identified—much less their functional roles in ecosystems—to within 10%, nor the overall species total to within an order-of-magnitude, we clearly cannot say much about how many species are likely to become extinct this century. We can note that the IUCN Red Data Books (2004), using specific and sensible criteria, estimated 20% of recorded mammal species are threatened with extinction, and likewise 12% of birds, 4% of reptiles, 31% of amphibians, 3% of fish, and 31% of the 980 known species of gymnosperms. However, when these figures are re-expressed in terms of the number of species whose status has been evaluated (as distinct from dividing the number known to be threatened by the total number known—however slightly—to science), the corresponding numbers are 23%, 12%, 61%, 31%, 26%, and 34%, respectively. This says a lot about how comparatively little attention reptiles and fish have received.

The corresponding figures for the majority of plant species, dicotyledons and monocotyledons, are respectively 4% and 1% of those known versus 74% and 68% of those evaluated. Most telling are the two numbers for the most numerous group, insects: 0.06% of all known species are threatened, compared with 73% of those actually evaluated. The same pattern holds true for other invertebrate groups. For these small things, which arguably run the world, we know too little to make any rough estimate of the proportions that have either become extinct or are threatened with it.

Perhaps surprisingly, we can nevertheless say some relatively precise things about current and likely future *rates* of extinction in relation to the average rates seen over the roughly 550 million years sweep of the fossil record (May, 1999; May et al., 1995). For bird and mammal species (a total of approximately 14,000), an average of about one certified extinction per year has occurred over the past century. This is a very conservative estimate of the true extinction rate, because many species receive little attention, even in this unusually well-studied group. Such a rate, if continued, translates into an average "species' life expectancy" of the order of 10,000 years. By contrast, the average life expectancy—from origination to extinction—of a species in the fossil record lies in the general range 1–10 million years, albeit with great variation both within and among groups.

So, if birds and mammals are typical—and no good reason exists to assume they are not—extinction rates in the twentieth century were higher, by a factor of 100–1,000, than the fossil record's average background rates. And four different lines of argument suggest a further 10-fold speeding up over the coming century. Such an acceleration in extinction rates is of the magnitude which characterized the Big Five mass extinction events in the fossil record (Raup, 1978; Sepkoski, 1992). These Big Five are used to mark changes from one geological epoch to the next. Although much need exists for further work to refine estimates of this kind, it does seem likely that we are standing on the breaking tip of a sixth great wave of mass extinctions. These facts and estimates are set out schematically in Figure 9.1 (Millenium Ecosystem Assessment, 2005).

The crucial difference between the impending Sixth Wave of mass extinction and the previous Big Five is that the earlier ones stemmed from external environmental events. The sixth, set to unfold over the next several centuries—seemingly long to us, but a blink of the eye in geological terms—derives directly from human impacts.

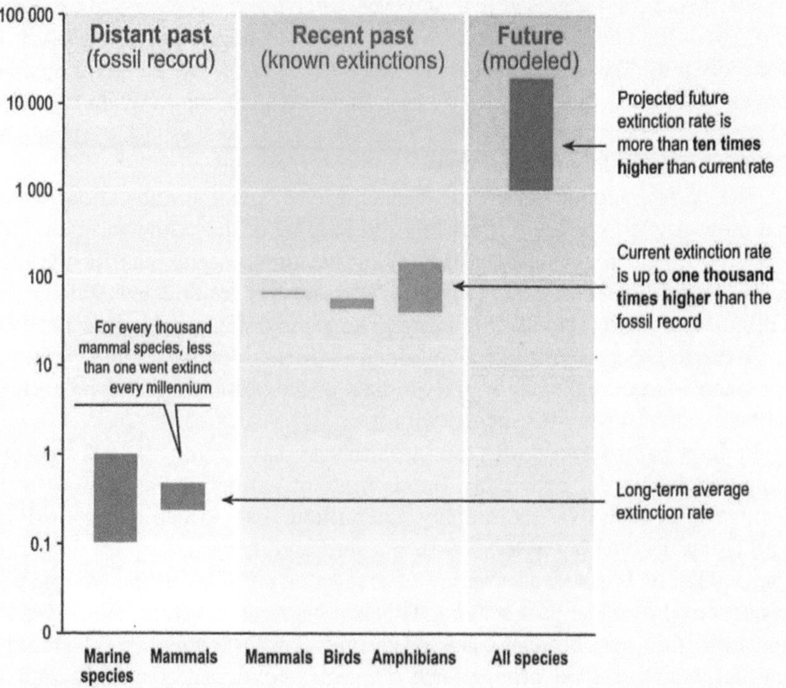

FIGURE 9.1 Extinctions per thousand species per millennium. *Source: Milennium Ecosystem Assessment.*

9.4 REASONS FOR CONCERN

The main causes of extinction are habitat loss, overexploitation, and introduction of alien species. Often two occur or all three combined. An increasing number of recent studies show, moreover, that the effects of climate change are compounding these more direct effects of human activities. Why should we worry about this accelerating loss of biological diversity? I think the reasons can be broadly grouped under three headings: narrowly utilitarian, broadly utilitarian, and ethical.

Taking narrowly utilitarian considerations first; it has been argued that plant and animal species—both known and yet undiscovered—are a precious resource of genetic novelties. They may well be the raw stuff of tomorrow's biotech revolution, producing new pharmaceutical products, new foodstuffs, and other products for the global economy. So let us not burn the books before we have read them.

Such efforts to move biological diversity into the gambit of conventional economics, if only to motivate political concern, would seem sensible. But I am skeptical of this argument. I think it more likely, with the pace of advances in understanding the molecular machinery of living things, that tomorrow's medicines will be designed from the molecules up, rather than emerging from high-tech bioprospecting.

But more generally and arguably more importantly, broadly utilitarian considerations suggest we do not yet know enough about the structure and function of ecosystems to be able to predict how much disturbance and species loss they can suffer yet still deliver ecosystem services upon which we depend.

This chapter began by showing just how little we know about how many species of animals, plants, and microbes are present on Earth today. Additionally, for most of those species which have been named and recorded—the majority of which are invertebrates—we know little or nothing about the roles they play in maintaining the ecosystems of which they are part. One estimate is that we have information about the behavior and ecology of fewer than 5% of all identified animal species (Raven, 2004). It is therefore not surprising that we are not yet very good at predicting the effects upon local or regional ecosystems of the loss of species as a consequence of habitat disturbance, or overexploitation, or introduction of alien species, or combinations of these extinction-causing perturbations.

The Millennium Ecosystem Assessment (2005), sponsored by the United Nations, involved some 1,360 scientists from 95 countries and was the first global assessment of the world's ecosystems. Despite many uncertainties, it gives a comprehensive appraisal of the condition of, and trends in, the world's ecosystems. These services are the benefit provided to humans as a result of species' interactions within the system. Some of these services are local (e.g., provision of pollinators for crops), others regional (e.g., flood control or water purification), and yet others global (e.g., climate regulation). In its massive report, the MEA identified 24 categories of such ecosystem services, broadly grouped under three headings: provisioning, regulating, and cultural.

Table 9.1 summarizes these 24 categories of services, along with indications of whether the service is being enhanced or degraded. Note that of the 24 categories of ecosystem services examined by the MEA, 15—roughly two-thirds—are being degraded or used unsustainably. Although 15 have suffered in this way, only 4 have been enhanced in the past 50 years, of which 3 involve food production: crops, livestock, and aquaculture. The status of the remaining five is equivocal or uncertain, as indicated in the notes in Table 9.1.

The way economists conventionally calculate gross domestic product (GDP) takes little or no account of the role of ecosystem services. For example, an oil tanker going aground, and wreaking havoc on the region's biota, will typically make a positive contribution to conventional GDP because cleanup costs are a plus whereas environmental damage is deemed not assessable! Constanza et al. (1997) have attempted to assess the "GDP equivalent" of the totality of the planet's ecosystem services. Their guesstimate is that such services have a value roughly equal to global GDP as conventionally assessed. Any calculation of this kind necessarily has many uncertainties, and some would argue that you simply cannot put a price upon a service which is essential to life. But I think it is helpfully indicative.

In essence, the broadly utilitarian argument recognizes that we do not know how much biological diversity we can lose, yet still keep ecosystem services upon which humans depend. In this situation, as emphasized by one of the founders of the Conservation Movement, Aldo Leopold, "the first rule of intelligent tinkering is to keep all the pieces." But maybe we could be clever enough to survive in a greatly biologically impoverished world. It would, very likely, be a world akin to that of the cult movie *Blade Runner.* The question arises, who would want to live in such a world? This takes us to our third consideration.

The ethical argument is simply put: we have a responsibility to hand on to future generations a planet as rich in natural wonders as the one we inherited. Narrowly utilitarian considerations urge us to preserve individual species, many of them not yet recorded much less studied, because tomorrow's biotechnology may find their genes useful. Broadly utilitarian considerations worry about preserving ecosystems because we depend upon them. Some would say ethical considerations are more vague, but I find them more compelling.

Some of the complexities of the ethical responsibilities of human stewardship were set out eloquently by Aldo Leopold. Mourning the death in the Cincinnati Zoo in 1917 of Martha, the last passenger pigeon, he wrote: "We grieve because no living man will see again the onrushing phalanx of victorious birds sweeping a path for Spring across the March skies, chasing the defeated winter from all the woods and prairies… Our grandfathers, who saw the glory of the fluttering hosts, were less well-housed, well-fed, well-clothed

TABLE 9.1 Global status of ecosystem services (Millennium Ecosystem Assessment, 2005)

Service	Status	Notes
Provisioning Services		
Food		
Crops	+	Substantial production increase
Livestock	+	Substantial production increase
Capture fisheries	–	Declining production due to overharvest
Aquaculture	+	Substantial production increase
Wild foods	–	Declining production
Fiber		
Timber	+/–	Forest loss in some regions, growth in others
Cotton, hemp, silk	+/–	Declining production of some fibers, growth in others
Wood fuel	–	Declining production
Genetic resources	–	Lost through extinction and crop genetic resource loss
Biochemicals, natural medicines, pharmaceuticals	–	Lost through extinction, overharvest
Freshwater	–	Unsustainable use for drinking, industry, and irrigation; amount of hydro energy unchanged, but dams increase ability to use that energy
Regulating Services		
Air quality regulation	–	Decline in ability of atmosphere to cleanse itself
Climate regulation		
Global	+	Net source of carbon sequestration since mid-century
Regional and local	–	Preponderance of negative impacts
Water regulation	+/–	Varies depending on ecosystem change and location
Erosion regulation	–	Increased soil degradation
Water purification and waste treatment	–	Declining water quality
Disease regulation	+/–	Varies depending on ecosystem change

Continued

TABLE 9.1 Global status of ecosystem services (Millennium Ecosystem Assessment, 2005)—cont'd

Service	Status	Notes
Pest regulation	–	Natural control degraded through pesticide use
Pollination	–	Apparent global decline in abundance of pollinators
Natural hazard regulation	–	Loss of natural buffers (wetlands, mangroves)
Cultural Services		
Spiritual and religious values	–	Rapid decline in sacred groves and species
Aesthetic values	–	Decline in quantity and quality of natural lands
Recreation and ecotourism	+/–	More areas accessible but many degraded

+, enhanced; –, degraded, in the senses defined in the main text.

than we are. The strivings by which they bettered our lot are also those which deprived us of pigeons. Perhaps we now grieve because we are not sure, in our hearts, that we have gained by the exchange… The truth is our grandfathers, who did the actual killing, were our agents. They were our agents in the sense they shared the conviction, which we have only now begun to doubt, that it is more important to multiply people and comforts than to cherish the beauty of the land in which they live." This not only gives poetic expression to how many of us feel, but I think it also raises the question of whether I would feel the same way if I were a poor farmer in a drought-stricken developing country, striving to feed my family.

REFERENCES

Clark, J.A., May, R.M., 2002. Taxonomic bias in conservation research. Science 297, 191–192.

Costanza, R., et al., 1997. The value of the world's ecosystem services and natural capital. Nature 387, 253–257.

Erwin, T.L., 1982. Tropical forests: their richness in coleoptera and other arthropod species. Coleopt. Bull. 36, 74–82.

Gaston, K.J., May, R.M., 1992. Taxonomy of taxonomists. Nature 356, 281–282. Reprinted in Italian: *Sapere*, No 59, 14–16 (1993).

Hamilton, A.J., et al., 2010. Quantifying incertainty in estimation of tropical arthropod species richness. Amer. Natur. 176, 90–98.

IUCN Red Data Book 2004. See www.redlist.org/info/tables.html.

May, R.M., 1999. The dimensions of life on Earth. In: Nature and Human Society. National Academy of Sciences Press, Washington DC, pp. 30–45.

May, R.M., Lawton, J.H., Stork, N.E., 1995. Assessing extinction rates. In: Lawton, J.H., May, R.M. (Eds.), Extinction Rates. Oxford University Press, pp. 1–24.

Millennium Ecosystem Assessment, 2005. Ecosystems and Human Well-being. Synthesis Island Press, Washington DC.

Prendagast, et al., 1993. Rare species, the coincidence of diversity hotspots and conservation strategies. Nature 365, 335–337.

Raup, D.M., 1978. Cohort analysis of genetic survivorship. Paleobiology 4, 1–15.

Raven, P.H., 2004. Taxonomy: where are we now? Proc. Roy. Soc. B 359, 729–730.

Sachs, J.D., 2008. Common Wealth: Economics for a Crowded Planet. Penguin Press, London.

Sepkoski, J.J., 1992. Phylogenetic and ecologic patterns in the phanerozoic history of marine biodiversity. In: Systematics, Ecology, and the Biodiversity Crisis.

Vitousek, P.M., Ehrlich, P.R., Ehlrich, A.H., Matson, P.A., 1986. Human appropriation of the products of photosynthesis. BioScience 36, 368–373.

Chapter 10

Chronic Environmental Diseases: Burdens, Causes, and Response

Kirsten M.M. Beyer
Division of Epidemiology, Institute for Health and Society, Medical College of Wisconsin, Milwaukee, WI, USA

ABSTRACT
Chronic diseases have been defined as diseases of long duration and slow progression, including well-known conditions such as heart disease, cancer, and diabetes. Environmental causes of chronic disease are increasingly implicated, as elements of the built and social environments are generally recognized, along with environmental contaminants in air, soil, and water, as important influences on chronic disease development. In this chapter, we consider the burdens, causes, and responses to major chronic environmental diseases. We begin with a review of recent epidemiological evidence for several categories of disease that place a heavy burden on the global population and are known to be influenced by environmental causes. We consider briefly the important ideas of inequality and disparity, which affect populations and diseases at a variety of spatial scales. We then turn to consider the mechanisms linking environments to chronic diseases, including exposure to contaminants, health behaviors, health care access issues, and stress. We conclude by discussing intervention and preparedness strategies that have been, or could be, employed to reduce burdens of chronic environmental diseases worldwide. Throughout the chapter, we present brief sidebars that highlight case studies or important processes complementary to the main text.

10.1 WHAT IS A CHRONIC ENVIRONMENTAL DISEASE?

Chronic diseases have been defined as diseases of long duration and slow progression, including well-known conditions such as heart disease, cancer, and diabetes (World Health Organization, 2013). Environmental causes of chronic disease are increasingly implicated, as elements of the built and social environments are generally recognized, along with environmental contaminants in air, soil, and water, as important influences on chronic disease development.

Biological and Environmental Hazards, Risks, and Disasters. http://dx.doi.org/10.1016/B978-0-12-394847-2.00013-9

Here, we define chronic environmental diseases as chronic diseases for which a recognized environmental component of risk exists.

Several types of environments have been implicated in causing and perpetuating chronic diseases. Traditionally, environmental health focused primarily on contaminants or toxicants present in the natural environment (air, water, soil), and their potential to effect ill health. More recently, scholars have begun to recognize the importance of the social and built environments in affecting disease and injury risk, as well as access to health care, health care-seeking behavior, quality of life, long-term survival, and equity. Elements of this larger conception of environment have now been empirically linked with a wide range of chronic diseases, and efforts to intervene to reduce burdens of these diseases increasingly address or focus on environmental interventions.

As recently as 2010, the Global Burden of Disease study identified high blood pressure, tobacco smoking, and second-hand smoke, and household air pollution from solid fuels as the three leading risk factors contributing to the global burden of disease (Lim et al., 2013). All three factors have significant environmental components and can be minimized by environmental interventions. In addition, major chronic diseases that place a heavy burden on the global population count among their causes, health behaviors (diet, physical activity), and exposures (air pollution, urban green space) that are intimately linked with elements of the social and built environments, and require new, macrolevel approaches to effect change.

In this chapter, we consider the burdens, causes, and responses to major chronic environmental diseases. We begin with a review of recent epidemiological evidence for several categories of disease that place a heavy burden on the global population and are known to be influenced by environmental causes. We consider briefly the important idea of inequality and disparity, which affect populations and diseases at a variety of spatial scales. We then turn to consider the mechanisms linking environments to chronic diseases, including behaviors, issues of health care access, and stress. We conclude by discussing intervention and preparedness strategies that have been, or could be, employed to reduce burdens of chronic environmental diseases worldwide. Throughout the chapter, we present brief sidebars that highlight case studies or important processes complementary to the main text.

10.2 THE GLOBAL BURDEN OF CHRONIC ENVIRONMENTAL DISEASES

Chronic diseases are responsible for 63% of global mortality, (World Health Organization, 2013) and most of these diseases have an environmental component of risk. According to the Global Burden of Disease Study (2010)—an effort that brings together nearly 500 researchers from more than 300 institutions in 50 countries to quantify the comparative magnitude of health loss due to diseases, injuries, and risk factors by age, sex, and geography worldwide (Murray et al., 2012)—an epidemiologic transition is underway, whereby the burden of disease is shifting from communicable, maternal neonatal, and nutritional causes toward noncommunicable diseases, including chronic illnesses such as cancer, cardiovascular diseases (CVDs), chronic obstructive pulmonary disease (COPD), and diabetes (Lozano et al., 2013). This epidemiologic transition is due to a number of factors including an aging global population, and reductions in the burdens of some types of diseases, such as diarrheal diseases, which largely affect younger populations (Lozano et al., 2013).

The impact of this shift can be detected in examining several types of disease burdens, including measures of death, disability, and years of life lost. The top 20 causes of global death and disability, as measured by the Global Burden of Disease Study, are shown in Table 10.1 (Murray et al., 2013).

TABLE 10.1 Leading Causes of Global Death and Disability, 2010

	Leading Causes of Death	Leading Causes of Disability
1	Ischemic heart disease	Ischemic heart disease
2	Stroke	Lower respiratory infections
3	Chronic obstructive pulmonary disease	Stroke
4	Lower respiratory infections	Diarrheal disease
5	Lung cancer	HIV/AIDS
6	HIV/AIDS	Low back pain
7	Diarrheal disease	Malaria
8	Road injury	Preterm birth complications
9	Diabetes	Chronic obstructive pulmonary disease
10	Tuberculosis	Road injury
11	Malaria	Major depressive disorder
12	Cirrhosis	Neonatal encephalopathy
13	Self-harm	Tuberculosis
14	Hypertensive heart disease	Diabetes
15	Preterm birth complications	Iron-deficiency anemia
16	Liver cancer	Neonatal sepsis
17	Stomach cancer	Congenital anomalies
18	Chronic kidney disease	Self-harm
19	Colorectal cancer	Falls
20	Other cardiovascular and circulatory diseases	Protein-energy malnutrition

Chronic diseases contribute significantly to death and disability globally. There have been some important changes in these rankings since the 1990 disease burden study. Of particular note, HIV/AIDS has risen from a rank of 33 to a rank of 5, whereas protein energy malnutrition, measles, and meningitis have all fallen significantly in the rankings. Ischemic heart disease is now ranked the number one disease burden, followed by lower respiratory infections, stroke, and diarrhea. Diabetes has climbed from a rank of 21 to a rank of 14. In this section, we highlight some of the major chronic diseases for which a significant environmental role has been established. We begin with a consideration of the major causes of global death and disability.

10.2.1 Cardiovascular Diseases

CVDs, diseases of the heart and blood vessels, are the number one cause of death globally (World Health Organization, 2013b). CVDs include diseases of the blood vessels supplying the heart (coronary heart disease), brain (cerebrovascular disease), and extremities (peripheral artery disease), in addition to diseases directly affecting the heart (rheumatic heart disease, congenital heart disease), and diseases involving blood clots in the veins (thrombosis, embolism) (World Health Organization, 2013b). Ischemic heart disease (also called coronary heart disease) is ranked as the number one cause of global mortality (Lozano et al., 2013) and disability (Murray et al., 2013). Although some individuals with CVDs are identified before a culminating event, others become aware of the disease when affected by a heart attack or stroke, caused by restricted blood flow to the heart (heart attack) or brain (stroke). Stroke is currently the second-leading cause of death globally (Lozano et al., 2013). Environmental factors are known to affect CVDs, primarily through their linkages with behavioral factors such as physical activity, healthy diet, and tobacco use, and exposure to tobacco smoke. As the number one cause of death and disability globally, CVDs demand attention and intervention, including efforts to modify environments to increase risk and encourage healthy behaviors.

10.2.2 Cancer

Cancer is a leading cause of death globally and the burden continues to rise (Jemal et al., 2011). Cancers are commonly considered independently when measuring their global burden; lung, liver, stomach, and colorectal cancers were all top causes of mortality in 2010 (see Table 10.1; Lozano et al., 2013). Breast and cervical cancer also contribute significantly to the global burden of cancer (World Health Organization, 2013a). Cancers are generally characterized by the rapid proliferation of abnormal cells in the body, resulting in tumors. These tumors can spread to other organ systems and parts of the body, in a process called metastasis. The severity of a cancer is generally measured based on the tumor's size, location, and the degree to which it has spread.

For many cancers, early stage diagnosis can greatly improve survival. Cancer etiology is complicated, but numerous environmental causes of cancer exist, including exposures to biological agents such as human papillomavirus (Bouvard et al., 2009), chemical agents such as formaldehyde and benzene (Baan et al., 2009), radiation (El Ghissassi et al., 2009), hormonal treatments such as estrogen replacement therapies (Grosse et al., 2009), and other environmental contaminants, including smoke from tobacco and coal (Secretan et al., 2009), heavy metals such as chromium and cadmium (Straif et al., 2009), and asbestos (Straif et al., 2009). In addition, factors important for other chronic diseases, including nutrition and physical activity, are known to be associated with cancer risk, as well as survival. A growing body of literature examines the associations between social and built-environmental factors and cancer outcomes, including incidence, late-stage diagnosis, and mortality, implicating factors such as neighborhood poverty, stress, and spatial access to care in delaying cancer diagnoses, and limiting treatment and survival (Henry et al., 2009; Lian et al., 2008; Pruitt et al., 2009; Schootman et al., 2009, 2006).

The Rising Toll of Cancer in Africa

For years, medical and public health efforts in Africa have focused on diseases attributed to infectious agents, malnutrition, or contamination of food and water. However, cancer is a commonly diagnosed disease in Africa with significant variation across the continent (Parkin et al., 2008) as illustrated in Figure 10.1. In addition, as some parts of the continent experience economic development, diseases associated with unhealthy lifestyle factors such as poor diet, sedentary lifestyles, and tobacco smoking are becoming more of a concern. Generally speaking, the cancer burden in Africa looks quite different from that which is observed in highly developed nations. In fact, many top cancers found in Africa are linked to infectious agents (e.g., human papilloma viruses and cervical cancer), with the most prominent example being the increase in Kaposi's sarcoma associated with the HIV/AIDS epidemic (Parkin et al., 2008). Kaposi's sarcoma is now the most common cancer in males and the third most common in females (Parkin et al., 2008). Although it is known that cancer presents an important and growing burden and environmental factors contribute to risk, limitations in surveillance, cancer registration, and research capacity result in an insufficient knowledge base for the development of cancer prevention and control strategies; more empirical evidence is needed (Parkin et al., 2008; Sitas et al., 2008).

10.2.3 Diabetes

An estimated 347 million or more people globally have diabetes (Danaei et al., 2011; Whiting et al., 2011), and that number is expected to rise to 552 million people by 2030 (Whiting et al., 2011). Diabetes is an illness related to the control of blood sugar by the hormone insulin. When the pancreas does not

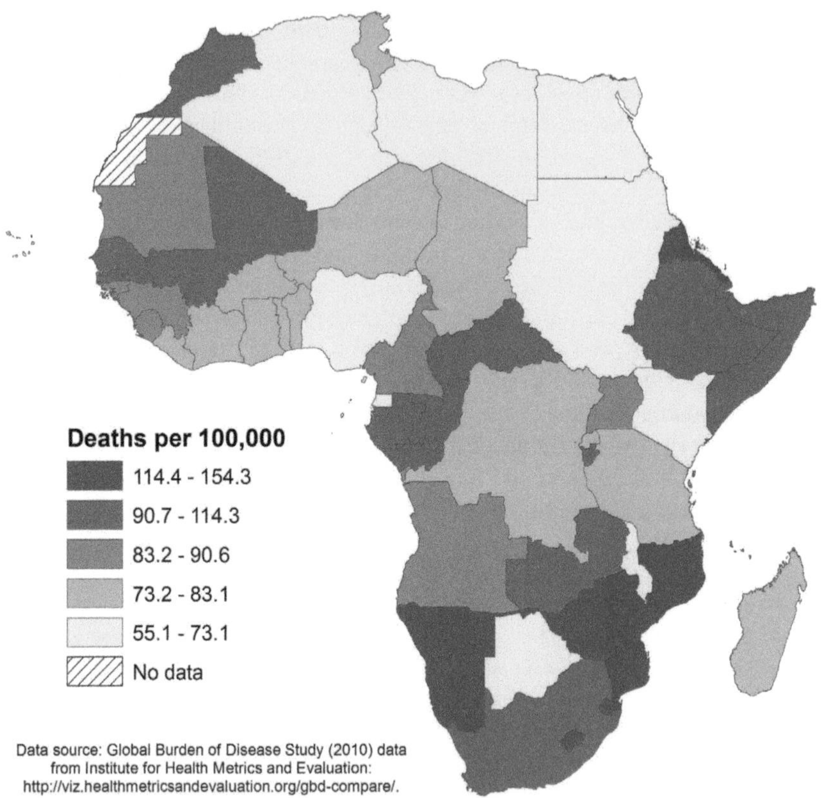

FIGURE 10.1 Age-adjusted cancer mortality rates in Africa by country (2010).

produce enough insulin or the body has difficulty using the insulin to regulate blood sugar, a condition known as hyperglycemia (high blood sugar) can result, ultimately damaging many of the body's organ systems if left uncontrolled. Individuals with diabetes must engage in significant self-management of the disease, including restricting diet, measuring blood sugar, and administering insulin. Two main types of diabetes occur. Type 1 diabetes is characterized by the body's deficient insulin production, requires daily insulin intake, and is not amenable to prevention. However, Type 2 diabetes—characterized by the body's ineffective use of insulin—is thought to be largely preventable through diet and nutrition, physical activity, and maintenance of a healthy weight. Type 2 diabetes was for many years referred to as "adult-onset diabetes"; however, increasing numbers of younger people, including children, have been diagnosed with Type 2 diabetes—leading to the declaration of an "epidemic"(Amos et al., 1997; King et al., 1998; Wild et al., 2004) and calls for more efforts to prevent and delay onset of the disease, including a focus on environmental causes of the disease (Zimmet et al., 2001). Diabetes has risen in the ranks and is now considered the 9th leading global cause of death (Lozano et al., 2013) and the 11th leading global cause of disability (Murray. et al., 2013).

10.2.4 Chronic Obstructive Pulmonary Disease

COPD is a disease of the lungs, characterized by a chronic blockage of air flow through the lungs, interfering with normal breathing. Symptoms of COPD include breathlessness and a chronic cough, and due to breathing difficulties, COPD can interfere with many functions of daily living. COPD is caused predominantly by smoking or inhaling second-hand tobacco smoke, with additional risks due to indoor and outdoor air pollution and some occupational exposures, thus ensuring its place as a chronic environmental disease. According to the World Health Organization, deaths from COPD are projected to increase over 30% in the next 10 years, emphasizing the need for urgent action to reduce tobacco use and minimize other exposures. From 1990 to 2010, COPD moved from the 4th to the 3rd leading cause of death globally (Lozano et al., 2013) and now affects over 65 million people worldwide, although it is known to be underdiagnosed (World Health Organization, 2013c). Other respiratory conditions such as asthma and emphysema also affect large numbers of people and have similar environmental causes.

10.2.5 Mental Illness

Major depressive disorder is the 11th leading cause of global disability (Murray et al., 2013), and self-harm is the 13th leading cause of global mortality (Lozanoi et al., 2013). Links between mental illness and self-harm, including suicide, are well established (World Health Organization, 2013d). Despite the burden of mental illness and its recognition as a top cause of death and disability globally, including close linkages with alcohol and substance abuse (Prince et al., 2007), mental illnesses are often stigmatized (Sartorius, 2007) and, in some countries, continue to struggle to be recognized as equally important as other types of disease, particularly through mental health "parity" legislation (Jeste, 2012; US Department of Labor (2010)). Dr Margaret Chan, the WHO Director-General, in October of 2013 released a Mental Health Action Plan 2013–2020 to focus global attention on issues of mental health and associated problems of stigma and discrimination; the plan was adopted in May 2013 by the World Health Assembly (World Health Organization, 2013e). The plan recognizes the importance of environmental factors in contributing to and perpetuating mental illnesses, and acknowledges the links between mental illness and other diseases such as cancer and CVDs (World Health Organization, 2013e). Although more attention has been paid to environmental characteristics that may exacerbate burdens of mental illness, some recent work has considered aspects of the environment that might benefit mental health. A relatively recent body of work examines the mental health benefits of access to green space and nature, particularly in urban environments (Alcock et al., 2013; Beyer et al., 2014; Maas et al., 2009; van den Berg et al., 2010). Work in this area has also found an added benefit for physical activity in natural areas (Barton and Pretty, 2010; Pretty et al., 2005), above and beyond the benefit to mental health from physical activity, which is well established (Penedo and Dahn, 2005).

10.2.6 HIV

Developments in treatments for HIV/AIDS—particularly antiretroviral therapies that suppress the HIV virus to limit progression of the disease—have led some to question whether the end of AIDS is possible, and whether living with HIV infection might be considered as a chronic disease (Deeks et al., 2013). Life expectancies for those diagnosed with HIV have increased significantly in recent years; one study determined that the life expectancy for a person diagnosed with HIV, particularly if they are diagnosed early before the need for antiretroviral therapies are indicated, now rivals the life expectancy for an individual not infected with HIV (Nakagawa et al., 2013). HIV has been found to be associated with an increased risk for CVDs (Triant, 2013), as well as psychosocial sequelae (Battles and Wiener, 2002), indicating the importance of considering the co-occurrence, and causal relationships, among various chronic conditions. Although HIV/AIDS is largely considered to be an infectious disease resulting primarily from human behavior, a growing evidence base implicates the role for neighborhood environment and place of residence in affecting sexual risk behaviors (Burns and Snow, 2012; Frye et al., 2010) and HIV infection (Maas et al., 2007), suggesting an important role for environmental conditions in affecting risk.

Violence: A Chronic Environmental Disease?

It has taken a long time for violence to be widely recognized as a public health issue (Krug et al., 2002; Mercy et al., 1993). Given evidence supporting environmental causes of violence (Beyer et al., 2013; Caetano et al., 2010; Cunradi et al., 2011; Frye et al., 2008; Li et al., 2010), trans-generational transmission of violence (Cordero et al., 2012; Gómez, 2011), and chronic suffering and health consequences often associated with domestic and intimate partner violence, child abuse and exposure to violence (Gass et al., 2010; Wuest et al., 2010), violence—like the evolving burden of HIV infection—challenges the definition of a chronic environmental disease. One might also ask, given recent work illustrating the contribution of violence and crime to other diseases, whether violence in fact may contribute most through shaping unhealthy environments within which other conditions, such as chronic stress, develop.

10.2.7 Health Inequalities and Vulnerable Populations

Of the 36 million people who died from chronic disease in 2008, 9 million were under age 60 years and 90% of these premature deaths occurred in low- and middle-income countries (World Health Organization, 2013). Globally, 80% of cardiovascular deaths occur in low- and middle-income countries (World Health Organization, 2013b). Individuals in low- and middle-income countries are more exposed to risk factors such as tobacco, which is more heavily regulated in higher income countries, as well as having less access to programs to combat the effect of these risk factors, such as tobacco cessation, and to health care resources to enable management and treatment of disease (World Health Organization, 2013b). A similar pattern is observable for other chronic diseases. Most of the individuals with diabetes live in low- and middle-income countries, which are expected to see the biggest increases in number with diabetes in the years to come (Whiting et al., 2011). In addition, 90% of all COPD deaths occur in low- and middle-income countries (World Health Organization, 2013c). Linkages between global wealth and disease and disability cannot be ignored, and these patterns of health disparity linked to wealth are observable at national, regional, and local scales as well. Cumulatively, it is clear that reducing the global burden of chronic environmental diseases will require a concerted effort to address inequalities.

10.3 CAUSES OF CHRONIC ENVIRONMENTAL DISEASES

Although some risk for chronic disease is hereditary, research has shown that a large proportion of risk can be attributed to other factors, such as behaviors and environmental exposures. Important behavioral risk factors for chronic diseases include tobacco use, nutrition, physical activity, and alcohol abuse—all of which are known to have important relationships with environmental conditions including wealth, access to built environments conducive to activity, enactment of tobacco policies, and spatial access to healthy foods. In addition, exposure to conditions of poverty, chronic stress, and environmental contaminants, as well as underexposure to salutogenic environmental features and health care, can increase chronic disease risk or complicate management and control of chronic conditions. As such, a growing body of research examines the contributions of (primarily residential) environments to chronic disease risk. (Diez Roux, 2003, 2009; Diez-Roux et al., 1999, 2000, 2001; Nash et al., 2011) In the sections below, we examine the ways in which environmental factors can contribute to chronic diseases.

10.3.1 Environment and Disadvantage

As discussed above, the burden of chronic disease falls heavily on lower and middle-income countries. At regional, national, and municipal levels, similar patterns of spatial disparity can be observed, with spatial patterns of disease commonly linked to spatial patterns of wealth and poverty. Globally, an understanding now exists that low-socioeconomic status is linked with a wide range of disease and injury outcomes, and that disparities based on socioeconomic disadvantage are, in some cases, widening (Marmot et al., 1997; Marmot, 2005; Marmot and McDowall, 1986; Orsi et al., 2010; Singh et al., 2013). Socioeconomic disadvantage can lead to health disparities and increased burdens of chronic disease specifically, through a number of pathways, including those discussed below—behaviors, exposure to chronic stress, reduced health care access, and increased exposure to environmental contaminants.

In some regions of the world, patterns of socioeconomic disadvantage are linked with other demographic characteristics—namely, constructs of race, ethnicity, and class. Segregation by race, ethnicity, and other demographic factors is an important problem in many nations, with perhaps the most extreme example occurring in the United States. Work has begun to focus specifically on the nature of residential racial segregation and the neighborhood environmental characteristics associated with segregation patterns, following Williams and Collins' (2001) notion that segregation creates "distinctive ecological environments for African-Americans" with significant health consequences (Collins and Williams, 1999; Williams and Collins, 2001). Addressing the potential impact of residential racial segregation for chronic disease prevention and control requires additional research that disentangles components of the "socioenvironmental milieu" (Russell et al., 2011) that may ultimately contribute to racial and ethnic cancer disparities, so that those components can be targeted by interventions to improve outcomes.

10.3.2 Environment and Behavior

Important behavioral risk factors for many chronic diseases include physical activity, diet/nutrition, health care-seeking behaviors, and alcohol and tobacco use. Poor diet and lack of physical activity can manifest as metabolic syndrome, overweight and obesity, and conditions such as high blood pressure (hypertension). Poor diet and physical inactivity were responsible for 10% of disability life years lost in 2010 (Lim et al., 2013), largely due to their impact on chronic disease burdens worldwide.

A rapidly growing body of work has examined the important role for built-environmental factors in affecting physical activity (Brown et al., 2013a; Jackson, 2003; Sallis et al., 2012). Studies have found links between built-environment infrastructure—such as street network connectivity or walkability, availability of parks and open space, and availability of amenities such as bike paths—and levels of physical activity (Humpel et al., 2002). Studies have more recently begun to emphasize important complexities of measurement (Brownson et al., 2009) and uncertainty related to the problem of neighborhood selection bias or self-selection in establishing causality (Frank et al., 2007; Handy et al., 2006), particularly with regard to drawing conclusions through analyses of observational data. A smaller number of studies have measured the impact of built-environment interventions including installation and/or modification of environmental features such as bike paths, trails, and playgrounds (Gustat et al., 2012). Some recent research has also considered the role for physical activity in different environments, with particular emphasis on types of outdoor environments (Aspinall et al., 2013).

A large body of work now describes the problem of "food deserts," and food insecurity more broadly, emphasizing the important impacts of spatial food access on community capacity to obtain and consume foods necessary to maintain a healthy diet and weight. A recent review of 49 studies in 5 countries found clear links between demographic characteristics such as income and race with disparities in food access, emphasizing that the evidence was strongest for the United States (Beaulac et al., 2009). Some research has examined linkages among food access and food shopping, eating behaviors, and weight (Cummins et al., 2005; Jennings et al., 2011; Skidmore et al., 2010), although more research is needed in this area to clearly establish links among food environments, behaviors, and outcomes.

Health care-seeking behaviors—such as going to a physician regularly and getting recommended screening tests—are important in both preventing chronic diseases from developing and identifying chronic diseases early, when they can be controlled and/or treated. Evidence has demonstrated that health-seeking behaviors are not attributable only to individuals, but have roots in the environmental context within which decisions are made. Cancer-screening behavior, for example, has been linked to aspects of neighborhood, employment, and social environments; the decision to seek care also must be seen as one that could be a decision to forgo some other benefit, such as additional income, and that competes with a long list of additional priorities that must be considered (Beyer et al., 2011b; Farley et al., 2002; Pruitt et al., 2009; Salant and Gehlert, 2008; Ward et al., 2004; Weitzman et al., 2001). Particularly within environments of socioeconomic disadvantage, these threats of financial loss or competing priorities can be very difficult to overcome.

Alcohol, tobacco, and drug-use behaviors have also been linked with environmental factors, with work emphasizing environmental exposure to opportunities (alcohol outlets, tobacco sales) as well as perceptions of neighborhood environments (Brown et al., 2013b; Datta et al., 2006; Furr-Holden et al., 2010; Milam et al., 2012; Oman et al., 2013; Reitzel et al., 2012; Tanjasiri et al., 2013). Substance-use behaviors (initiating substance use and/or failing to cease substance use) are known to be used as coping mechanisms to deal with stressors (Lopez et al., 2011; Siahpush et al., 2009)—including those generated by environmental conditions. Some work has shown that interventions to reduce behaviors such as tobacco smoking may benefit from multilevel structures that target individual decision making and behavior within the environmental context (Engbers et al., 2005; Frieden, 2010; Sorensen et al., 2002).

10.3.3 Environment and Stress

A growing body of work demonstrates an important role for the human-stress response, and its connections to socioeconomic status and environments, in contributing to chronic disease (Baum et al., 1999; Brody et al., 2013; Gruenewald et al., 2012; Jackson et al., 2010; McEwen and Gianaros, 2010; Schulz et al., 2012; Wallace et al., 2013). Stress and disruption of hormone levels related to exposure to chronic stress, are linked to a wide range of disease states, including heart disease, hypertension, diabetes, cognitive function, and depression, as well as conditions such as obesity (Juster et al., 2010; McEwen and Gianaros, 2010; Schulz et al., 2012). Limiting the body's exposure to stressful situations thus becomes an important way to reduce chronic disease.

Scholars increasingly recognize the important role that neighborhood environments can play in human stress levels. Environments characterized by low-socioeconomic status are commonly those characterized by an overabundance of stressors. Long-term habitation in these environments can thus lead to chronic stress and the development of numerous disease states. In addition, stress can make it more difficult to manage conditions already acquired, and can contribute to the worsening of conditions (Schootman et al., 2009). A new body of research has begun to examine correlations between neighborhood environments and allostatic load, finding empirical evidence of higher burdens of physiological stress for residents of lower income environments (Karb et al., 2012; Schulz et al., 2012). Some very recent work has found correlations among levels of green space (e.g., trees) and cortisol levels (Roe et al., 2013; Ward Thompson et al., 2012), suggesting a potential role for urban planning approaches such as neighborhood greening in reducing the physiological impacts of stressful environments.

The Human Stress Response

The human stress response, or "fight or flight" response, primarily involves the hypothalamic-pituitary-adrenal (HPA) axis, as shown in Figure 10.2. The hypothalamus—sometimes referred to as the "command center" of the brain—manages the body's response to stressful stimuli via the autonomic nervous system (the system responsible for involuntary body functions such as blood pressure). When a stressor is identified, the hypothalamus releases a hormone called corticotropin-releasing hormone (CRH), which signals to the pituitary gland to release an additional hormone—adrenocorticotropic hormone (ACTH). ACTH signals to the adrenal glands to release hormones including cortisol and adrenaline. Cortisol is characteristically referred to as the "stress hormone," and increases levels of sugar in the blood, as well as limiting body functions not essential during a threatening situation. Once the stressful situation has passed, a negative feedback loop should result in the body's hormone levels returning to normal, or allostasis.

The Human Stress Response—cont'd

However, when the body perceives threats more consistently—such as the experience one might have in living in a high crime neighborhood when shots fired more commonly occur (Browning et al., 2013)—allostasis is difficult to achieve. Instead, the fight or flight response—including the presence of cortisol in the blood stream—stays turned on. Overexposure of cortisol and other stress hormones can disrupt normal bodily functions and can lead to the development of chronic diseases. This "wear and tear" is often referred to as allostatic load, whereby allostasis—similar to homeostasis—is not achieved (Brody et al., 2013; Gruenewald et al., 2012; Schulz et al., 2012; Wallace et al., 2013).

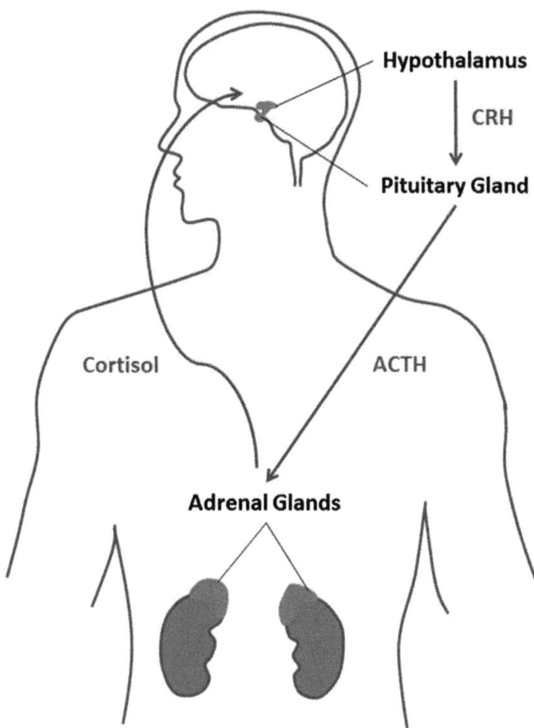

FIGURE 10.2 Hypothalamic pituitary adrenal axis.

10.3.4 Environment and Access to Care

A large body of evidence supports the notion that spatial or environmental access to health care affects care utilization. Spatial inaccessibility (long distance or time traveled to reach a service) has been found to be associated with reductions in regular physical exams, chronic disease care (for conditions such as diabetes, cancer, and heart disease), and number of health care visits in a given time period (Arcury et al., 2005; Nemet and Bailey, 2000). Distance has also been found to influence the type of care individuals receive—for instance, whether a woman chooses breast-conserving surgery for early stage breast cancer (Celaya et al., 2006; Nattinger et al., 2001; Schroen et al., 2005). Recent work has examined health care access at the neighborhood level, finding deficits in health care access for some regions as compared to others, and have begun to more carefully consider the interplay between spatial and aspatial aspects of accessibility (Bell et al., 2013; Bissonnette et al., 2012). Methods for measuring spatial and environmental aspects of health care accessibility have developed in recent years, considering impacts of different georeferencing, distance estimation, and catchment area methods on measures of access (Bell et al., 2012; Beyer et al., 2011a; Delamater, 2013; McLafferty, 2003; Rushton et al., 2006).

10.3.5 Environment and Contaminants

Exposure to contaminants in both outdoor and indoor environments is linked to chronic disease. In particular, environmental exposure to contaminants is linked with development of numerous cancers. In an October 17, 2013 news release, the International Agency for Research on Cancer (the agency of the World Health Organization focused on cancer) declared that outdoor air pollution is a human carcinogen (International Agency for Research on Cancer, 2013). Upon reviewing available evidence, the group concluded that sufficient evidence exists to say that outdoor air pollution causes lung cancer, and is positively associated with bladder cancer. In addition, they evaluated particulate matter—particle pollution that can be easily inhaled by humans—and classified it as carcinogenic to humans. These pollution sources are also implicated as causes of COPD and other respiratory ailments. Significant outdoor sources of air pollution, including particulate matter, include motor vehicle emissions and industrial effluent. In places where environmental regulations are not well developed, the release of air pollutants can pose significant population health danger. This is particularly troublesome in rapidly industrializing nations such as China (Zhang et al., 2010). In addition, given the recent increase in interest in community gardening as a way to overcome problems of food insecurity and inaccessibility in urban centers (Carney et al., 2012; Wakefield et al., 2007), it is worthwhile to note a significant literature on the subject of possible uptake of contaminants such as lead (Pb) in plants grown in urban soils (Clark et al., 2006; Davies, 1978; Mielke et al., 1983; Sangster et al., 2012; Spittler and Feder, 1979).

Indoor environments can also contribute to chronic disease risk—particularly to respiratory diseases and some cancers. In particular, a global need exists to limit inhalation of indoor smoke from the burning of fossil fuels and tobacco smoking, which contribute significantly to respiratory ailments, particularly in low- to middle-income countries. Other indoor exposures include molds, asbestos, lead from paint and radon, among others. Finally, it is well established that geographies of contamination are commonly interwoven with geographies of justice; an environmental-justice framework questions the equitable distribution of benefits and risks associated with the spatial distribution of pollution sources (Gilbert and Chakraborty, 2011; Miranda et al., 2011; Su et al., 2010).

10.4 PUBLIC HEALTH RESPONSE

10.4.1 Public Health Models to Guide Chronic Environmental Disease Response

The public health approach to addressing burdens of disease is commonly displayed in a simplified form, summarized in Figure 10.3. The public health process begins with identifying and defining health problems. Generally this is achieved through monitoring or surveillance of health-related datasets, including data from hospitals and clinics, population surveys, and administrative records from entities such as pharmacies and public health departments. On the global scale, the World Health Organization and other nongovernmental agencies play a large role in monitoring public health data and defining problems to make them actionable. An example relevant to the current chapter is the 2005 release of the World Health Organization report *Preventing chronic diseases: a vital investment* (Abegunde and Vita-Finzi, 2005). In this report, the authors follow a public-health process model, first outlining the burden of chronic diseases, then providing an overview of risk and protective factors, then identifying evidence-based interventions to prevent and control chronic diseases; they conclude with a call for action, outlining "essential steps for success" in addressing the global burden of chronic disease (Abegunde and Vita-Finzi, 2005). Reports such as these are meant to set the stage for action, and are commonly provided in multiple languages, accompanied by media kits to catalyze distribution, and summarized in fact sheets useful for communicating with policy makers and other stakeholders with the influence to effect change. It is well recognized that the public-health process is not immune

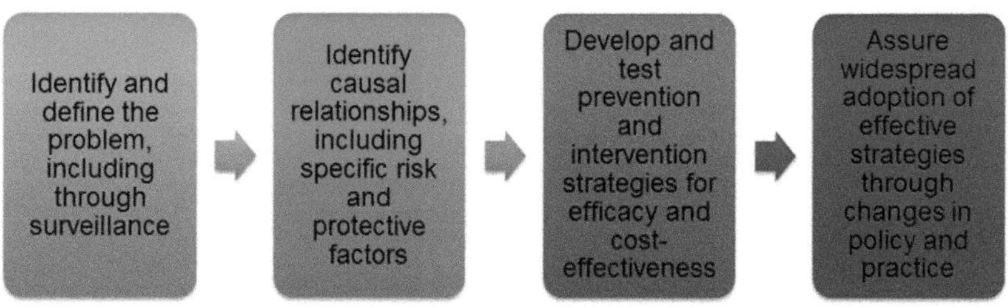

FIGURE 10.3 The public-health process model.

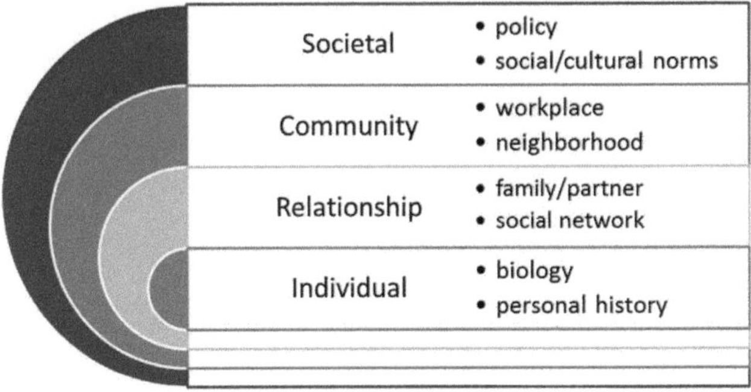

FIGURE 10.4 The social–ecological model of disease.

from political constraints, and that advocates for public-health policy and program change are commonly in competition with advocates for other issues to raise the profile and priority of their relative causes (Geneau et al., 2010). As goes policy change in other arenas, generally public-health policy change proceeds in an incremental or stepwise fashion, with gradual improvements taking place over time (Epping-Jordan et al., 2005).

Likely candidates for focused prevention and intervention strategies targeted at chronic disease prevention and control in the years to come include those focusing on diet, physical activity, pollution, and tobacco control. It is increasingly recognized that effecting healthy behavior change such as encouraging individuals to be physically active, eat healthy foods, and refrain from smoking means more than targeting those individuals. Instead, public health programs and policies should emphasize creating environments conducive to healthy decision-making, including making better choices available and affordable (Frieden, 2010; World Health Organization, 2013b). Similarly, interventions to reduce exposure to indoor air pollutants resulting from the burning of fossil fuels include not only the encouragement of reduced burning, but also environmental modifications such as chimneys, smoke hoods, eaves spaces, and windows, as well as ensuring wood is dried before use to minimize smoke (World Health Organization, 2013f). These multilevel strategies draw from the ideas of the social–ecological model of disease (see Figure 10.4), which emphasizes multilevel contributors to disease risk. Commonly, a social–ecological framework will include emphasis on individual, family and social network, neighborhood, and larger societal/policy level influences.

10.4.2 Key Targets for Prevention and Control of Chronic Environmental Disease

Two recent international efforts highlight the adoption of a social–ecological or multilevel model of chronic-disease causation—The Global Strategy on

Diet, Physical Activity and Health (Waxman and World Health Assembly, 2004) and the WHO Framework Convention on Tobacco Control (World Health Organization, 2003; World Health Organization, 2013g). Both specify interventions at multiple levels of influence, inherently recognizing the strong environmental contributions to chronic diseases worldwide.

Endorsed in 2004 by the 57th World Health Assembly, the Global Strategy on Diet, Physical Activity and Health is clearly informed by a socioecological or multilevel model of prevention and control of chronic disease, noting as its overall goal to guide "the development of an enabling environment for sustainable actions at individual, community, national and global levels that, when taken together, will lead to reduced disease and death rates related to unhealthy diet and physical inactivity" (Waxman and World Health Assembly, 2004). Among the many interventions specified in the Strategy are those that advocate environmental modifications to make healthy behaviors easier to achieve. Targeting physical activity, the strategy states that "environments should be promoted that facilitate physical activity, and supportive infrastructure should be set up to increase access to, and use of, suitable facilities" and "Schools… should be equipped with appropriate facilities and equipment" to provide students with daily physical activity (Waxman and World Health Assembly, 2004). Targeting healthy nutrition, the strategy advocates for schools to "limit the availability of products high in salt, sugar, and fats" and suggests that schools should consider contracting with local food growers to supply healthy foods for students (Waxman and World Health Assembly, 2004).

In addition, in recognition of the heavy burden caused by tobacco, global leaders have attempted to curtail the use of tobacco products, passing the WHO Framework Convention on Tobacco Control. This treaty seeks to "protect present and future generations from the devastating health, social, environmental, and economic consequences of tobacco consumption and exposure to tobacco smoke" (World Health Organization, 2003) through efforts such as regulation of tobacco pricing, taxing, packaging, and labeling, among other efforts. This treaty was signed by 168 countries before the signature closing date on June 29, 2004, and has (as of November 2013) 177 states who are parties to the convention (World Health Organization, 2013g). While the United States signed the treaty in 2004, it has not ratified the treaty (World Health Organization, 2013g). Advocacy groups have called on American Presidential administrations to send the FCTC to the senate for ratification (Action on Smoking and Health, 2012; American Lung Association, 2013). Some research has shown that tobacco packaging and labeling could be improved in a number of countries to more effectively align with the goals of the FCTC (Awopegba and Cohen, 2013). More recently, on November 12 2012, Parties to the FCTC also adopted the Protocol to Eliminate Illicit Trade in Tobacco Products, which is currently open for signature and ratification by all FCTC parties (World Health Organization, 2013g).

Can Nature Cure our Ills?

At the 2013 Annual Meeting of the American Public Health Association, the following resolution (20137 Nature, health and wellness) was adopted:

"To aid in promoting healthy and active lifestyles, encourages land use decisions that prioritize access to natural areas and green spaces for residents of all ages, abilities and income levels. Calls on public health, medical and other health professionals to raise awareness among patients and the public at-large about the health benefits of spending time in nature and of nature-based play and recreation. Also urges such professionals to form partnerships with relevant stakeholders, such as parks departments, school districts and nature centers. Calls for promoting natural landscaping" (American Public Health Association, 2013).

The adoption of this resolution by a major American public health organization, and recognition of the benefits of exposure to nature more generally, stem from a now well-established literature on the benefits of green space and exposure to nature for a range of conditions including increased levels of physical activity (Bell et al., 2008; Maas et al., 2008; Mitchell and Popham, 2008), stress reduction (Mitchell and Popham, 2008; Roe et al., 2013; van den Berg et al., 2010; Ward Thompson et al., 2012), and reduction of mental fatigue through attention restoration (Berman et al., 2008, 2012; Faber Taylor and Kuo, 2009; Hartig et al., 1991, 2003; Kaplan, 1995). The growing evidence base supporting a health promoting capacity for green space and nature may provide an opportunity to systematically deploy features known to be beneficial to human health into the environment – potentially even targeting vacant or ill-used lands for redevelopment as places to access nature's benefits (Branas, 2013; Branas et al., 2011; Hartig, 2008; Nordh et al., 2009).

10.5 CONCLUSIONS

Despite the promise of the interventions discussed above, other constraints remain, leaving health geographers, social epidemiologists, and others to ask questions such as: Can we achieve health improvement without addressing poverty? Can we reduce health inequalities without considering neighborhood environments? Are multilevel public health improvement strategies cost-effective and feasible? How do we adapt strategies for locations struggling with issues of racism or social class-based discrimination? These questions become more urgent as evidence continues to support the importance of stress in disease causation and perpetuation. Because environments characterized by economic distress, discrimination, and social problems such as racism and segregation are likely environments within which populations experience chronic stress, it is important to consider the complexity and primacy of problems such as un- and under-employment, lack of economic investment, under-education, and wealth disparities in generating disease risk and inequality. In addition, as built-environment interventions are considered, new problems such as gentrification and displacement of populations may arise, requiring forethought and planning to avoid negative consequences of well-intended interventions. Finally, as public health authorities move toward changes affecting the lives of individuals and populations, an emphasis on community engagement and participation is essential to ensure that interventions are likely to be effective, are responsive to community concerns, and are culturally appropriate. Chronic disease prevention and control is a complex and challenging endeavor; environmental interventions have the potential to create environments conducive to individual behavior change, while reducing harmful exposures to contaminants and stressors, and making health promoting environmental features equitably available to ultimately reduce the burden of chronic environmental diseases.

REFERENCES

Abegunde, D., Vita-Finzi, L., 2005. Preventing Chronic Diseases: A Vital Investment.

Action on Smoking and Health, 2012. ASH Calls on President Obama to Immediately Submit FCTC for Ratification, 8 Years after U.S. Signed Agreement, 2013.

Alcock, I., White, M.P., Wheeler, B.W., Fleming, L.E., Depledge, M.H., 2013. Longitudinal effects on mental health of moving to greener and less green urban areas. Environ. Sci. Technol. 48 (2), 1247–1255, 2014.

American Lung Association, 2013. Ratification of the Framework Convention on Tobacco Control (FCTC) Treaty.

American Public Health Association, 2013. American Public Health Association Adopts 17 New Policy Statements at Annual Meeting: New Policy Statements Address Paid Sick Leave, Access to Nature, Solitary Confinement and More.

Amos, A.F., McCarty, D.J., Zimmet, P., 1997. The rising global burden of diabetes and its complications: estimates and projections to the year 2010. Diabet. Med. 14, S7–S85.

Arcury, T.A., Gesler, W.M., Preisser, J.S., Sherman, J., Spencer, J., Perin, J., 2005. The effects of geography and spatial behavior on health care utilization among the residents of a rural region. Health Serv. Res. 40, 135–156.

Aspinall, P., Mavros, P., Coyne, R., Roe, J., 2013. The urban brain: analysing outdoor physical activity with mobile EEG. Br. J. Sports Med. http://bjsm.bmj.com/content/early/2013/03/05/bjsports-2012-091877.long.

Awopegba, A.J., Cohen, J.E., 2013. Country tobacco laws and article 11 of the WHO framework convention on tobacco control: a review of tobacco packaging and labeling regulations of 25 countries. Tob. Induc. Dis. 11, 23.

Baan, R., Grosse, Y., Straif, K., Secretan, B., El Ghissassi, F., Bouvard, V., Benbrahim-Tallaa, L., Guha, N., Freeman, C., Galichet, L., 2009. A review of human carcinogens—part F: chemical agents and related occupations. Lancet Oncol. 10, 1143–1144.

Barton, J., Pretty, J., 2010. What is the best dose of nature and green exercise for improving mental health? A multi-study analysis. Environ. Sci. Technol. 44, 3947–3955.

Battles, H.B., Wiener, L.S., 2002. From adolescence through young adulthood: psychosocial adjustment associated with long-term survival of HIV. J. Adolesc. Health 30, 161–168.

Baum, A., Garofalo, J., Yali, A., 1999. Socioeconomic status and chronic stress: does stress account for SES effects on health? Ann. N.Y. Acad. Sci. 896, 131–144.

Beaulac, J., Kristjansson, E., Cummins, S., 2009. Peer reviewed: a systematic review of food deserts, 1966–2007. Prev. Chronic Dis. 6.

Bell, J.F., Wilson, J.S., Liu, G.C., 2008. Neighborhood greenness and 2-year changes in body mass index of children and youth. Am. J. Prev. Med. 35, 547.

Bell, S., Wilson, K., Bissonnette, L., Shah, T., 2013. Access to primary health care: does neighborhood of residence matter? Ann. Assoc. Am. Geogr. 103, 85–105.

Bell, S., Wilson, K., Shah, T.I., Gersher, S., Elliott, T., 2012. Investigating impacts of positional error on potential health care accessibility. Spat. Spatiotemporal Epidemiol. 3, 17–29.

van den Berg, A.E., Maas, J., Verheij, R.A., Groenewegen, P.P., 2010. Green space as a buffer between stressful life events and health. Soc. Sci. Med. 70, 1203–1210.

Berman, M.G., Jonides, J., Kaplan, S., 2008. The cognitive benefits of interacting with nature. Psychol. Sci. 19, 1207–1212.

Berman, M.G., Kross, E., Krpan, K.M., Askren, M.K., Burson, A., Deldin, P.J., Kaplan, S., Sherdell, L., Gotlib, I.H., Jonides, J., November 2012. Interacting with nature improves cognition and affect for individuals with depression. J. Affect. Disord. 140 (3), 300–305.

Beyer, K.M., Kaltenbach, A., Szabo, A., Bogar, S., Nieto, F.J., Malecki, K.M., 2014. Exposure to neighborhood green space and mental health: evidence from the survey of the health of wisconsin. Int. J. Environ. Res. Public Health 11, 3453–3472.

Beyer, K.M., Layde, P.M., Hamberger, L.K., Laud, P.W., Summer 2013. Characteristics of the residential neighborhood environment differentiate intimate partner femicide in urban versus rural settings. J. Rural Health. 29 (3), 281–293.

Beyer, K.M., Saftlas, A.F., Wallis, A.B., Peek-Asa, C., Rushton, G., 2011a. A probabilistic sampling method (PSM) for estimating geographic distance to health services when only the region of residence is known. Int. J. Health Geogr. 10.

Beyer, K.M.M., Comstock, S., Seagren, R., Rushton, G., 2011b. Explaining place-based colorectal cancer health disparities: evidence from a rural context. Soc. Sci. Med. 72, 373–382.

Bissonnette, L., Wilson, K., Bell, S., Shah, T.I., 2012. Neighbourhoods and potential access to health care: the role of spatial and aspatial factors. Health Place 18, 841–853.

Bouvard, V., Baan, R., Straif, K., Grosse, Y., Secretan, B., Ghissassi, F.E., Benbrahim-Tallaa, L., Guha, N., Freeman, C., Galichet, L., 2009. A review of human carcinogens—Part B: biological agents. Lancet Oncol. 10, 321–322.

Branas, C.C., 2013. Safe and Healthy Places Are Made Not Born.

Branas, C.C., Cheney, R.A., MacDonald, J.M., Tam, V.W., Jackson, T.D., Ten Have, T.R., 2011. A difference-in-differences analysis of health, safety, and greening vacant urban space. Am. J. Epidemiol. 174, 1296–1306.

Brody, G.H., Yu, T., Chen, Y., Kogan, S.M., Evans, G.W., Beach, S.R., Windle, M., Simons, R.L., Gerrard, M., Gibbons, F.X., 2013. Cumulative socioeconomic status risk, allostatic load, and adjustment: a prospective latent profile analysis with contextual and genetic protective factors. Dev. Psychol. 49, 913.

Brown, B.B., Smith, K.R., Hanson, H., Fan, J.X., Kowaleski-Jones, L., Zick, C.D., 2013a. Neighborhood design for walking and biking: physical activity and body mass index. Am. J. Prev. Med. 44, 231–238.

Brown, Q.L., Milam, A.J., Smart, M.J., Johnson, R.M., Linton, S.L., Furr-Holden, C.D.M., Ialongo, N.S., 2013b. Objective and perceived neighborhood characteristics and tobacco use among young adults. Drug Alcohol Depend. 134 (1), 370–375. January 2014.

Browning, C.R., Cagney, K.A., Iveniuk, J., 2013. Crime rates, crime spikes and cardiovascular health in an urban population. In: Anonymous Crime, HIV and Health: Intersections of Criminal Justice and Public Health Concerns. Springer, pp. 187–205.

Brownson, R.C., Hoehner, C.M., Day, K., Forsyth, A., Sallis, J.F., 2009. Measuring the built environment for physical activity: state of the science. Am. J. Prev. Med. 36, S99–S123 e12.

Burns, P.A., Snow, R.C., 2012. The built environment & the impact of neighborhood characteristics on youth sexual risk behavior in cape town, south africa. Health Place 18, 1088–1100.

Caetano, R., Ramisetty-Mikler, S., Harris, T.R., 2010. Neighborhood characteristics as predictors of male to female and female to male partner violence. J. Interpers. Violence 25, 1986–2009.

Carney, P.A., Hamada, J.L., Rdesinski, R., Sprager, L., Nichols, K.R., Liu, B.Y., Pelayo, J., Sanchez, M.A., Shannon, J., 2012. Impact of a community gardening project on vegetable intake, food security and family relationships: a community-based participatory research study. J. Community Health 37, 874–881.

Celaya, M.O., Rees, J.R., Gibson, J.J., Riddle, B.L., Greenberg, E.R., 2006. Travel distance and season of diagnosis affect treatment choices for women with early-stage breast cancer in a predominantly rural population (United States). Cancer Causes Control 17, 851–856.

Clark, H.F., Brabander, D.J., Erdil, R.M., 2006. Sources, sinks, and exposure pathways of lead in urban garden soil. J. Environ. Qual. 35, 2066–2074.

Collins, C.A., Williams, D.R., 1999. Segregation and mortality: the deadly effects of racism? 14, 495–523.

Cordero, M.I., Poirier, G., Marquez, C., Veenit, V., Fontana, X., Salehi, B., Ansermet, F., Sandi, C., 2012. Evidence for biological roots in the transgenerational transmission of intimate partner violence. Transl. Psychiatry 2, e106.

Cummins, S., Petticrew, M., Higgins, C., Findlay, A., Sparks, L., 2005. Large scale food retailing as an intervention for diet and health: quasi-experimental evaluation of a natural experiment. J. Epidemiol. Community Health 59, 1035–1040.

Cunradi, C.B., Mair, C., Ponicki, W., Remer, L., 2011. Alcohol outlets, neighborhood characteristics, and intimate partner violence: ecological analysis of a California city. J. Urban Health 88, 191–200.

Danaei, G., Finucane, M.M., Lu, Y., Singh, G.M., Cowan, M.J., Paciorek, C.J., Lin, J.K., Farzadfar, F., Khang, Y., Stevens, G.A., 2011. National, regional, and global trends in fasting plasma glucose and diabetes prevalence since 1980: systematic analysis of health examination surveys and epidemiological studies with 370 country-years and 2·7 million participants. Lancet 378, 31–40.

Datta, G.D., Subramanian, S.V., Colditz, G.A., Kawachi, I., Palmer, J.R., Rosenberg, L., 2006. Individual, neighborhood, and state-level predictors of smoking among US black women: a multilevel analysis. Soc. Sci. Med. 63, 1034–1044.

Davies, B.E., 1978. Plant-available lead and other metals in british garden soils. Sci. Total Environ. 9, 243–262.

Deeks, S.G., Lewin, S.R., Havlir, D.V., 2013. The end of AIDS: HIV infection as a chronic disease. Lancet 382, 1525–1533.

Delamater, P.L., 2013. Spatial accessibility in suboptimally configured health care systems: a modified two-step floating catchment area (M2SFCA) metric. Health Place 24, 30–43.

Diez-Roux, A.V., 2009. The persistent puzzle of the geographic patterning of cardiovascular disease. Prev. Med. 49, 133–134.

Diez-Roux, A.V., 2003. Residential environments and cardiovascular risk. J. Urban Health 80, 569–589.

Diez-Roux, A.V., Link, B.G., Northridge, M.E., 2000. A multilevel analysis of income inequality and cardiovascular disease risk factors. Soc. Sci. Med. 50, 673–687.

Diez-Roux, A.V., Nieto, F.J., Caulfield, L., Tyroler, H.A., Watson, R.L., Szklo, M., 1999. Neighbourhood differences in diet: the atherosclerosis risk in communities (ARIC) study. J. Epidemiol. Community Health 53, 55–63.

Diez-Roux, A.V., Merkin, S.S., Arnett, D., Chambless, L., Massing, M., Nieto, F.J., Sorlie, P., Szklo, M., Tyroler, H.A., Watson, R.L., 2001. Neighborhood of residence and incidence of coronary heart disease. N. Engl. J. Med. 345, 99–106.

El Ghissassi, F., Baan, R., Straif, K., Grosse, Y., Secretan, B., Bouvard, V., Benbrahim-Tallaa, L., Guha, N., Freeman, C., Galichet, L., 2009. A review of human carcinogens—part D: Radiation. Lancet Oncol. 10, 751–752.

Engbers, L.H., van Poppel, M.N., Chin A Paw, M.J.M., van Mechelen, W., 2005. Worksite health promotion programs with environmental changes: a systematic review. Am. J. Prev. Med. 29, 61–70.

Epping-Jordan, J.E., Galea, G., Tukuitonga, C., Beaglehole, R., 2005. Preventing chronic diseases: taking stepwise action. Lancet 366, 1667–1671.

Faber Taylor, A., Kuo, F.E., 2009. Children with attention deficits concentrate better after walk in the park. J. Atten. Disord. 12, 402–409.

Farley, M., Golding, J.M., Minkoff, J.R., 2002. Is a history of trauma associated with a reduced likelihood of cervical cancer screening? J. Fam. Pract. 51, 827–830.

Frank, L.D., Saelens, B.E., Powell, K.E., Chapman, J.E., 2007. Stepping towards causation: do built environments or neighborhood and travel preferences explain physical activity, driving, and obesity? Soc. Sci. Med. 65, 1898–1914.

Frieden, T.R., 2010. A framework for public health action: the health impact pyramid. Am. J. Public Health 100, 590–595.

Frye, V., Koblin, B., Chin, J., Beard, J., Blaney, S., Halkitis, P., Vlahov, D., Galea, S., 2010. Neighborhood-level correlates of consistent condom use among men who have sex with men: a multi-level analysis. AIDS Behav. 14, 974–985.

Frye, V., Galea, S., Tracy, M., Bucciarelli, A., Putnam, S., Wilt, S., 2008. The role of neighborhood environment and risk of intimate partner femicide in a large urban area. Am. J. Public Health 98, 1473–1479.

Furr-Holden, C., Campbell, K., Milam, A., Smart, M., Ialongo, N., Leaf, P., 2010. Metric properties of the neighborhood inventory for environmental typology (NIfETy): an environmental assessment tool for measuring indicators of violence, alcohol, tobacco, and other drug exposures. Eval. Rev. 34, 159–184.

Gass, J.D., Stein, D.J., Williams, D.R., Seedat, S., 2010. Intimate partner violence, health behaviours, and chronic physical illness among south african women. S. Afr. Med. J. 100, 582–585.

Geneau, R., Stuckler, D., Stachenko, S., McKee, M., Ebrahim, S., Basu, S., Chockalingham, A., Mwatsama, M., Jamal, R., Alwan, A., 2010. Raising the priority of preventing chronic diseases: a political process. Lancet 376, 1689–1698.

Gilbert, A., Chakraborty, J., 2011. Using geographically weighted regression for environmental justice analysis: cumulative cancer risks from air toxics in florida. Soc. Sci. Res. 40, 273–286.

Gómez, A.M., 2011. Testing the cycle of violence hypothesis: child abuse and adolescent dating violence as predictors of intimate partner violence in young adulthood. Youth Soc. 43, 171–192.

Grosse, Y., Baan, R., Straif, K., Secretan, B., El Ghissassi, F., Bouvard, V., Benbrahim-Tallaa, L., Guha, N., Galichet, L., Cogliano, V., 2009. A review of human carcinogens—part A: pharmaceuticals. Lancet Oncol. 10, 13–14.

Gruenewald, T.L., Karlamangla, A.S., Hu, P., Stein-Merkin, S., Crandall, C., Koretz, B., Seeman, T.E., 2012. History of socioeconomic disadvantage and allostatic load in later life. Soc. Sci. Med. 74, 75–83.

Gustat, J., Rice, J., Parker, K.M., Becker, A.B., Farley, T.A., 2012. Effect of changes to the neighborhood built environment on physical activity in a low-income african american neighborhood. Prev. Chronic Dis. 9.

Handy, S., Cao, X., Mokhtarian, P.L., 2006. Self-selection in the relationship between the built environment and walking: empirical evidence from northern california. J. Am. Plan. Assoc. 72, 55–74.

Hartig, T., Evans, G.W., Jamner, L.D., Davis, D.S., Gärling, T., 2003. Tracking restoration in natural and urban field settings. J. Environ. Psychol. 23, 109–123.

Hartig, T., Mang, M., Evans, G.W., 1991. Restorative effects of natural environment experiences. Environ. Behav. 23, 3–26.

Hartig, T., 8 November 2008. Green space, psychological restoration, and health inequality. Lancet. 372 (9650), 1614–1615.

Henry, K.A., Niu, X., Boscoe, F.P., 2009. Geographic disparities in colorectal cancer survival. Int. J. Health Geogr. 8, 48.

Humpel, N., Owen, N., Leslie, E., 2002. Environmental factors associated with adults' participation in physical activity: a review. Am. J. Prev. Med. 22, 188–199.

International Agency for Research on Cancer, 2013. IARC: Outdoor Air Pollution a Leading Environmental Cause of cancer Deaths.

Jackson, J.S., Knight, K.M., Rafferty, J.A., 2010. Race and unhealthy behaviors: chronic stress, the HPA axis, and physical and mental health disparities over the life course. Am. J. Public Health 100, 933–939.

Jackson, R.J., 2003. The impact of the built environment on health: an emerging field. Am. J. Public Health 93, 1382.

Jemal, A., Bray, F., Center, M.M., Ferlay, J., Ward, E., Forman, D., 2011. Global cancer statistics. CA Cancer J. Clin. 61, 69–90.

Jennings, A., Welch, A., Jones, A.P., Harrison, F., Bentham, G., Van Sluijs, E.M., Griffin, S.J., Cassidy, A., 2011. Local food outlets, weight status, and dietary intake: associations in children aged 9–10 years. Am. J. Prev. Med. 40, 405–410.

Jeste, D.V., 2012. Mental health and the 2012 US election. Lancet 380, 1206–1208.

Juster, R., McEwen, B.S., Lupien, S.J., 2010. Allostatic load biomarkers of chronic stress and impact on health and cognition. Neurosci. Biobehav. Rev. 35, 2–16.

Kaplan, S., 1995. The restorative benefits of nature: toward an integrative framework. J. Environ. Psychol. 15, 169–182.

Karb, R.A., Elliott, M.R., Dowd, J.B., Morenoff, J.D., 2012. Neighborhood-level stressors, social support, and diurnal patterns of cortisol: the Chicago community adult health study. Soc. Sci. Med. 75, 1038–1047.

King, H., Aubert, R.E., Herman, W.H., 1998. Global burden of diabetes, 1995–2025: prevalence, numerical estimates, and projections. Diabetes Care 21, 1414–1431.

Krug, E.G., Mercy, J.A., Dahlberg, L.L., Zwi, A.B., 2002. The world report on violence and health. Lancet 360, 1083–1088.

Li, Q., Kirby, R.S., Sigler, R.T., Hwang, S.S., LaGory, M.E., Goldenberg, R.L., 2010. A multilevel analysis of individual, household, and neighborhood correlates of intimate partner violence among low-income pregnant women in Jefferson county, Alabama. Am. J. Public Health 100, 531–539.

Lian, M., Schootman, M., Yun, S., 2008. Geographic variation and effect of area-level poverty rate on colorectal cancer screening. BMC Public Health 8, 358.

Lim, S.S., Vos, T., Flaxman, A.D., Danaei, G., Shibuya, K., Adair-Rohani, H., Amann, M., Anderson, H.R., Andrews, K.G., Aryee, M., 2013. A comparative risk assessment of burden of disease and injury attributable to 67 risk factors and risk factor clusters in 21 regions, 1990–2010: a systematic analysis for the global burden of disease study 2010. Lancet 380, 2224–2260.

Lopez, W.D., Konrath, S.H., Seng, J.S., 2011. Abuse-related post-traumatic stress, coping, and tobacco use in pregnancy. J. Obstet. Gynecol. Neonatal Nurs. 40, 422–431.

Lozano, R., Naghavi, M., Foreman, K., Lim, S., Shibuya, K., Aboyans, V., Abraham, J., Adair, T., Aggarwal, R., Ahn, S.Y., 2013. Global and regional mortality from 235 causes of death for 20 age groups in 1990 and 2010: a systematic analysis for the global burden of disease study 2010. Lancet 380, 2095–2128.

Maas, B., Fairbairn, N., Kerr, T., Li, K., Montaner, J.S., Wood, E., 2007. Neighborhood and HIV infection among IDU: place of residence independently predicts HIV infection among a cohort of injection drug users. Health Place 13, 432–439.

Maas, J., Verheij, R.A., de Vries, S., Spreeuwenberg, P., Schellevis, F.G., Groenewegen, P.P., 2009. Morbidity is related to a green living environment. J. Epidemiol. Community Health 63, 967.

Maas, J., Verheij, R.A., Spreeuwenberg, P., Groenewegen, P.P., 2008. Physical activity as a possible mechanism behind the relationship between green space and health: a multilevel analysis. BMC Public Health 8, 206.

Marmot, M., 2005. Social determinants of health inequalities. Lancet 365, 1099–1104.

Marmot, M.G., McDowall, M.E., 1986. Mortality decline and widening social inequalities. Lancet 328, 274–276.

Marmot, M., Ryff, C.D., Bumpass, L.L., Shipley, M., Marks, N.F., 1997. Social inequalities in health: next questions and converging evidence. Soc. Sci. Med. 44, 901–910.

McEwen, B.S., Gianaros, P.J., 2010. Central role of the brain in stress and adaptation: links to socioeconomic status, health, and disease. Ann. N. Y. Acad. Sci. 1186, 190–222.

McLafferty, S.L., 2003. GIS and health care. Annu. Rev. Public Health 24, 25–42.

Mercy, J.A., Rosenberg, M.L., Powell, K.E., Broome, C.V., Roper, W.L., 1993. Public health policy for preventing violence. Health Aff. 12, 7–29.

Mielke, H.W., Anderson, J.C., Berry, K.J., Mielke, P.W., Chaney, R.L., Leech, M., 1983. Lead concentrations in inner-city soils as a factor in the child lead problem. Am. J. Public Health 73, 1366–1369.

Milam, A., Furr-Holden, C., Cooley-Strickland, M., Bradshaw, C., Leaf, P., 2012. Risk for exposure to alcohol, tobacco, and other drugs on the route to and from school: the role of alcohol outlets. Prev. Sci. 1–10.

Miranda, M.L., Edwards, S.E., Keating, M.H., Paul, C.J., 2011. Making the environmental justice grade: the relative burden of air pollution exposure in the United States. Int. J. Environ. Res. Public Health 8, 1755–1771.

Mitchell, R., Popham, F., 2008. Effect of exposure to natural environment on health inequalities: an observational population study. Lancet 372, 1655–1660.

Murray, C.J., Ezzati, M., Flaxman, A.D., Lim, S., Lozano, R., Michaud, C., Naghavi, M., Salomon, J.A., Shibuya, K., Vos, T., 2012. GBD 2010: design, definitions, and metrics. Lancet 380, 2063–2066.

Murray, C.J., Vos, T., Lozano, R., Naghavi, M., Flaxman, A.D., Michaud, C., Ezzati, M., Shibuya, K., Salomon, J.A., Abdalla, S., 2013. Disability-adjusted life years (DALYs) for 291 diseases and injuries in 21 regions, 1990–2010: a systematic analysis for the global burden of disease study 2010. Lancet 380, 2197–2223.

Nakagawa, F., May, M., Phillips, A., 2013. Life expectancy living with HIV: recent estimates and future implications. Curr. Opin. Infect. Dis. 26, 17–25.

Nash, S.D., Cruickshanks, K.J., Klein, R., Klein, B.E., Nieto, F.J., Ryff, C.D., Krantz, E.M., Shubert, C.R., Nondahl, D.M., Acher, C.W., 2011. Socioeconomic status and subclinical atherosclerosis in older adults. Prev. Med. 52, 208–212.

Nattinger, A.B., Kneusel, R.T., Hoffmann, R.G., Gilligan, M.A., 2001. Relationship of distance from a radiotherapy facility and initial breast cancer treatment. J. Natl. Cancer Inst. 93, 1344–1346.

Nemet, G.F., Bailey, A.J., 2000. Distance and health care utilization among the rural elderly. Soc. Sci. Med. 50, 1197–1208.

Nordh, H., Hartig, T., Hagerhall, C., Fry, G., 2009. Components of small urban parks that predict the possibility for restoration. Urban For. Urban Green. 8, 225–235.

Oman, R.F., Tolma, E.L., Vesely, S.K., Aspy, C.B., 2013. Youth gender differences in alcohol use: a prospective study of multiple youth assets and the neighborhood environment. Open. J. Prev. Med.

Orsi, J.M., Margellos-Anast, H., Whitman, S., 2010. Black-white health disparities in the United States and Chicago: a 15-year progress analysis. Am. J. Public Health 100, 349–356.

Parkin, D.M., Sitas, F., Chirenje, M., Stein, L., Abratt, R., Wabinga, H., 2008. Part I: cancer in indigenous Africans—burden, distribution, and trends. Lancet Oncol. 9, 683–692.

Penedo, F.J., Dahn, J.R., 2005. Exercise and well-being: a review of mental and physical health benefits associated with physical activity. Curr. Opin. Psychiatry 18, 189–193.

Pretty, J., Peacock, J., Sellens, M., Griffin, M., 2005. The mental and physical health outcomes of green exercise. Int. J. Environ. Health Res. 15, 319–337.

Prince, M., Patel, V., Saxena, S., Maj, M., Maselko, J., Phillips, M.R., Rahman, A., 2007. No health without mental health. Lancet 370, 859–877.

Pruitt, S.L., Shim, M.J., Mullen, P.D., Vernon, S.W., Amick III, B.C., 2009. Association of area socioeconomic status and breast, cervical, and colorectal cancer screening: a systematic review. Cancer Epidemiol. Biomarkers Prev. 18, 2579–2599.

Reitzel, L.R., Vidrine, J.I., Businelle, M.S., Kendzor, D.E., Cao, Y., Mazas, C.A., Li, Y., Ahluwalia, J.S., Cinciripini, P.M., Cofta-Woerpel, L., 2012. Neighborhood perceptions are associated with tobacco dependence among African American smokers. Nicotine Tob. Res. 14, 786–793.

Roe, J.J., Thompson, C.W., Aspinall, P.A., Brewer, M.J., Duff, E.I., Miller, D., Mitchell, R., Clow, A., 2013. Green space and stress: evidence from cortisol measures in deprived urban communities. Int. J. Environ. Res. Public Health 10, 4086–4103.

Rushton, G., Armstrong, M.P., Gittler, J., Greene, B.R., Pavlik, C.E., West, M.M., Zimmerman, D.L., 2006. Geocoding in cancer research: a review. Am. J. Prev. Med. 30, S16–S24.

Russell, E., Kramer, M.R., Cooper, H.L., Thompson, W.W., Arriola, K.R.J., 2011. Residential racial composition, spatial access to care, and breast cancer mortality among women in georgia. J. Urban Health 88, 1117–1129.

Salant, T., Gehlert, S., 2008. Collective memory, candidacy, and victimisation: community epidemiologies of breast cancer risk. Sociol. Health Illn. 30, 599–615.

Sallis, J.F., Floyd, M.F., Rodríguez, D.A., Saelens, B.E., 2012. Role of built environments in physical activity, obesity, and cardiovascular disease. Circulation 125, 729–737.

Sangster, J.L., Nelson, A., Bartelt-Hunt, S.L., 2012. The occurrence of lead in soil and vegetables at a community garden in Omaha, Nebraska. Int. J. Serv. Learn. Eng. Humanit. Eng. Soc. Entrepreneursh. 7, 62–68.

Sartorius, N., 2007. Stigma and mental health. Lancet 370, 810–811.

Schootman, M., Jeffe, D.B., Gillanders, W.E., Aft, R., 2009. Racial disparities in the development of breast cancer metastases among older women. Cancer 115, 731–740.

Schootman, M., Jeffe, D.B., Baker, E.A., Walker, M.S., 2006. Effect of area poverty rate on cancer screening across US communities. J. Epidemiol. Community Health 60, 202.

Schroen, A.T., Brenin, D.R., Kelly, M.D., Knaus, W.A., Slingluff, C.L., 2005. Impact of patient distance to radiation therapy on mastectomy use in early-stage breast cancer patients. J. Clin. Oncol. 23, 7074–7080.

Schulz, A.J., Mentz, G., Lachance, L., Johnson, J., Gaines, C., Israel, B.A., 2012. Associations between socioeconomic status and allostatic load: effects of neighborhood poverty and tests of mediating pathways. Am. J. Public Health 1–9.

Secretan, B., Straif, K., Baan, R., Grosse, Y., El Ghissassi, F., Bouvard, V., Benbrahim-Tallaa, L., Guha, N., Freeman, C., Galichet, L., 2009. A review of human carcinogens—part E: tobacco, areca nut, alcohol, coal smoke, and salted fish. Lancet Oncol. 10, 1033–1034.

Siahpush, M., Yong, H., Borland, R., Reid, J.L., Hammond, D., 2009. Smokers with financial stress are more likely to want to quit but less likely to try or succeed: findings from the international tobacco control (ITC) four country survey. Addiction 104, 1382–1390.

Singh, G.K., Azuine, R.E., Siahpush, M., 2013. Widening socioeconomic, racial, and geographic disparities in HIV/AIDS mortality in the united states. Adv. Prev. Med. 2013, 1987–2011.

Sitas, F., Parkin, D.M., Chirenje, M., Stein, L., Abratt, R., Wabinga, H., 2008. Part II: cancer in indigenous Africans—causes and control. Lancet Oncol. 9, 786–795.

Skidmore, P., Welch, A., van Sluijs, E., Jones, A., Harvey, I., Harrison, F., Griffin, S., Cassidy, A., 2010. Impact of neighbourhood food environment on food consumption in children aged 9–10 years in the UK SPEEDY (Sport, Physical Activity and Eating Behaviour: Environmental Determinants in Young people) study. Public Health Nutr. 13, 1022–1030.

Sorensen, G., Stoddard, A.M., LaMontagne, A.D., Emmons, K., Hunt, M.K., Youngstrom, R., McLellan, D., Christiani, D.C., 2002. A comprehensive worksite cancer prevention intervention: behavior change results from a randomized controlled trial (United States). Cancer Causes Control 13, 493–502.

Spittler, T.M., Feder, W.A., 1979. A study of soil contamination and plant lead uptake in Boston urban gardens. Commun. Soil Sci. Plant Anal. 10, 1195–1210.

Straif, K., Benbrahim-Tallaa, L., Baan, R., Grosse, Y., Secretan, B., El Ghissassi, F., Bouvard, V., Guha, N., Freeman, C., Galichet, L., 2009. A review of human carcinogens—part C: metals, arsenic, dusts, and fibres. Lancet Oncol. 10, 453–454.

Su, J.G., Larson, T., Gould, T., Cohen, M., Buzzelli, M., 2010. Transboundary air pollution and environmental justice: Vancouver and Seattle compared. GeoJournal 75, 595–608.

Tanjasiri, S.P., Lew, R., Mouttapa, M., Lipton, R., Lew, L., Has, S., Wong, M., September 2013. Environmental influences on tobacco use among Asian American and Pacific Islander youth. Health Promot. Pract. 14 (Suppl. 1), 40S–47S.

Triant, V.A., 2013. Cardiovascular disease and HIV infection. Curr. HIV/AIDS Rep. 10, 199–206.

US Department of Labor, 2010–2013. Mental Health Parity and Addiction Equity Act of 2008 (MHPAEA).

Wakefield, S., Yeudall, F., Taron, C., Reynolds, J., Skinner, A., 2007. Growing urban health: community gardening in south-east toronto. Health Promot. Int. 22, 92–101.

Wallace, M., Harville, E., Theall, K., Webber, L., Chen, W., Berenson, G., 2013. Neighborhood poverty, allostatic load, and birth outcomes in African American and white women: findings from the Bogalusa Heart Study. Health Place 24, 260–266.

Ward Thompson, C., Roe, J., Aspinall, P., Mitchell, R., Clow, A., Miller, D., 15 April 2012. More green space is linked to less stress in deprived communities: evidence from salivary cortisol patterns. Landscape Urban Plan. 105 (3), 221–229.

Ward, E., Jemal, A., Cokkinides, V., Singh, G.K., Cardinez, C., Ghafoor, A., Thun, M., 2004. Cancer disparities by race/ethnicity and socioeconomic status. CA Cancer J. Clin. 54, 78–93.

Waxman, A., World Health Assembly, 2004. WHO global strategy on diet, physical activity and health. Food Nutr. Bull. 25, 292–302.

Weitzman, E.R., Zapka, J., Estabrook, B., Goins, K.V., 2001. Risk and reluctance: understanding impediments to colorectal cancer screening. Prev. Med. 32, 502–513.

Whiting, D.R., Guariguata, L., Weil, C., Shaw, J., 2011. IDF diabetes atlas: global estimates of the prevalence of diabetes for 2011 and 2030. Diabetes Res. Clin. Pract. 94, 311–321.

Wild, S., Roglic, G., Green, A., Sicree, R., King, H., 2004. Global prevalence of diabetes estimates for the year 2000 and projections for 2030. Diabetes Care 27, 1047–1053.

Williams, D.R., Collins, C., 2001. Racial residential segregation: a fundamental cause of racial disparities in health. Public Health Rep. 116, 404.

World Health Organization, 2013a. Cancer: Fact Sheet.

World Health Organization, 2013b. Cardiovascular Diseases: Fact Sheet.

World Health Organization, 2013c. Chronic Obstructive Pulmonary Disease: Fact Sheet.

World Health Organization, 2013d. Depression: Fact Sheet.

World Health Organization, 2013e. Mental Health Action Plan 2013–2020. WHO Document Production Services, Geneva, Switzerland.

World Health Organization, 2003. WHO Framework Convention on Tobacco Control. WHO Document Production Services, Geneva, Switzerland.

World Health Organization, 2013. Chronic Diseases.

World Health Organization, 2013f. Interventions to Reduce Indoor Air Pollution.

World Health Organization, 2013g. WHO Framework Convention on Tobacco Control.

Wuest, J., Ford-Gilboe, M., Merritt-Gray, M., Wilk, P., Campbell, J.C., Lent, B., Varcoe, C., Smye, V., 2010. Pathways of chronic pain in survivors of intimate partner violence. J. Women's Health 19, 1665–1674.

Zhang, J., Mauzerall, D.L., Zhu, T., Liang, S., Ezzati, M., Remais, J.V., 2010. Environmental health in China: progress towards clean air and safe water. Lancet 375, 1110–1119.

Zimmet, P., Alberti, K., Shaw, J., 2001. Global and societal implications of the diabetes epidemic. Nature 414, 782–787.

Chapter 11

Land Degradation and Environmental Change

Paolo D'Odorico[1] and Sujith Ravi[2]

[1] *Department of Environmental Sciences, University of Virginia, Charlottesville, VA, USA,*

[2] *Department of Earth and Environmental Sciences, Temple University, Philadelphia, PA, USA*

ABSTRACT

Land degradation is a process resulting in a reduced ability of the land to provide ecosystem services. This phenomenon is affecting (directly or indirectly) billions of people around the world and may lead to the overexploitation of soil resources, loss of ecosystem productivity, shifts in vegetation composition, and/or loss of rural livelihoods. Here we review the main drivers of land degradation, evaluate its magnitude, impact, and significance, and discuss the conditions underlying the possible persistence and irreversibility of this phenomenon.

11.a INTRODUCTION

"Desertification" and "land degradation" are two terms commonly associated with losses of vegetation cover, ecosystem productivity, and soil resources that are crucial to rural livelihoods and the provision of ecosystem services (e.g., MEA, 2005). Different definitions have been provided for these terms, depending on the context in which they are used and the progress made in the understanding of the underlying processes. Desertification is a process of environmental "deterioration" (Dregne, 1977), leading to a persistent and even irreversible transition to "desert-like conditions" (MEA, 2005; D'Odorico et al., 2013). This process is generally identified with dryland degradation resulting from either natural or anthropogenic drivers (UNCCD, 1994).

The notions of desertification and land degradation have been criticized commonly on various grounds (Thomas, 1997; Eswaran et al., 2001), including the fact that many processes occur that may turn an ecosystem into a desert, and many types of desert environments with different landforms, climates, and biota (e.g., Nicholson, 2013). Thus, these two terms do not uniquely identify the driver of degradation, nor the state of the degraded system. Moreover, the definition of desertification as a loss of ecosystem

Biological and Environmental Hazards, Risks, and Disasters. http://dx.doi.org/10.1016/B978-0-12-394847-2.00014-0

productivity is in contrast with the fact that land degradation may result from (and contribute to) shifts in plant community composition characterized by an increase in productivity (e.g., Van Auken, 2000). Likewise, biodiversity may increase in the degraded state (Maestre et al., 2009; Eldridge et al., 2011). Therefore, land degradation could enhance erosion and soil losses, while improving the provision of other ecosystem services. Although the focus is here on dryland degradation, we need to stress that wetter climate zones are also prone to degradation. For instance, many regions across the dry subhumid and wet tropics are affected by losses of forest cover, soil fertility, and habitat (e.g., Runyan et al., 2012).

11.b INDICATORS OF LAND DEGRADATION

Indicators commonly used to characterize land degradation include loss of vegetation cover, increase in soil erosion, losses of soil nutrients and soil water-holding capacity, increase in soil salinity and sodicity, seed-bank depletion, degradation of biological crust, appearance of (and colonization by) invasive plant species, or shifts in plant community composition such as those associated with woody plant encroachment (Dregne, 1977; Glantz, and Orlovsky, 1983; Schlesinger et al., 1990; Ravi et al., 2009a,b; D'Odorico et al., 2012, 2013; Bhattachan et al., 2014). Decline in productivity—which often results from accelerated soil erosion and soil salinity—is also used as an indicator of land degradation. Other indicators include the economic and societal effects of decreasing crop yields, decline in rangeland productivity, and other losses of ecosystem services. For instance, increasing unemployment, land tenure changes, abandonment of agricultural land, rise in "environmental migrations," and increase in remittance flow to the affected region can be symptoms of societal restructuring in response to land degradation (*sensu*, Geist and Lambin, 2004; D'Odorico et al., 2013). Some of these societal responses can further enhance and sustain land degradation by weakening (or even dismantling) the institutions that traditionally enforced rules of resource governance ensuring environmental stewardship and sustainable use of the land.

11.c THE GLOBAL SIGNIFICANCE

Land degradation is happening at an alarming pace and is affecting regions inhabited by over one-third of the global population. This phenomenon contributes to a dramatic decline in the productivity of croplands and rangelands worldwide (Dregne, 1977; Reynolds and Stafford Smith, 2002, D'Odorico et al., 2012), thereby, threatening food security and environmental quality. Land degradation is, therefore, considered as a major global environmental issue of this century (Eswaran et al., 2001). Its environmental and socioeconomic–political effects involve a complex interplay of biophysical and anthropogenic factors acting at different spatial and temporal scales (Geist and Lambin, 2004). Considering the direct and indirect linkages of land degradation to poverty and environmental quality, the mitigation of desertification would considerably improve living standards in dryland areas, especially in the developing world (MEA, 2005).

The areal extent of global degraded areas varies depending on the definitions (Eswaran et al., 2001). Globally, about 24% of the global land area has been affected by degradation and over 1.5 billion people live on degraded lands (IFPRI, 2012). Using a comprehensive analysis to determine the extent, degree, and drivers of soil degradation, Oldeman et al. (1991) estimated that, globally, human-induced soil degradation has affected 1965 million ha (Table 11.1). In the case of the World's drylands, estimates by Dregne and Chou (1992) indicated that the continents of Africa and Asia are particularly affected by land degradation (Table 11.2). Unfortunately, more recent comprehensive global assessments of land degradation are missing due to challenges in defining land degradation and identifying appropriate indicators for quantifying the extent of land degradation. Global maps of land degradation can now be developed using vegetation indices from satellite observations. Their interpretation, however, is characteristically a difficult task because some areas appear to have "greened up" over the last few decades, which could reflect an increase in vegetation cover or a shift in plant community composition. Thus the greening and browning signals detected in satellite data require adequate investigation and interpretation before they can be used to infer patterns of land degradation (Bai et al., 2008). Dryland

TABLE 11.1 Degree of global human-induced soil degradation (In millions of hectare).

	Degree of Degradation			
In Millions of Hectare	Light	Moderate	Strong	Extreme
Africa	173	192	124	5
Asia	295	344	108	<1
Oceania	96	4	2	<1
Europe	60	144	10	4
North America	17	78	1	–
Central America	2	35	26	–
South America	105	113	25	–
Total	749	910	296	9

Modified from Bridges and Oldeman, 1999.

TABLE 11.2 Areal extent of dryland degradation, including arid, semiarid, and dry subhumid regions

	Irrigated Land		Rainfed Cropland		Rangeland	
In 1000 Hectare	Degraded	% of Total	Degraded	% of Total	Degraded	% of Total
Africa	1902	18	48,863	61	995,080	74
Asia	31,813	35	122,284	56	1,187,610	76
Australia and New Zealand	250	13	14,320	34	361,350	55
Europe	1905	16	11,854	54	80,517	72
North America	5860	28	11,611	16	411,154	85
South America	1417	17	6635	31	297,754	76
Total	43,147	30	215,567	47	3,333,465	73

Modified from Dregne and Chou, 1992.

systems worldwide are experiencing rapid population growth and increase in standard of living (MEA, 2005; Schultz and Prasad, 2008). The resulting increase in human pressure and overexploitation of ecosystem services is commonly manifested by the expansion and intensification of agriculture and livestock grazing, which may strongly contribute to land degradation (MEA, 2005). Moreover, climate warming and recurrent droughts are further increasing the pressure on soil and water resources of dryland regions.

11.d DRIVERS

Drivers of land degradation involve a number of biophysical and social factors, through a variety of possible interactions (Thomas, 1997; Geist and Lambin, 2004; Herrmann and Hutchinson, 2005; Verón et al., 2006). Anthropogenic land degradation is typically initiated by changes in land management associated with the intensification of either agricultural or livestock production. The human-induced soil degradation is commonly differentiated into degradation types (e.g., soil erosion, nutrient decline, salinization) and causative factors (e.g., deforestation, overgrazing) as shown in Figure 11.1 (Bridges and Oldeman, 1999).

Dryland cropping systems are generally rainfed and are typically left fallow (bare) during rainless periods, which favor a rapid degradation through soil erosion. On the other hand, in areas with limited access to freshwater, irrigated agriculture may accelerate soil salinization, which is a major hurdle for

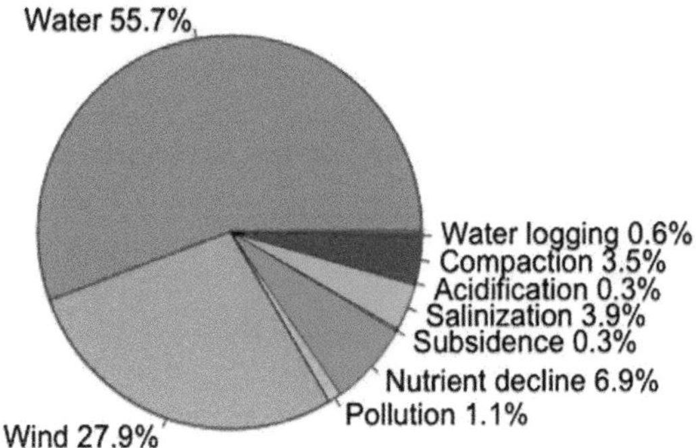

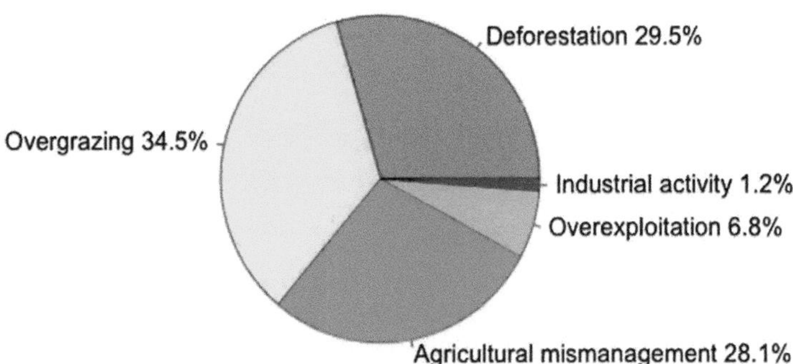

FIGURE 11.1 Drivers of global human-induced soil degradation by degradation type (top) and causative factors (bottom). *Based on data from Oldeman et al. (1991) and Bridges and Oldeman (1999).*

economic crop production and may lead to the abandonment of vast areas that remain prone to soil erosion. The degradation of grasslands and rangelands is occurring at an even greater extent than croplands (Dregne and Chou, 1992; Pagiola, 1999). The overexploitation of grazing lands—which comprises around 70% of the world's drylands—is typically due to overgrazing and conversion to croplands (Dregne and Chou, 1992). Overgrazing hinders vegetation recovery, alters species composition, and impacts soil hydrological properties. These changes may lead to irreversible declines in rangeland productivity by reducing the ecosystem carrying capacity and resilience. For instance, in the drylands of Mongolia and China, the most important anthropogenic factors contributing to land degradation are changes in animal husbandry (Batjargal, 1992; Zhao et al., 2005). The carrying capacities of these grazing systems are increasingly exceeded, resulting in vegetation degradation and enhanced soil erosion. Even in the case of monsoon deserts like the Thar in India, which turns lush green following precipitation events, the overexploitation of fodder and fuel wood has caused the ecological destruction of the desert ecosystem, resulting in slower rates of vegetation regeneration following precipitation (Sinha et al., 1999; Chauhan, 2003; Ravi and Huxman, 2009).

Potential degradation pathways may also be associated with changes in plant community composition acting in concert with climatic drivers (e.g., precipitation variability) (Schlesinger et al., 1990; Ravi et al., 2009a). Thus land degradation can result as an indirect effect of changes in vegetation compositions such as: (1) shifts from palatable perennial to unpalatable annual grass dominance as result of overgrazing; (2) exotic grass invasions; and (3) encroachment of woody plants due to grazers, changes in fire management, climate warming, or increase in atmospheric CO_2 concentrations (e.g., Van Auken, 2000; Ravi et al., 2009b). All of these changes in vegetation typically entail a severe loss of ecosystem services (e.g., rangeland productivity and soil sheltering against erosion). For instance, anthropogenic degradation of dryland ecosystems after the European colonization has been reported in the case of the Southwestern United States, Australia, Southern Africa, and South America. Large-scale commercial grazing in conjunction with management practices (e.g., fire suppression) led to an increase in woody plant density (or "shrub encroachment") and to the invasion of desert ecosystems by exotic grasses, with negative impacts on ecosystem function and services (Archer, 1989; D'Antonio and Vitousek, 1992; Pickup, 1998; Van Auken, 2000; Ravi et al., 2009a,b).

In many areas, cropland and rangeland degradation are interrelated. For example, decline in crop productivity resulting from several factors (e.g., soil erosion) may contribute to a shift to a livestock-based system in some rural agrarian communities. The introduction of new technology is also known for being a possible contributor to land degradation (e.g., Geist and Lambin, 2004). In some rural areas, the use of modern pumping technology has increased the availability of drinking water for livestock, thereby, resulting in overgrazing-induced degradation in the surroundings of water points (e.g., Wang et al., 2008). The increased pressure on drylands is also contributed by climatic changes, urbanization, and energy development (MEA, 2005; Hernandez et al., 2014). In the case of the sub-Saharan Africa, climatic disasters (most notably, the multidecadal drought from the late 1960s)—combined with weak economies and unsustainable use of marginal resources—increased the stress on dryland ecosystems (Glantz, 1987; Hutchinson, 1996), which were unable to meet the demands of the increasing human population (Darkoh, 1998). These factors caused famines and large-scale human migrations with important socioeconomic and political consequences (Glantz, 1977; Kassas, 1977; Mabbut and Wilson, 1980; Nnoli, 1990; Darkoh, 1998). Another "dramatic" example of anthropogenic dryland degradation occurred in the Great Plains of the United States in the 1930s and is known as "the Dust Bowl." In this region, dramatic soil loss and dust emissions were observed as a result of poor land management practices in conjunction with dry climatic conditions (Worster, 1979).

11.e CONCLUSIONS

One of the main questions arising from the analysis of land degradation is whether this shift to a degraded state is permanent and irreversible, at least within the timescales of few human generations. In other words, can the landscape recover its ecosystem functions and resume its provision of ecosystem services once the disturbance (e.g., grazers, cultivation) is removed? Some of the definitions of desertification imply that the process is persistent and permanent, which means that the underlying dynamics exhibit two alternative stable states, with the system being stable both in its degraded and nondegraded states (D'Odorico et al., 2013). Thus, once the disturbance is removed, the system does not spontaneously revert to the initial nondegraded conditions. Bistable dynamics are typically induced by positive feedback; in the case of land degradation a number of feedback mechanisms have been invoked to explain the (seemingly) irreversible transition to a degraded state (e.g., D'Odorico et al., 2013). These feedback involve interactions between vegetation and their physical environment, and the ability of vegetation to modify resource availability or disturbance regime to its own advantage. Although desertification feedback have been investigated both theoretically and experimentally, empirical studies assessing the reversibility or irreversibility of land degradation (e.g., Ravi et al., 2009b; Bhattachan et al., 2014) are needed to evaluate the factors underlying landscape recovery.

REFERENCES

Archer, S., 1989. Have southern Texas savannas been converted to woodlands in recent history. Am. Nat. 134 (4), 545–561.

Bai, Z.G., Dent, D.L., Olsson, L., Schaepman, M.E., 2008. Proxy global assessment of land degradation. Soil Use Manag. 24 (3), 223–234. http://dx.doi.org/10.1111/j.1475-2743.2008.00169.x.

Batjargal, Z., 1992. The climatic and man-induced environmental factors of the degradation of ecosystem in Mongolia. International Workshop on Desertification, Ulaanbaatar, Mongolia, pp. 19.

Bhattachan, A., D'Odorico, P., Dintwe, K., Okin, G.S., Collins, S.L., 2014. Resilience and recovery of the Kalahari dunes. Ecosphere 5 (2), 1–14.

Bridges, E.M., Oldeman, L.R., 1999. Global assessment of human-induced soil degradation. Arid Soil Res. Rehabilitat. 13, 319–325.

Chauhan, S.S., 2003. Desertification control and management of land degradation in the Thar desert of India. Environmentalist 23, 219–227.

Darkoh, M.B.K., 1998. The nature, causes and consequences of desertification in the drylands of Africa. Land Degrad. Dev. 9, 1–20.

D'Antonio, C.M., Vitousek, P.M., 1992. Biological invasion by exotic grasses, the grassfire cycle, and global change. Annu. Rev. Ecol. Syst. 23, 63–87.

D'Odorico, P., Okin, G.S., Bestelmeyer, B.T., 2012. A synthetic review of feedbacks and drivers of shrub encroachment in arid grasslands. Ecohydrology 5 (5), 520–530. http://dx.doi.org/10.1002/eco.259.

D'Odorico, P., Bhattachan, A., Davis, K.F., Ravi, S., Runyan, C.W., 2013. Global desertification: drivers and feedbacks. Adv. Water Resour. 51, 326–344. http://dx.doi.org/10.1016/j.advwatres.2012.01.013.

Dregne, H.E., 1977. Desertification of arid lands. Econ. Geogr. 53 (4), 322–331.

Dregne, H.E., Chou, N.T., 1992. Global desertification dimensions and costs. In: Dregne, H.E. (Ed.), Degradation and restoration of arid lands. Texas Tech. University, Lubbock, TX, pp. 249–282.

Eldridge, D.J., Bowker, M.A., Maestre, F.T., Roger, E., Reynolds, J.F., Whitford, W.G., 2011. Impacts of shrub encroachment on ecosystem structure and functioning: towards a global synthesis. Ecol. Lett. 14, 709–722.

Eswaran, H., Lal, R., Reich, P.F., 2001. Land degradation: an overview. In: Bridges, E.M., Hannam, I.D., Oldeman, L.R., Pening de Vries, F.W.T., Scherr, S.J., Sompatpanit, S. (Eds.), Responses to Land Degradation. Proc. 2nd. International Conference on Land Degradation and Desertification, Khon Kaen, Thailand. Oxford Press, New Delhi, India.

Geist, H.J., Lambin, E.F., 2004. Dynamic causal patterns of desertification. Bioscience 54 (9), 817–829.

Glantz, M.H. (Ed.), 1977. Desertification: Environmental Degradation in and Around Arid Lands. Westview Press, Boulder.

Glantz, M.H., 1987. Drought and Hunger in Africa: Denying Famine a Future. Cambridge University Press, pp. 457.

Glantz, M.H., Orlovsky, N.S., 1983. Desertification: a review of the concept. Desertification Control Bull. 9, 15–22.

Hernandez, R., Easter, S.B., Murphy-Mariscal, M.L., Maestre, F.T., Tavassoli, M., Allen, E.B., Barrows, C.W., Belnap, J., Ochoa-Hueso, R., Ravi, S., Allen, M.F., 2014. Environmental impacts of utility-scale solar energy. Renew. Sustain. Energy Rev. 29, 766–779.

Herrmann, S.M., Hutchinson, C.F., 2005. The changing contexts of the desertification debate. J. Arid Environ. 63, 538–555.

Hutchinson, C.F., 1996. The Sahelian desertification debate: a view from the American South West. J. Arid Environ. 33, 519–524.

IFPRI, 2012. 2011 Global Food Policy Report. International Food Policy Research Institute, Washington, DC.

Kassas, M., 1977. Arid and semi-arid lands: Problems and prospects. Agro-Ecosystems 3, 185–204.

Mabbutt, J.A., Wilson, A.W. (Eds.), 1980. Social and Environmental Aspects of Desertification. Proceedings of an Inter-Congress Meeting of the International Geographical Union Working Group on Desertification in and around Arid Lands; Held in conjunction with the Arid Lands Sub-programme of the United Nations University Natural Resources Programme, and the UNESCO/MAB Programme, 3–8 January 1979, Tuscon, Arizona, USA. NRTS-5/UNUP-127; The United Nations University, Tokyo, 40 pp.

MEA, Millennium Ecosystem Assessment, 2005. Ecosystems and Human Well-being: Desertification Synthesis. World Resources Institute, Washington, DC.

Maestre, F.T., Bowker, M.A., Puche, M.D., Hinojosa, M.B., Martinez, I., Garcia-Palacios, P., Castillo, A.P., Soliveres, S., Luzuriaga, A.L., Sanchez, A.M., Carreria, J.A., Gallardo, A., Escudero, A., 2009. Shrub encroachment can reverse desertification is semi-arid Mediterranean grasslands. Ecol. Lett. 12, 930–941.

Nicholson, S., 2013. Dryland Climatology. Cambridge University Press.

Nnoli, O., 1990. Desertification, refugees and regional conflict in West Africa. Disasters 14, 132–139.

Oldeman, R.L., Hakkeling, R.T.A., Sombroek, W.G., 1991. World Map of the Status of Human Induced Soil Degradation, second rev. ed. International Soil Reference and Information Centre, Wageningen, Netherlands.

Pagiola, S., 1999. The Global Environmental Benefits of Land Degradation Control on Agricultural Land: Global Overlays Program World Bank Environment Paper. No:16. World Bank.

Pickup, G., 1998. Desertification and climate change- the Australian perspective. Clim. Res. 11, 51–63. http://dx.doi.org/10.3354/ cr011051.

Ravi, S., Huxman, T.E., 2009. Land degradation in the Thar Desert. Front. Ecol. Environ. 7 (10), 517–518.

Ravi, S., D'Odorico, P., Collins, S.L., Huxman, T.E., 2009a. Can biological invasions induce desertification? New Phytol. 181 (3), 508–511.

Ravi, S., D'Odorico, P., Wang, L., White, C., Okin, G.S., Collins, S.L., 2009b. Post-fire resource redistribution in desert grasslands: a possible negative feedback on land degradation. Ecosystems 12 (3), 434–444.

Reynolds, J.F., Stafford Smith, D.M., 2002. Do humans cause deserts? In: Reynolds, J.F., Stafford Smith, D.M. (Eds.), Global Desertification. Do Humans Cause Deserts? Dahlem Workshop Series, vol. 88. Dahlem University Press, Berlin, pp. 1–21.

Runyan, C.W., D'Odorico, P., Lawrence, D., 2012. Physical and biological feedbacks on deforestation. Rev. Geophys. 50, RG4006. http://dx.doi.org/10.1029/2012RG000394, 201.

Schlesinger, W.H., Reynolds, J.F., Cunningham, G.L., Huenneke, L.F., Jarrell, W.M., Virginia, R.A., et al., 1990. Biological feedbacks in global desertification. Science 247 (4946), 1043–1048.

Schultz, B., Prasad, K., 2008. In: Lee, C., Schaaf, T. (Eds.), The Future of Drylands. Springer, UNESCO Publishing.

Sinha, R.K., Bhatia, S., Vishnoi, R., 1999. Desertification control and rangeland management in the Thar desert of India. RALA Report No. 200.

Thomas, D.S.G., 1997. Science and the desertification debate. J. Arid Environ. 37, 599–608.

UNCCD, 1994. United Nations Convention to Combat Desertification, Elaboration of an International Convention to Combat Desertification in Countries Experiencing Serious Drought And/or Desertification, Particularly in Africa (U.N. Doc. A/AC.241/27, 33 I.L.M. 1328, United Nations).

Van Auken, O.W., 2000. Shrub invasions of North American semiarid grasslands. Annu. Rev. Ecol. Syst. 31, 197–215.

Verón, S.R., Paruelo, J.M., Oesterheld, M., 2006. Assessing desertification. J. Arid Environ. 66, 751–763.

Wang, L., D'Odorico, P., 2008. The limits of water pumps. Science 321 (5885), 36–37. http://dx.doi.org/10.1126/science.321.5885.36c.

Worster, D., 1979. Dust Bowl: The Southern Plains of 1930s. Oxford University press, New York, pp. 277.

Zhao, H.L., Zhao, X.Y., Zhou, R.L., Zhang, T.H., Drake, S., 2005. Desertification processes due to heavy grazing in sandy rangeland, Inner Mongolia. J. Arid Environ. 62 (2), 309–319.

Chapter 11.1

Desertification

John Oswald and Sarah Harris
Department of Geography and Geology, Eastern Michigan University, MI, USA

ABSTRACT
Desertification, an intentionally foreboding term, is at its core land degradation in dryland environments. After nearly 40 years of international attention, the analysis of the causes and consequences of land degradation in dryland systems remains shrouded in the nuanced complexities of multiple interwoven variables (biophysical, anthropogenic, and climatic) working simultaneously to bring about change in these ecologically and socially vulnerable areas. In order to fully grasp the complexities of desertification, this chapter follows a fivefold approach. First, it provides a synthesis of the history and politicization of the term. Second, it establishes desertification's geospatial context. Third, it identifies the interrelated drivers of land degradation and their impacts, with special focus placed on North Africa, the Middle East, and Central Asia. Fourth, it outlines the current approaches to studying land degradation and desertification. Finally, it concludes by situating the desertification dynamic in the context of sustainability and poverty eradication.

Desertification, an intentionally ominous term, is a complex and multilayered phenomenon that simultaneously denotes both the process of landscape transformation, as well as the end state of degradation in dryland ecosystems (arid, semiarid, and dry-subhumid regions). As discussed in greater detail below, desertification as a process includes the combination of anthropogenic, biophysical, and climatic inputs that negatively impact these highly adapted, yet vulnerable, ecosystems around the world. Within these increasingly hydrologically stressed dryland regions, traditionally marginalized populations (ecologically, economically, and politically, or what Reynolds et al. (2007) term the "dryland syndrome") living in these areas face increasing challenges in their ability to produce food, generate income, and maintain their traditional lifestyles. As population numbers, currently at two billion people (MEA, 2005), in dryland regions continue to rise, a simultaneous amplification has occurred of human landscape exploitation. This intensification can push dryland systems beyond sustainable thresholds and when unsustainable resource exploitation continues unabated, the process can translate into ecological disequilibrium in the short term and landscape degradation over the long term, creating what has been termed "desert-like" conditions in these areas.

Biological and Environmental Hazards, Risks, and Disasters. http://dx.doi.org/10.1016/B978-0-12-394847-2.00015-2

Severe degradation in vulnerable dryland regions, the terminus of the desertification process, includes a panoply of negative biotic and pedogenic changes across the landscape. In general, these changes result in the limitation of a dryland system's resilience, in other words, its capability to absorb changes, to the relentless climatic and meteorological dynamics of more arid regions (limited, variable, and intense rainfall patterns, intense solar radiation, and wind). The diminishing capacity of these degraded landscapes, in turn, intensifies the socioeconomic stresses faced by the human inhabitants, creating a near self-perpetuating feedback loop, resulting in a multiplicity of social responses ranging from the abandonment of degraded lands, the expansion of unsustainable land-use practices into even more marginal areas, or to either voluntary or forced internal and/or external migration.

As seen above, human impacts, biophysical responses, and climatic variability can create an intricate interdigitation of compounding causal mechanisms, often building upon each other in a nonlinear fashion, that accelerate desertification/land degradation in dryland environments. As discussed throughout this chapter, this dynamic of "compounded desertification" involves transformative stresses that soon exceed the limited carrying capacity of the pedogenic and hydrologic realities in affected regions, thus leading to system-wide degradation in ecological diversity and agroeconomic potential. Furthermore, the compounding effects of one region's degradation can soon affect social and ecological transformation in nearby, possibly hitherto unaffected, regions, thus further reinforcing the spiraling effect of desertification and its role as a supraregional and even global issue.

As will be seen throughout this chapter, the desertification process is a complex, multicausal, and dynamic process in which feedback loops between humans and the environment transform local and regional agroecological systems in numerous and often distinct ways. Along with making it difficult to pinpoint a definitive set of primary drivers or metavariables of desertification (Thomas, 1997; Reynolds and Stafford-Smith, 2002; Geist, 2005; Lambin et al., 2009), the multivariate and dynamic nature of desertification has also created a level of controversy and uncertainty surrounding the term.

To best address this dynamic nature of desertification and the difficulty of identifying the causes of it, this chapter follows a fivefold approach. First, it provides a synthesis of the history and politicization of the term "desertification." Second, it establishes the specific geospatial context of desertification, namely drylands, by defining physical parameters of this type of diverse ecosystem. Third, this chapter identifies the interrelated human, biophysical, and meteorological drivers that are key to understanding the processes and impacts of dryland degradation, with special attention being given to examples from North Africa, the Middle East, and Central Asia. Fourth, this chapter turns to an analysis of the current methodological approaches for studying land degradation and desertification. Finally, it concludes by revisiting the compounding effects of the desertification dynamic and its significance for the global community, especially in terms of human development and poverty and regional and international stability, with special emphasis being placed on the principles of sustainability, community involvement, and degradation mitigation and remediation (Reynolds et al., 2007; Geist, 2005; Imeson, 2012).

11.1.1 HISTORICAL ROOTS, EVOLVING DEFINITIONS, AND CRITIQUES OF DESERTIFICATION

Although the term "desertification" does not formally appear in publication until 1949 (Aubreville, 1949), French environmentalists and colonial officers, as well as British colonial officers and American scholars, were utilizing the concept of desertification to describe how the ruinous use of the land by humans could lead to the transformation of once productive land into a desert since the mid-to-late nineteenth century (Davis, 2004; Harris, 2012; Butzer and Harris, 2007). These individuals frequently applied this concept of human-induced degradation to many of the same regions that are the focus of desertification discussions today, such as West and North Africa, the Eastern Mediterranean, the Levant, and the Middle East.

George Perkins Marsh, an American diplomat and environmentalist traveling abroad, provided one of the most frequently cited examples of this belief that humans were causing great damage to the landscape when he argued in 1864 that within the arid regions of the world, human activities, especially perceived overgrazing, were ruining the landscape by disallowing the growth of forests. He believed that if humans were removed from the landscape and forests were allowed to grow, rainfall amounts would increase and the arid regions would become lush and green with vegetation (he even suggested that the Sahara would become forest covered). He, as well as other authors, found support for this belief in the writings of classical authors (e.g., Strabo and Pliny), who described areas such as North Africa as verdant and productive during their time period, even though modern archaeological work has shown that the climate/landscape in this region has not dramatically changed over the past 3,000 years (Barker, 2002).

Moving into the latter twentieth century, the term once again appears in print, now as "desertization," to describe the spread of desert conditions into regions that previously had none through human misuse (Le Houérou, 1977; Rapp, 1974; Rapp et al., 1976 also include climatic forces as part of the definition). This term, "desertization," served as the foundation for the United Nations Conference on Desertification (UNCOD) definition, marking the first time that the concept of "desertification" was internationalized as a prescient issue for global human development and ecological stability (UNCOD, 1977). As defined by the UNCOD, "Desertification is the diminution or destruction of the biological potential of land, and can lead ultimately to desert-like conditions." The UNCOD's (1977) publication *Desertification: Its Causes and Consequences* and the subsequent creation of a 1:25,000,000 *World Map of Desertification* (UN, 1977) visualized the problem at a global scale and accentuated human overexploitation of environmental resources as one of the main causes of desertification. Its *Plan of Action* (UNCOD, 1977) identified three main objectives—to halt desertification and reverse its impacts, to establish sustainable land uses, and to ensure the socioeconomic development of people impacted by desertification—and introduced a number of monitoring programs and a series of recommendations and technical fixes to remediate the negative impacts of desertification (Thomas and Middleton, 1994, p. 31).

Although the UNCOD brought the issue of desertification to the international forefront, much uncertainty still existed as to what desertification actually entailed. The definition was ambiguous, the inputs were difficult to quantify, and more attention was being paid to the consequences rather than to a better definition and description of the causes (see Drenge, 1983). The ambiguity of the UNCOD's definition was also furthered by the fact that by the early 1980s dozens of different formal definitions existed, which unfortunately led to miscommunication between scholars and policy (Glantz and Orlovsky, 1983).

Thanks in part to these uncertainties and ambiguities, a "crisis narrative" (Thomas and Middleton, 1994) concerning the ill effects of desertification soon developed. The negative impacts of desertification became popularized in the media (see Veron et al., 2006, p. 753), and politicians, both within and outside affected areas, began to utilize the threat of desertification to further their goals in a number of different ways (Thomas and Middleton, 1994). In some instances, desertification was used as a clarion call for opposition to the government in power by those who desired power for themselves. In other cases, governments used the threat of desertification as a means to secure new streams of international funding that might otherwise have not been made available to them, although there was no guarantee that said funding did fully go toward combating desertification. As noted at the beginning of this section, blaming indigenous people for deleterious land-use practices was nothing new, but this loosely defined threat of desertification made these claims even easier for governments to make and garner support in terms of public opinion and funding for settling and/or moving social groups who they perceived as having detrimental land-use practices (see Davis, 2004, 2007 for discussion of the treatment of the nomadic and seminomadic Berbers in North Africa; see Rosen, 2000; Schaldach et al., 2013; Leu et al., 2014 for the Negev in Israel). In short, the vagaries of the definition and the creation of the "crisis narrative" of social and ecological consequences of desertification allowed national governments to manipulate the term for political expediency in whatever ways they saw fit.

Beyond the use and abuse of the term by national actors and the international community, criticism was also levied at the type of research and monitoring and remediation activities that were actually carried out to address desertification. Numerous works were published attempting to define the term and codify the processes and consequences of the phenomenon, but little progress was made in the creation of a comprehensive assessment strategy of desertification (Veron et al., 2006). Apart from research shortcomings, in terms of actual work on the ground, many of the UNCOD-sponsored programs were short-term fixes with no proposals for any long-term social and economic development initiatives. Further, critics argued that many of these fixes were based on "Western" land management systems, such as large landowner, sedentary range management, and often did not conform to traditional land-use patterns in dryland regions (Thomas and Middleton, 1994), nor did they take into account the social complexities in the affected areas (Spooner and Mann, 1982; see also ICIHI, 1986). Because of this lack of attention to social complexities, the human role was misunderstood, but it was decidedly not ignored. Indeed, many projects focused heavily on human drivers and significantly downplayed the role of climatic/meteorological variation and impacts, and thus never fully understood what was happening on the ground.

To try to counter the critiques outlined above, especially those regarding the human role, the UNEP/FAO (1984) proposed another definition that aimed to better address both human and natural induced causes of desertification. It reads "A comprehensive expression of economic and social processes as well as those natural and induced ones which destroy the equilibrium of soil, vegetation, air and water, in the areas subject to edaphic and/or climatic aridity." It follows this with "Continued deterioration leads to a decrease in, or destruction of the biological potential of the land, deterioration of living conditions and an increase of desert landscape." However, the 1990 UNEP Consultative Meeting on the Assessment of Desertification in Nairobi soon returned the focus to the human causes. According to this revised definition "Desertification is land degradation in arid, semi-arid and dry Sub-humid areas resulting from adverse human impact" (Thomas and Middleton, 1994). In this definition, humans are considered the main driver of land degradation in dryland areas.

By 1992 the focus had shifted once again back to both climatic and human drivers. As developed during UNEP's 1992 Conference on Environment and Development (UNCED) in Rio de Janeiro (also known as the Earth Summit), the new definition read "land degradation in arid, semi-arid and dry sub-humid areas, resulting from various factors, including climatic variations and human activities" (see Chapter 12 of the Agenda 21). This definition was formally adopted by the United Nations Convention to Combat Desertification (UN, 1994a,b and UNCCD, 2000). Although this definition remains arguably overgeneralized and open-ended (causing even further issues for a comprehensive assessment and mitigation strategy), it now serves as the primary definition used in most publications and policies—though this wide acceptance may simply reflect a history of definition and issue "fatigue" (see discussion section and Stringer, 2008).

Although the definition has not noticeably changed, the turn of the century, specifically the Rio+20 conference, the UN General Assembly and the 2011 Council of Parties (COP 11), brought with it an interest in situating desertification in the context of sustainable development and poverty reduction as the cornerstone of the new green economy, with an overarching goal of zero-net land degradation in countries experiencing desertification and drought (see Imeson, 2012 for further discussion). This was to be accomplished by working with national governments in affected areas to create National Action Plans (NAPs). The implementation strategies created by these plans aims include all affected stakeholders, from national officials to local community members and NGOs. On a different front, emphasis has also been placed on public awareness campaigns such as the designation of 2006 as the *International Year of Deserts and Desertification* to further popularize social and ecological issues surrounding desertification; however, as Tal (2007) and Stringer (2008) have both stated, limited measurable success has stemmed from this awareness campaign.

11.1.2 DRYLANDS AND THEIR VARIABILITY

As mentioned above, at its core desertification is a spatially situated umbrella term for dryland ecosystem degradation (slight to very severe). One might rightfully ask the question: what constitutes a "dryland" region? According to various UNEP assessments, dryland ecosystems constitute between 47% (Middleton and Thomas, 1997) and 41% (MEA, 2005) of the terrestrial landmass, making these regions the world's largest biome. In general, the defining characteristic of these regions is aridity, or the lack or deficit of water relative to climatic conditions that impacts human land use. More specifically, drylands are characterized by high summer temperatures, high solar radiation intensity (no cloud cover for over 70% of the year (Leity, 2008)), erratic and high interannual precipitation variability, low relative humidity rates, high evapotranspiration rates, and a soil structure dominated by aridisols (high mineral and low organic matter content) and entisols (sandy, undifferentiated soils).

Dryland soils are defined as having low organic matter (limiting microbial processing of nutrients for plants), weak structure and high salt content, and limited moisture retention capabilities (see Laity, 2008). The low nutrient availability coupled with high average temperatures and low moisture availability lends itself to sparse vegetation cover mostly consisting of drought-tolerant (xerophytic) plant species. The land surface in these regions is exposed to a number of simultaneous climatological/meteorological stresses. High winds (owing to a relatively flat topography) introduce erosion in the form of deflation and transportation of dust (fine silts and aeolian sediments), whereas concentrated (high magnitude, low frequency) interannual rainfall events introduce soil loss through water erosion (sheet and rill) and surface compaction from raindrops.

After millennia of long-term desiccation and decadal and interannual variations, dryland landscapes react by forming either inorganic or biological/cryptobiotic surface crusts (Laity, 2008). In drier regions, the inorganic crusts (fine-textured soils, salts, and rocks) act as a sealant, which reduces water infiltration, inhibits plant growth, and increases surface reflection (albedo). In wetter regions with more organic surface cover, biological crusts of nonvascular plants (algae, fungi, lichens) form on exposed surfaces that serve as a more permeable sealant allowing for greater soil stability, increased water infiltration and storage, and plant growth. These crusts are the results of long-term systemic adaptations, but are easily disturbed by human land-use intensification (such as agriculture, pastoralism, or even tourism).

The combined physical, biological, and meteorological features of drylands culminate in an overall heightened susceptibility to degradation and drought. Societies that live in these dryland regions have established over time a series of coping mechanisms such as advanced systems of water collection and drought-tolerant, agricultural practices and polyculture to adapt to the severe water and pedogenic stresses in these regions. As addressed in the next section, however, regional political structures, the intensification of unsustainable land-use practices for regional and global markets, and growing populations within the region (around two billion people living within global drylands according to the MEA (2005)) will likely overpower these coping mechanisms and intensify existing issues as well as potentially introduce new stressors.

Drylands are subdivided into hyperarid deserts (7.5% of the total global landmass), arid (12%), semiarid (18%), and dry-subhumid regions (10%) (MEA, 2005; see also Spellman, 2000; Middleton and Thomas, 1997; Heathcote, 1983). Hyperarid or "true" deserts are areas of extreme dryness with less than 80 mm (<3 in) annual precipitation; by definition they must also have at least 12 consecutive months with no precipitation. These regions also have a P/PET ratio of less than 0.05 (also known as the index of aridity; in drylands precipitation (P) is less than two-thirds of the potential evapotranspiration (PET), which includes soil evaporation and plant transpiration). Interestingly, although mobilization of sand dunes and desert expansion does occur at the desert margins, the geographical distribution of deserts around the globe has remained fairly static since their formation in response to climate conditions in the Quaternary period (Spellman, 2000).

Arid regions, on the other hand, can have anywhere between 25 and 350 mm of rainfall annually, with 50–100% of that total falling at one time during the year, and a P/PET ratio of 0.05–0.20. Semiarid regions, on the other hand, receive a mean annual rainfall of between 250 and 500 mm and have a P/PET ratio of 0.20–0.50. These regions, commonly referred to as savannas, are dominated by grasslands and shrubs and are suited to pastoralism and limited rain-fed, and irrigated agricultural intensification. The final dryland classification includes the dry-subhumid regions. These regions can receive up to 700 mm of annual precipitation and have a P/PET ratio of 0.50–0.65. In these latter two dryland regions, between 20% and 50% of the total rainfall amounts occurs during a single time of the year.

The biotic and abiotic landscapes of the combined dryland biome represent dynamic, highly adapted, and resilient ecosystems. Dryland regions in North Africa and the Sahel, the Middle East, and Central Asia have experienced both long- and short-term climatic fluctuations over the past 20,000 years, especially in terms of rainfall amounts and wind patterns. According to Barker and Gilbertson (2000), the Late Pleistocene (18,000–10,000 years ago) Sahara region was hyperarid and extended much further into nearby regions than it does today. However, the archaeological record from the Quaternary to Holocene transition (10,000 and 8,000 BP) shows a notably wetter climate in North Africa, the Middle East, and Central Asia.

The first signs of regional desiccation appear in the archaeological record of North Africa and the Middle East at about 6,000 years BP with its current aridification dynamic occurring by 4,500 BP (concerning the Sahel see Falkenmark and Rockstrom, 2008). According to Issar et al. (2012), a recognizable shift toward greater humidity occurred around 5,000 BP in the Middle East, and the archaeological record indicates that since then the dryland regions from the North African Maghreb to the deserts and steppes of Central Asia have experienced cycles of wetter and drier climates over centuries with recognizable periods of long droughts. As an example of these fluctuations, the Dead Sea Salt Cave and sediment layer analysis shows higher rainfall levels during the first two centuries CE in the Negev region, with a shift to greater aridity in the middle of the first millennium CE, which remained the norm until the early 1900s (Rosen, 2000).

The understanding that dryland regions experience changes in their prevailing climatic patterns at the millennial and century scale assists archaeologists and paleoclimatologists alike in understanding the expansion and contraction of human settlements and agricultural intensification patterns within these regions (see Barker and Gilbertson, 2000; Barker, 2002; Rosen, 2000, and the 2012 special issue of the *Journal of Arid Environments*). An archaeologically based understanding of drought and humid cycles is a significant aid for our understanding of long-term human settlement patterns across a landscape, as it is not until the late nineteenth century that temperature and rainfall amounts were precisely recorded, giving researchers a much finer resolution for understanding annual and decadal lengths of drought cycles.

The African Sahel is a good example of a region with this type of well-documented temporal sequencing of drought cycles and their human impacts. According to Swift (1977), the Sahel experienced frequent droughts and famines during the first half of the twentieth century (1913–1914, 1930–1932, and the early 1940s). From the latter half of the 1940s until the late 1960s, however, the region experienced higher than average rainfall amounts. The wetter climate pattern allowed inhabitants to expand agriculture and pastoral ranges deeper into traditionally non- or underutilized lands. Beginning in 1968 and lasting through 1973, however, the Sahel experienced a severe shortage of rainfall leading to an extended drought cycle. This drought triggered a catastrophic famine in the region that claimed the lives of approximately 250,000 people and countless numbers of domesticated animals, ushered in a massive migration of millions of people throughout the region, and caused a general collapse of the agricultural economy. This drought and famine in the dryland region of the Sahel garnered international attention, and in response to this catastrophe, the world was officially introduced to the term "desertification" (see previous section for a fuller description of the history and meaning of the term).

11.1.3 PHYSICAL, CLIMATIC, AND ANTHROPOGENIC DRIVERS OF DESERTIFICATION AND THEIR IMPACTS

It is estimated that drylands encompass nearly five billion hectares of land (MEA, 2005) and, excluding hyperarid regions, anywhere from 10% to 20% (MEA, 2005) up to as high as 70% (UNEP, 1992) of these lands have experienced some form of degradation. As addressed above, degradation is amplified by continuous population growth in the world's dryland regions. Between 1990 and 2000 an 18.5% population growth rate occurred across all global drylands (MEA, 2005). In North Africa, the Middle East, and Central Asia (the geographical focus of this chapter), 2013 estimates place the total population at around 550 million with a 3.7% annual growth rate (PRB, 2013). Given these facts, it becomes clear that a detailed understanding of human–environment interaction in these vulnerable regions is needed. As humans continue to amplify their agropastoral land uses, other mechanisms (both biophysical and climatic) begin to build upon each other within the system and quicken the pace of land degradation. This section identifies the multiplicity of interlocking drivers (biophysical, climatic, and anthropogenic) that exacerbate degradation as well as outlines the compounded social and ecological impacts of these drivers present in affected areas through examples and case studies from North Africa through Central Asia.

11.1.3.1 Integrating Driving, Underlying Driving Forces, and Proximate Causes

It is helpful to build upon Geist's (2005, p. 26) analysis of "typical pathways," or the "chain of events and sequences of cause and effects" that drive dryland change when examining the cumulative impacts of how human land-use modifications and biophysical processes interact to heighten degradation susceptibility. Geist identified *driving forces*, *underlying driving forces*, *proximate causes*, and *trigger events* in his analysis. Driving forces include the slow and fast biophysical and climatic variables (biotic and abiotic changes in a region's natural capacity for certain land uses), whereas underlying driving forces include macro-stresses such as population increases, socioeconomic dynamics, and government policies, as well as long-term climatic conditions that contribute to a region's aridity. Proximate causes include human land use (grazing and agricultural intensification) as well as regional climate variations. Meteorological/climatic variables, such as short-term changes in rainfall amounts and intensities, temperature changes, and socioeconomic stresses and/or crises that force a further intensification of land exploitation serve as "triggers" that push a system beyond its carrying capacity thresholds, thus heightening the potential for land degradation.

The main *driving forces* of pedogenic degradation in drylands include exposure of the surface to high wind velocity and heavy, intermittent rainfall events. As discussed in the drylands section, aridisols and entisols, with their low organic matter and nutrient-cycling capabilities, are predominated by diffuse xerophytic vegetation patterns, which offer little coverage and are susceptible to erosion in the absence of either inorganic or biological surface crusts. When these crusts are disturbed and the vegetation removed, the weakly structured soils can be exposed to substantial wind and water erosion. Resultant soil erosion through deflation, transportation, and compaction causes a lessening of organic matter within the soil and higher salt contents (primary salinization and alkalinization (Thomas and Middleton, 1994)). This erosion, in turn, further increases nutrient loss and lowers the biological productivity capacity (soil exhaustion). This normally occurs over an extended period of time. Lower biological potential leads to less ground cover, which, in turn, leads to higher surface exposure and solar reflection rates (albedo). Surface albedo, it is hypothesized, decreases the region's atmospheric convective properties thereby leading to a more stable regional atmosphere with lower relative humidity levels and a reduction in rainfall potential, thus facilitating the onset of a positive feedback (see Charney et al., 1975).

Long-term climatic variability can be viewed both as a *driving force* and as an *underlying driver* of land degradation. Shifts toward greater aridification and desiccation result in higher temperatures and lower regional moisture regimes (lessening groundwater quantity and availability through reduced river flows, aquifer recharge rates, and so on) which compounds the pedogenic degradation processes addressed above. In the context of long-term, global climate change as Spellman (2000) estimates that for every 1° in temperature rise there is an estimated PET increase of 72 mm. Similarly, other climatic/meteorological factors that affect immediate geomorphological stability include periodic meteorological droughts (see Wilhite and Glantz, 1985) and episodic rainfall inundations that can quicken the reduction of surface vegetation cover and heighten the pace of soil erosion and flash-floods in affected regions (see Lavee et al., 1998).

Human exploitation of the natural system, as well, can be classified as a *driving force* and as an *underlying driver*. The stresses placed upon the environment stem from interwoven processes of direct and indirect macro-level forces (socioeconomic and political stability, national agricultural policies, and economic integration in the supraregional and global systems, to name a few), as well as more micro-level forces such as cultural associations with certain land-use types (traditional nomadic pastoralism, for example) and individual and regional land management practices to fulfill familial and communal subsistence and economic development needs.

One *underlying driver* of human-induced, degradation/desertification is the continued integration of dryland regions into the global economic system, the process identified as globalization. In terms of economic development potential, historically marginalized groups in dryland regions now have greater access to global markets for their land-based products. Similarly, they also have greater access to modern agricultural technologies leading to higher crop yields (mechanization, hybrid or genetically modified seeds, and fertilizers and pesticides). This access, however, can have two negative impacts. First, it can stimulate further agricultural expansion into traditional un- or underutilized areas (allowing for increased slope cultivation due to mechanized terracing as well as movement into drier and more nutrient-challenged areas where traditional, rain-fed farming has little success). Second, access to modern agricultural techniques can ultimately result in unsustainable practices such as overintensification of crops unfit for the dryland climate, as well as the further marginalization and displacement of people who cannot afford such technologies (Barbier, 2000).

Barbier (2000) argued that integration in the global system often leads these regions to intensify production to meet global and supraregional agricultural demands (commercialization), rather than focus on meeting local subsistence needs. Thus, the local agricultural system is weakened in its ability to buffer itself from market fluctuations in agricultural commodity pricing. It also has difficulties protecting itself from trends in market liberalization such as lessened import quotas and tariffs and state subsidies, among many others (see also Adhikari, 2013 for further discussion), under the present globalization paradigm. The end result is further human-induced degradation driven by agricultural expansion, often perpetuated by a society that has become, for better or for worse, intrinsically linked with the global economic system.

National development policies, commonly in conjunction with the increased integration of dryland systems into the globalized system, can also serve as *underlying drivers* of land degradation, as is the case across North Africa, the Middle East, and Central Asia (see Nielsen and Adriansen, 2005). Governmental policies aimed at national economic growth and stability, or to expand their global access, can direct the intensification of land exploitation into otherwise marginal areas. The spread of irrigated agriculture in Tunisia and Egypt, where it is known as “land reclamation,” has moved agricultural lands into desert areas fed by oases in both countries as well as intensively along the Nile River in the latter. (King and Thomas, 2014), whereas agricultural expansion in the arid regions of Jordan and Saudi Arabia necessitates the use of fossil waters deep underground, which requires advanced oil-drilling technologies.

Other techniques used to increase the agroeconomic development of a country are the building of multipurpose hydroelectric dam projects in underdeveloped regions of a nation to create new irrigated agricultural lands extending off the reservoir, as well as purportedly creating new and stable sources of income. The Aswan High Dam in Egypt (see Abu-Zeid and El-Shibini, 1997), the GAP Project in Turkey (see Harris, 2002), and the Tabqa Dam project in Syria (see Altinbilek, 2004; Shapland, 1997), serve to accentuate this point. Millions of acres of newly irrigated agricultural lands in these regions are dedicated to cotton production to meet national and global demands, with little land being used for subsistence crops. This pattern is repeated in Iran and in the Aral Sea Basin, both of which are discussed in more detail below. In all of these cases, new irrigation potential not only transforms former pastoral areas into higher-profit cropland, at least in the short-term, but it also displaces local inhabitants. As might be expected, this agricultural intensification can lead to degradation of the landscape. Agricultural movement into these marginal areas is often predicated on the maximization of production with the goal of national self-sufficiency. Thus, agricultural-sector policies that are growth oriented can lead to regional overcultivation and overexploitation of the natural resources (soil and water), leading to nutrient depletion, waterlogging, salinization, and subsequent abandonment.

As noted above, agricultural expansion and intensification (the *proximate cause* of overcultivation) can lead to a range of negative impacts across the social and biophysical landscape. More specifically, agricultural expansion in dryland areas generally requires the transformation of traditional pastoral lands to crop productions (displacing or sedentarizing those communities and their animals, gradually or by force). This land alteration, especially in the case of large fields, commonly requires deep plowing that exposes the soil surface to wind and water erosion. Investments in larger field systems in dryland environments require a number of insurance strategies such as the increased and sustained use of hybrid seeds and inputs of fertilizers, pesticides, and water. Access to these expensive inputs squeeze poor farmers out of the production system, therefore forcing their displacement to other marginal lands or migration to urban areas. Short or nonexistent fallow periods, coupled with the already low nutrient availability, quickly deplete the soils, which then require increased use of agrochemicals. Increased chemical usage and its runoff soon degrade water quality across the system.

Water requirements for agricultural intensification are met by the expansion of irrigation systems (including surface water and subterranean sources—aquifers and fossil waters). Not only does the overexploitation of water soon challenge availability, introducing scarcity, but inefficient practices such as channel and flood irrigation systems lead to two deleterious issues—waterlogging and salinization. First, overirrigation leads to waterlogging which not only diminishes the downsystem water availability, but also raises the overall water table within affected agricultural regions. This, in turn, means fewer salts are flushed through the system leaving higher levels of these and other minerals close to the surface, impacting the biotic potential of the soil, a process known as secondary salinization and alkalinization (Thomas and Middleton, 1994). Overirrigated fields in these arid regions also are exposed to rapid evaporation, which not only removes water from the system, but also quickens the pace of the infusion of salts and minerals into the topsoil and, in extreme cases, leaves salt crusts on top of the fields. Because few salt-tolerant crop species exist (date palms and licorice being two exceptions), the presence of high concentrations of sodium in the topsoil quickly creates a toxic environment and diminishes the productive capacity of the farm lands, thus leading to field abandonment.

The social and environmental stresses introduced by agricultural production in drylands are often underscored by national policies oriented around achieving sociopolitical stability through land-use intensification and lifestyle change (from a pastoral to sedentary dynamic). It is this national, political orientation toward agricultural self-sufficiency and lifeway change (to one that is easier to control and monitor) that Geist (2005) argued can be viewed as *underlying driving forces*. Across North Africa and the Middle East, national governments, building upon the aforementioned colonial perceptions of desertification being driven by deleterious indigenous land-use practices, continue to press for the sedentarization of traditional nomadic pastoralists. Recent studies continue to document the impact grazing has on localized regions within countries such as Algeria (Hirche et al., 2011), Morocco (Zucca et al., 2011), and Israel (Leu et al., 2014; Schaldach et al., 2013). Similarly, national policies that call for a decrease in rangeland in favor of sedentary crop production, and concurrently restrict pastoralist movement due to the enforcement of national borders, lead to a severe restriction in the pastoralists' horizontal ranges (between fields and areas within a region/nation) and vertical ranges (from low to high altitudes in the summer months). As demand for meats, milk and cheese, and wool increases, so too do the herd-stocking demands. These limitations on the movement of large herds of animals result in a number of biophysical impacts associated with overstocking and overgrazing, two of Geist's (2005) *proximate causes*.

The last examples of Geist's (2005) typical pathways presented in this section, overstocking and overgrazing, are the traditional culprits in desertification. As noted above, they lead to a number of negative biophysical impacts that challenge the potential carrying capacity of the region—in this case, the number of livestock per unit of land that can be raised without detrimental effects on the environment. Large herds occupying a region for an extended period of time (multiseasonal sedentarization) both affect the crusts that develop in certain regions (exposing the surface to erosion) as well as compact the soils due to trampling pressures from the animals' hooves, leading to increased water runoff. Concerning the compaction of soils by grazing animals, according to Heathcote (1983, p. 98) sheep exert a hoof pressure of 0.65 kg/cm^2, whereas cattle exert 1.7 kg/cm^2—he stated that comparably a crawler and a two-wheeled tractor exert 0.63 and 2.0 kg/cm^2 respectively. In the end, limited ranges lead to increased herd concentrations and increased foraging, and this, in turn, removes plant cover (increasing surface albedo) and/or allows for the intrusion of inedible species. Providing water for these herds through the drilling of bore-holes, compounds the negative effects of overstocking and overgrazing within the affected regions (Thomas and Middleton, 1994).

11.1.3.2 Case Studies of the Compounded Effects of the Drivers of Desertification: *Iran* and the *Aral Sea Basin*

Irrigation, agricultural development, and salinization in drylands have a long history dating back many millennia. Water exploitation and (mis)management are also a common denominator of agricultural intensification. Today, however, overextension of water resources is practiced at a much greater scale. Water usage in Iran is an example of the impacts of overexploitation leading to systemic degradation. According to Amiraslani and Dragovich (2010), much of Iran's 1.64 million km^2 geography is classified as arid and semiarid lands. They state that around 20% of the total landmass is desert areas, 55% is rangeland, and 11% is agricultural land. In addition, around 85% of the country's water usage (over 100 billion cubic meters) is dedicated to agricultural production, making water overexploitation a key systemic issue of compounding desertification in the country. According to Zafarnejad (2009), Iran's agricultural development policies have led to the damming of almost every major river for hydroelectric production and increased agricultural potential (especially the Karaj, Sefidrud, Jiroft, and Tajan rivers). On account of this, Iran has suffered from increased salinization in its irrigated fields and heightened rural depopulation.

The overexploitation of Iran's water systems, coupled with regional rainfall variability, has led to aquifers being drained to the extent that most of the country's 31,000 *qanats* (the traditional chain of wells linking aquifers in higher regions with drier, lower regions) are drying up (Zafarnejad, 2009; see also Amiraslani and Dragovich, 2010). Whereas salinization has been a major problem, another impact of hydrologic overexploitation is severe ecological degradation of downsystem riverine and riparian habitats through decreased flows and increased pollution contents. Large water bodies have seen severe retractions of their shorelines; Maharlu Lake volume has decreased by 90%, Bakhtegan Lake by 95%, and Parishan Lake by 80%. Lake Urmia, which has decreased from an average depth of 12 m in 2000 to just 6 m in 2014, has garnered international attention due to its drying, and recently a 1.3 billion dollar restoration project planned by the United Nations Development Programme (UNDP) and the Iranian government has been commenced

(UNDP, 2014). With the reduction in the volume of these lakes due to decreases in water inflow based on regional droughts and overextraction by humans, the salt- and agrochemical-infused lakebeds become exposed and dry up contributing to aerosol loading and leading to downwind deposition of these particles on agricultural and pastoral lands. Thus the degradation of these lake systems compounds the regional impacts of desertification by introducing salinization and chemical pollution into hitherto unaffected across the larger regional system. Unfortunately the process of hydrologic overexploitation leading to the drying of entire lake systems during long-term drought cycles is not just limited to Iran.

Another example of the regional impacts of human overextraction of water resources can be seen in the Aral Sea Basin of Central Asia, where there has been unprecedented land degradation based on agroeconomic intensification. The deterioration of the Aral Sea system is well published, so to avoid redundancies only the most significant aspects of the system's demise are presented here (see Kaplan et al., 2014; Micklin, 2010; Oren et al., 2010; MacKay, 2009; Oberhansli et al., 2007; Lioubimstva and Cole, 2006; Waltham and Sholji, 2001; among many others). The region, once under the control of the Soviet Union, was master planned as a cotton-growing region, and from the 1960s through today, the production of raw cotton for export has dominated the agriculture sectors of Kazakhstan, as well as Uzbekistan and Turkmenistan. To meet the production demands, Moscow facilitated the creation and expansion of irrigation systems along the Syr Dar'ya and Amu Dar'ya, the two principal water sources for the inland Aral Sea. The Soviets also constructed the 1,400-km-long Karakum Canal off of the Amu Dar'ya, which diverts nearly 13 km^3 of the river's flow and provides water to irrigate around one million hectares in the Karakum desert of Turkmenistan (Nesbit and O'Hara, 2000).

Not surprisingly, since the 1960s the Aral Sea has experienced a precipitous decline in water levels (by 26 m) and a decrease in overall volume (around 92%) and surface area (around 88%) (Micklin, 2010). These changes also mean that around 7.5 million acres of salt- and agrochemical-infused (also radioactive and biological weapons residue-infused) seafloor is now exposed. The wind activates the dry materials on the seafloor and blows them hundreds of kilometers downwind, depositing the dust, salts, and chemicals across agricultural and pastoral areas surrounding the desiccated sea. By the 1990s, between 40 and 150 million tons of material had been removed from the exposed seafloor and this figure has continued to rise, as exemplified by the 2008 dust storm that stretched 600 km downwind of the seafloor (Micklin, 2010).

The compounded nature of desertification drivers and impacts in the Aral Sea region is obvious. Overcultivation of cotton for colonial and global markets in Central Asia's dryland region (especially in the Karakum and Kyzylkum deserts) epitomizes the cumulative impacts of land mismanagement. Waterlogging remains a major problem throughout the irrigated fields along the Amu Dar'ya and Syr Dar'ya systems' Karakum Canal, where the water table in the Merv region is up by 20 m (Nesbit and O'Hara, 2000). Secondary salinization continues to degrade the productivity of the extant cropland, leading to an increased rate of abandonment of salinized fields and the conversion of pasture land to irrigated agricultural areas (Qadir et al., 2009). As one example, in Turkmenistan 4500 km^2 of former pasture land was converted to agricultural production between 1974 and 2004, representing an 86% increase in irrigated areas (Kaplan et al., 2014). Issues with a diminished supply, infused with salt and agrochemical runoff, have severely impacted the Aral Sea and its estuaries (one can clearly see from recent remotely sensed images that the Amu Dar'ya no longer even flows into the Aral Sea). The ecological impacts of this decreased flow to the Aral Sea include the division of the sea into four separate bodies of water, the loss of most fish species outside of the North Aral Sea (destroying the once prolific fish industry), and the extreme loss of bird species (from 319 to 160) and mammal species (from 70 to 32) (see Micklin, 2010, p. 203). Simultaneously, the ever-increasing seafloor exposure has increased regional salinization and human respiratory ailments due to aerosol loading, thus affecting land productivity and human health in regions with no direct causal link to the Aral Sea's demise.

In conclusion to this section, human-induced degradation is characteristically a reflection of mismanagement and nonsustainability (e.g., implementation of unsustainable agricultural uses, greater movement into marginal areas, and inefficient use of irrigation systems, among others). Unpredictable climate-triggering events that occur accentuate vulnerable, human-manipulated landscapes, leading to heightened degradation potential across regional landscapes. Soil structures, already facing issues of weak structure and cohesion, become further exposed to the wind and rain and, once activated, cause sheet and rill erosion (and in its most extreme form, gullying). The impacts of this soil erosion include increased loss of organic matter, deflation and the blowing of sands and fine particles (dune formation/movement and dust storms), and the deposition of eroded sediments in local water bodies and/or outwash areas. The loss of soil cover increases surface albedo and the loss of nutrients challenges the biological potential of the affected regions, which, in turn, leads to a decrease in the natural and agricultural carrying capacity of an affected region.

Trigger events as identified by Geist (2005) accentuate the diminishing capacity of degraded landscapes, and this diminished capacity, in turn, intensifies the socioeconomic stresses faced by the human inhabitants and can result in a range of responses from the benign—such as a transfer of uses (agriculture to pastoralism); to the more severe—field and/or regional abandonment (due to loss of biomass in rangelands or salinization or waterlogging of fields) and migration, both internal and external (see Leighton, 2006; Bettini and Anderson, 2014 for further discussion); to the acute—environmental security issues (perceived and real threats) of systemic collapse, famine, and geopolitical turmoil (for example see Kepner, 2006; Brauch et al., 2003).

11.1.4 APPROACHES TO IDENTIFYING DESERTIFICATION

Due to the complex interweaving of desertification drivers and their scalar interchangeability (subregional to global), as seen following Geist's (2005) *underlying driving forces*, *driving forces*, *proximate causes*, and *trigger events*, it is perhaps not surprising that no single assessment process has been adopted by researchers or policy makers. A variety of methodological approaches and a dizzying array of acronyms have been developed to assess the drivers and impacts of land degradation across dryland ecosystems. Some are macro-level studies that attempt to address the correlations between general atmospheric circulation models (especially the North Atlantic Oscillation and El Niño Southern Oscillation), rainfall and temperature variability, and vegetation coverage to understand the role climatic variations play as an underlying driving force for desertification in terms of vegetation productivity/coverage (see Alkama et al., 2012; Wang, 2003; for a palaeoclimate modeling of Central Asia see Jin et al., 2012), whereas others take a more micro-level approach, aiming to understand land degradation in a specific region (for example, Lavee et al., 1998).

Remotely sensed image analysis, aggregated with annual rainfall amounts and temperature trends, remains the backbone of most biophysical analysis of the drivers of desertification. Turning first to remotely sensed image analysis, this research frequently utilizes data from the NOAA Advanced Very High Resolution Radiometer (AVHRR) sensor array, Landsat ETM, and Landsat ETM+, other earth resource satellites such as SPOT, IKONOS, and Quickbird, and aerial photos. If possible, a Normalized Difference Vegetation Index (NDVI) value typically is calculated from the sensor data, specifically the infrared (NIS) and visible red (VIS) bands, as a measure of biomass productivity. These bands in particular are compared as healthy plants reflect more NIS and absorb more VIS, whereas unhealthy or water-stressed plants do the opposite (Kundu and Dutta, 2011; see also Symeonakis and Drake, 2004), and therefore patterns of change in the condition of the region's biomass can be monitored over time.

These remotely sensed sources, however, are not without their issues. The drawback to AVHRR technology is that the resolution is coarse (1.1 km^2 pixel size) and poses accuracy issues for analysis of small regions and the farm/plot level. In terms of Landsat ETM and Landsat ETM+, although it does provide data as far back as the late 1970s, thus allowing for a time series analysis of land-use/land-cover change (LULCC), it too suffers from resolution limitations (30 m by 30 m resolution), issues of cloud cover (tropospheric masking), and seasonal availability issues for image sequencing (drylands appear much more desiccated in summer months than during interannual rain events). Image costs remain high for other earth resource satellites such as SPOT, IKONOS, and Quickbird, and, although they do offer much finer multispectral resolutions (from 20 to 2.5 m, 3.3 m, and 2.5 m respectively), they also suffer from limitations such as image cloud cover and seasonal variability in imaging sequences. Aerial photos can be helpful for temporal sequencing, but they remain limited by coverage extents and availability within the dryland areas and, as Hirche et al. (2011) have noted, low-lying and sparse dryland vegetation is difficult to identify even on these images.

Turning now to the study of annual rainfall amounts and temperature trends, NDVI data are used by some as proxy-indicators for meso-scale climate and vegetation changes (see Kucharski et al., 2013; Lauwaet et al., 2009 for testing of the Charney Hypothesis (1975)), whereas other regional analyses attempt to identify and explain trends in the expansion and contraction of plant species in dryland areas, based on the principle of rain use efficiency (RUE). RUE is the ratio of aboveground natural primary production and the levels of annual precipitation. Dryland moisture regimes generally dictate what plants can survive in these arid regions and any departures or variations from these species-specific norms are considered a sign of human influence and/or potential degradation (Veron et al., 2006; see also Linstadter and Baumann, 2013; Prince et al., 2007; Hein and De Ridder, 2006). Critiques of the RUE process revolve around the resolution of the base data, as well as uncertainties concerning natural cycles of vegetation and climate versus the onset of degradation. Concerning the latter issue, Veron et al. (2006) have stated that the uncertainty is based on lack of historic reference situations and to remedy this, protected areas and national parks should be used as a control.

In addition to remotely sensed image analysis and climate and vegetation modeling, other studies seek to identify the physical processes of desertification following a geomorphological approach. Identification of the physical geography of a study area serves as the foundation for the main field-based, analytical structures. Building upon the techniques outlined by the UNCOD in 1977, the UNEP/FAO (1984) established a 22-part assessment matrix used to evaluate the desertification status in affected areas. As a means to measure the scale of the impact of degradation, this matrix focuses primarily on identifying the levels of erosion and salinization in a region (from slight to very severe, such as evidence of gullies for wind and water erosion and salt crusts in salinized areas (see Veron et al., 2006)), while also examining the percentage of plant cover. As another example of a geomorphological-focused approach, Lavee et al. (1998) analyzed the physical geography at the plot level along a single climatic transect in Israel and underscored the variability between landform types in the same transect and their differences in sensitivity to degradation.

Building upon the UNEP/FAO matrix, the Mediterranean Desertification Land Use Survey (MEDALUS) technique is also used by some desertification scholars as a robust ecological assessment tool. Environmental indicators are measured and classified according to the Soil Quality Index, the Climate Quality Index, Vegetation Quality Index, and Management Quality Index. This adaptable matrix allows for a detailed assessment of the environmental sensitivity of a region (see Bakr et al., 2012). Variations of the MEDALUS approach serve as the underlining assessment strategy for current studies that combine the environmental indicators with other variables, such as topography, slope and aspect, land use, soil types, and land cover, among others, within a GIS to create composite maps of multilayered desertification-hazard zonation (see Barzani and Khairulmaini, 2013; Mashayekhan and Honardoust, 2011). Though field analysis components are time-consuming and the data recorded in the geodatabases can be limited by the availability and quality from the source (national governments and the international community), the process holds much promise for the synthesis of the multivariate nature of desertification vulnerability analysis and the identification of areas suffering from potentially severe issues of degradation, or desertification "hot spots."

Regardless of which approach is used, as can be seen, a robust, interdisciplinary methodology that includes both human and biophysical analytical techniques (for example see Owusu et al., 2013; Reynolds et al., 2007) must be followed to understand the multiple variables working simultaneously to affect system vulnerability within dryland systems. The Desertification Development Paradigm (DDP) outlined by Reynolds et al. (2007) attempts to do this by developing a framework for integrated human–environment interaction analysis as it relates to the causes and consequences of desertification. They outlined five guiding "principles" that seek to link human livelihoods and ecosystem management (2007: 848 ff): (1) dryland human–environment systems are dynamic, coadapting, with no single target equilibrium point; (2) dryland dynamics are influenced by long-term (slow) drivers as well as intermittent and/or immediate (fast) changes in a system, and though they work in tandem not all variables carry equivalent weight; (3) dryland systems have multiple social and ecological thresholds that when crossed, change regions often in nonlinear ways; (4) multiple stakeholders and multiple hierarchical scales and networks must be considered; and (5) to maintain a functional and sustainable human–environment system, a much greater value must be placed on indigenous, local "hybrid" environmental knowledge (LEK).

In summary, a multiplicity of analytical methods, models, and monitoring techniques have been utilized to identify degradation throughout dryland regions, but they have been unable thus far to overcome the lack of a unifying set of input parameters, issues with biophysical data quantity and quality, remotely sensed data interoperability and resolution limitations (including the "people to pixel" problem discussed in more detail below), and the lack of relevant time series, or seasonal data. In response to these concerns, and with a goal of better achieving the reversal and/or remediation of desertification in the context of poverty alleviation and sustainable development (the revised tenants of the post-Rio+20 Convention to Combat Desertification [CCD]), scholars continue to work toward better interdisciplinary approaches that successfully emphasize the dynamic, coevolving nature of human environment interaction within dryland systems. Future integrated studies can help this process by better identifying the main drivers of desertification, as well as calibrating remediation processes that will achieve the goals of sustainable socioeconomic development.

11.1.5 DISCUSSION

After nearly 40 years of critical (re)evaluations, the analysis of desertification remains a complex, nuanced topic rife with intentional and unintentional obfuscation. No comprehensive or widely accepted assessment strategy has percolated to the surface, nor has a common mitigation and rehabilitation strategy been defined (both social and ecological). Due to the dynamic nature of the battery of potential drivers and their compounding effects, as well as the wide array of regional human–environment interaction variations, it may be naïve to think that there is a panacea for desertification/dryland degradation analysis.

In general, macro-level approaches such as described above are critiqued for glossing over regional complexities, whereas studies that take a more micro-level, inductive regional approach (such as advocated by Geist, 2005) are also critiqued because an assessment strategy such as this requires a fluid, multivariate assessment structure that may or may not be duplicable between regions as there is no standard baseline or control by which to compare and/or aggregate the analyses. Further, a multivariate assessment dynamic is difficult to "sell" at the international and national levels. Veron et al. (2006, p. 752) summarized it best: from action-oriented politicians to indigenous communities facing impending socioeconomic and ecological deterioration to (semi)informed donors, "the coexistence of conflicting definitions and divergent estimates negatively affects societal perceptions, leading to skepticism and, ultimately, to the delay of eventual solutions." The threat is that a general malaise fueled by "issue fatigue" (Stringer, 2008) will disrupt the development and implementation of future projects and, in the end, could threaten to remove desertification/land degradation from its rightful place in the international discussion of significant human–environment issues such as poverty, sustainability, resource scarcity, and potential socioeconomic–political instability.

As noted above, many studies rely on remote sensing technologies and multivariate GIS geodatabases. Positive ongoing trends with these technologies are occurring that allow for increasingly more detailed assessment levels than in the past (for example, improved resolutions, greater availability, and lower costs). Unfortunately, however, a number of significant limitations to the technologies exist, such as temporal scale, data quality deficiencies, and resolution issues. Perhaps most significantly, remote sensing monitoring and modeling techniques are commonly plagued with the inability to link "people to pixels" (Lambin et al., 2009), reducing our ability to understand sustainability and poverty eradication in the context of land degradation. Hazard-zonation maps of desertification may highlight degradation in certain regions, but pixel sizes and lines on the maps may cross between poorly defined private landholdings and even political boundaries, thus increasing the scale of stakeholder involvement and delaying any actions to reverse the degradation. Although human land-use practices may be one of the main drivers of degradation, humans also suffer from the most recognizable impacts in the context of the desertification dynamic, namely socioeconomic stresses/collapse, famines, and migration. Therefore, methods must be developed and utilized that better detail the human-economic-political landscape as it interacts with a region's ecology, and, to this end, methods that focus on highlighting local, socioecological systems that accentuate the importance of local environmental knowledge (see Easdale and Domptail, 2014; Reynolds et al., 2007) are increasingly being developed and utilized.

Another issue associated with these technologies is that of recognizing the importance of map currency, i.e., ensuring that the most up-to-date maps possible are used. Online searches today return links to the older USDA/NRCS *Global Desertification Vulnerability* map (1998) or derivations of the UNEP *World Map of Desertification* (1992), and many regional maps are derived from the 1997 *World Atlas of Desertification* indicating that need exists for a widely accessible, updated desertification map. In respect to this deficiency, the Joint Research Center of the European Commission and the UNEP currently are in the process of creating a new, revised World Atlas of Desertification accessible in print and in a limited form online (JRC, 2015).

Although this is a positive step, taking into account the current advancements in Web GIS mapping and publically accessible, high-resolution mapping platforms such as Google Earth, it seems almost a disservice to desertification studies to not have an official online mapping portal whereby continuously updated, dynamic regional maps of land degradation and desertification can be queried, manipulated, and produced by any interested or vested party. This portal could also provide public access to a comprehensive database of social, economic, and ecological statistics, and fill a much needed gap in the information relevant to land degradation. Such a portal would include a country by country list of current and historic statistics concerning the total area (ideally in a consistent form of measure such as km^2—which is lacking on many available aggregate data sets) of land classified as vulnerable, degraded, and/or desertified. This database would help numerically reinforce claims of increasing degradation in certain regions as well as celebrate successful reversal campaigns in other areas. The one major limitation plaguing this searchable map and database system, however, is the continued lack of a concrete system of measures for assessing the extent and impact of desertification. Currently those interested in the topic only have access to aggregate data, both online and in print, with limited ability to corroborate or verify any statistics presented. This introduces another avenue for potential skepticism in a contentious topic that is already suffering from the aforementioned "issue fatigue."

Finally, as noted above, perhaps the most significant issue with techniques of remote sensing monitoring and modeling is its inability to link "people to pixels." Images and maps may identify degradation or its vulnerability, but the lines or pixels may cross between individual/communal holdings and/or national boundaries. One way to bridge the gap between remote sensing strengths at macro-level analysis and the micro-level analysis of affected communities within or near desertification hot spots is to stimulate more public participatory mapping (PPM) and participatory GIS (PGIS) projects such as those utilized heavily by geographers and other field researchers in the social sciences (Dunn, 2007; see also Bisaro et al., 2013; Hessel et al., 2009; Raddaoui, 2009). When and where available, derivations on PPM and PGIS can also incorporate current social media outlets and online mapping sites for real-time interaction between stakeholders, thus further increasing community engagement within the process, as well as providing an opportunity for community members in affected regions to identify areas of degradation when direct field observation by international and government scientists is challenged by climatic or sociopolitical stresses. The data collected from these participatory mapping projects could also be collected and displayed on any official online Web-mapping portal.

11.1.6 CONCLUSIONS

An unfortunate truism of desertification analysis is that the forces that drive land degradation in dryland regions have only increased in recent years. Production of food and fiber in dryland regions has intensified not only to meet the needs of growing regional populations, but also to meet the global and interregional demands for these land-based products. Expanded cultivation and grazing ranges and the overextraction of water will continue to negatively impact the populations residing in these ecologically stressed areas, and though these populations carry with them a long history of coping mechanisms, the severity of macro-level stresses bearing down upon these regions, triggered by climatic variations or socioeconomic or political upheaval, may soon prove too much for the system to withstand.

Early on in the debate over the drivers of desertification, human (over)use of the natural resources in an area was seen as most detrimental, whereas ecological inputs were downplayed. Local communities were either considered active agents of degradation or passive actors who were being exposed to system-wide degradation, and therefore projects to reverse and/or remediate desertification were instituted in a top-down fashion, generally with parameters derived from Western or developed agricultural and pastoral structures. This approach was unfortunate. Although the emphasis on human actions is perhaps inevitable, since, as Thomas and Middleton (1994, p. 84) have stated, "irrespective of whether human action or nature, or a combination of the two, is believed to be responsible for desertification, it is the human role that we have the ability to modify...," we must keep in mind that the victims of desertification should not be made the culprits. Commonly communities residing in ecologically stressed areas have been displaced by landscape conversion to more intensive yet higher-income-generating practices, reflecting the pressures of national government policies and/or global forces.

The realization of the need to better contextualize the human role appeared in the 1990s, when it became apparent that UNCOD's approach was having little success, and it has continued into the present, with an emphasis on investigating human and climatic inputs, as well as working with local communities to develop a more regionally tailored remediation framework. The principles of the DDP (Reynolds et al., 2007) clearly illustrate this new approach. As they have argued, human–environment systems are adaptive, coevolving, and have multiple thresholds. Also, multiple stakeholders exist across multiple hierarchical scales in regions affected by desertification, and so any project to halt and reverse the impacts of dryland degradation needs to create a functional and sustainable structure aimed at long-term social and ecological stability. To accomplish this goal, local environmental knowledge needs to play a larger role in the decision-making process.

The principles of the DDP align with the UN's call for zero-net land degradation, sustainability, and poverty alleviation in dryland and drought-stricken areas. In this context, another example of a framework for poverty alleviation based on alternative livelihoods is addressed by Adhikari (2013). These alternatives include expanding successful local environmental knowledge techniques already implemented, such as intercropping strategies for animal fodder or for traditional high-value, drought-tolerant species (olives, cumin, dates, among others), encouraging multiple uses for irrigated areas, such as aquaculture in reservoirs and water-catchment areas, and facilitating other potential income-generating activities such as bee-keeping and solar-power generation, to all dryland regions. Other more controversial suggestions have been circulated, such as the creation of regional ecotourism industries or a greater push to utilize genetically modified, highly drought-tolerant seeds. Building upon Adhikari's (2013) suggestions, it must be clear that any project aimed at developing alternative livelihoods requires community acceptance and must remain a viable income generator to prevent the common occurrence of communities returning to previous livelihood activities once project funding runs dry.

Desertification remains a highly complex and nuanced topic. The generally incalculable variations in regional biophysical processes, social-economic-political volatility, and climatic dynamics result in a concrete approach to analyzing and remediating the impacts of land degradation in dryland areas remaining elusive. The multivariate complexities of human–environment interaction in these dryland areas continue to confound scientists, practitioners, and policy makers, and the lack of accessible data continues to shroud the causes and consequences of desertification in mystery. Researchers, officials, and local communities need to continue to interact to find viable, sustainable solutions to land degradation in these regions because pressures from food and cash-crop production, water scarcity, political instability, globalization, and climatic variations will continue to compound the processes of degradation currently at work within these vulnerable dryland areas.

REFERENCES

Abu-Zeid, M.A., El-Shibini, F.Z., 1997. Egypt's High Aswan Dam. Int. J. Water Resour. Dev. 13 (2), 209–218.

Adhikari, B., 2013. Poverty reduction through promoting alternative livelihoods: implications for marginal drylands. J. Int. Dev. 25, 947–967.

Alkama, R., Kageyama, M., Ramstein, G., 2012. A sensitivity study to global desertification in cols and warm climates: results from the IPSL OAGCM model. Clim. Dyn. 38, 1629–1647.

Altinbilek, D., 2004. Development and management of the Euphrates–Tigris basin. Int. J. Water Resour. Dev. 20 (1), 15–33.

Amiraslani, F., Dragovich, D., 2010. Cross-sectoral and participatory approaches to combating desertification: the Iranian experience. Nat. Resour. Forum 34, 140–154.

Aubreville, A., 1949. Climats, Forêts et Désertification de l'Afrique Tropicale. Société d'Éditions Géographiques Maritimes et Coloniales, Paris.

Bakr, N., Weindorf, D.C., Bahnassy, M., El-Badawi, M.M., 2012. Multi-temporal assessment of land sensitivity to desertification in a fragile agro-ecosystem: environmental indicators. Ecol. Indic. 15, 271–280.

Barbier, E.B., 2000. Links between economic liberalization and rural resource degradation in the developing regions. Agric. Econ. 23, 299–310.

Barker, G., Gilbertson, D., 2000. Living at the margin: themes in the archaeology of drylands. In: Barker, Gilbertson (Eds.), The Archaeology of Drylands: Living at the Margins. Routledge, London.

Barker, G., 2002. A tale of two deserts: contrasting desertification histories on Rome's desert frontiers. World Archaeol 33 (3), 488–507.

Barzani, N.M., Khairulmaini, O.S., 2013. Desertification risk mapping of the Zayandeh Rood Basin in Iran. J. Earth Syst. Sci. 122 (5), 1269–1282.

Bettini, G., Andersson, E., 2014. Sand waves and human tides: exploring environmental myths on desertification and climate-induced migration. J. Environ. Dev. 23 (1), 160–185.

Bisaro, A., Kirk, M., Zdruli, P., Zimmerman, W., 2013. Global drivers setting desertification research priorities: insights from a stakeholder consultation forum. Land Degrad. Dev. 25 (5), 5–16.

Brauch, H., Liotta, P., Marquina, A., Rogers, P., El-Sayed Selim, M. (Eds.), 2003. Security and Environment in the Mediterranean; Conceptualising Security and Environmental Conflicts. Springer-Verlag, Berlin.

Butzer, K.W., Harris, S.E., 2007. Geoarchaeological approaches to the environmental history of Cyprus. J. Archaeol. Sci. 34 (11), 1932–1952.

Charney, J., Stone, P.H., Quirk, W.J., 1975. Drought in the Sahara: a biogeophysical feedback mechanism. Science 187, 434–435.

Davis, D.K., 2004. Desert wastes of the Maghreb: desertification narratives in French colonial environmental history of North Africa. Cult. Geogr. 11, 359–387.

Davis, D.K., 2007. Resurrecting the Granary of Rome: Environmental History and French Colonial Expansion in North Africa. Ohio University Press, Athens.

Drenge, H.E., 1983. Desertification of Arid Lands. Hardwood Academic Publishers, Chur.

Dunn, C.E., 2007. Participatory GIS—a people's GIS? Prog. Hum. Geogr. 31 (5), 616–637.

Easdale, M.H., Domptail, S.E., 2014. Fates can be changed! Arid rangeland in a globalizing world—a complementary co-evolution perspective on the current 'desert syndrome'. J. Arid Environ. 52, 52–62.

Falkenmark, M., Rockstrom, J., 2008. Building resilience to drought in desertification-prone savannas in sub-Saharan Africa: the water perspective. Nat. Resour. Forum 32, 93–102.

FAO/UNEP, 1984. Provisional Methodology for Assessment and Mapping of Desertification. Food and Agriculture Organization of the United Nations, United Nations Environmental Programme, Rome.

Geist, H., 2005. The Causes and Progression of Desertification. Ashgate, Burlington.

Glantz, M.H., Orlovsky, N.S., 1983. Desertification: a review of the concept. Desertification Control Bull. 9, 15–22.

Harris, L., 2002. Water and conflict geographies of the Southeastern Anatolia project. Soc. Nat. Resour. 15 (8), 743–759.

Harris, S.E., 2012. Cyprus as a degraded landscape of resilient environment in the wake of colonial intrusion. Proc. Natl. Acad. Sci. U.S.A. 109 (10), 3670–3675.

Heathcote, R.L., 1983. The Arid Lands: Their Use and Abuse. Longman, London.

Hein, L., De Ridder, N., 2006. Desertification in the Sahel: a reinterpretation. Glob. Change Biol. 12, 751–758.

Hessel, R., van den Berg, J., Kaboré, O., van Kekem, A., Verzandvoort, S., Dipama, J.-M., Diallo, B., 2009. Linking participatory and GIS-based land use planning methods: a case study from Burkina Faso. Land Use Policy 26, 1162–1172.

Hirche, A., Salamani, M., Abdellaoui, A., Benhouhou, S., Martinez Valderrama, J., 2011. Landscape changes of desertification in arid areas: the case of south-west Algeria. Environ. Monit. Assess. 179, 403–420.

ICIHI (Independent Commission on International Humanitarian Issues), 1986. The Encroaching Desert: The Consequence of Human Failure. Zed Books Ltd, London.

Imeson, A., 2012. Desertification, Land Degradation and Sustainability: Paradigms, Processes, Principles and Policies. John Wiley and Sons, Ltd, West Sussex.

Issar, A.S., Ginat, H., Zohar, M., 2012. Shifts from deserted to inhabited terrain in the arid part of the Middle East, a function of climate changes. J. Arid Environ. 86, 5–11.

Jin, L., Chen, F., Morrill, C., Otto-Bliesner, B.L., Rosenbloo, N., 2012. Causes of early Holocene desertification in arid central Asia. Clim. Dyn. 38, 1577–1591.

Joint Research Center of the European Commission (JRC) and the United Nations Environmental Programme (UNEP), 2015. World Atlas of Desertification. Active link: http://wad.jrc.ec.europa.eu/ (last accessed 24.06.15.).

Kaplan, S., Blumberg, D.G., Mamedov, E., Orlovsky, L., 2014. Land-use change and land degradation in Turkmenistan in the post-Soviet era. J. Arid Environ. 103, 96–106.

Kepner, W.G., 2006. Desertification in the Mediterranean Region: A Security Issue. Guildford, London.

King, C., Thomas, D.S.G., 2014. Monitoring environmental change and degradation in the irrigated oases of the Northern Sahara. J. Arid Environ. 103, 36–45.

Kucharski, F., Zeng, N., Kalnay, E., 2013. A further assessment of vegetation feedback on decadal Sahel rainfall variability. Clim. Dyn. 40, 1453–1466.

Kundu, A., Dutta, D., 2011. Monitoring desertification risk through climate change and human interference using remote sensing and GIS techniques. Int. J. Geomatics Geosci. 2 (1), 21–33.

Laity, Julie, 2008. Deserts and Desert Environments. Wiley-Blackwell, Oxford.

Lambin, E.F., Geist, H., Reynolds, J.F., Stafford-Smith, D.M., 2009. Coupled human-environment system approaches to desertification: linking people to pixels. In: Roder, Hill (Eds.), Recent

Advances in Remote Sensing and Geoinformation Processing for Land Degradation Assessment. CRC Press, Boca Raton.

Lauwaet, D., van Lipzig, N.P.M., De Ridder, K., 2009. The effect of vegetation changes on precipitation and Mesoscale Convective Systems in the Sahel. Clim. Dyn. 33, 521–534.

Lavee, H., Imeson, A.C., Sarah, P., 1998. The impact of climate change on geomorphology and desertification along mediterranean-arid transect. Land Degrad. Dev. 9, 407–422.

Le Houérou, H.N., 1977. The nature and causes of desertization. In: Glantz (Ed.), Desertification: Environmental Degradation in and around Arid Lands. Westview Press, Boulder.

Leighton, M., 2006. Desertification and migration. In: Johnson, Mayrand, Paquin (Eds.), Governing Global Desertification: Linking Environmental Degradation, Poverty and Participation. Ashgate, Aldershot.

Leu, S., Mor Mussery, A., Budovsky, A., 2014. The effects of long time conservation of heavily grazed shrubland: a case study in the Northern Negev, Israel. Environ. Manage. 54, 309–319.

Linstädter, A., Baumann, G., 2013. Abiotic and biotic recovery pathways of arid rangelands: lesson from the High Atlas Mountains, Morocco. Catena 103, 3–15.

Lioubimtseva, E., Cole, R., 2006. Uncertainties of climate change in arid environments of Central Asia. Rev. Fish. Sci. 14, 29–49.

MacKay, J., 2009. Running dry: international law and the management of the Aral Sea. Cent. Asian Surv. 28 (1), 17–27.

Marsh, G.P., 1864. In: Lowenthal, D. (Ed.), Man and Nature. University of Washington Press, Seattle, 2003.

Mashayekhan, A., Honardoust, F., 2011. Multi-criteria evaluation model for desertification hazard zonation mapping using GIS. J. Appl. Biol. Sci. 5 (3), 49–54.

MEA Safriel, U., Adeel, Z., 2005. Dryland systems (Chapter 22). In: Millennium Ecosystem Assessment. World Resources Institute, Island Press, Washington, DC.

Micklin, P., 2010. The past, present, and future Aral Sea. Lakes Reser. Res. Manage. 15, 193–213.

Middleton, N., Thomas, D., 1997. World Atlas of Desertification, second ed. United Nations Environmental Program, New York.

Nesbitt, M., O'Hara, S., 2000. Irrigation agriculture in Central Asia: a long-term perspective from Turkmenistan. In: Barker, Gilbertson (Eds.), The Archaeology of Drylands: Living at the Margins. Routledge, London.

Nielsen, T.T., Adriansen, H.K., 2005. Government policies and land degradation in the Middle East. Land Degrad. Dev. 16, 151–161.

Oberhansli, H., Boroffka, N., Sorrel, P., Krivonogov, S., 2007. Climate variability during the past 2,000 years and past economic and irrigation activities in the Aral Sea Basin. Irrig. Drain. Syst. 21, 167–183.

Oren, A., Plotnikov, I.S., Sokolov, S., Aladin, N., 2010. The Aral Sea and the Dead Sea: disparate lakes with similar histories. Lakes Reser. Res. Manag. 15, 223–236.

Owusu, A.B., Cervone, G., Luzzadder-Beach, S., 2013. Analysis of desertification in the Upper East Region (UER) of Ghana using remote sensing, field study, and local knowledge. Cartographica 48 (1), 22–37.

PRB (Population Reference Bureau) Data Finder: International Profiles, 2013. Active link: http://www.prb.org/DataFinder/Geography.aspx?loct=3 (last accessed 01.08.14.).

Prince, S.D., Wessels, K.J., Tucker, C.J., Nicholson, S.E., 2007. Desertification in the Sahel: a reinterpretation of a reinterpretation. Glob. Change Biol. 13, 1308–1313.

Qadir, M., Noble, A.D., Qureshi, A.S., Gupta, R.K., Yuldashev, T., Karimov, A., 2009. Salt-induced land and water degradation in the Aral Sea basin: a challenge to sustainable agriculture in Central Asia. Nat. Resour. Forum 33, 134–149.

Raddaoui, B., 2009. Participatory monitoring and evaluation of a project to combat desertification in drylands (case study in Centre western Tunisia). In: Marini, Talbi (Eds.), Desertification and Risk Analysis Using High & Medium Resolution Satellite Data. Springerlink, Dordrecht.

Rapp, A., Le Houérou, H.N., Lundholm, B. (Eds.), 1976. Can Desert Encroachment Be Stopped? A Study with Emphasis on Africa. Swedish Natural Science Research Council, Stockholm.

Rapp, A., 1974. A Review of Desertization in Africa: Water, Vegetation and Man. Secretariat for International Ecology, Stockholm.

Reynolds, J.F., Stafford-Smith, D.M. (Eds.), 2002. Global Desertification: Do Humans Cause Deserts? Dahlem University Press, Berlin.

Reynolds, J.F., Stafford-Smith, D.M., Lambin, E.F., Turner II, B.L., Mortimore, M., Batterbury, S.P.J., Downing, T.E., Dowlatabadi, H., Fernández, R.J., Herrick, J.E., Huber-Sannwald, E., Jiang, H., Leemans, R., Lynam, T., Maestre, F.T., Ayarza, M., Walker, B., 2007. Global desertification: building a science for dryland development. Science 316, 847–851.

Rosen, S.A., 2000. The decline of desert agriculture: a view from the classical period of the Negev. In: Barker, Gilbertson (Eds.), The Archaeology of Drylands: Living at the Margins. Routledge, London.

Schaldach, R., Wimmer, F., Koch, J., Volland, J., Geißler, K., Martin, K., 2013. Model-based analysis of the environmental impacts of grazing management on Eastern Mediterranean ecosystems in Jordan. J. Environ. Manag. 127, 84–95.

Shapland, G., 1997. Rivers of Discord: International Water Disputes in the Middle East. Palgrave Macmillan, New York.

Spellman, G., 2000. The dynamic climatology of drylands. In: Barker, Gilbertson (Eds.), The Archaeology of Drylands: Living at the Margins. Routledge, London.

Spooner, B., Mann, H.S. (Eds.), 1982. Desertification and Development: Dryland Ecology in Social Perspective. Academic Press, London.

Stringer, L.C., 2008. Reviewing the International Year of Deserts and Desertification 2006: what contribution towards combating global desertification and implementing the United Nations Convention to Combat Desertification. J. Arid Environ. 72, 2065–2074.

Swift, J., 1977. Sahelian pastoralists: underdevelopment, desertification, and famine. Annu. Rev. Anthropol. 6, 457–478.

Symeonakis, E., Drake, N., 2004. Monitoring desertification and land degradation over sub-Saharan Africa. Int. J. Remote Sens. 25 (3), 573–592.

Tal, A., 2007. Degraded commitments: reviving international efforts to combat desertification. Brown J. World Aff. 13 (2), 187–197.

Thomas, D.S., Middleton, N.J., 1994. Desertification: Exploding the Myth. John Wiley and Sons, Chichester.

Thomas, David. S., 1997. Science and the desertification debate. J. Arid Environ. 37 (4), 599–608.

UN (United Nations), 1977. World map of desertification. In: UN Conference on Desertification, Nairobi, Kenya. Document A/CONF.74/2.

UN (United Nations), 1994a. UN earth Summit. Convention on desertification. In: UN Conference in Environment and Development. Rio de Janeiro, Brazil. June 3–14, 1992. DPI/SD/1576. United Nations, New York.

UN (United Nations), 1994b. United Nations Convention to Combat Desertification in Countries Experiencing Serious Drought and/or Desertification, Particularly in Africa (U.N. Doc. A/AC.241/27, 33 I.L.M. 1328). New York: United Nations.

UNCCD (United Nations Convention to Combat Desertification), 2000. Assessment of the Status of Land Degradation in Arid, Semi-arid and Dry Sub-humid Areas. United Nations Convention to Combat Desertification, Bonn.

UNCED (United Nations Conference on Environment and Development), 1992. Managing Fragile Ecosystems: Combating Desertification and Drought. Agenda 21, New York: United Nations.

UNCOD (United Nations Conference on Desertification), 1977. Desertification: Its Causes and Consequences. Pergamon Press, Oxford.

UNDP (United Nations Development Programme), 2014. Towards a Solution for Iran's Drying Wetlands. International Technical Round Table 16-18 March 2014 Conclusions and Recommendations. Active link: http://www.ir.undp.org/content/dam/iran/docs/Publications/E&SD/WIRT%20Conclusions%20and%20Recommendations.pdf (last accessed 20.06.14.).

UNEP (United Nations Environmental Program), 1984. General Assessment of Progress in the Implementation of the Plan of Action to Combat Desertification, 1978–1984. GC-12/9 United Nations Environmental Programme.

UNEP (United Nations Environmental Program), 1992. World Atlas of Desertification. Arnold, London.

Veron, S.R., Paruelo, J.M., Oesterheld, M., 2006. Assessing desertification. J. Arid Environ. 66, 751–763.

Waltham, T., Sholji, I., 2001. The demise of the Aral Sea—an environmental disaster. Geol. Today 17 (6), 218–224.

Wang, G., 2003. Reassessing the impact of North Atlantic oscillation on sub-Saharan vegetation patterns. Glob. Change Biol. 9, 493–499.

Whilhite, D.A., Glantz, M.H., 1985. Understanding the drought phenomenon: the role of definitions. Water Int. 10, 111–120.

Zafarnejad, F., 2009. The contribution of dams to Iran's desertification. Int. J. Environ. Stud. 66 (3), 327–341.

Zucca, C., Julitta, F., Franco, P., 2011. Land restoration by fodder shrubs in a semi-arid agropastoral area of Morocco. Effects on soils. Catena 87, 306–312.

Chapter 11.2

Grassland Degradation

Abbey F. Wick [1], Benjamin A. Geaumont [2], Kevin K. Sedivec [1] and John R. Hendrickson [3]

[1] *North Dakota State University, School of Natural Resource Sciences, Fargo, ND, USA,* [2] *North Dakota State University, Hettinger Research Extension Center, Hettinger, ND, USA,* [3] *United States Department of Agriculture, Agricultural Research Service, Mandan, ND, USA*

ABSTRACT

Grassland biomes occur naturally worldwide with an estimated 1.5 million square kilometers in North America alone. The three grassland types occurring in North American are tall, mixed and short grass prairies, each of which provide valuable ecosystem services to humans. Ecological services provided by grasslands include but are not limited to, carbon storage, habitat for pollinators, and forage for livestock and wild herbivores. While once prevalent, grassland ecosystems have and continue to be degraded throughout much of North America. The degradation of grasslands is affecting their ability to function properly and is hindering the ability of grasslands to provide the full suite of ecological services they could offer. Causes of degradation are many and may vary by region. Much of the tall grass prairie has been plowed for crop production. In the mixed grass prairie, recent technological innovations in oil production (fracking) has increased oil production but has also resulted in degradation of the grassland environment in multiple ways including habitat fragmentation. All three grassland types have suffered from livestock overgrazing and the disappearance of fire from the landscape. The true value of ecological services provided by grasslands is unknown; however, many of those provided may be critical for the long-term survival and health of humans. With the number of intact grasslands continuing to decline in North America in coincident with their degradation, care needs to be taken when developing effective management plans and government policy for grassland preservation.

11.2.1 INTRODUCTION

The Society for Range Management and the United States Department of Agriculture define grasslands as land on which the vegetation is dominated by grasses, grass-like plants, and/or forbs (Jacoby, 1989). Grasslands naturally occur worldwide and are often thought of as having the greatest diversity of grazing animals and predators on the planet (White et al., 2000). Because of their extent, grasslands are one of the largest ecosystems in the world, making

Biological and Environmental Hazards, Risks, and Disasters. http://dx.doi.org/10.1016/B978-0-12-394847-2.00016-4

up about 40.5% of the earth's terrestrial surface excluding Greenland and Antarctica (FAO, 2005). However, throughout the world, grasslands are under threat of degradation through human activities.

In wetter areas, such as the North American prairie and South American Pampas, grasslands have been converted to annual agriculture (FAO, 2005). Much of the grassland conversion occurred during European settlement of the Americas; however, the conversion process is still continuing. Between 2006 and 2011, Wright and Wimberly (2013) found grasslands declined by 530,000 ha in the US states of North Dakota, South Dakota, Nebraska, Minnesota, and Iowa. Baldi and Paruelo (2008) recorded a 6% decrease in grassland cover in the Rio de la Plata biogeographical region of South America between 1989 and 2002. Changes in land use also threaten European grasslands (Toland et al., 2008).

Although land use change directly and quickly alters grassland, grasslands can also be degraded by management practices. Currently, grazing pressures have reduced vegetative biomass on grass steppes in Mongolia (Liu et al., 2013). Overgrazing was recognized as a threat to Great Plains grasslands in the United States (Weaver, 1954). Large scale overgrazing by domestic livestock in North America began at the end of the nineteenth century and continued into the early twentieth century, following the near eradication of bison and rapid expansion of domestic livestock (Vermeire et al., 2004). The effects of overgrazing can have major and long-term impacts on plant communities (Laycock, 1991).

In this chapter, we focus on historic native grasslands in the Great Plains of North America and human-based processes, which alter the function previously provided by grassland systems. However, some attention is brought to less diverse grasslands which have replaced native grassland systems.

An estimated 1.5 million square kilometers of prairie communities occur in North America, which contain a majority of the native grasslands (Mac et al., 1998). Each grassland can typically be described as one of the three categories: (1) tallgrass prairie, (2) mixed-grass prairie, and (3) short-grass prairie. Each native grassland type was formed based on ecological drivers that include climate; frequency of fire; and the frequency, density, and duration of visitation by native grazers (Samson et al., 2004). Within each grassland category, plant diversity and community dynamics differ among regions, landscape position, and soil types (Barbour et al., 1998). The transition in type of grassland across the Great Plains is shown in Figure 11.2.1.

Tallgrass prairie develops under higher levels of precipitation than mixed- and short-grass prairies. Historically, tallgrass prairie dominated 677,300 km^2 and has declined in area by 96.8% to current estimates of 21,548 km^2 (Samson et al., 1998). This is largely attributed to the inherent fertility of these soils and suitability for conversion to agricultural production (Samson et al., 2004). Mixed-grass prairie historically dominated 628,000 km^2 of the Great Plains and has declined to 225,803 km^2, which is a 64% decline. Short-grass prairie has seen similar declines of 65.8% from 181,790 to 62,115 km^2

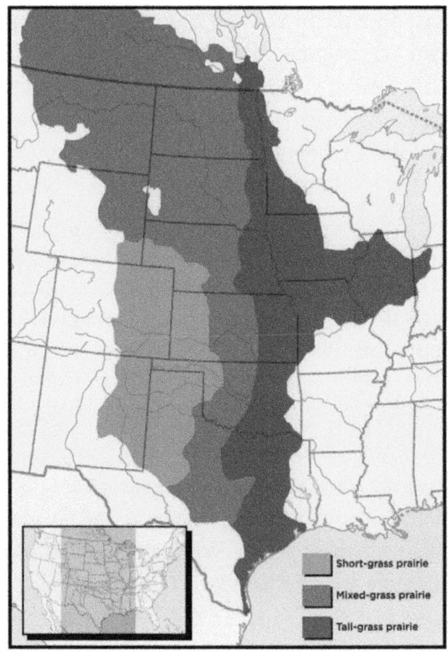

FIGURE 11.2.1 Extent of grassland categories in North America (Samson et al., 1998).

(Samson et al., 1998). Native grasslands are at risk of degradation for various human services. Additionally, many of the native grasslands have been replaced with less diverse grassland communities following their conversion to some other land use (cropping is an example).

Grassland species composition varies by geographic region. The southern Great Plains region is dominated with warm-season grasses. Cultivated grasslands may or may not contain forbs and legumes whereas native grasslands typically do contain flowering plant species. Common grassland types of the southern Great Plains include a mixture of big bluestem (*Andropogon gerardii* Vitman), switchgrass (*Panicum virgatum* L.), little bluestem (*Schizachyrium scoparium* (Michx.) Nash), and Indian grass (*Sorghastrum nutans* (L.) Nash) with and without a mixture of forbs and legumes. The nonnative Bermudagrass (*Cynodon dactylon* L.) also commonly occurs in our southern Great Plains as a grassland type from Texas to Kansas. Bahiagrass (*Paspalum notatum* Flugge) is another nonnative species occurring in the southern Great Plains and is commonly planted to perennial grasslands in southern Texas.

The central Great Plains region grasslands, both cultivated and native, are comprised of a mixture of cool- and warm-season grasses. This region is best described as that area found within Nebraska and Kansas, and may include western Iowa and Missouri, and eastern Colorado and New Mexico. A common grassland can comprise a mixture of big bluestem, switchgrass, little bluestem, Indian grass, western wheatgrass (*Pascopyrum smithii* (Rydb.) Á. Löve), and needlegrasses (*Heterostipa* spp.). Nonnative smooth bromegrass-type (*Bromus inermis* Leyss.) grasslands can occur in the northern areas of this region whereas nonnative crested wheatgrass-type (*Agropyron cristatum* (L.) Gaertn.) grasslands occur in the western areas. Intermediate wheatgrass (*Thinopyrum intermedium* (Host) Barkworth & D.R. Dewey), another nonnative type grassland are common throughout the central Great Plains region.

The grasslands of northern Great Plains region are dominated by cool-season grasses with forbs included in native grasslands and cultivated grasslands either with or without forbs and legumes. This region includes North and South Dakota, and may include western Minnesota and eastern Montana. However, switchgrass and big bluestem grassland types are scattered throughout the northern Great Plains. The most common cultivated grassland type is wheatgrass-dominated grasslands that are dominated by introduced grasses including intermediate and tall wheatgrass (*Thinopyrum ponticum* (Podp.) Z.-W. Liu & R.-C. Wang) followed by crested wheatgrass and smooth bromegrass type grasslands. The native grass, slender wheatgrass (*Elymus trachycaulus* (Link) Gould ex Shinners) is also a common species occurring in cultivated grasslands. Introduced meadow bromegrass (*Bromus biebersteinii* Roem. & Schult.) has become a popular grass planted for pasture and hay in recent years. Alfalfa (*Medicago sativa* L.) and yellow sweetclover (*Melilotus officinalis* (L.) Lam.) are common forbs occurring in cultivated grasslands of the northern Great Plains. Purple coneflower (*Echinacea purpurea* (L.) Moench) and yellow prairie coneflower (*Ratibida columnifera* (Nutt.) Woot. & Standl.) are common forbs established in native grasslands of the region.

In this chapter, we discuss the benefits and ecological services provided by historically native grasslands and how degradation through land use conversion and management practices impacts those services. At the same time, we acknowledge the services provided by less diverse grasslands that have replaced the native grassland systems.

11.2.2 BENEFITS AND ECOLOGICAL SERVICES

Grasslands contribute to a suite of ecological services including, but not limited to, soil and water resources, biological diversity and habitat, forage and recreation.

11.2.2.1 Soil and Water Resources

One of the most identifiable ecological services of the soil resource under grasslands is the carbon (C) storage capabilities. The third largest pool of carbon (C) is contained in the soil resource (estimated at 1220–1550 Pg of organic C and 685–748 Pg of inorganic C to 1 m depth; Eswaran et al., 1995; Batjes, 1996), with grasslands containing up to 12% of the global pool of soil organic C (Schlesinger, 1997). Soil C in grasslands provides a suite of environmental services, including support of plant nutrition through nutrient supply and retention, improved water retention, building of aggregation to facilitate water movement through soils, and erosion reduction (Conant et al., 2001). Management practices utilized in grasslands play a large role in the preservation or degradation of soil C, primarily organic C. Practices that preserve or enhance soil C and maintain nutrient cycling include controlled or rotational grazing (Schuman et al., 2009), soil conservation practices of reduced disturbance (Lal et al., 1998; Franzluebbers, 2009), prescribed burning or natural fire regime (DeBano et al., 1998; Vermeire et al., 2011), and maintenance of native plant composition (Sanderson et al., 2013).

An important hydrologic link has been identified for grassland establishment and management (Fiedler et al., 2003). Changes in hydrologic services may occur as the result of shifts in vegetation diversity and density, canopy structure, and soil health parameters in addition to climatic factors (precipitation and temperature; Frank, 2003). Therefore, a management practice influencing vegetation or soil health, such as grazing intensity or fire regime, will ultimately influence the hydrologic functions performed by grasslands. Additionally, grasslands surrounding wetlands play an important role for wetland function (Van der Kamp et al., 2003).

11.2.2.2 Biological Diversity and Habitat

North American grasslands are rich in biological diversity as a result of the previously described categories (tall-, mixed- and short-grass prairies). Due in part to this heterogeneity across categories, North American grasslands provide habitat for a variety of animal species.

Biological diversity in grasslands facilitates pollination. Grassland flora and fauna have coevolved leading in many instances to specific flower characteristics in plants and corresponding anatomy or appendages in pollinators (Kiester et al., 1984; Ricklefs and Miller, 1999). Animal pollinators play an important role in the production of numerous crops that are essential for human nutrition (Daily, 1997; Klein et al., 2007). Not all crops which provide human nourishment are pollinated by animals; however, it is plausible that most humans could not adequately fulfill their nutritional needs on nonanimal pollinated foods alone (Free, 1993; Steffan-Dewenter et al., 2005; Eilers et al., 2011).

Introduced honeybees (*Apis mellifera*) have been the most important and economically viable pollinator for most crop monocultures occurring throughout North America and worldwide (Southwick and Southwick, 1992; Williams, 1994; Klein et al., 2007). In recent years, honeybee populations have experienced significant declines (Kremen and Ricketts, 2000; Kluser and Peduzzi, 2007). The diversity of native pollinators and their importance to human food production has been ignored for years, but as honeybee populations decline, their relevance to human food production is being considered (Kremen et al., 2004; Morandin and Winston, 2005). Wild pollinators may not currently exist in large enough numbers to pollinate the extensive agricultural crops that exist across today's agricultural landscape; however, they can provide an important ecological service by increasing the production of certain crops, fill the void of population declines associated with introduced honeybees, and maintain plant diversity within remaining grassland habitats (Kearns et al., 1998; Kremen et al., 2002; Klein et al., 2007; Winfree et al., 2007; Potts et al., 2010).

Historically, mega-fauna including grizzly bears (*Ursus arctos*), American bison (*Bison bison*), and elk (*Cervus canadensis*) were prevalent within the prairie ecosystem along with smaller mammals including black-tailed prairie dogs (*Cynomys ludovicianus*) and black-footed ferrets (*Mustela nigripes*). Today, the great majority of the mega-fauna have been removed from the landscape and populations of important ecosystem engineers such as the black-tailed prairie dogs have been greatly reduced (Miller et al., 1994). Nonetheless, North American grasslands continue to provide habitat for a whole suite of animals. Numerous grassland bird species occur within the grasslands each selecting for specific attributes within the short-, tall-, or mixed-grass prairies (Knopf, 1996). Along with bison, deer, elk, rodents, and birds, grasslands are home to insects, butterflies, bats, amphibians, reptiles, and more.

11.2.2.3 Forage

Grasslands of the Great Plains provide herbage for livestock in the form of grazing land or hay land, and often times used as both. Traditionally, native grasslands provide extensive diversity for continuous grazing activities. Many of these native grasslands have been replaced with less diverse grasslands containing both native and introduced grasses used for grazing. It is important to recognize how grasslands were replaced throughout history because the services provided by native grasslands are very different from those provided by "replacement" grasslands.

"Replacement" grasslands within the Great Plains region were developed on previously cultivated lands. Most of the Great Plain's grasslands were established on degraded cropland in the 1930s following legislation associated with the dust bowl as a depression-era solution to lower crop prices (United States Department of Agriculture, 1985). The Soil Bank Act of 1956, which was part of the Agricultural Act of 1956, created the second large conversion of farm land to grasslands, with 11.6 million hectares nationwide converted from crop production to conservation uses (Helms, 1985). The final legislative act that converted almost 16.2 million hectares to grasslands was the 1985

Farm Bill that created the Conservation Reserve Program (United State Department of Agriculture, 2002). Although many of these conservation program lands have been converted back to farm land, much of the today's grasslands of the Great Plains are a product of past and current legislative actions.

Much of the grasslands replacing native grasslands in the Great Plains were developed for pasture and hay land, and conservation use. Early, developed grasslands provided feed for livestock, either as grazing lands or for hay production. The most recent grasslands were also developed to create wildlife habitat as nesting and brood-rearing habitat, and winter cover; as well as recreational areas for hunting and other outdoor activities. Forbs and legumes were added to most perennial plant mixtures starting in the 1980s to provide high quality feed for livestock and wildlife, diversity for wildlife habitat, and pollinator species for insects.

11.2.2.4 Recreation

Grasslands provide substantial recreational opportunities for the public. From bird watching to hunting, humans find great solace in spending time afield in grasslands. Bird watching may be the fastest growing form of ecotourism in the United States with approximately 33% (or 70.4 million people over the age of 16 years) taking part in the activity in 2001 (Cordell and Herbert, 2002). The grasslands of North America are a common destination for birders in search of endemic species to add to their life list including among others the Sprague's Pipit (*Anthus spragueii*), Baird's sparrow (*Ammodramus bairdii*), and ferruginous hawk (*Buteo regalis*; Knopf, 1996).

Hunting is another common form of recreation undertaken on America's grasslands. North American grasslands provide habitat for numerous game species, which attract sportspeople from around the globe. Although not a direct recreational activity afforded to humans by grasslands, gardening continues to be a popular past time in North America with many of today's floristic cultivars originating from native prairie species (Armitage, 2006). Most forms of grassland recreation can be considered ecotourism and can provide an important source of income to local communities and people through the expenditures associated with such activities (Diamantis, 1999; Hodur et al., 2004). The true recreational value of grasslands is unknown; however, the opportunities are countless and it is clear that these opportunities bring great joy to many who participate in them providing yet another suit of ecological services to society.

11.2.3 GRASSLAND DEGRADATION

For the purpose of this paper, grassland degradation is defined as a reduction in the ability of grassland to provide ecosystem goods and services (FAO, 2014).

Several identified practices or conversions occur that lead to grassland degradation. Some of that which have been identified include conversion to agricultural or urban lands and fragmentation, energy development, exotic species invasion, fire dynamics, and grazing intensity.

11.2.3.1 Conversion and Fragmentation

The continued loss due to conversion and degradation of grasslands to alternate uses has negatively affected several ecological services listed in the previous section. Soil resources are particularly impacted through grassland conversion to cropland and urbanization, where as much as 90% of soil structure that facilitates water movement through soils (infiltration) and water storage (water holding capacity) in soils is lost (Jastrow and Miller, 1998), C storage potential is reduced (Liebig et al., 2009), and soil function is impeded. Disturbances, which reduce soil organic C and alter nutrient cycling include overgrazing (Conant et al., 2001), biomass removal (Lal, 2004), invasion of undesirable species, cultivation (Liebig et al., 2009; Sanderson et al., 2013), and drainage/changes in hydrology.

Additionally, many animal pollinators are also drastically reduced through the conversion of grasslands to a monoculture of crops in agricultural lands or urbanization because of the removal of the flowering diversity pollinators require to survive (Grixti et al., 2009; Winfree et al., 2009; Potts et al., 2010). Fragmentation of grasslands, the introduction of exotic plants and pollinators, intensive grazing, and increased use of pesticides and herbicides have all been found to reduce wild pollinator populations (Kearns and Inouye, 1997; Kearns et al., 1998; Grixti et al., 2009; Winfree et al., 2011). Habitat fragmentation can isolate both pollinators and flowering plants straining pollinator populations by potentially lowering their food source and reducing genetic transfer both among and within populations (Kearns et al., 1998). Strong support also exists for the loss of bee diversity and abundance as a result of habitat degradation and fragmentation (Ricketts et al., 2008; Winfree et al., 2009).

Grassland conversion to crop production has had a profound effect on wildlife habitat, reducing grasslands by millions of hectares and commonly fragmenting those that remain (Samson et al., 2004; Wright and Wimberly, 2013). Although the result of habitat loss is somewhat clear, "no home, no wildlife," the effect of fragmentation on grassland wildlife appears somewhat species- and scale dependent (Fahrig, 2003; Stephens et al., 2003). Grassland birds as a whole have declined steadily for over 45 years (Sauer et al., 2013). Many grassland birds are sensitive to patch size and these populations typically show sharper declines than less area sensitive birds (Herkert, 1994; Bender et al., 1998; Johnson and Igle, 2001; Sauer et al., 2013). Fragmentation and urbanization have influenced predator populations and at times reduced nest success of birds (Crooks and Soule, 1999; Chalfoun et al., 2002; Stephens et al., 2003). Smaller grassland fragments can be more efficiently covered by predators resulting in more lost nests whereas urbanization may reduce mesopredator numbers leading to an altered predator community that may be more specialized at feeding on birds and their eggs (Crooks and Soule, 1999).

11.2.3.2 Energy Development

Energy development in the United States is a potential cause of grassland degradation. In the search for energy independence, the United States has ramped up energy production with many of these efforts occurring on grasslands (McDonald et al., 2009). The recent development and refinement of fracking has facilitated the drilling of thousands of new wells within the mixed-grass prairie region of North Dakota, Montana, and Canada. The oil play in this region, known as the Bakken Formation, has resulted in the fragmentation of millions of hectares of grasslands potentially affecting wildlife and exposing much of the area to invasive species. Due to the limited infrastructure in much of the region, road construction and truck traffic has greatly increased, raising the risk of grassland degradation from increases in road dust. Road dust is known to affect plant growth by decreasing their ability to photosynthesize by reducing stomata function and through the potential accumulation of heavy metals (Farmer, 1993; Lokeshwari and Chandrappa, 2006).

The increase in oil production has led to the construction of thousands of kilometers of pipelines. Pipeline construction requires the disturbance of vegetation and topsoil and has the potential to further degrade the local plant community. The associated disturbance opens up the plant community to invasion by unwanted weeds, many of which may be nonnative or considered invasive. Further, following pipeline construction, native and nonnative plants may be sowed back into the disturbed areas, but these plantings may provide an additional avenue for nonnative plants to encroach on the surrounding plant communities (Desserud et al., 2010).

Wind energy is a growing industry in the United States with many wind farms being constructed on grassland habitat (McDonald et al., 2009). Government incentives have helped facilitate the recent growth in wind energy (U.S. Department of Energy, 2014). Wind energy has the potential to degrade existing grasslands by further fragmenting grasslands and opening them up to invasion by nondesirable plants. The actual footprint of a wind farm may be small, but the associated transmission lines and additional infrastructure can greatly expand the impacted area (McDonald et al., 2009). Wind farms are commonly located in some of the more windy areas within a region and typically occur on hill tops. In many regions, these hill tops consist of the last remaining prairie fragments and may further degrade existing grasslands (Pruett et al., 2009). Wind farms can be devastating to populations of certain wildlife species whereas others avoid such areas all together (Robel et al., 2004; Pitman et al., 2005; Drewitt and Langston, 2006; Kunz et al., 2007).

The Energy Independence and Security Act of 2007 strived to bring the United States to energy independence partially through the use of clean renewable biofuels. Similar to other forms of energy development, the federal government has put incentives in place to help stimulate the development of new bioenergy sources and secure the future of existing ones (Yacobucci, 2010). The production of corn for use in ethanol has resulted in the turning over of many grassland hectares in the United States (Fargione et al., 2009). The increased demand for corn has played a significant role in the loss of Conservation Reserve Program lands and evidence exists that the demand for corn has also led to the destruction of native grassland hectares (Secchi et al., 2009; Wright and Wimberly, 2013). The loss of grasslands to ethanol production is a cause for concern; however, the act of doing so fragments the existing grassland area causing further degradation.

11.2.3.3 Exotic Species

Invasions of exotic species can have above- and below-ground impacts on grasslands resulting in degradation. Exotic species have been reported to increase biomass, increase nitrogen availability, change nitrogen fixation rates, and increase decomposition rates by increasing litter quality (Ehrenfeld, 2003). Both native and exotic species characteristically have favorable responses to increased fertility (Stohlgren et al., 1999). However, Wedin and Tilman (1990) indicated that strong feedbacks occur between species composition and nitrogen cycling and plots dominated by nonnative cool-season grasses had lower nitrogen retention than plots dominated by native warm-season grasses (Wedin and Tilman, 1996). Dominance of smooth bromegrass in tallgrass prairies has been linked to being able to compete more effectively for nitrogen and cycling nitrogen more rapidly (Vinton and Goergen, 2006; Figure 11.2.2).

FIGURE 11.2.2 Smooth brome invasion into native grasslands in Mandan, ND. *Photo taken by John Hendrickson.*

Hendrickson et al. (2001) reported greater rates of decomposition for smooth bromegrass and crested wheatgrass compared to two native grasses, which could result in more rapid cycling and the potential for loss of nitrogen from the system.

Invasive plants have also been reported to facilitate plant invasion (Jordan et al., 2008). This study indicated that two common invasive grasses, smooth bromegrass and crested wheatgrass suppressed growth on two of three native forbs tested (Jordan et al., 2008) although the impact was less on native grasses. Sites invaded by crested wheatgrass have been reported to be lower in species diversity, although not species richness, compared to uninvaded sites (Krzic et al., 2000) and have also been reported to have reduced forb richness and diversity (Henderson and Naeth, 2005). Although native forbs are reduced by invasive grasses, invasive forbs also reduce species richness and relative cover (Pritekel et al., 2006). Although exotic species may have similar impacts on grasslands, it is important not to treat them as a single entity. Hendrickson and Lund (2010) pointed out the importance of understanding both the target species and the plant community in developing restoration for development of adaptive management strategies.

11.2.3.4 Fire Dynamics

The role of fire in the formation of Northern Great Plains plant communities may be considered secondary to climate, but still substantial (Singh et al., 1983; Higgins, 1986). Historical accounts of fire prior to European settlement and charcoal studies depict numerous fires occurring across the Great Plains that were typically caused by Native Americans and lightning strikes (Higgins, 1986; Umbabhowar, 1996). The seasonality, intensity, and return rate of fire altered the plant community in different ways (Kucera and Koelling, 1964; Ewing and Engle, 1988; Biondini et al., 1989). Furthermore, recently burned areas typically result in nutritional regrowth that attracts large ungulates potentially further altering species composition (Milchunas et al., 1988; Hartnett et al., 1996). Along with European settlement of the Great Plains, came a large effort to suppress wildfires across the landscape in an attempt to protect crops, livestock, and settlers. The subsequent removal of Native Americans from vast areas further reduced the occurrence of fire across the Great Plains.

Grasslands are considered disturbance-dependent ecosystem and it is reasonable to believe that the century-long reduction in fire has degraded many grasslands throughout the Great Plains. The overall effect of fire reduction may differ within tall-, mixed-, and short-grass prairies as the frequency of fire differed within these ecosystems (Frost, 1998). The general knowledge regarding the true effect of fire reduction across all grasslands is limited due in part to few and often vague historical accounts of community composition prior to European settlement. Nonetheless, in areas of tallgrass prairie, woody species have encroached onto grasslands changing the general plant community, often increasing woody cover at the expense of grasses (Abrams et al., 1986; Briggs et al., 2002). In other areas, the accumulation of litter has affected species composition and production through the loss of certain grasses and forbs (Weaver and Rowland, 1952; Knapp and Seastedt, 1986; Hobbs and Huenneke, 1992). Further, the loss of fire and shifts in species diversity may open up grasslands to invasion by exotic species (Hobbs and Huenneke, 1992; Smith and Knapp, 1999). In recent years, much knowledge has been gleaned through scientific research regarding the effect of reintroducing fire onto the grassland ecosystems.

11.2.3.5 Grazing

Grassland herbivory has long occurred affecting vegetation in various ways (Anderson, 2006). Large bison herds once moved freely scouring the grasslands in search of forage. Movement patterns of foraging bison were typically driven by rain and fire events with return intervals of grazing bison varying from one location to the next. The preference of bison for certain vegetative species (C_4 grasses) over others created competition among plants that influenced the species composition present prior to their extermination (Schwartz and Ellis, 1981; Anderson, 2006). Numerous other herbivores with different forage preferences, including pronghorn (*Antilocapra americana*), occurred on the landscape with bison (Schwartz and Ellis, 1981). The selection of certain plant types by these additional herbivores may have further impacted species composition, albeit to a more localized and lesser extent than bison.

The removal of bison changed the disturbance regime experienced throughout North America for hundreds of years and in their place domesticated livestock were introduced. The introduction of domestic livestock brought about changes to the grazing regime of grasslands in North America. At first, the grasslands of the west were relatively free of private ownership and treated like the commons where livestock were allowed to roam and graze freely. In 1862, the passage of the Homestead Act ignited settlement and subsequent ownership of the grasslands. Landownership brought about the establishment of fences used to keep domesticated cattle and other livestock in designated areas. Unlike bison that grazed freely across the landscape, domesticated livestock were under private ownership and generally restricted to certain areas. Limited landownership, in part due to the inability to produce cattle on the 160 acres deeded to producers by the Homestead Act, resulted in overuse of grassland resources and at times altered the grassland community (Fleischner, 1994).

In addition to changes in the disturbance and distribution regime between cattle and bison, differences occur in foraging habits between the two. Cattle similar to bison prefer graminoids over forbs and shrubs, but generally have a slightly different foraging preference and social structure that may affect plant community dynamics in different ways (Schwartz and Ellis, 1981; Plumb and Dodd, 1993; Towne et al., 2005). Undoubtedly, the replacement of free-ranging bison with confined domestic livestock has led to grassland degradation in certain areas, but the extents to which these changes have occurred or continue to occur are unclear. However, grazing by domestic livestock remains a dominated land use of intact grasslands today (Lubowski et al., 2006).

Two additional considerations occur for the historic grazing shifts on grasslands—one with respect to overgrazing and the other pertaining to species availability in the less diverse, replacement grasslands relative to native grasslands.

Overgrazing occurs when plants are exposed to heavy grazing pressure for extended periods of time, or without sufficient recovery periods (Jacoby, 1989). Overgrazing can occur by livestock in poorly managed grazing operations and by overpopulations of native or nonnative wild animals. Overgrazing reduces the usefulness, productivity, plant vigor, and biodiversity of the land and is typically considered the number one cause of desertification and erosion. It can also lead to the spread of invasive species of nonnative plants and weeds. Overgrazing can occur under continuous or rotational grazing and is typically caused by having too many animals on the grasslands over a period of time or by not properly controlling their grazing activity. Palatable plant leaf areas can be reduced under overgrazing, which subsequently reduces the interception of sunlight and plant growth. Plants become weakened, lose vigor, and have reduced root length (Parker and Sampson, 1931; Crider, 1955). The reduced root length makes the plants more susceptible to death during drought conditions and harsh, open winters. Low vigor perennial grasses allow weed seeds to germinate and grow.

With the replacement of high diversity of cool- and warm-season native grasslands with less diverse communities (Tomanek and Albertson, 1957), they have a marked seasonal variation in production and quality. Cool-season grass mixtures tend to maximize herbage production in early summer with nutritional quality adequate for grazing animals through late spring or prior to seed production (Whitman et al., 1951; Sedivec et al., 2010). Warm-season grass mixtures maximize herbage production in mid to late summer, while nutritional quality adequate for grazing livestock through mid-summer or prior to heading (Whitman et al., 1951; Sedivec et al., 2009). Grasslands with a mixture of cool- and warm-season grasses retain maximum production levels longer, creating a sustained peak in herbage production from early to late summer. Cool- and warm-season grass mixtures will also have an adequate level of nutritional quality throughout the grazing period, including spring, summer, and fall periods, depending on amount of regrowth and available moisture.

Understanding the growth phenology of cool- and warm-season grasses will allow livestock growers to best match the grazing season or hay

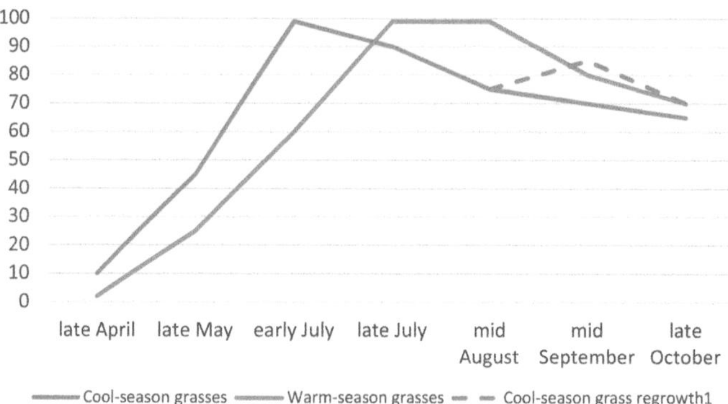

FIGURE 11.2.3 Percentage of average peak biomass for cool- and warm-season grasses in the Great Plains. Regrowth of cool-season grasses will occur when late summer moisture is available and the plants have been previously grazed or hayed during the active growing period (May–June; Sedivec et al., 2009, 2010).

harvesting period with the relationship between nutritional quality and available biomass. Cool-season grasses reach peak production approximately 2–3 weeks earlier than warm-season grasses (Figure 11.2.3; Sedivec et al., 2009, 2010). Both cool- and warm-season grasses lose standing crop after peak production due to senescence or leaf loss, with about 70% of forage produced remaining by late October, early November in most years. If moisture is available in September and October, cool-season grass leaf area will elongate and new tillers will develop, creating a seasonal increase in plant growth during this time of year. In semiarid regions where evaporation is greater than precipitation in most years, this secondary growth of cool-season is minimal. However, in the eastern regions of the Great Plains, this regrowth is common most years.

Nutritional content will vary between grass species, varieties, and growth physiology, especially in warm-season grasses (Sedivec et al., 2009). Managing grasslands to meet the nutritional demands of domestic livestock and wild herbivores will differ based on age and class of the grazing animal. It is important to understand that as grass matures, crude protein and digestibility decline and fiber fractions increase. Once most grass species reach the heading stage, nutritional quality can become deficient for many classes of livestock. Effective management reduces the risk for degradation.

11.2.4 CONCLUSIONS

Grasslands serve ecological functions that cannot be replaced by other land uses. The true value of the ecological services provided by grasslands is unknown, but in many instances may be priceless. Degradation is a concern as we look to maintain or improve soil resources and hydrologic cycling, biodiversity and pollination, livestock, and recreational uses. Land use conversion, invasive species, fire regime alteration, energy development, and grazing management are some of the processes that can lead to degradation and care needs to be taken when developing effective management plans for grassland preservation.

REFERENCES

Abrams, M.D., Knapp, A.K., Hulbert, L.C., 1986. A ten year-record of aboveground biomass in a Kansas tallgrass prairie: effects of fire and topographical position. J. Bot. 73, 1509–1515.

Anderson, R.C., 2006. Evolution and origin of the Central Grassland of North America: climate, fire, and mammalian grazers. J. Torrey Bot. Soc. 133, 626–647.

Armitage, A.M., 2006. Armitage's Native Plants for North American Gardens. Timber Press, 452 pp.

Baldi, G., Paruelo, J.M., 2008. Land-use and land cover dynamics in South American temperate grasslands. Ecol. Soc. 13, 6.

Barbour, M.G., Burk, J.H., Pitts, W.D., Gilliam, F.S., Schwartz, M.W., 1998. Terrestrial Plant Ecology, third ed. Benjamin/Cummings, Menlo Park, CA. 649 pp.

Batjes, N.H., 1996. Total C and N in the soils of the world. Eur. J. Soil Sci. 47, 151–163.

Bender, D.J., Contreras, T.A., Fahrig, L., 1998. Habitat loss and population decline: a meta-analysis of the patch size effect. Ecology 79, 517–533.

Biondini, M.E., Steuter, A.A., Grygiel, C.E., 1989. Seasonal fire effects on the diversity patterns, spatial distribution and community structure of forbs in the Northern Mixed Prairie, USA. Vegetatio 85, 21–31.

Briggs, J.M., Knapp, A.K., Brock, B.L., 2002. Expansion of woody plants in tallgrass prairie: a fifteen-year study of fire and fire-grazing interactions. Am. Midl. Nat. 147, 287–294.

Chalfoun, A.D., Thompson III, F.R., Ratnaswamy, M.J., 2002. Nest predators and fragmentation: a review and meta-analysis. Conserv. Biol. 16, 306–318.

Conant, R.T., Paustian, K., Elliott, E.T., 2001. Grassland management and conversion into grassland: effects on soil carbon. Ecol. Appl. 11, 343–355.

Cordell, H.K., Herbert, N.G., February 2002. The popularity of birding is still growing. Birding 54–61.

Crider, F.J., 1955. Root Growth Stoppage Resulting from Defoliation of Grass, vol. 1102. U. S. Dept. Agri. Tech. Bull., 23 pp.

Crooks, K.R., Soule, M.E., 1999. Mesopredator release and avifaunal extinctions in a fragmented system. Nature 400, 563–566.

Daily, G., 1997. Nature's Services: Societal Dependence on Natural Ecosystems. Island Press, 391 pp.

DeBano, L.F., Neary, D.G., Ffolliott, P.F., 1998. Fire's Effects on Ecosystems. John Wiley and Sons, New York, NY, p. 330.

Desserud, P., Gates, C.C., Adams, B., Revel, R.D., 2010. Restoration of foothills rough fescue grassland following pipeline disturbance in southwestern Alberta. J. Environ. Manage. 97, 2763–2770.

Diamantis, D., 1999. The concept of ecotourism: evolution and trends. Curr. Issues Tour. 2, 93–122.

Drewitt, A.L., Langston, R.H.W., 2006. Assessing the impacts of wind farms on birds. Ibis 148, 29–42.

Ehrenfeld, J.G., 2003. Effects of exotic plant invasions on soil nutrient cycling processes. Ecosystems 6, 503–523.

Eilers, E.J., Kremen, C., Greenleaf, S.S., Garber, A.K., Klein, A.M., 2011. Contribution of pollinator-mediated crops to nutrients in the human food supply. PLoS One 6, 1–6.

Eswaran, H., Van den Berg, E., Reich, P., Kimble, J.M., 1995. Global soil C resources. In: Lal, R., Kimble, J., Levine, E., Stewart, B.A. (Eds.), Soils and Global Change. Lewis Publishers, Boca Raton, FL, pp. 27–43.

Ewing, A.L., Engle, D.M., 1988. Effects of late summer fire on tallgrass prairie microclimate and community composition. Am. Midl. Nat. 120, 212–223.

Fahrig, L., 2003. Effects of habitat fragmentation on biodiversity. Annu. Rev. Ecol. Evol. Syst. 34, 487–515.

FAO, 2005. Grasslands of the World. No. 34. Suttie, J.M., Reynolds, Stephen G., Batello, C. (Eds.). Food and Agricultural Organization of the United Nations, Rome, Italy.

FAO, 2014. Land Degradation Assessment. http://www.fao.org/nr/land/degradation/en/ (accessed 11/2014).

Farmer, A.M., 1993. The effects of dust on vegetation—a review. Environ. Pollut. 79, 63–75.

Fargione, J.E., Cooper, T.R., Flaspohler, D.J., Hill, J., Lehman, C., Tilman, D., McCoy, T., McLeod, S., Nelson, E.J., Oberhauser, K.S., 2009. Bioenergy and wildlife: threats and opportunities for grassland conservation. Bioscience 59, 767–777.

Fiedler, F.R., Frazier, G.W., Ramirez, J.A., Ahuja, L.P., 2003. Hydrologic response of grasslands: effects of grazing, interactive infiltration and scale. J. Hydrol. Eng. 7, 293–301.

Fleischner, T.L., 1994. Ecological costs of livestock grazing in western North America. Conserv. Biol. 8, 629–644.

Frank, A.B., 2003. Evapotranspiration from northern semiarid grasslands. Agron. J. 95, 1504–1509.

Franzluebbers, A., 2009. Linking soil organic carbon and environmental quality through conservation tillage and residue management. In: Lal, R., Follet, R.F. (Eds.), Soil Carbon Sequestration and the Greenhouse Effect, SSSA Special Publication 57, second ed., Madison, WI, pp. 263–290.

Free, J.B., 1993. Insect Pollination of Crops, second ed. Academic Press, London. 684 pp.

Frost, C.C., 1998. Pre-settlement fire frequency regimes of the United States: a first approximation. In: Pruden, T., Brennan, L.A. (Eds.), Fire in Ecosystem Management: Shifting the Paradigm from Suppression to Prescription. Tall Timbers Fire Ecology Conference Proceedings, No. 20. Tall Timbers Research Station, Tallahassee, FL, pp. 70–81.

Gritxti, J.C., Wong, L.T., Cameron, S.A., Favret, C., 2009. Decline of bumble bees (*Bombus*) in the North American Midwest. Biol. Conserv. 142, 75–84.

Hartnett, D.C., Hickman, K.R., Walter, L.E.F., 1996. Effects of bison grazing, fire, and topography on floristic diversity in tallgrass prairie. J. Range Manag. 49, 413–420.

Helms, J.D., 1985. Brief History of the USDA Soil Bank Program. USDA, Natural Resource Conservation Services. http://www.nrcs.usda.gov/Internet/FSE_DOCUMENTS/stelprdb1045666.pdf (accessed 18.07.14.).

Henderson, D.C., Naeth, M.A., 2005. Multi-scale impacts of crested wheatgrass invasion in mixed-grass prairie. Biol. Invasions 7, 639–650.

Hendrickson, J.R., Wienhold, B.J., Berdahl, J.D., 2001. Decomposition rates of native and improved cultivars of grasses in the Northern Great Plains. Arid Land Resour. Manag. 15, 347–357.

Hendrickson, J.R., Lund, C., 2010. Plant community and target species affect responses to restoration strategies. Rangel. Ecol. Manag. 63, 435–442.

Herkert, J.R., 1994. The effects of habitat fragmentation on Midwestern grassland bird communities. Ecol. Appl. 4, 461–471.

Higgins, K.F., 1986. Interpretation and Compendium of Historical Fire Accounts in the Northern Great Plains. U.S. Fish and Wildlife Service. Resource Publication 161, 39 pp.

Hobbs, R.J., Huenneke, L.F., 1992. Disturbance, diversity, and invasion: implications for conservation. Conserv. Biol. 6, 324–337.

Hodur, N.M., Bangsund, D.A., Leistritz, F.L., 2004. Characteristics of Nature-Based Tourism Enterprises in North Dakota. Department of Agribusiness and Applied Economics, Agricultural Experiment Station, North Dakota State University. Report #537. 58 pp.

Jacoby, P.W., 1989. A Glossary of Terms Used in Range Management: A Definition of Terms Commonly Used in Range Management. Society for Range Management, Denver, CO, 20 pp.

Jastrow, J.D., Miller, R.M., Lussenhop, J., 1998. Contributions of interacting biological mechanisms to soil aggregate stabilization in restored prairie. Soil Biol. Biochem. 30, 905–916.

Johnson, D., Igl, L., 2001. Area requirements of grassland birds: a regional perspective. The Auk 118, 24–34.

Jordan, N.R., Larson, D.L., Huerd, S.C., 2008. Soil modification by invasive plants: effects on native and invasive species of mixed-grass prairies. Biol. Invasions 10, 177–190.

Kearns, C.A., Inouye, D.W., 1997. Pollinators, flowering plants, and conservation biology. Bioscience 47, 297–307.

Kearns, C.A., Inouye, D.W., Waser, N.M., 1998. The conservation of plant-pollinator interactions. Annu. Rev. Ecol. Syst. 29, 83–112.

Kiester, A.R., Lande, R., Schemske, D.W., 1984. Models of coevolution and speciation in plants and their pollinators. Am. Nat. 124, 220–243.

Klein, A.M., Vaissiere, B.V., Cane, J.H., Steffan-Dewenter, I., Cunningham, S.A., Kremen, C., Tscharntke, T., 2007. Importance of pollinators in changing landscapes for world crops. Proc. R. Soc. B 274, 303–313.

Kluser, S., Peduzzi, P., 2007. Global Pollinator Decline: A Literature Review. United Nations Environmental Program/GRID-Europe. WWW.grid.unep.ch (accessed 18.07.14.).

Knapp, A.K., Seastedt, T.R., 1986. Detritus accumulation limits productivity of tallgrass prairie. BioScience 36, 662–668.

Knopf, F.L., 1996. Prairie Legacies—Birds. Prairie Conservation: Preserving North America's Most Endangered Ecosystem. Island Press, Covelo, CA, 135–148.

Kremen, C., Ricketts, T., 2000. Global perspectives on pollination disruptions. Conserv. Biol. 14, 1226–1228.

Kremen, C., Williams, N.M., Thorp, R.W., 2002. Crop pollination from native bees at risk from agricultural intensification. Proc. Natl. Acad. Sci. 99, 16812–16816.

Kremen, C., Williams, N.M., Bugg, R.L., Fay, J.P., Thorp, R.W., 2004. The area requirements of an ecosystem service: crop pollination by native bee communities in California. Ecol. Lett. 7, 1109–1119.

Krzic, M., Broersma, K., Thompson, D.J., Bomke, A.A., 2000. Soil properties and species diversity of grazed crested wheatgrass and native rangelands. J. Range Manag. 53, 353–358.

Kucera, C.L., Koelling, M., 1964. The influence of fire on composition of central Missouri prairie. Am. Midl. Nat. 72, 142–147.

Kunz, T.H., Arnett, E.B., Erickson, W.P., Hoar, A.R., Johnson, G.D., Larkin, R.P., Strickland, M.D., Thresher, R.W., Tuttle, M.D., 2007. Ecological impacts of wind energy development on bats: questions, research needs, and hypotheses. Front. Ecol. Environ. 5, 315–324.

Lal, R., 2004. Agricultural activities and the global carbon cycle. Nutr. Cycl. Agroecosys. 70, 103–116.

Lal, R., Kimble, J.M., Follett, R.F., Cole, C.V., 1998. The potential of U.S. Cropland to sequester carbon and mitigate the greenhouse effect. Sleeping Bear Press Inc., Chelsea, MI.

Laycock, W.A., 1991. Stable states and thresholds of range condition on North American rangelands: a viewpoint. J. Range Manag. 44, 427–433.

Liebig, M.A., Mikha, M.M., Potter, K.N., 2009. Management of dryland cropping systems in the U.S. Great Plains: effects on soil organic carbon. In: Lal, R., Follet, R.F. (Eds.), Soil Carbon Sequestration and the Greenhouse Effect, SSSA Special Publication 57, second ed., Madison, WI, pp. 97–114.

Liu, Y.Y., Evans, J.P., McCabe, M.F., AM de Jeu, R., IJM van Dijk, A., Dolman, A.J., Saizen, I., 2013. Changing climate and overgrazing are decimating Mongolian steppes. PLoS One 8, e57599.

Lokeshwari, H., Chandrappa, G.T., 2006. Impact of heavy metal contamination of Bellandur Lake on soil and cultivated vegetation. Curr. Sci. 91, 622–627.

Lubowski, R.N., Vesterby, M., Bucholtz, S., Baez, A., Roberts, M.J., 2006. Major Uses of Land in the United States, 2002. U.S. Department of Agriculture, Economic Research Service, Economic Information Bulletin Number 14, 54 pp.

Mac, M.J., Opler, P.A., Puckett Haecker, C.E., Doran, P.D., 1998. Status and Trends of the Nation's Biological Resources. U.S. Department of the Interior, U.S. Geological Survey, Reston, VA, 964 pp.

McDonald, R.I., Fargione, J., Kiesecker, J., Miller, W.M., Powell, J., 2009. Energy sprawl or energy efficiency: climate policy impacts on natural habitat for the United States of America. PLoS One 4, e6802. http://dx.doi.org/10.1371/journal.pone.0006802.

Milchunas, D.G., Sala, O.E., Lauenroth, W.K., 1988. A generalized model of the effects of grazing by large herbivores on grassland community structure. Am. Nat. 132, 87–106.

Miller, B., Ceballos, G., Reading, R., 1994. The prairie dog and biotic diversity. Conserv. Biol. 8, 677–681.

Morandin, L.A., Winston, M.L., 2005. Wild bee abundance and seed production in conventional, organic, and genetically modified canola. Ecol. Appl. 15, 871–881.

Parker, K.W., Sampson, A.W., 1931. Growth and yield of certain *Gramineae* as influenced by reduction of photosynthetic tissue. Hilgardia 5, 361–381.

Pitman, J.C., Hagen, C.A., Robel, R.J., Loughin, T.M., Applegate, R.D., 2005. Location and success of lesser prairie-chicken nests in relation to vegetation and human disturbance. J. Wildl. Manag. 69, 1259–1269.

Plumb, G.E., Dodd, J.L., 1993. Foraging ecology of bison and cattle on a mixed prairie: implications for natural area management. Ecol. Appl. 3, 631–643.

Potts, S.G., Biesmeijer, J.C., Kremen, C., Neumann, P., Schweiger, O., Kunin, W.E., 2010. Global pollinator declines: trends, impacts and drivers. Trends Ecol. Evol. 25, 345–353.

Pritekel, C., Wittemore-Olson, A., Snow, N., Moore, J.C., 2006. Impacts from invasive plant species and their control on the plant community and belowground ecosystem at Rocky Mountain National Park, USA. Appl. Soil Ecol. 32, 132–141.

Pruett, C.L., Patten, M.A., Wolfe, D.H., 2009. It's not easy being green: wind energy and a declining grassland bird. BioScience 59, 257–262.

Ricketts, T.H., Regetz, J., Steffan-Dewenter, I., Cunningham, S.A., Kremen, C., Bogdanski, A., Gemmill-Herren, B., Greenleaf, S.S., Klein, A.M., Mayfield, M.M., Morandin, L.A., Ochieng, A., Viana, B.F., 2008. Landscape effects on crop pollinations services: are there general patterns? Ecol. Lett. 11, 499–515.

Ricklefs, R.E., Miller, G.L., 1999. Ecology, fourth ed. W.H. Freeman and Company, New York, USA. 822 pp.

Robel, R.J., Harrington Jr., J.A., Hagen, C.A., Pitman, J.C., Reker, R.R., 2004. Effect of energy development and human activity on the use of sand sagebrush habitat by lesser prairie-chickens in southwestern Kansas. Trans. N. Am. Nat. Resour. Conf. 69, 251–266.

Samson, F.B., Knopf, F.L., Ostlie, W.R., 1998. Grasslands. In: Mac, M.J., Opler, P.A., Puckett Haecker, C.E., Doran, P.D. (Eds.), Status and Trends of the Nation's Biological Resources, vol. 2. Northern Prairie Wildlife Research Center Online, Jamestown, ND, pp. 437–472.

Samson, F.B., Knopf, F.L., Ostlie, W., 2004. Great Plains ecosystems: past, present, and future. Wildl. Soc. Bull. 32, 6–15.

Sanderson, M.A., Archer, D., Hendrickson, J., Kronberg, S., Liebig, M., Nichols, K., Schmer, M., Tanaka, D., Aguilar, J., 2013. Diversification and ecosystem services for conservation agriculture: outcomes from pastures and integrated crop-livestock systems. Renew. Agric. Food Syst. 28, 129–144.

Sauer, J.R., Link, W.A., Fallon, J.E., Pardieck, K.L., Ziolkowski Jr., D.J., 2013. The North American breeding bird survey 1966–2011: summary analysis and species accounts. North Am. Fauna 79, 1–32.

Schlesinger, W.H., 1997. Carbon balance in terrestrial detritus. Annu. Rev. Ecol. Syst. 8, 51–81.

Schuman, G.E., Ingram, L.J., Stahl, P.D., Derner, J.D., Vance, G.F., Morgan, J.A., 2009. Influence of management on soil organic carbon dynamics in northern mixed-grass rangeland. In: Lal, R., Follet, R.F. (Eds.), Soil Carbon Sequestration and the Greenhouse Effect, SSSA Special Publication 57, second ed., Madison, WI, pp. 169–180.

Schwartz, C.C., Ellis, J.E., 1981. Feeding ecology and niche separation in some native and domestic ungulates on the shortgrass prairie. J. Appl. Ecol. 343–353.

Sedivec, K.K., Tober, D.A., Duckwitz, W.L., Dewald, D.D., Printz, J.L., Craig, D.J., 2009. Grasses for the Northern Plains Growth Patterns, Forage Characteristics and Wildlife Values. Vol. II – Warm-Season Grasses. Circ. R-1390. North Dakota State University Extension Service, Fargo, ND, 67 pp.

Sedivec, K.K., Tober, D.A., Duckwitz, W.L., Dewald, D.D., Printz, J.L., 2010. Grasses for the Northern Plains Growth Patterns, Forage Characteristics and Wildlife Values. Vol. 1 – Cool-season Grasses. Circ. R-1323. North Dakota State University Extension Service, Fargo, ND, 90 pp.

Secchi, S., Gassman, P.W., Williams, J.R., Babcock, B.A., 2009. Corn-based ethanol production and environmental quality: a case of Iowa and the conservation reserve program. Environ. Manag. 44, 732–744.

Singh, J.S., Lauenroth, W.K., Milchunas, D.G., 1983. Geography of grassland ecosystems. Prog. Phys. Geogr. 7, 46–80.

Smith, M.D., Knapp, A.K., 1999. Exotic plant species in a C_4-dominated grassland: invisibility, disturbance, and community structure. Oecologia 120, 605–612.

Southwick, E.E., Southwick Jr., L., 1992. Estimating the economic value of honey bees (*Hymenoptera apidae*) as agricultural pollinators in the United States. J. Econ. Entomol. 85, 621–633.

Steffan-Dewenter, I., Potts, S.G., Packer, L., 2005. Pollinator diversity and crop pollination services at risk. Trends Ecol. Evol. 20, 651–652.

Stephens, S.E., Knoons, D.N., Rotella, J.J., Willey, D.W., 2003. Effects of habitat fragmentation on avian nesting success: a review of the evidence at multiple scales. Biol. Conserv. 115, 101–110.

Stohlgren, T.J., Binkley, D., Chong, G.W., Kalkhan, M.A., Schell, L.D., Bull, K.A., Otsuki, Y., Newman, G., Bashkin, M., Son, Y., 1999. Exotic plant species invade hot spots of native plant diversity. Ecol. Monogr. 69, 25–46.

Toland, J., Jones, W., Eldridge, J., Thorpe, E., O'hara, E., 2008. LIFE and Europe's Grasslands: Restoring a Forgotten Habitat. Office for Official Publications of the European Communities, Luxembourg.

Tomanek, G.W., Albertson, F.W., 1957. Variation in cover, composition, production, and roots of vegetation on two prairies in western Kansas. Ecol. Monogr. 27, 267–281.

Towne, E.G., Hartnett, D.C., Cochran, R.C., 2005. Vegetation trends in tallgrass prairie from bison and cattle grazing. Ecol. Appl. 15, 1550–1559.

Umbanhowar, C.E., 1996. Recent fire history of the northern Great Plains. Am. Midl. Nat. 135, 115–121.

United States Department of Agriculture, 1985. History of Agricultural Price-Support and Adjustment Programs 1933–84. Agriculture Information Bulletin Number 485. United States Department of Agriculture, Economic Research Service. http://naldc.nal.usda.gov/naldc/download.xhtml?id=CAT10842840&content=PDF.

United State Department of Agriculture, 2002. USDA 2002 Farm Bill Conservation Provisions: Summary. USDA, Washington, DC, 14 pp.

United State Department of Energy, 2014. Energy Efficiency and Renewable Energy: Wind Program. US Department of Energy, Washington, DC. DOE/EE-1071. 2 pp.

Van der Kamp, G., Hayashi, M., Gallen, D., 2003. Comparing the hydrology of grassed and cultivated catchments in the semi-arid Canadian prairies. Hydrol. Process. 17, 559–575.

Vermeire, L.T., Crowder, J.L., Wester, D.B., 2011. Plant community and soil environmental response to summer fire in the Northern Great Plains. Rangel. Ecol. Manag. 64, 37–46.

Vermeire, L.T., Heitschmidt, R.K., Johnson, P.S., Sowel, B.F., 2004. The prairie dog story: Do we have it right? Bioscience 54, 689–695.

Vinton, M.A., Goergen, E.M., 2006. Plant–soil feedbacks contribute to the persistence of *Bromus inermis* in tallgrass prairie. Ecosystems 9, 967–976.

Weaver, J.E., 1954. North American Prairie. Johnsen Publishing Company, Lincoln, NE.

Weaver, J.E., Rowland, N.W., 1952. Effects of excessive natural mulch on development, yield, and structure of native grassland. Bot. Gaz. 114, 1–19.

Wedin, D.A., Tilman, D., 1990. Species effects on nitrogen cycling: a test with perennial grasses. Oecologia 84, 433–441.

Wedin, D.A., Tilman, D., 1996. Influence of nitrogen loading and species composition on the carbon balance of grasslands. Science 274, 1720–1723.

White, R.P., Murray, S., Rohweder, M., 2000. Pilot Analysis of Global Ecosystems: Grassland Ecosystems. World Resources Institute, 81 pp.

Williams, I.H., 1994. The dependence of crop production within the European Union on pollination by honey bees. Agric. Zool. Rev. 6, 229–257.

Winfree, R., Williams, N.M., Dushoff, J., Kremen, C., 2007. Native bees provide insurance against ongoing honey bee loses. Ecol. Lett. 10, 1105–1113.

Winfree, R., Auguilar, R., Vazquez, D.P., LeBuhn, G., Aizen, M., 2009. A meta-analysis of bees' responses to anthropogenic disturbance. Ecology 90, 2068–2076.

Winfree, R., Bartmoeus, I., Cariveau, D.P., 2011. Native pollinators in anthropogenic habitats. Annu. Rev. Ecol. Evol. Syst. 42, 1–22.

Whitman, W.C., Bolin, D.W., Klostermann, E.W., Ford, K.D., Moomaw, L., Hoag, D.G., Buchanan, M.L., 1951. Caroten, protein, phosphorus in grasses of western North Dakota. Agri. Exp. Sta., North Dakota Agri. College. North Dakota Agri. Exp. Sta. Bull. No. 370.

Wright, C.K., Wimberly, M.C., 2013. Recent land use change in the western corn belt threatens grasslands and wetlands. Proc. Natl. Acad. Sci. 110, 4134–4139.

Yacobucci, B.D., 2010. Biofuels Incentives: A Summary of Federal Programs. Congressional research service reports. Paper 8.

Chapter 11.3

Land Degradation in Rangeland Ecosystems

Jay P. Angerer, William E. Fox and June E. Wolfe
Texas A&M AgriLife Research, Blackland Research and Extension Center, Temple, TX, USA

ABSTRACT
Rangelands provide an array of ecosystem services such as food, fiber, water, recreation, minerals, and are important to the livelihoods of people across the globe, especially in developing countries. Competing land uses, overgrazing, extreme climate events, and socioeconomic changes are resulting in rangeland degradation in many parts of the world. Given our reliance on rangelands, degradation of this resource can have far-reaching effects. In this chapter, causes of rangeland degradation are examined. Indicators that can be used to identify degradation and methods for assessing the degree of degradation in rangeland ecosystems are discussed. Options and considerations for restoring disturbed rangeland are presented, in addition to future directions in rangeland degradation monitoring and assessment.

11.3.1 INTRODUCTION

Over 40% of the Earth's terrestrial land cover is occupied by rangelands, which are defined as lands characterized by native plant communities, with grazing generally being a major land use, and are managed by ecological, rather than agronomic methods (SRM, 2002). With regard to grazing, rangelands provide almost 75% of the forage used by domesticated livestock (Brown and Thorpe, 2008) and livestock production on rangelands can contribute significantly to the overall gross domestic product (GDP), especially in developing countries.

Rangelands provide an array of ecosystem services such as food, fiber, water, recreation, minerals, and medicinal plants to both rural and urban populations (Havstad et al., 2007). They also store 10−30% of global soil organic carbon (Scurlock and Hall, 1998). On a majority of rangelands in developing countries, livestock production is a primary land use, and many people depend on rangeland livestock production for their livelihoods. In these countries, livestock provides food and income for the majority of these

Biological and Environmental Hazards, Risks, and Disasters. http://dx.doi.org/10.1016/B978-0-12-394847-2.00017-6

individuals who live on less than 1 USD per day (FAO, 2008). Demand for livestock products is increasing worldwide (Delgado et al., 1999; FAO, 2008); however, the land area available for livestock production is decreasing due to competing interests, including mining, oil and gas exploration/production, expansion of crop production, biofuel schemes, and urban sprawl (Estell et al., 2012; Herrick et al., 2012). On communal rangelands, the governments and policy makers that allow expansion into these areas do not always take into consideration the needs of pastoralists whose livelihoods depend on these lands. Competing uses can reduce access to key resources such as water, and shrink the land area available for grazing; thus, increasing the potential for overgrazing and conflict for limited resources (Little and McPeak, 2014). Given our reliance on rangelands for ecosystem services, and the dependency of livestock producers on the rangeland for their livelihoods, degradation of this resource can have profound ecological and societal effects. The large areas occupied by rangelands, the remote locations in which they are located, and the diverse mix of species that graze these lands, pose challenges for monitoring, assessing, and mitigating the causes of rangeland degradation.

The aim of this chapter is to: (1) provide an overview of the causes of rangeland degradation; (2) illustrate indicators that can be used to identify degradation; and (3) discuss methods for assessing the degree of degradation in rangeland ecosystems. In addition, managing and repairing rangeland degradation, and future efforts for responding to rangeland degradation are discussed.

11.3.2 DEFINITION AND EXTENT OF RANGELAND DEGRADATION

Degradation on rangelands can generally be described as a reduction in biological and economic productivity that occurs over a sustained period of time and is linked to improper or unsustainable human land uses, and the subsequent impacts of this unsustainable use on vegetation composition, hydrology, and soil processes (Bedunah and Angerer, 2012). An important consideration in this definition is the period of time that the reduction in productivity occurs. The reduction must be persistent (i.e., the amount of time it has remained below a corresponding baseline) and not the result of short-term climate variation (Safriel, 2007). In some instances, confusion exists between degradation and desertification, and the words are sometimes used interchangeably to describe degradation. Although desertification is a form of degradation, desertification is defined by the United Nations Convention to Combat Desertification (UNCCD) as "land degradation in arid, semi-arid and dry subhumid areas resulting from various factors, including climatic variations and human activities" (UNCCD, 1994). The desertification definition applies to rangelands that occur in dryland regions (i.e., areas with an aridity index of less than 0.65), but not rangelands that occur in more humid areas; therefore the terms are not always interchangeable.

In a global assessment of land degradation conducted by the United Nations (UN), degradation of pastures and rangelands was estimated at 20% of the total land area and up to 73% in dry areas (Steinfeld et al., 2006). In reviewing the UN and other global land degradation assessments, Safriel (2007) states that problems exist with these global assessments because of the difficulties in measuring the attribute used to represent land degradation and a general lack of field data to verify degradation.

11.3.3 CAUSES OF RANGELAND DEGRADATION

Given that land degradation can occur across a range of time scales, combined with the interaction and interrelatedness of climate and human impacts, efforts to determine the specific causes of degradation are a challenge (Henry et al., 2007). The causes of rangeland degradation are complex, and are often combinations of factors rather than a single, identifiable factor. Land and livestock management, in conjunction with climate are overarching causes of rangeland degradation. These are commonly interrelated, and are driven by contributing factors, such as government policy, land use changes, collection of fuelwood, land tenure, and economic conditions (Bedunah and Angerer, 2012; Sivakumar and Stefanski, 2007).

11.3.3.1 Land and Livestock Management

Historically, most discussions of rangeland degradation have centered on effects of livestock grazing, especially overgrazing (e.g., Asner et al., 2004; Ayoub, 1998; Dregne, 2000; Eswaran et al., 2001; Hess and Holechek, 1995; Horn et al., 2002; Steinfeld et al., 2006; Wilcox, 2007, 2010; Wilcox and Thurow, 2006; Zhang et al., 2014). Overgrazing occurs when the consumption of vegetation biomass by livestock and other grazers (e.g., wildlife) exceeds the vegetation's ability to recover in a timely fashion, thus exposing the soil and reducing the vegetation's productive capacity. The soil exposure can accelerate the potential for water and wind erosion processes, and can lead to soil losses. Accelerated erosion processes, along with soil compaction by the grazing animals, can reduce the amount of water infiltrating into the soil, which can lead to reduced plant growth. Continued overgrazing reduces inputs of soil organic matter because less plant biomass is available as litter, which in turn, reduces soil organic matter, nutrients, and biotic activity. This leads to deteriorated soil structure, which increases the potential for erosion and reduces water-holding capacity of soil. This process has been described as a "degradation spiral," a set of stepwise processes with positive feedback that leads to land degradation (King and Hobbs, 2006; Whisenant, 1999, 2002).

Stocking rates (i.e., density of livestock per unit area for a given period of time) that exceed the carrying capacity of the land (i.e., the maximum number of animals a unit of land can sustain without depleting the vegetation and soil resources) can result in overgrazing (Holechek et al., 1995). If overgrazing on a site continues for a long period of time, it can spiral into degradation. Generally, it is recommended that stocking rates be adjusted according to forage availability to avoid conditions that lead to degradation. However, this is not always done, especially in developing countries. The reasons that livestock producers do not adjust stocking rates are varied. Collecting data for stocking rate assessments can be time-consuming, expensive, and many producers may lack the tools to do this efficiently for a given land area (Angerer, 2012). Also, for communal lands, a lack of infrastructure occurs with which to support assessments or the inability to enforce proper stocking rates or grazing schedules (Steinfeld et al., 2006). In addition, expansion of competing land uses into communal rangelands can remove key areas (e.g., dry season grazing), fragment the landscape, and create barriers that restrict mobility, thus reducing pastoralist's ability to escape low-forage availability. The result is vegetation overuse.

The expansion of woody plants on rangelands in many parts of the world has been seen as a form of rangeland degradation (Archer, 2010; Asner et al., 2004; Huxman et al., 2005; Wilcox, 2007). In arid and semiarid rangelands, overgrazing by livestock has been a primary factor driving the expansion of woody plants. Added to this, drought and fire suppression can significantly affect the degradation process (Archer, 2010). In the more arid dryland regions, woody-plant encroachment causes desertification where mesophytic grasses are replaced by xerophytic shrubs (Archer, 2010; Browning and Archer, 2010), which leads to greater spatial and temporal heterogeneity in soil resources and exposure of the soil to processes of wind and water erosion (Schlesinger et al., 1990). In semiarid regions with higher rainfall, woody-plant encroachment can displace grasses and reduce carrying capacity for livestock that are primarily grass grazers; however, primary production and nutrient cycling in the system may be enhanced (Archer, 2010). Encroachment of woody plants can result in changes to the hydrological processes in a landscape. Increases in the shrub component can increase evaporation from bare soils relative to evapotranspiration on landscapes where grass cover is reduced at the expense of shrub expansion. This can potentially lead to streamflow reductions in landscapes where subsurface flow is predominant (Huxman et al., 2005). Woody-plant encroachment also has implications on the global carbon cycle and carbon sequestration. Increases in woody plants generally lead to increases in aboveground and belowground carbon, although there is a substantial lag in the buildup of belowground carbon (Hibbard et al., 2003). Although the increase in carbon pools associated with woody plants can be viewed as a form of carbon sequestration, they are at risk for rapid loss in the event of fire or extreme drought (Archer, 2010).

From the socioeconomic perspective of livestock production, the increase in woody plants is perceived as degradation because it generally reduces the number of livestock that can graze an area due to decreases in herbaceous production. Mechanical treatments (root plowing, chaining, grubbing, etc.), herbicidal treatments, and prescribed burning, have been used to try to eradicate or reduce woody plants on the landscape to promote increased grass and livestock production. However, in many cases where treatments were applied indiscriminately, the desired effects were often short-lived because woody plants increased in dominance within 5–10 years after treatment, which in turn, affected wildlife and biological diversity (Archer et al., 2011; Archer, 2010). Since the 1980s, a more integrated approach to woody-plant management has been encouraged that takes into consideration multiple uses and values of rangelands, the major ecosystem components (plants, animals, and soils), and the response of these to woody-plant management (Archer et al., 2011; Archer, 2010; Scifres et al., 1985). Archer and Stokes (2000) have noted that although some may consider the progressive encroachment of woody plants as a degraded state because of limits on livestock production, some of these landscapes can be ecologically productive and diverse, and have other socioeconomic values, such as wildlife hunting or ecotourism.

11.3.3.2 Climate

Given that the majority of rangelands are in arid and semiarid areas, the vegetation biomass production is driven by precipitation, and drought is a common occurrence (Figure 11.3.1). The reluctance or inability to adjust stocking rates during drought periods when livestock forage is below average

FIGURE 11.3.1 Wind erosion and dust storms on drought-stricken rangeland in Northeastern Uvurkhangai province, Mongolia. Heavy grazing pressure removes vegetation ground cover and leaves the soil exposed, making it susceptible to wind and water erosion.

can result in conditions that lead to the downward spiral of degradation. However, it has been hypothesized that in areas of high rainfall variability, plant species composition and vegetation production are primarily controlled by precipitation rather than grazing pressure (Behnke and Scoones, 1992; Ellis and Swift, 1988). Frequent droughts keep animal numbers below equilibrium densities, which decreases the likelihood of overgrazing that could result in degradation. These systems have been described as nonequilibrium systems and pastoralists in these systems rely on mobility to exploit larger geographic areas during periods of low-forage availability (Ellis and Swift, 1988). Others have challenged the assertion that livestock grazing, without regard to stocking rate in these nonequilibrium systems, has little long-term impact on vegetation productivity (Illius and O'Connor, 1999; Müller et al., 2007; Vetter, 2005; Wessels et al., 2007b). For example, Wessels et al. (2007b) examined nonequilibrium rangelands in South Africa that had been grazed at high stocking rates and found that degradation had reduced long-term productivity of the sites. Miehe et al. (2010), in a study of different grazing intensities on Sahelian rangeland in Senegal, found that the sites shifted between a period of nonequilibrium and equilibrium conditions over a 27-year period. During the drier, nonequilibrium period, degradation was not evident, whereas during the wetter, equilibrium period, changes in species composition indicated degradation under high grazing intensities. The authors concluded that longer periods of monitoring are required to detect degradation in these systems.

With climatic change, extreme climate events on most continents are predicted to increase with greater numbers of heat waves and heavy precipitation occurrences (Easterling et al., 2000; IPCC, 2012), and in some areas of the world, droughts are expected to become more intense and frequent (e.g., Southern Europe and West Africa) (IPCC, 2012). With these extreme climate events, conditions on rangelands can become more favorable for loss of plant species and cover, exposure of the soil to wind and water erosion, and susceptible to invasion by plant species. Disentangling the impacts of climate and land-management effects on rangeland degradation will likely not be straightforward. Even when areas are properly managed, extreme climate events such as drought can lead to lower productivity and loss of plant species. For example, Young (1956) found that lightly grazed rangeland near Sonora, Texas (USA), had only slightly higher survival of dominant, perennial grasses (15%) than heavily grazed areas (9%) on the same ecological site after an extreme drought in the early 1950s. In southern Queensland, Australia, Clarkson and Lee (1988) found that extreme drought reduced ground cover by 6.4% on native pastures, whereas grazing reduced ground cover by 3.7%. The authors also noted that the effects of heavy grazing on ground cover were exacerbated by the extreme drought conditions. Large die-offs of woody plants on rangelands during extreme drought events have also been observed (Fensham et al., 2009; Twidwell et al., 2014), which can alter plant community composition and carbon sinks associated with the woody-plant biomass.

Miehe et al. (2010) described the need for longer monitoring periods to capture natural fluctuations in drier and wetter precipitation periods in order to identify long-term degradation processes and to inform management policies.

11.3.4 INDICATORS OF RANGELAND DEGRADATION

11.3.4.1 Purpose of Indicators

Degradation can be viewed from the perspective of rangeland health in that degraded areas represent "unhealthy" rangeland or rangeland with a series of symptoms that are affecting the overall health of the land. Rangeland health has been defined by the National Research Council as "the degree to which the integrity of the soil and ecological processes of rangeland ecosystems are maintained" (NRC, 1994). In 1995, the Task Group on Unity of Concepts and Terminology of the Society for Range Management (SRM, 1995), further refined the definition as "the degree to which the integrity of the soil, vegetation, water, and air, as well as the ecological processes of the rangeland ecosystem are balanced and sustained." Although these definitions provide an excellent framework for evaluating rangeland health, it has been difficult to quantify levels of degradation primarily due to the fact that monitoring is either limited or nonexistent in many of the world's rangeland systems. In addition, degradation is characteristically an "eye of the beholder" phenomenon, where individuals or groups may have competing concepts of degradation due to the application of values to the systems in question.

The complexity of these issues, and the competing values that stakeholders have about rangelands, provides a critical challenge to the art and science of sustainable rangeland management. Holling et al. (2001) submitted that the issues associated with the complexity of sustainable development are not singularly limited to social, economic, or ecological paradigms, but must be integrated across all three components of an ecosystem. However, integrating these factors across rangeland systems that cross multiple political boundaries and jurisdictions adds to the complexity and commonly exceeds the capacity for assessment in efforts to quantify rangeland degradation. Schlesinger (2010) has suggested that unless science can be brought to society through meaningful translations, landscapes will continue to progress down the path of degradation. Sustainability is more than just the aggregation of issues; it is the integration and interrelationships of dynamic systems (Singh et al., 2009). Fox et al. (2009) developed a conceptual framework (integrated social, economic and ecological concept) for illustrating the interrelationships of biophysical and socioeconomic processes on rangeland systems as a tool for evaluating the utility of indicators for assessing these ecosystems (Figure 11.3.2).

In the past, rangeland monitoring or assessment efforts have tended to center on one of the three components of rangeland ecosystems (i.e., social, economic, or ecological) or on a specific issue; thus, providing a skewed view

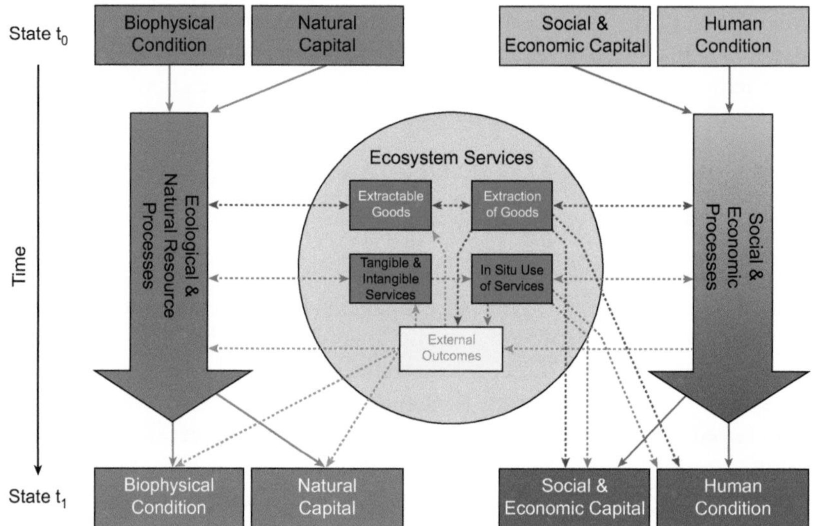

FIGURE 11.3.2 Conceptual framework for the Integrated Social, Economic and Ecological Concept. In this framework, the world is categorized into four states: (1) current biophysical conditions; (2) natural resource capital; (3) social capacity and economic capital; and (4) current human condition. As indicated by the large downward arrows, the four states are acted upon by ecological and natural resource processes and by social and economic processes. The outcomes of the various processes acting upon the states and conditions existing at time period_0 result in a modified set of states and conditions existing at time period_1. The ecological and natural resource processes and the social and economic processes then act upon those modified states and conditions, resulting in new states and conditions in the next time period. *From Fox et al. (2009).*

of potential degradation processes taking place. To address this challenge, numerous groups of rangeland and environmental scientists from around the world have attempted to develop "indicators" of rangeland sustainability. "Indicators," as a tool for measuring, tend to develop from values (i.e., the choice to measure what "we" care about), and result in the creation of values (i.e., the choice to care about what we measure) (Singh et al., 2009). More recent efforts seek to standardize an approach to quantifying the biophysical and socioeconomic processes ongoing in rangeland ecosystems in an effort to understand trends and changes that are taking place.

The use of the term "indicator" provides land managers with a solid basis for assessment, tracing its origins back to the Latin *indicare* meaning to "disclose or point out, to announce or make publicly known, or to estimate or put a price on." Hammond et al. (1995) have defined an "indicator" as something that "provides a clue to a matter of larger significance or makes perceptible a trend or phenomenon that is not immediately detectable." As a result, an indicator points out something "larger" than the actual measurement provides. The use of indicators helps to visualize and highlight trends; thus, providing a means to communicate otherwise complicated information (Warhurst, 2002). Through the continued development of the science of natural resource management, society has obtained the ability to apply indicators as tools for understanding complex processes ongoing in nature and the information required to affect decision-making, and ultimately land management. This critical information provision is pointed out by Kates et al. (2001) stating that "the purpose for sustainability assessment is to provide decision-makers with an evaluation of global to local integrated nature-society systems in short- and long-term perspectives in order to assist them to determine which actions should or should not be taken in an attempt to make society sustainable."

Below are descriptions of selected rangeland indicator systems that are currently being applied around the globe. These systems have been implemented to provide a better understanding of processes and provide information for decision-makers and stakeholders in order to more efficiently understand degradation risk. This information can then be used in efforts to restore or stabilize these ecosystems for provision of the goods and services they provide.

11.3.4.2 International Indicator Programs

11.3.4.2.1 Australia

In 2011, the Australian government published "Sustainable Australia – Sustainable Communities: A Population Strategy for Australia" in recognition that to build a sustainable Australia, critical information on the economy, environment, and society and the interrelationships between them were needed to more effectively inform decision-makers (National Sustainability Council, 2013). Under this program, sustainability indicators were defined within the criteria of social/human capital, natural capital, and economic capital. The effort recognized the need to look beyond traditional approaches of economic indicators (i.e., GDP) to assess well-being. Additionally, expanding assessment beyond economic indicators to include social and environmental parameters provides the ability to see the full impact of policy and decision-making on the systems of interest.

Australia is unique in that approximately 75% of the terrestrial systems in the country are classified as rangeland. Initial indicator work in Australia found that the extent of native vegetation, country wide, had declined by 14% since 1750, with an additional 62% subjected to varying degrees of disturbance (National Sustainability Council, 2013). Disturbance and/or rangeland degradation accounted for approximately 12% of arid zone mammal extinctions (11 species) and 7% of flowering plant extinctions (6 species) across rangeland systems (Australian Department of Environment, 1999).

11.3.4.2.2 United States—Sustainable Rangeland Roundtable

Since 2001, the Sustainable Rangeland Roundtable (SRR), currently directed by the University of Wyoming, has endeavored to develop, illustrate, and integrate into rangeland-management policy the use of criteria and indicators of rangeland sustainability. The SRR's mission is to "promote ecological, economic and social sustainability of rangelands through the widespread use of criteria and indicators for rangeland assessment, and by providing a forum for dialogue on rangeland sustainability." From 2001 to 2003, rangeland scientists, environmental biologists, economists, sociologists, legal professionals, and nongovernmental organizations, spanning the realm of rangeland interests, developed a standard suite of indicators across five major criteria for assessing rangeland sustainability (Mitchell, 2010).

Bringing together a diverse group of experts, with varying sets of values, requires considerable effort in developing a consensus. Each expert brings to the table their individual expertise and experiences, but also the biases and values developed over their careers. In addition, the broad scope of expertise adds an additional level of challenge to indicator development. These varying expertise and personal opinions can pose a significant challenge to the process of determining indicators for rangeland sustainability. The SRR applied a collaborative Delphi methodology for systematically gathering and integrating the informed judgment of a group of experts concerning rangeland assessment. The Delphi process has been used to establish research results where traditional data-driven methods are not feasible, to aid in policy decision-making, to resolve environmental disputes, and to facilitate economic planning (Miller and Cuff, 1986; Smit and Mason, 1990). For the SRR, the results of the Delphi and its consensus building, along with consensus that was gained during the SRR quarterly meetings, allowed for the development of a well-vetted set of potential indicators for assessing rangeland systems.

Within each criteria, participants developed specific indicators that can be measured in an effort to understand the impacts of rangeland use and management on sustainability. The SRR effort culminated with the development of five criteria groups that contained 64 indicators of sustainability. The five criteria groups and their associated indicators, which fell under two criteria categories (biophysical and socioeconomic), included the following:

11.3.4.2.3 Biophysical Criteria

- soil and water conservation on rangelands (10 indicators)
- indicators for conservation and maintenance of plant and animal resources on rangelands (10 indicators)
- maintenance of rangeland productive capacity (6 indicators)

11.3.4.2.4 Socioeconomic Criteria

- social and economic indicators of rangeland sustainability (28 indicators)
- legal, institutional, and economic framework for rangeland conservation and sustainable management (10 indicators)

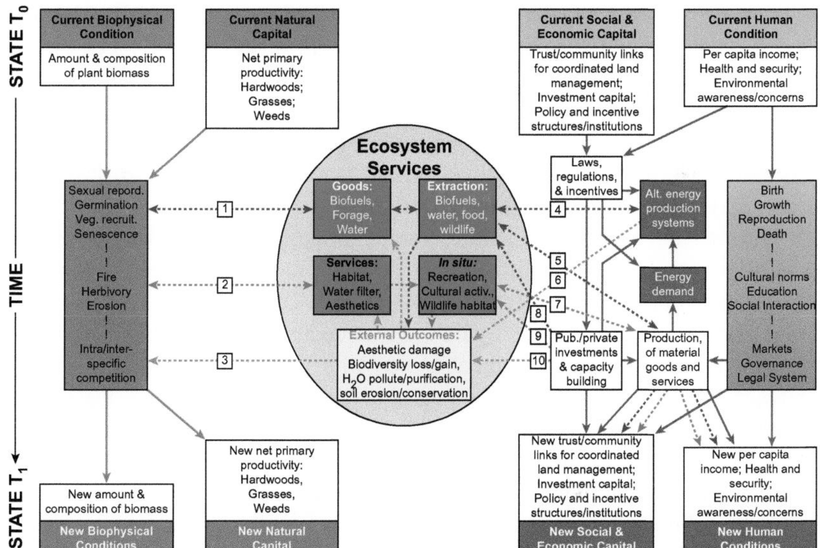

FIGURE 11.3.3 An example application of the Integrated Social, Economic and Ecological Concept framework to identify key biophysical–socioeconomic linkages that impact the delivery of ecosystem goods and services, and that affect or are affected by biofuel production from rangeland ecosystems. Numbered boxes indicate linkages between biophysical and socioeconomic subsystems that affect delivery of ecosystem services. *Adapted from Kreuter et al. (2012).*

In reality, the collection of data for all 64 indicators is not likely; however, the intention is for users to be able to agree that the indicators are valid and viable, and that a suite of indicators can be established based upon the region/locale of interest. Kreuter et al. (2012) demonstrated the utility of this approach through an investigation of the potential impacts of energy production on Western rangelands in the US (Figure 11.3.3).

11.3.4.3 Interpreting Indicators of Rangeland Health

An interagency collaboration was conducted by four U.S. land management/scientific teams including the U.S. Department of Agriculture's (USDA) Natural Resource Conservation Service and Agricultural Research Service (ARS), along with the U.S. Department of Interior's Bureau of Land Management (BLM) and the U.S. Geological Survey. The goal of this effort was to produce a user-friendly standard for land managers to conduct rangeland health assessment. The most recent version of the process was completed in 2005 (Pellant et al., 2005). The use of "qualitative assessments" of rangeland health is able to provide land managers and technical assistance specialists with a communication tool for use with interested stakeholders. The technique, along with quantitative monitoring and inventory information, is designed to provide early warnings on upland rangelands (Pellant et al., 2005). The process is not designed to provide diagnostic capabilities or understand cause and effect of degradation processes. Instead, through assessment, it provides information regarding ecosystems that may be experiencing degradation and in need of additional quantitative assessment in order to understand ongoing processes. The rangeland health assessment attempts to establish how well ecological processes on a site are functioning.

Through the definition of attributes of rangeland health, the assessment process introduces users to three cornerstones of the system: soil/site stability, hydrologic function, and biotic integrity. The 17 qualitative indicators of rangeland health are contained within the three attribute areas included in the assessment process (some indicators may be present in multiple attribute areas). Soil stability includes 11 assessment indicators (e.g., rills, water flow patterns, gullies, bare ground, etc.). The hydrologic function attribute includes 10 indicators such as presence/absence of pedestals, gullies, surface erosion resistance to erosion, compaction layer, etc. Last, the biotic integrity attribute includes nine indicators including functional/structural groups, plant mortality, litter amount, invasive plants, etc. The rangeland health approach suggests a series of key quantitative assessment indicators that can be obtained if the qualitative assessment indicates concerns along with methodologies suitable for generating the quantitative measurements.

The utility of the assessment process of rangeland health has been widely utilized by land-management agencies within the United States and has been adapted and applied on rangelands around the world. This approach provides an efficient and effective means to assess trends or change that is taking place on rangeland ecosystems and can inform land managers on the need to increase the assessment process by collection of quantitative data.

11.3.5 MONITORING RANGELAND DEGRADATION

In the assessment of indicators for rangeland health, monitoring data need to be collected to inform land managers and stakeholders on the condition of the indicators. Monitoring may be defined as the process of observing and tracking the progress or quality of resources, attributes, or their proxies over a period of time (West, 2003); in other words, to keep attributes under systematic review. Thus, monitoring provides a methodology to evaluate trend or change relative to a management objective or resource potential. The emphasis on change distinguishes monitoring from a single, point-in-time, survey or inventory (Ruyle and Dyess, 2010). The evaluation of trend in the time series allows assessment of decreasing, increasing, or stable trajectories in the items being monitored (West, 2003). Time series data provide this necessary reference and offer the best means available to evaluate and quantify positive or negative trends of management strategies or range improvements. Unfortunately, long-term data sets are rare due to their expense.

11.3.5.1 Ground Monitoring

Ground monitoring is needed to provide the inventory and trend data for indicators of rangeland health, and to examine trends toward degradation or improvement. Measurements represent indicators of chemical, physical, or biological processes and the linkages between human activity and environmental change with consideration of societal responses. The SRR notes that conservation and maintenance of soil and water resources are too general for direct monitoring, but may be characterized by a set of indictors focused on soils, vegetation, invasive species, capacity, and social commodities (Mitchell, 2010). Each of the indicators has specific methods or protocols for data collection with considerations for spatial and temporal scales, and statistical soundness. Spatial and temporal scales are important for rangeland monitoring because ecological and socioeconomic indicators may show different characteristics at different temporal and spatial scales (Mitchell, 2010). With regard to spatial scale, patterns at one level of scale that are identified through monitoring may not be evident at lower or higher levels of resolution. On temporal scales, longer time periods with repeated measurements are needed to identify actual change in rangeland condition (e.g., degradation) rather than natural fluctuations (e.g., from climate) (Mitchell, 2010; West, 2003).

Methods and protocols for rangeland monitoring and assessment are available to assist in planning and implementing field-monitoring programs. Several practical how-to manuals discuss monitoring of basic rangeland attributes including soil and site stability, watershed function, and biotic integrity. Reasons, methods, experimental designs, analyses, and interpretations are discussed in detail (Herrick et al., 2005a,b; Mitchell, 2010; Pellant et al., 2005; Riginos and Herrick, 2010). A general overview of indicators and ground-monitoring methods is provided below.

11.3.5.1.1 Soil Attributes

Maintaining and improving the quality of soil resource base is a prerequisite for sustainable rangeland use. Establishing permanent reference sites or areas of interest should be a consideration of a monitoring program. Soil degradation, through erosion, is visually evaluated by matching the condition of the rangeland to one of the three erosion forms, sheet, rill, and gully. These basic indicators provide information such as loss of fertility, increased runoff, poor soil structure, reduced rainfall infiltration, and poor vegetation cover. Soil properties indicating degradation include increases in salinity, depletion of nutrients, compaction and crusting, loss of organic matter, and reduced biodiversity (Kellner and Moussa, 2014).

Soil aggregate stability is an important measure of soil resistance to erosion and an important indicator for assessing soil erosion risk (Herrick et al., 2010). Soils with higher aggregate stability are less prone to water erosion because the larger soil aggregates improve resistance to soil detachment and generally higher water infiltration rates (Mitchell, 2010). The larger aggregates also reduce risk for wind erosion because of the increased weight of the aggregate. From the standpoint of rangeland degradation, activities or practices that result in changes to the plant community, break up soil crusts, break up or loosen aggregates, or impact mycorrhizal fungal communities can result in soil degradation. Aggregate stability can be measured using standard laboratory methods from samples collected in the field. However, because stability of soil aggregates can vary spatially and with soil texture, a large number of samples are required for a representative sample, thus increasing costs and soil transport considerations (Herrick et al., 2001). A field kit has been developed that can be used to gather a large number of samples to test for aggregate stability in a short amount of time (Herrick et al., 2001).

Change in the extent of bare ground is another important soil indicator for degradation. As the soil becomes exposed due to loss of vegetation by fire, grazing, drought, or other means, it becomes susceptible to wind and water erosion. The connectivity of the patches of bare ground and the size of the patches influence the speed of water and wind as it travels across the surface (Mitchell, 2010). Large patches of bare ground have a larger fetch area, thus reducing the opportunities for vegetation barriers to reduce the velocity of water and wind. For example, Bartley et al. (2006) found that the increases in bare patches on hillslopes in Australian savanna rangelands were significant contributors to runoff and soil loss, and that medium to high cover of vegetation at the bottom of hillslopes reduced the amount of sediment reaching the streams. Field measurement of bare ground can be conducted using transect methods that measure basal gaps (i.e., distances between plant spaces) and line-point intercept where the intercepts of bare ground can be compared to that of vegetation and litter (Herrick et al., 2005b). This type of monitoring should be continued over a long period of time in order to detect trends in the contraction or expansion of bare ground.

11.3.5.1.2 Biotic Attributes

Several important indicators have been identified for monitoring rangeland plant and animal resources that have implications for rangeland degradation. These include land area occupied by rangeland vegetation and rangeland plant communities, degree of rangeland and rangeland plant community fragmentation, intensity of human uses, changes in the fire regime, and displacement of native plants with invasive or non-native plants (Mitchell, 2010). The changes in the land area occupied by rangeland vegetation are important for assessing land area changes that could influence ecosystem services, plants and animals, and livestock stocking rates on rangelands. Because of the competing land uses that are shrinking the area of rangeland and fragmenting the landscape (Alkemade et al., 2013; Estell et al., 2012; Herrick et al., 2012), assessing the area of rangeland is important to reduce overgrazing (especially in developing countries) and for preventing loss of critical habitat or species. However, determining the total land area can be difficult because of the changing land uses and the ability to reliably map the land area on a timely basis. For national and global assessments, this would be logistically difficult and expensive if field methods were used. Therefore, remote sensing approaches are more feasible (see Section 11.3.5.2).

On degraded rangelands, the productive capacity has generally been reduced due to factors such as soil loss, loss of plant species, fragmentation, and other factors. Several indicators have been identified for monitoring rangeland productivity with regard to degradation including standing aboveground biomass, amount of litter, and numbers of livestock on rangeland (Mitchell, 2010; Pyke et al., 2002). Aboveground biomass measurements is one of the most common indicators examined on rangelands. Methods for measurement of aboveground biomass include harvesting, estimation, or a combination of these two methods (double sampling) (USDA, 2003). The harvesting method generally involves direct measurement of standing biomass from a set of plots of known area. Vegetation biomass within the plots is clipped and removed for weighing. For herbaceous plants (grasses, grass-like plants, and forbs), aboveground plant parts would be harvested by clipping/removing the aboveground biomass from within the plot (USDA, 2003). For trees and shrubs, entire plants may be harvested, but may require separation into leaf and wood components. Catchpole and Wheeler (1992) have cautioned that plot sampling for trees and large shrubs may not be practical where spatial variability is high, and that other techniques such as estimation of biomass may be more useful. After harvesting biomass from plots, samples are generally oven-dried to remove water so that biomass can be expressed on a dry-matter basis and converted to weight per unit area. Biomass measurements should take into consideration stratification of the land area for collecting representative samples for the variability in the plant communities being monitored, in addition to statistical methods to insure that adequate samples are collected to detect change (Herrick et al., 2005b).

The amount of plant litter is an important indicator as it reduces water runoff (Hart and Frasier, 2003; Thurow et al., 1988), protects the soil from erosion (Benkobi et al., 1993; Ludwig et al., 2004) and returns nutrients and organic matter to the soil via decomposition (Whitford, 1988). Plant litter is defined as dead plant material that is detached from the plant, and the amount is affected by the utilization of vegetation by grazers and climate variability (Pellant et al., 2005). Ground cover of litter can be measured using techniques similar to that used for bare ground such as line-point intercept. Litter biomass can be monitored using methods similar to aboveground biomass measurements where the litter biomass is removed from a known plot area, dried, weighed, and scaled by the appropriate weight per unit area.

Numbers of livestock on rangeland provide an indicator of the secondary production capacity of the rangeland (Mitchell, 2010), and when expressed as the number of livestock per unit area, it reflects the grazing pressure that the landscape is receiving. As discussed previously, sustained heavy grazing pressure can lead to overgrazing, and subsequently lead to degradation. Monitoring numbers of livestock can be conducted at local, regional, and national levels. Data are generally collected by census at the county (district) level and then aggregated to state (province) and national levels. Livestock numbers can be separated by species of livestock (e.g., numbers of cattle, sheep, and goats). However, because the forage intake and weights vary between different species of livestock, numbers are often normalized and expressed as animal units (AUs). Each species of livestock has an AU equivalent. In the United States, an AU is generally defined as a 454 kg cow with a forage dry matter intake of 9.5 kg/hd/day (Holechek, 1988). In countries where sheep and goats are predominant, a sheep unit is commonly used (a 50 kg sheep with 1 kg/hd/day intake). In Africa, tropical livestock units are used which represents one cow with a body weight of 250 kg and a dry matter intake of 5.2 kg/day. Changes in livestock numbers over time, both from an AU and a livestock-species perspective, can be a method of evaluating grazing pressure and livestock management changes over time. For example, with the collapse of the Soviet Union in 1991, Mongolia transitioned to a market economy and rangelands became freely available to grazing with little or no control over timing of grazing or animal numbers (Bedunah and Schmidt, 2004; Fernández-Giménez, 1997; Humphrey and Sneath, 1999; Mearns, 1996). From the period after 1991, livestock numbers peaked to numbers higher than those observed in the last 50 years, and are most likely higher than those observed during the last 100 years (Figure 11.3.4). The availability of livestock-density information such as this can be extremely helpful in assessing degradation impacts that may have resulted because of socioeconomic changes.

11.3.5.1.3 Water and Hydrological Attributes

Rangeland degradation, in the form of desertification, woody species encroachment, and nonnative invasion appears to be affecting regional streamflow, and may ultimately affect the global water cycle (Wilcox, 2007). Numerous examples are available, but sustained monitoring is needed to determine trends. Rangeland degradation affects water quality because streams, rivers, and lakes are strongly influenced by landscape characteristics of their watershed. Core monitoring topics in evaluation of rangeland watersheds include erosion, hydrology, vegetation, stream channel, water quality, species, habitat, and human use responses (Lewis et al., 2006; Moore et al., 2012; Wilkinson et al., 2013).

Water quality is strongly coupled to the landscape. Land use, cover type, and spatial configuration are strong predictors of nutrient and sediment loading. Vegetation type and location are also strongly correlated with

FIGURE 11.3.4 Changes in the numbers of livestock (converted to sheep units) over time in Mongolia. This type of data provides important information on the grazing pressure that has occurred on the landscape and informs trend analysis to assess livestock impacts on degradation. *Livestock number data from 1910 to 1960 are from Suttie (1999) and data from 1970 to 2013 are from National Statistical Office of Mongolia (2014).*

stream-nutrient concentrations and make good predictors (Bateni et al., 2013). Ellison et al. (2009) found improvements in water quality, as reduction in total dissolved solids and increases in macroinvertebrate diversity, were related to the best management practices of livestock grazing used in the Muddy Creek Basin of Wyoming.

Water environments are some of the most spatially and temporally heterogeneous spaces on Earth, making representative sampling difficult (Keith, 1991). Modern instrumentation and communications networks provide continuous and immediately available water-quality/quantity information that can be used directly or indirectly to compute concentrations of many water-quality constituents. Sensors most commonly available for directly measuring water parameters include specific conductance, pH, temperature, turbidity, fluorescence, and flow (acoustic Doppler). These in-stream measurements may be used as surrogates (or indicators) for estimating other water constituents. Because measurements are automated and recorded in small time intervals, they are often referred to as "continuous" or "real-time" measurements. Continuously recorded data associated with data from manually collected water-quality samples may be used to develop statistical relations between surrogate sensor measurements and constituents of concern that cannot be measured continuously. For example, turbidity threshold sampling is used to accurately calculate suspended sediment loads by measuring turbidity (an optical parameter) as a surrogate for suspended sediment measurements (Lewis, 1996). An in-stream turbidimeter is installed in conjunction with an automated water sampler. The sampler is activated at specified rising and falling turbidity thresholds. Water samples are analyzed for suspended sediment concentration and a regression relating concentration to optical turbidity is calculated. Using this regression, the continuous turbidity data accurately represent the suspended sediment loads in the runoff, which reflects erosion and ultimately reflects rangeland condition. A similar approach is used for other water-quality constituent concentrations such as bacteria, nutrients, metals, and organics. Computed estimates of water quality can be made available in near real-time at significantly decreased costs, compared to manual sample collection and analysis. These computed or estimated values are being used by water-management agencies for a wide variety of issues including water regulation, total maximum daily loads, recreation, and water treatment. Application of this technology to rangeland resource management is apparent.

11.3.5.2 Remote Sensing Assessments

The use of remote sensing imagery for assessing rangeland degradation is attractive because of the large areal coverage it provides, the ability to examine remote areas that may be inaccessible, and the ability to receive information at greater temporal frequencies. A variety of approaches have been used to develop ways of examining conditions on rangelands using remotely sensed data. Vegetation indices are one of the more popular and extensively studied products of remote sensing, and use transformations of spectral bands of the electromagnetic spectrum that are measured as reflectance from the Earth's surface by satellites. These indices allow the spatial and temporal variation, along with the relative contribution of vegetation properties, such as photosynthetic activity and canopy structure to be examined (Huete et al., 2002). A variety of vegetation indices have been proposed (see Tucker [1979] and Huete et al. [2002] for a review of indices), which generally involve some combination of the red and near infrared (NIR) portions of the electromagnetic spectrum, specifically wavelengths in the 0.6–0.7 μm (red) and 0.75–1.1 μm (NIR) ranges (Tucker et al., 1983). Of the vegetation indices that have been derived in the past 40 years, the Normalized Difference Vegetation Index (NDVI) is the most used and accepted (Cracknell, 2001).

Currently, multiple satellite platforms exist that are producing remote sensing imagery in a variety of spatial resolutions, and many of the products are freely available. Below is a description of some remote sensing analyses and products that have been developed and used for rangeland monitoring or to provide information on indicators of rangeland health and degradation.

11.3.5.2.1 Land Cover and Change

Several remote sensing products have been developed to assess land cover and land-cover change. The Moderate Resolution Imaging Spectroradiometer (MODIS) Land Cover product provides a yearly assessment of land cover with global coverage (Friedl et al., 2002). The product has five different land-cover classification schemes available, so users must decide which scheme to use and which classes to designate as rangeland. The product was recently enhanced to provide 500-m resolution for land areas across the globe (Friedl et al., 2010). In the United States, a National Land Cover Database (NLCD) was developed in 2001 (Homer et al., 2007), in 2006 (Fry et al., 2011), and in 2011 (Jin et al., 2013). Using Landsat satellite data, the NLCD uses a 16-class scheme of land-cover classification applied consistently across the coterminous United States at a spatial resolution of 30 m. Land-cover change products were also developed to compare changes from 2006 to 2011 and 2001 to 2011 (Jin et al., 2013). As with the MODIS land-cover data, users need to select which classes to designate as rangeland for assessing total rangeland area and change in land cover for rangeland.

11.3.5.2.2 Aboveground Biomass Estimation

Assessment of aboveground biomass using remote sensing generally uses two approaches (1) empirical models that use a statistical relationship between ground locations of aboveground biomass matched with the spectral bands (or some combination of bands) in pixels of remote sensing images; and (2) process models that use remote sensing data as inputs to predict aboveground biomass. Empirical approaches generally involve using a regression relationship between the remote sensing imagery and biomass data collected from field campaigns (Dungan, 1998). Tucker et al. (1983) used both a linear and logarithmic regression between Advanced Very High Resolution Radiometer (AVHRR) NDVI (8-km resolution) and ground-collected, biomass data to predict biomass on a regional scale in the Sahel region of Senegal. In the Xilingol Steppe of Inner Mongolia, Kawamura et al. (2005) used the MODIS Enhanced Vegetation Index (500-m pixels) to predict live biomass and total biomass of livestock forage with linear regression models with accuracy of almost 80%. Other examples include Sannier et al. (2002), Frank and Karn (2003), Kogan et al. (2004), Xu et al. (2008), and Yu et al. (2010). One consideration for using these empirical relationship models is that the extrapolation of these relationships to new areas is not always feasible or recommended. Generally, additional field data would be required for expanding or improving the prediction equations for new areas or regions.

Process models that use remote sensing variables as inputs are another approach for predicting aboveground biomass on a spatial and temporal basis. Reeves et al. (2001) used products from the MODIS imagery suite to estimate fraction of absorbed photosynthetically active radiation (fAPAR) which fed into a light-use efficiency model (Montieth, 1972) to estimate aboveground net primary productivity (ANPP) on a weekly basis at 1-km resolution for rangeland in the Northwestern USA. Hunt and Miyake (2006), using a light-use efficiency model and AVHRR data to estimate fAPAR, predicted aboveground biomass for estimating stocking rates within 1 km^2 cells for the state of Wyoming, USA. Modeling using remote sensing data can be integrated with geographic information systems (GIS) to produce comprehensive outputs for early warning systems on rangelands. In the African Sahel regions, Ham and Fillol (2012) described an early warning system that incorporates a light-use efficiency model with remote sensing inputs from the SPOT platform to predict forage biomass at 1-km resolution. An integrated approach to biomass quantity assessment on a regional level is the Livestock Early Warning System in East Africa (Stuth et al., 2003, 2005) and Mongolia (Angerer, 2012). This system uses process modeling and vegetation index data (AVHRR and MODIS NDVI) to provide near real-time estimates of biomass and deviation from average conditions (anomalies) on a bimonthly basis at 8-km resolution.

11.3.5.2.3 Degradation Assessments

Because of the high spatial and temporal frequency of remote sensing data, studies have been conducted with the intent of mapping areas of degraded rangeland for local, regional, and global assessments. Many of the degradation assessments that have been conducted to date make use of vegetation indices as a proxy for vegetation biomass or net primary productivity. Several of the initial degradation assessments using remote sensing time series involved evaluating relationships between vegetation indices and rainfall. Rainfall use efficiency (RUE), defined as the ratio of ANPP to annual rainfall (Le Houerou, 1984), was evaluated as an indicator for degradation. For the RUE method, ANPP is calculated from the seasonal integral of NDVI (i.e., the sum of the NDVI from the start to the end of the season) and it is colocated with the appropriate rainfall for calculation of RUE for a given pixel (Prince et al., 1998). Downward trends in RUE indicate potential degradation and upward trends indicate improved vegetation conditions. In an RUE study in the Sahel using AVHRR-NDVI, Prince et al. (1998) found no significant downward trends in RUE in this region that would indicate widespread degradation, and instead found a slight positive trend in RUE.

Using NDVI as a proxy for production, Evans and Geerken (2004) examined trends in the residuals of regressions using annual maximum NDVI and the accumulated rainfall that triggered the maximum NDVI in an effort to detect human- and climate-induced degradation in Syrian drylands. The authors concluded that negative trends in residuals would indicate human-induced degradation since the climate signal was removed from the analysis. In South Africa, Wessels et al. (2007a) conducted a comparison of the RUE method and a modification of the residual trends analysis (RESTREND) proposed by Evans and Geerken (2004) to examine if these methods could distinguish human-induced degradation from rainfall variability by comparing the results of these methods to degraded areas identified in field surveys. Results indicated that degraded areas did exhibit lower RUE (i.e., degraded areas had less-biomass production than nondegraded areas receiving the same amount of rainfall); however, RUE was highly correlated with rainfall, and therefore was not recommended for use in detecting degradation. The RESTREND method performed better in identifying degraded areas, but had several weakness that included: (1) it likely underestimates the magnitude of degradation; (2) the point in the time series when the degradation takes place influences the trend of the residuals with degradation in the middle of the time series having more negative trends in the residuals; and (3) the RESTREND method can only detect degradation that has occurred during the time series.

In a study of land degradation in Zimbabwe, Prince (2004) examined the use of scaling accumulated NDVI (as a proxy for ANPP) in individual pixels to that of a maximum accumulated NDVI of all pixels falling within a defined land type. The land-type boundaries are based on similar soils, climate, topography, and land cover. This technique is described as Local NPP scaling (LNS) and it attempts to define the relative difference in production of an individual pixel to that of the potential production in the entire land unit. Individual pixels having a large relative difference from the potential production (i.e., the maximum accumulated NDVI) in the land type would be designated as potentially degraded, thus requiring follow-up assessments (Prince, 2004). This technique has been applied in South Africa (Wessels et al., 2008), in Italy (Fava et al., 2012) and a follow-up analysis in Zimbabwe by Prince et al. (2009). Wessels et al. (2008) has stated that the LNS method is a valuable tool for mapping land degradation, but cautions that the performance of the method is strongly dependent on the quality of the land-unit stratifications.

Reeves and Baggett (2014) developed a degradation assessment protocol that compares the vegetation state and the temporal trend of a pixel to an objectively derived reference condition. Using MODIS NDVI (250 m) for the Great Plains region of the US, they compared each pixel to a reference condition that was determined from surrounding pixels having similar vegetation species, structure, and productivity. By statistically comparing the slopes of the linear trends in annual mean NDVI and annual maximum NDVI for each pixel against the reference condition, they found that 9% of the rangeland in the Northern Great Plains and 16% in the Southern Great Plains were degraded from their reference conditions.

A global assessment of land degradation was conducted by Bai et al. (2008) using modifications of the RUE methodology with AVHRR-NDVI for the period of 1981–2003. They found that almost 24% of the global land area was degraded. For land-cover classes that can be grouped together as rangelands (e.g., shrub cover, closed-open, evergreen; shrub cover, closed-open, deciduous; and herbaceous cover, closed-open), global land area classified as degraded ranged from 3% to 8% in their global assessment.

Although remote sensing protocols can be useful in regional assessments for defining land areas having characteristics of degradation, Wessels et al. (2007a) pointed out the need for field assessments to verify the areas mapped as degraded, and for long-term field-monitoring programs to improve degradation detection and to verify degradation indicators, such as species change and erosion that are not easy to detect with presently existing, remote sensing data sets over the long term. Another consideration is that many of the remote sensing analyses can only detect degradation that has occurred during the period of the satellite record, not degradation that occurred prior to the dates of the imagery (Wessels et al., 2008). A tremendous need exists for research and policy changes that provide for long-term monitoring programs that are able to capture the vegetation and soil indicators of degradation (Miehe et al., 2010).

11.3.6 RESTORATION AND THE DEGRADATION SPIRAL

Restoration of degraded semiarid and arid rangelands challenges the art and science of rangeland science and management. Rangelands that have encountered degrees of perturbation above and beyond their resilience thresholds (i.e., their capacity to absorb a disturbance without fundamental changes to characteristic processes and feedback [Holling, 1973]) are destined, without action, to spiral downward along a gradient of degradation. Therefore, losing the ability to provide the ecosystem goods and services that human, animal, and vegetation communities require for sustaining function within the ecosystem. "Change" processes are continual and part of the natural cycle on all rangelands of the world. Processes such as soil erosion, species immigration/emigration, variations in species, and other factors are part of every rangeland ecosystem, but when they expand beyond the normal range of variation, the perceived value of the ecosystem may be diminished. At this point, human values may determine that the system has sustained a disturbance beyond what is considered normal. It is important to remember that land degradation is a human value, driven by what society deems should be the function of the ecosystem in question; that is, the ecosystem will continue to function at some level regardless the expanse of disturbance. However, human perceptions of the "values" of ecosystems provide us with a reference of what "we" desire for any given ecosystem; therefore, we are able to assign a value of degradation and, if desired, a value of "restoring" the ecosystem to maintain the values society has assigned to the system. Assuming all ecosystems retain a level of inherent value, we are afforded the opportunity to recognize degradation and, in many cases, the opportunity to employ efforts to restore/rehabilitate the systems to a desired level of function.

Whisenant (1999) illustrated land degradation as a stepwise process of feedback initiated by some disturbance action, and which continues through time and space in a downward spiral away from a point that is considered as valuable, either to human or other species (see Section 11.3.3). It is important to recognize throughout the degradation-spiral process, a combination of biotic and abiotic processes are being affected across the ecosystem, as well as a change in structure and function. King and Hobbs (2006) have suggested that this combination truly drives the degradation of an ecosystem. Focusing on one process alone can lead to management decisions that can drastically impact the success/failure of any restoration programs that is being considered.

To better understand this critical interaction between structural and functional views of land degradation, one needs to understand the difference between the two. A structural view of degradation focuses on: (1) static patterns of change; (2) structural attributes of the system; (3) relies on mechanical manipulation; and (4) restores on a "quick fix" basis (King and Hobbs, 2006). A functional view centers on assessing: (1) dynamic processes; (2) functional systems; (3) a manipulation of interactions; and (4) results in "autogenic recovery." Either view, on its own, diminishes the potential for successful restoration of degraded ecosystems. In a true restoration program, the structural and functional aspects of both biotic and abiotic components must be addressed to initiate a program that has the potential for lasting effects. For those active in the restoration process, a desire should occur to reverse the downward spiral of degradation through restoring functions/processes in both the biotic and abiotic components of an ecosystem. The ultimate goal of restoration programs should be the generation of positive feedback loops that promote a continued improvement of the ecosystem components as a whole. Achieving this is a challenge and requires the input of experience and expertise across a wide variety of disciplines; however, when planned well, this has the potential to significantly improve the chances of success in returning to a desired system.

A question arises as to when a site has already begun the spiral, when can it be deemed degraded. The answer to this question lies within both the value system that is placed upon the ecosystem and the biophysical thresholds that are attributed to varying states of ecosystems. This interaction of human values and biophysical thresholds challenges land managers and scientists alike as they seek to identify the point at which an ecosystem is degrading. Therefore, it is important to consider the concept of thresholds when examining a degradation/restoration phenomenon. Wallington et al. (2005) have pointed out that over the recent past, conservation and restoration scientists have recognized that ecosystem dynamics are considerably more complex, nonlinear, and unpredictable than previous generations understood.

Thresholds can be viewed as points where a small change in environmental condition can lead to significant changes in ecosystem "state" variables. Considerable efforts have been undertaken to define and identify ecological thresholds while describing the processes of change once a "breaking point" has been reached (Bestelmeyer et al., 2003; Carpenter, 2001; Vandekerckhove et al., 2000). Once a threshold has been crossed, it is likely that alternative states may be established (Scheffer et al., 1993, 2001). The Resilience Alliance, made up of scientists from the biophysical and socioeconomic disciplines around the world, has defined ecological thresholds as a "bifurcation point between alternate states which, when passed, causes the system to 'flip' to a different state" (Meyers and Walker, 2003). Building upon this definition, Bennett and Radford (2003) submitted that "ecological thresholds" are points or zones within space or time where a rapid change takes place resulting in a change from one ecological condition to another. Archer (1989) illustrated a conceptual rangeland system and the states within which it can function. In this illustration, he described two states, one herbaceous-driven, and the other shrub-driven (Figure 11.3.5), with a threshold separating the two states. Within each state were several levels of function, with these levels of function being either herbaceous-, or shrub-driven. In the Figure 11.3.5 example, the inflection points between A and B can be viewed as "point" thresholds. Within each of these phases, those inflection points could be considered "managerial" thresholds. This would be an indication that the system has the potential to degrade to a less-desirable herbaceous-driven system, but through proper land management, the downward spiral could be reversed. However, upon reaching the inflection point "C," significant biotic and abiotic change has likely taken place and the system could be considered entering an "ecological/biological" threshold, where less-intrusive land management options are no longer capable of reversing the degradation spiral. These concepts are relatively easy to

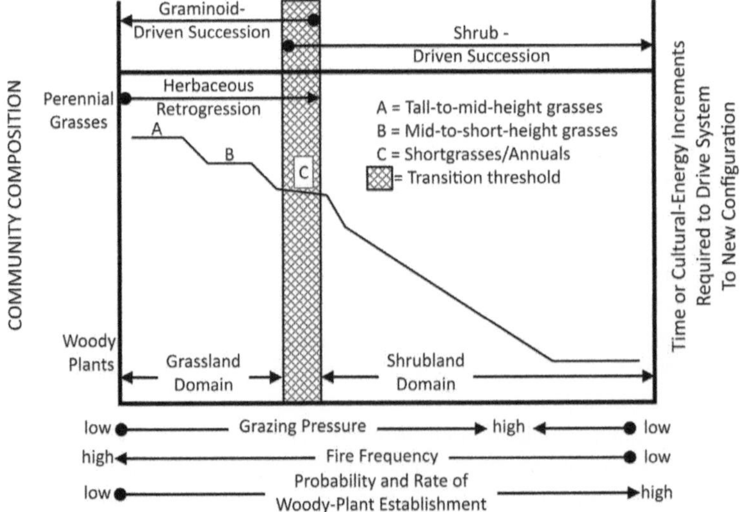

FIGURE 11.3.5 A conceptual model of factors influencing a change in state from grassland to woodland. *From Archer (1989).*

understand, but how can the land manager/restoration conservationist implement these approaches in the development of a restoration program to combat the degradation process? Huggett (2005) provided some insight to this question by defining four areas of application for "threshold" theory in biodiversity conservation. There were as follows:

1. If "sensitivity thresholds" are known for species of conservation or commercial significance, then cost-effective management policies and plans that strategically target these thresholds may be developed and implemented (Bayliss and Choquenot, 1998; Choquenot and Parkes, 2001);
2. If thresholds can be identified in natural or modified systems, it can help in understanding and managing trade-offs between biodiversity conservation and production (Roe and Van Eeten, 2001; Wallace, 2003);
3. Understanding ecological thresholds can potentially assist land managers in setting targets for habitat retention and restoration, and where possible, landscape recovery (Huggett, 2005); and
4. Knowledge of thresholds can be utilized in the development of landscape designs for biodiversity conservation (Brooker and Brooker, 2003; Brooker and Lefroy, 2004).

Managing within the realm of the degradation/restoration interface can challenge the most astute land managers. Many times an ecosystem may appear resilient, only to have some level of perturbation take place that rapidly transitions the system across an ecological threshold and requiring energy inputs to address (Suding and Hobbs, 2009). Arid and semiarid rangeland systems, worldwide, are highly diverse and complex systems with dynamic ecological and socioeconomic drivers moving the ecosystems. However, these systems maintain a relatively stable state until disturbance regimes are changed to the point that the inherent resilience of the system is compromised. At this point, the potential to cross ecological thresholds exists, and the possibility of initiating the downward degradation spiral may begin. Understanding the concepts of rangeland degradation and the ability to apply threshold theories can provide a powerful tool for the rangeland manager and significantly increase the ability to sustain these critical ecosystems for the goods and services that can be provided based upon societies values.

11.3.7 CONCLUSIONS AND FUTURE DIRECTIONS

Rangelands provide an array of ecosystem services to both rural and urban populations such as food, fiber, water, recreation, minerals, and medicinal plants. Therefore, the degradation of rangelands can have far-reaching effects. The causes of rangeland degradation are complex, and are often combinations of factors rather than a single, identifiable factor. Overarching causes are land management and extreme climate events, but these can be driven by factors such as land tenure, government policies, loss of rangeland to other uses (e.g., energy, mining, and biofuels), and economic conditions.

Given that livestock production is one of the main land uses and economic activities on rangelands, herd management can have tremendous impacts on rangeland degradation. Management that results in grazing more animals that the land can support over time (i.e., overgrazing) exposes the soil and reduces the productive capacity of the vegetation. This can begin the stepwise process of the degradation spiral where soil loss and vegetation productivity continue to decline. Reversal of the degradation spiral can come through improved management that reduces stocking rates. However, if the degradation spiral continues long enough, the site may cross a threshold where human and energy inputs would be required to move it back to its original state.

Climate change is expected to increase heat waves, heavy precipitation events, and the intensity and frequency of drought. These extreme climate events could increase degradation in vulnerable areas of the world, leading to plant species and cover loss, increased wind and water erosion, and susceptibility to invasive plant species (e.g., wood plant encroachment). Monitoring of rangeland systems will be extremely important in order to understand the impacts of climate and land-management effects on rangeland ecosystems and to develop appropriate adaptive strategies.

Formalized and consistent monitoring programs that focus on indicators for rangeland health will assist in understanding the processes that drive rangeland degradation and in efforts to restore or stabilize these systems. Ground-monitoring programs that highlight data collection for soil and site stability, hydrologic functions, and biotic integrity that are carried out over appropriate time and spatial scales can provide the necessary information to more efficiently understand degradation risk and develop restoration strategies. Remote sensing assessments are also useful in monitoring for degradation by providing spatially dense data sets at high temporal frequencies. These assessments can define "hot spots" or land areas having characteristics of degradation, but ground-monitoring assessments are needed to verify the degradation indicators that are not easy to detect with remote sensing.

With changing climate and competing land uses for rangelands, people who depend upon rangelands for their livelihoods will face many challenges that create greater uncertainty and increased risk in the future. Long-term monitoring programs that use common methodologies and statistically valid sampling schemes will be needed to provide the necessary data for identifying thresholds and providing information for adaptive management and policy changes in the coming years. Identification of important indicators of rangeland health and protocols for ground and remote sensing assessments have been developed, and implementation of these programs across the globe would greatly improve our ability to detect degradation and provide solutions.

Remote sensing and ground-based monitoring systems will continue to improve, along with telecommunication systems to allow information delivery and dissemination. Advances in high-resolution imagery obtained from unmanned aerial vehicles can provide the ability to quantify plant cover, composition, and structure at multiple spatial scales (Rango et al., 2009). LIght Detection And Ranging (LIDAR) technology can provide high-resolution quantification for biomass and land-cover assessments (Bork and Su, 2007; Ku et al., 2012). Bestelmeyer and Briske (2012), in an examination of grand challenges that need to be addressed for adoption of resilience-based rangeland management, expressed the need to develop knowledge systems to support adaptation and transformation. They defined knowledge systems as "technologies and institutions that bring together and mobilize diverse sources of information to support decision-making." The development of a rangeland degradation knowledge system that incorporate monitoring data, early warning and trend assessment, and local knowledge would be extremely beneficial to reducing rangeland degradation and improving management. Herrick et al. (2013) described an innovative system that combines mobile phone technology, cloud-based computing, and crowd-sourcing to collect data and analyze data from field sources, models, and databases to define land potential and provide users with critical information for land use planning and land management decision-making. A knowledge system also connects users to provide information and data sharing on best practices. The integration of remote sensing, ground monitoring, GIS, and simulation modeling provides avenues to connect to rangeland managers and policy makers to provide early warning of conditions that can lead to degradation. Rangeland scientists can continue to work with land managers and policy makers to communicate and translate information on the threats of degradation in terms of both ecological and societal factors to assist in educating land users and guide policy and regulation (Bedunah and Angerer, 2012).

REFERENCES

Alkemade, R., Reid, R.S., van den Berg, M., de Leeuw, J., Jeuken, M., 2013. Assessing the impacts of livestock production on biodiversity in rangeland ecosystems. Proc. Natl. Acad. Sci. U.S.A. 110, 20900–20905.

Angerer, J.P., 2012. Gobi forage livestock early warning system. In: Coughenour, M.B., Makkar, H.P.S. (Eds.), Conducting National Feed Assessments. Food and Agriculture Organization, Rome, Italy, pp. 115–130.

Archer, S., Stokes, C., 2000. Stress, disturbance and change in rangeland ecosystems. In: Arnalds, O., Archer, S. (Eds.), Rangeland Desertification. Springer, Netherlands, pp. 17–38.

Archer, S., Davies, K.W., Fulbright, T.E., McDaniel, K.C., Wilcox, B.P., Predick, K., Briske, D., 2011. Brush management as a rangeland conservation strategy: a critical evaluation. In: Conservation Benefits of Rangeland Practices: Assessment, Recommendations, and Knowledge Gaps. USDA-NRCS, Washington, DC, USA, pp. 105–170.

Archer, S., 1989. Have southern Texas savannas been converted to woodlands in recent history? Am. Nat. 134, 545–561.

Archer, S.R., 2010. Rangeland conservation and shrub encroachment: new perspectives on an old problem. In: Wild Rangelands. John Wiley & Sons, Ltd, pp. 53–97.

Asner, G.P., Elmore, A.J., Olander, L.P., Martin, R.E., Harris, A.T., 2004. Grazing systems, ecosystem responses, and global change. Annu. Rev. Environ. Resour. 29, 261–299.

Australian Department of Environment, 1999. Australia's Rangelands. Available at: http://www.environment.gov.au/land/publications/australias-rangelands (accessed 30.10.14.).

Ayoub, A.T., 1998. Extent, severity and causative factors of land degradation in the Sudan. J. Arid Environ. 38, 397–409.

Bai, Z.G., Dent, D.L., Olsson, L., Schaepman, M.E., 2008. Proxy global assessment of land degradation. Soil Use Manage. 24, 223–234.

Bartley, R., Roth, C.H., Ludwig, J., McJannet, D., Liedloff, A., Corfield, J., Hawdon, A., Abbott, B., 2006. Runoff and erosion from Australia's tropical semi-arid rangelands: influence of ground cover for differing space and time scales. Hydrol. Process. 20, 3317–3333.

Bateni, F., Fakheran, S., Soffianian, A., 2013. Assessment of land cover changes & water quality changes in the Zayandehroud river basin between 1997–2008. Environ. Monit. Assess. 185, 10511–10519.

Bayliss, P., Choquenot, D., 1998. Contribution of modelling to definition and implementation of management goals for overabundant marsupials. Managing Marsupial Abundance for Conservation Benefits. In: Proceedings of a Symposium Held at the 1998 Society for Conservation Biology Annual Meeting (Occasional Papers of the Marsupial CRC No. 1). Macquarie University, Sydney, Australia, pp. 69–75.

Bedunah, D.J., Angerer, J.P., 2012. Rangeland degradation, poverty, and conflict: how can rangeland scientists contribute to effective responses and solutions? Rangeland Ecol. Manage. 65, 606–612.

Bedunah, D.J., Schmidt, S.M., 2004. Pastoralism and protected area management in Mongolia's Gobi Gurvansaikhan National Park. Dev. Change 35, 167–191.

Behnke, R.H., Scoones, I., 1992. Rethinking Range Ecology: Implications for Rangeland Management in Africa. IIED, London, 43 p. En IIED Dryland Networks Programme Issues Paper.

Benkobi, L., Trlica, M.J., Smith, J.L., 1993. Soil loss as affected by different combinations of surface litter and rock. J. Environ. Qual. 22, 657–661.

Bennett, A., Radford, J., 2003. Know your ecological thresholds. In: Thinking Bush (Native Vegetation Research and Development Program, Land and Water Australia, Canberra), vol. 2, pp. 1–3.

Bestelmeyer, B.T., Briske, D.D., 2012. Grand challenges for resilience-based management of rangelands. Rangeland Ecol. Manage. 65, 654–663.

Bestelmeyer, B.T., Brown, J.R., Havstad, K.M., Alexander, R., Chavez, G., Herrick, J.E., 2003. Development and use of state-and-transition models for rangelands. J. Range Manage. 56, 114–126.

Bork, E.W., Su, J.G., 2007. Integrating LIDAR data and multispectral imagery for enhanced classification of rangeland vegetation: a meta analysis. Remote Sens. Environ. 111, 11–24.

Brooker, L., Brooker, M., 2003. Local distribution, metapopulation viability and conservation of the Blue-breasted Fairy-wren in fragmented habitat in the Western Australian wheatbelt. Emu 103, 185–198.

Brooker, L., Lefroy, T., 2004. Habitat Neighbourhoods for Conserving Viable Populations of Birds. CSIRO Sustainable Ecosystems, Perth, Australia.

Brown, J.R., Thorpe, J., 2008. Climate change and rangelands: responding rationally to uncertainty. Rangelands 30, 3–6.

Browning, D.M., Archer, S.R., 2010. Protection from livestock fails to deter shrub proliferation in a desert landscape with a history of heavy grazing. Ecol. Appl. 21, 1629–1642.

Carpenter, S.R., 2001. Alternate states of ecosystems: evidence and its implications. In: Press, M.C., Huntly, D., Levin, S. (Eds.), Ecology: Achievement and Challenge. Blackwell, London, pp. 357–383.

Catchpole, W.R., Wheeler, C.J., 1992. Estimating plant biomass: a review of techniques. Aust. J. Ecol. 17, 121–131.

Choquenot, D., Parkes, J., 2001. Setting thresholds for pest control: how does pest density affect resource viability? Biol. Conserv. 99, 29–46.

Clarkson, N., Lee, G., 1988. Effects of grazing and severe drought on a native pasture in the traprock region of southern Queensland. Trop. Grassl. 22, 176–183.

Cracknell, A.P., 2001. The exciting and totally unanticipated success of the AVHRR in applications for which it was never intended. Adv. Space Res. 28, 233–240.

Delgado, C., Rosegrant, M., Steinfeld, H., Ehui, H., Courbois, C., 1999. Livestock to 2020: The Next Food Revolution. International Food Policy Research Institute, Washington, D.C, USA, p. 83.

Dregne, H.E., 2000. Desertification: problems and challenges. Ann. Arid Zone 39, 363–371.

Dungan, J., 1998. Spatial prediction of vegetation quantities using ground and image data. Int. J. Remote Sens. 19, 267–285.

Easterling, D.R., Meehl, G.A., Parmesan, C., Changnon, S.A., Karl, T.R., Mearns, L.O., 2000. Climate extremes: observations, modeling, and impacts. Science 289, 2068–2074.

Ellis, J.E., Swift, D.M., 1988. Stability of African pastoral ecosystems: alternate paradigms and implications for development. J. Range Manage. Arch. 41, 450–459.

Ellison, C.A., Skinner, Q.D., Hicks, L.S., 2009. Assessment of best-management practice effects on rangeland stream water quality using multivariate statistical techniques. Rangeland Ecol. Manage. 62, 371–386.

Estell, R.E., Havstad, K.M., Cibils, A.F., Fredrickson, E.L., Anderson, D.M., Schrader, T.S., James, D.K., 2012. Increasing shrub use by livestock in a world with less grass. Rangeland Ecol. Manage. 65, 553–562.

Eswaran, H., Lal, R., Reich, P., 2001. Land degradation: an overview. In: Responses to Land Degradation, pp. 20–35.

Evans, J., Geerken, R., 2004. Discrimination between climate and human-induced dryland degradation. J. Arid Environ. 57, 535–554.

Fava, F., Colombo, R., Bocchi, S., Zucca, C., 2012. Assessment of Mediterranean pasture condition using MODIS normalized difference vegetation index time series. J. Appl. Remote Sens. 6, 063530-063531-063530-063512.

Fensham, R.J., Fairfax, R.J., Ward, D.P., 2009. Drought-induced tree death in savanna. Glob. Change Biol. 15, 380–387.

Fernández-Giménez, M.E., 1997. Landscapes, Livestock, and Livelihoods: Social, Ecological, and Land-use Change Among the Nomadic Pastoralists of Mongolia. University of California, Berkeley.

Food and Agriculture Organization [FAO], 2008. Livestock Policy and Poverty Reduction. Food and Agriculture Organization, Rome, Italy, p. 8.

Fox, W.E., McCollum, D.W., Mitchell, J.E., Swanson, L.E., Kreuter, U.P., Tanaka, J.A., Evans, G.R., Theodore Heintz, H., Breckenridge, R.P., Geissler, P.H., 2009. An integrated social, economic, and ecologic conceptual (ISEEC) framework for considering rangeland sustainability. Soc. Nat. Resour. 22, 593–606.

Frank, A.B., Karn, J.F., 2003. Vegetation indices, CO_2 flux, and biomass for northern plains grasslands. J. Range Manage. 56, 382–387.

Friedl, M.A., McIver, D.K., Hodges, J.C.F., Zhang, X.Y., Muchoney, D., Strahler, A.H., Woodcock, C.E., Gopal, S., Schneider, A., Cooper, A., Baccini, A., Gao, F., Schaaf, C., 2002. Global land cover mapping from MODIS: algorithms and early results. Remote Sens. Environ. 83, 287–302.

Friedl, M.A., Sulla-Menashe, D., Tan, B., Schneider, A., Ramankutty, N., Sibley, A., Huang, X., 2010. MODIS Collection 5 global land cover: algorithm refinements and characterization of new datasets. Remote Sens. Environ. 114, 168–182.

Fry, J.A., Xian, G., Jin, S., Dewitz, J.A., Homer, C.G., Limin, Y., Barnes, C.A., Herold, N.D., Wickham, J.D., 2011. Completion of the 2006 national land cover database for the conterminous United States. Photogramm. Eng. Remote Sens. 77, 858–864.

Ham, F., Fillol, E., 2012. Pastoral surveillance system and feed inventory in the Sahel. In: Coughenour, M.B., Makkar, H.P.S. (Eds.), Conducting National Feed Assessments. Food and Agriculture Organization, Rome, Italy, pp. 83–94.

Hammond, A., Adriaanse, A., Rodenburg, E., Bryant, D., Woodward, R., 1995. Environmental Indicators: A Systematic Approach to Measuring and Reporting on Environmental Policy Performance in the Context of Sustainable Development. World Resources Institute, Washington, DC.

Hart, R.H., Frasier, G.W., 2003. Bare ground and litter as estimators of runoff on short-and mixed-grass prairie. Arid Land Res. Manage. 17, 485–490.

Havstad, K.M., Peters, D.P.C., Skaggs, R., Brown, J., Bestelmeyer, B., Fredrickson, E., Herrick, J., Wright, J., 2007. Ecological services to and from rangelands of the United States. Ecol. Econom. 64, 261–268.

Henry, B., McKeon, G., Syktus, J., Carter, J., Day, K., Rayner, D., 2007. Climate variability, climate change and land degradation. In: Sivakumar, M.V.K., Ndiang'ui, N. (Eds.), Climate and Land Degradation. Springer Berlin Heidelberg, pp. 205–221.

Herrick, J.E., Whitford, W.G., de Soyza, A.G., Van Zee, J.W., Havstad, K.M., Seybold, C.A., Walton, M., 2001. Field soil aggregate stability kit for soil quality and rangeland health evaluations. Catena 44, 27–35.

Herrick, J., Van Zee, J., Havstad, K., Burkett, L., Whitford, W., 2005a. Monitoring Manual for Grassland, Shrubland, and Savanna Ecosystems. Jornada Experimental Range, USDA. The University of Arizona Press, 236 p.

Herrick, J., Van Zee, J., Havstad, K., Burkett, L., Whitford, W., 2005b. Monitoring Manual for Grassland, Shrubland, and Savanna Ecosystems, Volume II. Design, Supplementary Methods and Interpretation. Jornada Experimental Range, USDA. The University of Arizona Press. p. 236.

Herrick, J.E., Lessard, V.C., Spaeth, K.E., Shaver, P.L., Dayton, R.S., Pyke, D.A., Jolley, L., Goebel, J.J., 2010. National ecosystem assessments supported by scientific and local knowledge. Front. Ecol. Environ. 8, 403–408.

Herrick, J.E., Brown, J.R., Bestelmeyer, B.T., Andrews, S.S., Baldi, G., Davies, J., Duniway, M., Havstad, K.M., Karl, J.W., Karlen, D.L., Peters, D.P.C., Quinton, J.N., Riginos, C., Shaver, P.L., Steinaker, D., Twomlow, S., 2012. Revolutionary land use change in the 21st century: is (rangeland) science relevant? Rangeland Ecol. Manage. 65, 590–598.

Herrick, J.E., Urama, K.C., Karl, J.W., Boos, J., Johnson, M.-V.V., Shepherd, K.D., Hempel, J., Bestelmeyer, B.T., Davies, J., Guerra, J.L., Kosnik, C., Kimiti, D.W., Ekai, A.L., Muller, K., Norfleet, L., Ozor, N., Reinsch, T., Sarukhan, J., West, L.T., 2013. The global Land-Potential Knowledge System (LandPKS): supporting evidence-based, site-specific land use and management through cloud computing, mobile applications, and crowdsourcing. J. Soil Water Conserv. 68, 5A–12A.

Hess Jr., K., Holechek, J., 1995. Policy roots of land degradation in the arid region of the United States: an overview. Environ. Monitor. Assess. 37, 123–141.

Hibbard, K., Schimel, D., Archer, S., Ojima, D., Parton, W., 2003. Grassland to woodland transitions: integrating changes in landscape structure and biogeochemistry. Ecol. Appl. 13, 911–926.

Holechek, J., Pieper, R.D., Herbel, C.H., 1995. Range Management: Principles and Practices. Prentice Hall, Englewood Cliffs, NJ, xvi, 526 pp.

Holechek, J.L., 1988. An approach for setting the stocking rate. Rangelands 10, 10–14.

Holling, C., Gunderson, L.H., Ludwig, D., 2001. In quest of a theory of adaptive change. In: Holling, C., Gunderson, L.H. (Eds.), Panarchy: Understanding Transformations in Human and Natural Systems. Island Press, New York, pp. 4–22.

Holling, C.S., 1973. Resilience and stability of ecological systems. Ann. Rev. Ecol. Syst. 4, 1–23.

Homer, C., Dewitz, J., Fry, J., Coan, M., Hossain, N., Larson, C., Herold, N., McKerrow, A., VanDriel, J.N., Wickham, J., 2007. Completion of the 2001 national land cover database for the counterminous united states. Photogramm. Eng. Remote Sens. 73, 337.

Horn, B.E., Hart, C.R., Paisley, S.I., 2002. Management of rangeland livestock under drought. Ann. Arid Zone 41, 1–23.

Huete, A., Didan, K., Miura, T., Rodriguez, E.P., Gao, X., Ferreira, L.G., 2002. Overview of the radiometric and biophysical performance of the MODIS vegetation indices. Remote Sens. Environ. 83, 195–213.

Huggett, A.J., 2005. The concept and utility of 'ecological thresholds' in biodiversity conservation. Biol. Conserv. 124, 301–310.

Humphrey, C.A., Sneath, D., 1999. The End of Nomadism? Society, State and the Environment in Inner Asia. Duke University Press, Durham, NC.

Hunt, E.R., Miyake, B.A., 2006. Comparison of stocking rates from remote sensing and geospatial data. Rangeland Ecol. Manage. 59, 11–18.

Huxman, T.E., Wilcox, B.P., Breshears, D.D., Scott, R.L., Snyder, K.A., Small, E.E., Hultine, K., Pockman, W.T., Jackson, R.B., 2005. Ecohydrological implications of woody plant encroachment. Ecology 86, 308–319.

Illius, A.W., O'Connor, T.G., 1999. On the relevance of nonequilibrium concepts to arid and semiarid grazing systems. Ecol. Appl. 9, 798–813.

Intergovernmental Panel on Climate Change [IPCC], 2012. Managing the Risks of Extreme Events and Disasters to Advance Climate Change Adaptation. A Special Report of Working Groups I and II of the Intergovernmental Panel on Climate Change. Cambridge University Press, New York, NY, 582 p.

Jin, S., Yang, L., Danielson, P., Homer, C., Fry, J., Xian, G., 2013. A comprehensive change detection method for updating the national land cover database to circa 2011. Remote Sens. Environ. 132, 159–175.

Kates, R.W., Clark, W.C., Corell, R., Hall, J.M., Jaeger, C.C., Lowe, I., McCarthy, J.J., Schellnhuber, H.J., Bolin, B., Dickson, N.M., Faucheux, S., Gallopin, G.C., Grübler, A., Huntley, B., Jäger, J., Jodha, N.S., Kasperson, R.E., Mabogunje, A., Matson, P., Mooney, H., Moore, B., O'Riordan, T., Svedin, U., 2001. Sustainability science. Science 292, 641–642.

Kawamura, K., Akiyama, T., Yokota, H., Tsutsumi, M., Yasuda, T., Watanabe, O., Wang, G., Wang, S., 2005. Monitoring of forage conditions with MODIS imagery in the Xilingol steppe, Inner Mongolia. Int. J. Remote Sens. 26, 1423–1436.

Keith, L.H., 1991. Environmental Sampling and Analysis: A Practical Guide. Lewis Publishers, Chelsea, MI.

Kellner, K., Moussa, A., 2014. A conceptual tool for improving rangeland management decision-making at grassroots level: the local-level monitoring approach. Afr. J. Range Forage Sci. 26, 139–147.

King, E.G., Hobbs, R.J., 2006. Identifying linkages among conceptual models of ecosystem degradation and restoration: towards an integrative framework. Restor. Ecol. 14, 369–378.

Kogan, F., Stark, R., Gitleson, A., Jargalsaikhan, C., Dugrajav, C., Tsooj, S., 2004. Derivation of pasture biomass in Mongolia from AVHRR-based vegetation health indices. Int. J. Remote Sens. 25, 2889–2896.

Kreuter, U., Fox, W., Tanaka, J., Maczko, K., McCollum, D., Mitchell, J., Duke, C., Hidinger, L., 2012. Framework for comparing ecosystem impacts of developing unconventional energy resources on western US rangelands. Rangeland Ecol. Manage. 65, 433–443.

Ku, N.-W., Popescu, S.C., Ansley, R.J., Perotto-Baldivieso, H.L., Filippi, A.M., 2012. Assessment of available rangeland woody plant biomass with a terrestrial LIDAR system. Photogramm. Eng. Remote Sens. 78, 349–361.

Le Houerou, H.N., 1984. Rain use efficiency: a unifying concept in arid-land ecology. J. Arid Environ. 7, 213–247.

Lewis, D.J., Singer, M.J., Dahlgren, R.A., Tate, K.W., 2006. Nitrate and sediment fluxes from a california rangeland watershed. J. Environ. Qual. 35, 2202–2211.

Lewis, J., 1996. Turbidity-controlled suspended sediment sampling for runoff-event load estimation. Water Resour. Res. 32, 2299–2310.

Little, P.D., McPeak, J.G., 2014. Resilience and pastoralism in Africa south of the Sahara with a particular focus on the horn of Africa and the Sahel, West Africa. In: 2020 Conference. International Food Policy Research Institute, Addis Ababa, Ethiopia.

Ludwig, J.A., Tongway, D.J., Bastin, G.N., James, C.D., 2004. Monitoring ecological indicators of rangeland functional integrity and their relation to biodiversity at local to regional scales. Austral Ecol. 29, 108–120.

Mearns, R., 1996. Community, collective action and common grazing: the case of post-socialist Mongolia. J. Dev. Stud. 32, 297–339.

Meyers, J., Walker, B.H., 2003. Thresholds and Alternate States in Ecological and Socio-ecological Systems: Thresholds Database. Available at: http://www.resalliance.org.au.

Miehe, S., Kluge, J., Von Wehrden, H., Retzer, V., 2010. Long-term degradation of Sahelian rangeland detected by 27 years of field study in Senegal. J. Appl. Ecol. 47, 692–700.

Miller, A., Cuff, W., 1986. The Delphi approach to the mediation of environmental disputes. Environ. Manage. 10, 321–330.

Mitchell, J.E. (Ed.), 2010. Criteria and Indicators of Sustainable Rangeland Management. University of Wyoming, Cooperative Extension Service. No. SM-56, p. 227.

Montieth, J.L., 1972. Solar radiation and productivity in tropical ecosystems. Appl. Ecol. 9, 747–766.

Moore, G.W., Barre, D.A., Owens, M.K., 2012. Does shrub removal increase groundwater recharge in Southwestern Texas semiarid rangelands? Rangeland Ecol. Manage. 65, 1–10.

Müller, B., Frank, K., Wissel, C., 2007. Relevance of rest periods in non-equilibrium rangeland systems – a modelling analysis. Agric. Syst. 92, 295–317.

National Research Council [NRC], 1994. Rangeland Health: New Methods to Classify, Inventory, and Monitor Rangelands. National Academy Press, Washington, D.C., 180 p.

National Statistical Office of Mongolia, 2014. Mongolia Livestock Statistical Data. Available at: http://www.1212.mn/en/contents/stats/contents_stat_fld_tree_html.jsp (accessed 25.10.14.).

National Sustainability Council, 2013. Sustainable Australia Report 2013, Conversations with the Future. DSEWPaC, Canberra, p. 263.

Pellant, M., Shaver, P., Pyke, D.A., Herrick, J.E., 2005. Interpreting Indicators of Rangeland Health. version 4. Technical Reference 1734-6. US Department of the Interior, Bureau of Land Management: National Science and Technology Center, Denver, CO. BLM/WO/ST-00/001+1734/REV05. p. 122.

Prince, S.D., De Colstoun, E.B., Kravitz, L.L., 1998. Evidence from rain-use efficiencies does not indicate extensive sahelian desertification. Glob. Change Biol. 4, 359–374.

Prince, S.D., Becker-Reshef, I., Rishmawi, K., 2009. Detection and mapping of long-term land degradation using local net production scaling: application to Zimbabwe. Remote Sens. Environ. 113, 1046–1057.

Prince, S., 2004. Mapping desertification in Southern Africa. In: Gutman, G., Janetos, A., Justice, C., Moran, E., Mustard, J., Rindfuss, R., Skole, D., Turner II, B., Cochrane, M. (Eds.), Land Change Science. Springer, Netherlands, pp. 163–184.

Pyke, D., Herrick, J., Shaver, P., Pellant, M., 2002. Rangeland health attributes and indicators for qualitative assessment. J. Range Manage. 55, 584–597.

Rango, A., Laliberte, A., Herrick, J.E., Winters, C., Havstad, K., Steele, C., Browning, D., 2009. Unmanned aerial vehicle-based remote sensing for rangeland assessment, monitoring, and management. J. Appl. Remote Sens. 3, 033542-033542-033515.

Reeves, M.C., Baggett, L.S., 2014. A remote sensing protocol for identifying rangelands with degraded productive capacity. Ecol. Indic. 43, 172–182.

Reeves, M.C., Winslow, J.C., Running, S.W., 2001. Mapping weekly rangeland vegetation productivity using MODIS algorithms. J. Range Manage. 54, 90–105.

Riginos, C., Herrick, J.E., 2010. Monitoring Rangeland Health: A Guide for Pastoralists and Other Land Managers in Eastern Africa. Version II. Nairobi, Kenya.

Roe, E., Van Eeten, M., 2001. Threshold-based resource management: a framework for comprehensive ecosystem management. Environ. Manage. 27, 195–214.

Ruyle, G.B., Dyess, J., 2010. Rangeland Monitoring and the Parker 3-step Method: Overview, Perspectives and Current Applications. College of Agriculture and Life Sciences, University of Arizona, Tucson, AZ.

Safriel, U.N., 2007. The assessment of global trends in land degradation. In: Sivakumar, M.K., Ndiang'ui, N. (Eds.), Climate and Land Degradation. Springer Berlin Heidelberg, pp. 1–38.

Sannier, C.A.D., Taylor, J.C., Du Plessis, W., 2002. Real-time monitoring of vegetation biomass with NOAA-AVHRR in Etosha National Park, Namibia, for fire risk assessment. Int. J. Remote Sens. 23, 71–89.

Scheffer, M., Hosper, S., Meijer, M., Moss, B., Jeppesen, E., 1993. Alternative equilibria in shallow lakes. Trends Ecol. Evol. 8, 275–279.

Scheffer, M., Carpenter, S., Foley, J.A., Folke, C., Walker, B., 2001. Catastrophic shifts in ecosystems. Nature 413, 591–596.

Schlesinger, W.H., Reynolds, J.F., Cunningham, G.L., Huenneke, L.F., Jarrell, W.M., Virginia, R.A., Whitford, W.G., 1990. Biological feedbacks in global desertification. Science 247, 1043–1048.

Schlesinger, W.H., 2010. Translational ecology. Science 329, 609.

Scifres, C.J., Rasmussen, G.A., Hamilton, W.T., Smith, R.P., Conner, J.R., Stuth, J.W., Inglis, J.M., Welch, T.G., 1985. Integrated Brush Management Systems for South Texas: Development and Implementation. Texas A&M University, College Station, Texas, p. 76.

Scurlock, J., Hall, D., 1998. The global carbon sink: a grassland perspective. Glob. Change Biol. 4, 229–233.

Singh, R.K., Murty, H.R., Gupta, S.K., Dikshit, A.K., 2009. An overview of sustainability assessment methodologies. Ecol. Indic. 9, 189–212.

Sivakumar, M.K., Stefanski, R., 2007. Climate and land degradation — an overview. In: Sivakumar, M.V.K., Ndiang'ui, N. (Eds.), Climate and Land Degradation. Springer, Berlin, pp. 105–135.

Smit, J., Mason, A., 1990. A Policy Delphi study in the socialist Middle East. Public Adm. Dev. 10, 453–465.

Society for Range Management [SRM]: Task Group on Unity in Concepts and Terminology Committee Members, 1995. New concepts for the assessment of rangeland condition. J. Range Manag. 48(3):271–282.

Society for Range Management [SRM], 2002. Policy Statement, Position Statements, and Resolutions. Society For Range Management, Denver, CO, p. 9.

Steinfeld, H., Gerber, P., Wassenaar, T.D., Castel, V., de Haan, C., 2006. Livestock's Long Shadow: Environmental Issues and Options. Food and Agriculture Organization of the United Nations, Rome, Italy.

Stuth, J., Angerer, J., Kaitho, R., Zander, K., Jama, A., Heath, C., Bucher, J., Hamilton, W., Conner, R., Inbody, D., 2003. The Livestock Early Warning System (LEWS): Blending Technology and the Human Dimension to Support Grazing Decisions. Available at: http://cals.arizona.edu/OALS/ALN/aln53/stuth.html (accessed 25.10.14.).

Stuth, J.W., Angerer, J., Kaitho, R., Jama, A., Marambii, R., 2005. Livestock early warning system for Africa rangelands. In: Boken, V.K., Cracknell, A.P., Heathcote, R.L. (Eds.), Monitoring and Predicting Agricultural Drought: A Global Study. Oxford University Press, New York, p. 472.

Suding, K.N., Hobbs, R.J., 2009. Threshold models in restoration and conservation: a developing framework. Trends Ecol. Evol. 24, 271–279.

Suttie, J.M., 1999. Grassland and Pasture Crops: Country Pasture/forage Resource Profile-Mongolia. Available at: http://www.fao.org/ag/AGP/AGPC/doc/Counprof/mongolia (accessed 01.10.14.).

Thurow, T.L., Blackburn, W.H., Taylor Jr., C.A., 1988. Infiltration and interrill erosion responses to selected livestock grazing strategies, Edwards Plateau, Texas. J. Range Manage. 41, 296–302.

Tucker, C.J., Vanpraet, C., Boerwinkel, E., Gaston, A., 1983. Satellite remote-sensing of total dry-matter production in the Senegalese Sahel. Remote Sens. Environ. 13, 461–474.

Tucker, C.J., 1979. Red and photographic infrared linear combinations for monitoring vegetation. Remote Sens. Environ. 8, 127–150.

Twidwell, D., Wonkka, C.L., Taylor, C.A., Zou, C.B., Twidwell, J.J., Rogers, W.E., 2014. Drought-induced woody plant mortality in an encroached semi-arid savanna depends on topoedaphic factors and land management. Appl. Veg. Sci. 17, 42–52.

United Nations Convention to Combat Desertification [UNCCD], 1994. Article 2 of the Text of the United Nations Convention to Combat Desertification, p. 58.

USDA., 2003. National Range and Pasture Handbook. U.S. Dept. of Agriculture, Natural Resources Conservation Service, Grazing Lands Technology Institute, Fort Worth, Tex.

Vandekerckhove, L., Poesen, J., Oostwoud Wijdenes, D., Nachtergaele, J., Kosmas, C., Roxo, M.J., de Figueiredo, T., 2000. Thresholds for gully initiation and sedimentation in Mediterranean Europe. Earth Surf. Proc. Land. 25, 1201–1220.

Vetter, S., 2005. Rangelands at equilibrium and non-equilibrium: recent developments in the debate. J. Arid Environ. 62, 321–341.

Wallace, K.J., 2003. Managing Natural Biodiversity in the Western Australian Wheatbelt: A Conceptual Framework. K.J. Wallace, B.C. Beecham, B.H. Bone. [Crawley, W.A.]. Dept. of Conservation and Land Management.

Wallington, T.J., Hobbs, R.J., Moore, S.A., 2005. Implications of current ecological thinking for biodiversity conservation: a review of the salient issues. Ecol. Soc. 10, 16.

Warhurst, A., 2002. Sustainability indicators and sustainability performance management. Report to the Project: Mining, Minerals and Sustainable Development (MMSD). International Institute for Environment and Development (IIED), Warwick, England.

Wessels, K., Prince, S., Malherbe, J., Small, J., Frost, P., VanZyl, D., 2007a. Can human-induced land degradation be distinguished from the effects of rainfall variability? A case study in South Africa. J. Arid Environ. 68, 271–297.

Wessels, K.J., Prince, S.D., Carroll, M., Malherbe, J., 2007b. Relevance of rangeland degradation in semiarid northeastern south africa to the nonequilibrium theory. Ecol. Appl. 17, 815–827.

Wessels, K.J., Prince, S.D., Reshef, I., 2008. Mapping land degradation by comparison of vegetation production to spatially derived estimates of potential production. J. Arid Environ. 72, 1940–1949.

West, N.E., 2003. Theoretical underpinnings of rangeland monitoring. Arid Land Res. Manage. 17, 333–346.

Whisenant, S., 1999. Repairing Damaged Wildlands: A Process-Orientated, Landscape-Scale Approach. Cambridge University Press.

Whisenant, S.G., 2002. Terrestrial systems. In: Perrow, M.R., Davy, A.J. (Eds.), Handbook of Ecological Restoration. Cambridge University Press, Cambridge, pp. 83–105.

Whitford, W.G., 1988. Decomposition and nutrient cycling in disturbed arid ecosystems. In: Allen, E.B. (Ed.), The Reconstruction of Disturbed Arid Lands. Westview Press, Boulder, Colorado, pp. 136–161.

Wilcox, B.P., Thurow, T.L., 2006. Emerging issues in rangeland Ecohydrology: vegetation change and the water cycle. Rangeland Ecol. Manage. 59, 220–224.

Wilcox, B.P., 2007. Does rangeland degradation have implications for global streamflow? Hydrol. Process. 21, 2961–2964.

Wilcox, B.P., 2010. Transformative ecosystem change and ecohydrology: ushering in a new era for watershed management. Ecohydrology 3, 126–130.

Wilkinson, S.N., Hancock, G.J., Bartley, R., Hawdon, A.A., Keen, R.J., 2013. Using sediment tracing to assess processes and spatial patterns of erosion in grazed rangelands, Burdekin River basin, Australia. Agric. Ecosyst. Environ. 180, 90–102.

Xu, B., Yang, X.C., Tao, W.G., Qin, Z.H., Liu, H.Q., Miao, J.M., Bi, Y.Y., 2008. MODIS-based remote sensing monitoring of grass production in China. Int. J. Remote Sens. 29, 5313–5327.

Young, V.A., 1956. The effect of the 1949–1954 drought on the ranges of Texas. J. Range Manage. 9, 139–142.

Yu, L., Zhou, L., Liu, W., Zhou, H.K., 2010. Using remote sensing and GIS technologies to estimate grass yield and livestock carrying capacity of alpine grasslands in Golog Prefecture, China. Pedosphere 20, 342–351.

Zhang, J.P., Zhang, L.B., Liu, W.L., Qi, Y., Wo, X., 2014. Livestock-carrying capacity and overgrazing status of alpine grassland in the Three-River Headwaters region, China. J. Geogr. Sci. 24, 303–312.

Chapter 12

Deforestation

Richard A. Houghton
Woods Hole Research Center, Falmouth, MA, USA

Until recently, the clearing of forests for settlements and agriculture was viewed as a good thing. Not only were developed lands more useful for human needs, but forests were wild, dangerous, and evil places. The Grimm's fairy tale, "Hänsel and Gretel," captures this sentiment, beginning, "Hard by a great forest dwelt a poor wood-cutter…." Only in recent decades, as a result of an appreciation of biodiversity and a concern about climate change, have forests been seen as beneficial and has deforestation been seen as having unwanted consequences or trade-offs.

This brief introduction considers three aspects of deforestation. It begins with definitions and ways of measuring deforestation, then discusses past and present rates of deforestation, and finally summarizes some of its effects.

12.a DEFINITIONS

Conceptually, the definition of deforestation is straightforward: it is the conversion of forest cover to nonforest cover. Operationally, the definition is not so simple for at least two reasons. First, this simple definition of deforestation requires a definition for forest. Without a clear definition of "forest," it is often difficult to assess whether a forest has been converted to another cover type. For example, the definition of forest used by the UN Food and Agriculture Organization (FAO) considers a land to be forest even if it has no trees; for example, if the trees have been clear-cut or burned but are expected to return to forest. The implicit notion of intent in this definition makes it troublesome to measure deforestation, especially from satellites. The recent global estimate of deforestation rates based on Landsat data by Matt Hansen and colleagues, for example, reports losses of forest area rather than rates of deforestation because the measurement approach does not distinguish between permanent (different cover type) and temporary removal of trees.

Shifting cultivation, where cropping alternates with fallow, is also ambiguous. From a timber perspective, a forest cleared for shifting cultivation has been deforested, but the land is only temporarily used for agriculture, and its fallow will return to forest if it is not recleared. Satellite data can be used to

Biological and Environmental Hazards, Risks, and Disasters. http://dx.doi.org/10.1016/B978-0-12-394847-2.00018-8

distinguish between temporary and permanent removal of forest cover if lands are monitored frequently enough.

12.b RATES AND EXTENT OF DEFORESTATION

Deforestation probably originated with the use of fire, and estimates are that 40–50% of the Earth's original forest area has been lost. Some of that loss happened before settled agriculture began, approximately 10,000 years ago, but only in recent decades is there reliable information on rates of deforestation. Before about 1950, most deforestation occurred in the temperate zones of Europe, Russia, China, North America, and Australia as agricultural lands in those regions expanded. Thereafter, deforestation and agricultural expansion on those regions slowed and even reversed. The US is probably the prime example of forest expansion, as farms moved west in the middle of the twentieth century and forests regrew on abandoned croplands in the east. But deforestation continues today in the Eastern US, now for residential and commercial land. The notion that deforestation waxes and wanes with economic development may be generally true but is not absolute.

In contrast to nineteenth- and twentieth-century deforestation in the world's temperate zones, deforestation in the tropics did not begin in earnest until after 1950, reaching rates of 12 million hectares per year in the 1990s. Rates have slowed in the twenty-first century, but not by much. The exception is Brazil, where the annual rate of deforestation in Amazonia declined by nearly 80% since a peak in 2004. Deforestation in the rest of the tropics continues, and the rate overall for the tropics is still about 9 million hectares per year. This rate is offset to some extent by increases in planted forests or plantations, for example in Brazil, India, and Vietnam.

12.c EFFECTS OF DEFORESTATION

Loss of forests is a loss of habitats, and the consequences for biodiversity, especially in tropical forests, are huge. The effects on climate are also huge, from local and regional effects on precipitation and drought to global effects on climate through emissions of greenhouse gases. Deforestation releases the carbon stored in trees to the atmosphere, mostly as carbon dioxide. Cultivation of native soils also releases carbon dioxide as soil organic matter decays. These emissions of carbon from deforestation accounted for about 10% of the global emissions of carbon from human activity in the decade 2000–2010, but the fraction was closer to 20% in 1990. Most of the relative decrease is the result of ever-growing emissions of carbon from fossil fuels, but some is from a decline in the rate of tropical deforestation.

The amount of carbon released to the atmosphere since 1850 as a result of land use, including agricultural expansion as well as forestry, has been nearly 50% of fossil fuel emissions over the same 165 years. The historical patterns vary by region, paralleling the history of agricultural expansion. Most emissions from deforestation today are from the tropics; before 1950 most emissions were from outside the tropics. It is important to recognize that the emissions of carbon from deforestation are not equivalent to the emissions of carbon from changes in land use, which include emissions from grasslands converted to crops as well as emissions from harvested wood (not considered deforestation).

Concern about climate change has brought deforestation (and carbon) to the public's attention. Deforestation is the first "D" in Reduced Emissions from Deforestation and forest Degradation (REDD+), which is a mechanism to reward developing countries for reducing rates of deforestation and increasing carbon stocks in forests. Indeed, stopping deforestation, allowing degraded and secondary forests to accumulate carbon, and expanding the areas of forests could all play a major role in stabilizing the concentration of carbon dioxide in the atmosphere, temporarily, while fossil fuels are replaced with renewable fuels. Many concerns exist with using forest restoration in such a global strategy, but doing so is cheaper than alternative approaches for taking carbon out of the atmosphere and would have a number of benefits aside from its climatic effects.

One concern that must be noted here is that forests have an effect on climate besides their carbon storage. They also affect the exchanges of heat, moisture, and momentum between the land and atmosphere. These biophysical effects also affect climate, offsetting the biogeochemical effects (carbon) in boreal forests but enhancing them in tropical forests. No one is seriously proposing deforestation in boreal forests to combat global warming, but simply growing more forests, everywhere, would not have the desired effect on climate.

Chapter 12.1

Deforestation in Southeast Asia

Edgar C. Turner[1] and Jake L. Snaddon[2]
[1]*Insect Ecology Group, Department of Zoology, University of Cambridge, Cambridge, UK,*
[2]*Centre for Biological Sciences, University of Southampton, Southampton, UK*

ABSTRACT
Despite housing exceptionally high biodiversity, Southeast Asia has suffered rapid rates of forest loss in recent decades, with forest cover in the region reported to have declined by over 13% between 1990 and 2010. In this section, we briefly consider drivers for this rapid decline and consequences of forest loss for biodiversity and important ecosystem services, such as carbon sequestration. We particularly highlight the impact of expanding oil palm cultivation in the region, as two Southeast Asian countries, Indonesia and Malaysia, are the world's largest palm oil producers. In addition, recorded changes in the environment as a result of oil palm expansion in the region are likely to be experienced in other areas of the world in the future, as oil palm cultivation expands in Africa and South America. In the last section, we describe some key factors and options for the conservation of Southeast Asia's remaining forest areas.

12.1.1 LONG-TERM CHANGES IN LAND USE IN SOUTHEAST ASIA

Southeast Asia houses an exceptionally high biodiversity, is rich in endemics (Sodhi et al., 2004), and includes 4 of the 25 global biodiversity hotspots, originally described by Myers et al. (2000). It has also experienced some of the highest reported levels of forest loss and agricultural expansion of any tropical region. Less than half of the original forest cover now remains in Southeast Asia (Sodhi et al., 2004), with large reductions reported in both primary and selectively logged forest (Wilcove et al., 2013). According to the Food and Agriculture Organization of the United Nations (FAOSTAT, 2015), the total forest cover across the 11 Southeast Asian countries declined from 247 million hectares in 1990 to 214 million hectares in 2010; a drop of over 13%. However, this change is variable between countries (Figure 12.1.1), depending on the country's past and current socioeconomic and environmental status. For example, Cambodia, Timor-Leste, and Indonesia show high levels of forest

Biological and Environmental Hazards, Risks, and Disasters. http://dx.doi.org/10.1016/B978-0-12-394847-2.00019-X

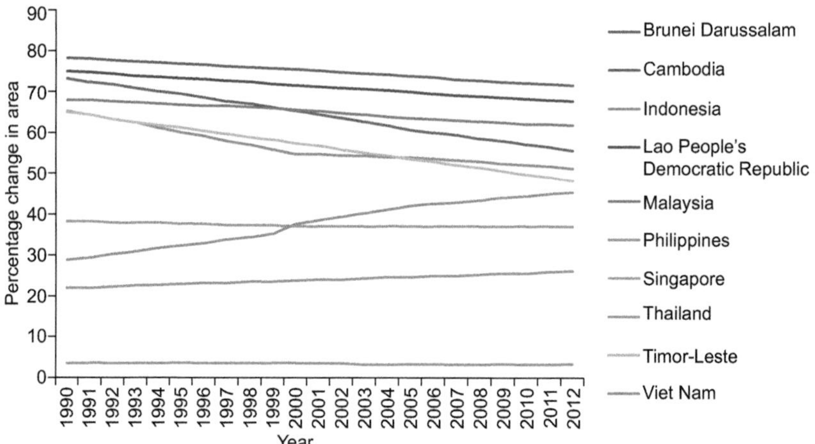

FIGURE 12.1.1 Percentage change in forest area from 1990 to 2012 for 10 Southeast Asian countries (FAOSTAT, 2015).

loss. In contrast, other countries, such as Singapore, show no or limited change, probably due to the historical loss of nearly all forest outside protected areas, whereas others, such as Vietnam, report an increase, probably due to an expansion in the area of planted forest (Food and Agricultural Organization of the United Nations, 2012). However, due to inaccuracies in reporting, the comparability of data, and definitions of "forest," the exact levels of forest cover, forest loss, and forest disturbance are uncertain and may be higher than figures suggest (Matthews and Grainger, 2002). For example, a recent remote sensing study revealed higher than expected levels of forest degradation in the Malaysian states of Sabah and Sarawak with almost 80% of the land area found to be impacted by previously undocumented logging and land clearance (Bryan et al., 2013).

The rapid reduction in forest cover across the region has resulted in a sharp decline in biodiversity (Sodhi et al., 2010), with high rates of extinction predicted in the future (Brooks et al., 2002). This is dramatically demonstrated by Singapore, which has experienced deforestation levels of more than 95% and has lost at least 881 of its 3196 recorded species (28%), but may have lost up to 73%, when species which were unrecorded, but likely to have been present before clearance, are taken into account (Brook et al., 2003).

The two main drivers for forest clearance in Southeast Asia are clear-felling and conversion to agricultural land and tree plantations (Gibbs et al., 2010). Agricultural land area has expanded dramatically in many of the Southeast Asian countries (Figure 12.1.2). Between 1980 and 2000, approximately 60% of this new agricultural land was established on intact forest areas and another 30% on disturbed forest. Up until 2000, new tree plantations were also mostly established on previously forested areas (Gibbs et al., 2010), but

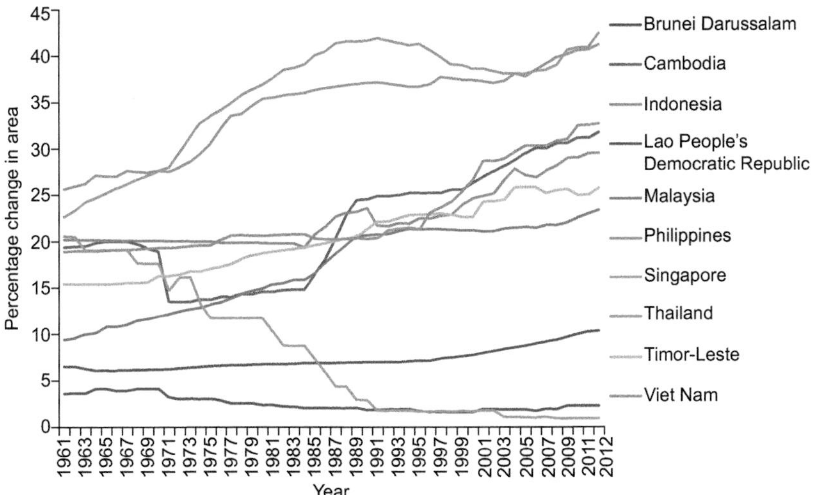

FIGURE 12.1.2 Percentage change in agricultural area from 1961 to 2012 for 10 Southeast Asian countries (FAOSTAT, 2015).

the source of land for later expansions is less consistently reported, with some studies citing continued forest replacement (Koh and Wilcove, 2008), but others suggesting expansion onto established agricultural land (Gibbs et al., 2010). In many cases, it is therefore clear that forest has been converted to industrial plantations and that these intensively managed areas now make up a significant proportion of the land area. For example, Gaveau et al. (2014), studying forest cover change in Borneo, reported that 10% of land in 2010 was covered by industrial-scale monoculture plantations, including oil palm, and timber plantations such as acacia. Abood et al. (2014), quantifying the level of forest loss experienced across different concession types in Indonesia from 2000 to 2010, found that, of a total of 14.7 million hectares of forest loss recorded in the country, 12.8% occurred in fiber plantation concessions, 12.5% in logging concessions, and 11% in oil palm plantation concessions, with a further 6.3% being allocated to mixed concessions and 2.1% to mining concessions.

Oil palm is, therefore, among the most important drivers of forest loss in the region and, owing to widely reported impacts on biodiversity, receives considerable media attention. Per area, oil palm is the highest-yielding vegetable oil crop. Its uses are diverse: from edible oil, to the production of soaps and cosmetics (Corley, 2009), and increasingly as a feedstock for biofuel production (Eynck et al., 2013). The two highest-producing palm oil countries worldwide are Indonesia and Malaysia, recorded as growing 7.1 and 4.5 million hectares of palm oil in 2013 (FAOSTAT, 2015). Between 1990 and 2005, a reported 55–59% of oil palm expansion in Malaysia, and 56% of oil palm expansion in Indonesia, replaced previously forested land (Koh and Wilcove, 2008). Although detrimental to the countries' forests and the biodiversity they contain, this rapid expansion has brought economic benefits to the countries involved including impoverished rural communities (Eynck et al., 2013), although it is argued that more still needs to be done to enhance the benefits of this expansion for smallholder farmers (Rist et al., 2010).

12.1.2 IMPACTS OF LOGGING ON THE ENVIRONMENT

Since the 1950s, an increasing global demand for tropical timber has resulted in widespread logging in parts of Southeast Asia (Miettinen et al., 2014). This expansion has been driven partly by the high number of commercially viable timber species that occur at high densities in Southeast Asian forests and has resulted in a relatively intensive timber harvest. For example, extraction estimates for Southeast Asia are between 8 and 15 trees removed per hectare compared to only 1–5 from South American or African forests (Putz et al., 2008). This is partly due to the dominance of the single tree family, the Dipterocarpaceae, in Southeast Asian forests (Corlett and Primack, 2005). These are very productive and make up about 25% of the global consumption of tropical hardwoods (Kettle, 2010).

Although logging has been widespread and relatively intense, an increasing number of studies have demonstrated that the process of logging and relogging of rain forests in Southeast Asia has had a surprisingly limited impact on forest biodiversity, with the majority of animal species continuing to be found in logged forest habitats (Didham, 2011). However, logging does affect the distribution and abundance of many taxa and may detrimentally impact associated ecosystem processes. For example, a study in Sabah, Malaysia found lower species richness of dung beetles and lower rates of dung removal and associated seed dispersal by the beetles in logged forest compared to undisturbed forest (Slade et al., 2011). Similarly, another recent study on termites, reported a large reduction in termite occurrence in logged compared to primary forest and a shift in the composition of termite functional groups, with soil-feeding termites being particularly badly affected by forest disturbance (Luke et al., 2014). Another recent large-scale study in Sabah, Malaysia found shifts in three important ecosystem processes (litter decomposition, seed predation and removal, and invertebrate predation) between primary forest and logged forest sites, with the contribution of invertebrates compared to vertebrates decreasing in logged forest and associated declines in the relevant functional groups (Ewers et al., 2015). Logging can also have an effect on ecosystem functions operating at larger scales. For example, logging has been documented to have a large negative impact on carbon storage in rain forest plots and to result in shifts in the proportion of living and deadwood-stored carbon (Pfeifer et al., 2015), with implications for climate change.

Fairly few studies have investigated the impacts of multiple rounds of logging, although evidence is mounting that even severe disturbance of this kind results in a surprisingly limited change in biodiversity. Another study in Sabah, Malaysia, found that 75% of primary forest bird and dung beetle species were still present in twice-logged forest, although there was a drop in overall richness of dung beetle species (Edwards et al., 2011). In the same area, 80% of leaf-litter-dwelling ant species, occurring in unlogged forest plots, were still recorded in twice-logged forest, although a lower species richness occurred at small spatial scales and a change in species composition in twice-logged plots occurred (Woodcock et al., 2011).

Realization of the value of disturbed forest habitats in Southeast Asia, has precipitated productive debates in the scientific community about the direction and focus of conservation messages and practice (Didham, 2011). Such a debate is particularly timely, given limited conservation funds and the potentially cost-effective option of conserving large areas of comparatively cheap logged forest, where commercially valuable timber species have already been harvested (Fisher et al., 2011a). The value of reduced impact logging techniques versus conventional logging in conserving forest species has also been studied, but the benefits of such management options for biodiversity are not clear (Edwards et al., 2012a), partly owing to the variably reported effects of traditional logging on biodiversity.

12.1.3 IMPACTS OF FOREST FRAGMENTATION ON THE ENVIRONMENT

During the process of forest conversion to agriculture, remaining forest areas characteristically become fragmented, leading to isolated forest patches of varying sizes within an agricultural matrix. These forest areas may remain because they receive special conservation protection, are under different ownership, are protected by law along riparian margins, or are unsuitable for agricultural production. Smaller forest areas generally contain a lower diversity of species than continuous forest. For example, a study investigating forest fragments isolated by oil palm plantations found a 60-fold decrease in the abundance of priority birds and a 1.8-times decrease in birds overall in forest fragments compared to continuous forest (Edwards et al., 2010). The management history of forest fragments can also be important in determining the quality of forest and species diversity remaining. For example, a study comparing privately and publically managed forest fragments found that privately managed forests contained smaller trees at lower densities with reduced canopy cover and shallower leaf litter compared to publically managed forest. Probably as a result of this, these privately managed fragments only contained about half the species richness of ants in publically managed forest patches, which in turn contained about 70% of the species occurring in continuous forest areas (Tawatao et al., 2014). Although they may contain fewer species than continuous forest areas, fragments can still be important habitats for forest species within agricultural landscapes. For example, although lower in biodiversity than continuous forest, riparian reserves within oil palm plantations contain a higher species richness and functional diversity of dung beetles than the surrounding agricultural land, although this is not associated with higher dung removal rates (Gray et al., 2014).

Fragment area, fragment isolation, characteristics of the taxa present, and the nature of the surrounding habitat type are all known to play a role in determining the number of species remaining in forest fragments in the long term (Laurance et al., 2011). For example, in the study of birds found in forest fragments in oil palm mentioned above, the abundance of priority bird species increased with fragment area, although even the largest fragments contained fewer species than continuous forest habitat (Edwards et al., 2010). In a study investigating fruit-feeding butterflies and moths and leaf-litter ants, forest fragment area was related positively to species richness for all three taxa, and patch isolation related negatively to species richness for butterflies alone (Hill et al., 2011). Another study investigating ant diversity in fragments ranging from 5 ha to 500 ha, found that ant species richness increased with area of fragment (Lucey et al., 2014). In a study in Peninsular Malaysia, forest fragment size was positively related to the abundance and species richness of bats. However, this relationship varied between different functional groups, with cavity- and foliage-roosting bat species being most strongly affected by fragment size, but edge and open-habitat specialists and cave-roosting species being unaffected (Struebig et al., 2008). Fragmentation can also reduce the genetic diversity of species isolated within forest patches, potentially compromising the ability of such species to persist in the long term in isolated fragments. For example, in the same study location, the most disturbance-sensitive of three bat species investigated showed lower genetic diversity in populations isolated in smaller forest patches (Struebig et al., 2011). Similarly, another study in Sabah, Malaysia, found lower genetic diversity in two species of rain forest ants in forest fragments compared to continuous forest (Bickel et al., 2006).

Forest fragments can sometimes have spillover effects on the surrounding habitat type, as species migrate out of forest fragments to forage within the matrix, or to migrate between patches. Sometimes this spillover can increase the biodiversity of the surrounding land use, with benefits for important ecosystem functions, such as pest control and pollination (for example, see Garibaldi et al. (2011) for a meta-analysis, investigating the role of noncrop habitats in supporting pollination). Oil palm plantations with larger native forest patches have been found to contain more mammal species (Azhar et al., 2014) and forest areas left in plantations can significantly increase the species richness of birds and butterflies in the plantation area (Koh, 2008). A study investigating butterflies and ants along an oil palm forest boundary in Sabah, Malaysia, found that butterfly diversity in oil palm areas increased at closer distances to the forest edge, but that both taxa had a lower species richness in plantations compared to forest (Lucey and Hill, 2012). Such enriching effects may also be more dramatic where larger areas of forest are maintained within an agricultural landscape. For example, another study investigating spillover effects on oil palm ants found increased species richness at both continuous forest and forest fragment edges, and that this effect was greater the larger the forest size (Lucey et al., 2014). As both birds and ants are important predators of crop pests in plantations, it is possible that forest fragments within the plantation landscape could enhance pest control. However, the only study so far to investigate the impact of forest fragments across a large area of oil palm plantation in Sabah, Malaysia (<23,000 ha), found no effect of forest patches on oil palm yield (Edwards et al., 2014b).

12.1.4 IMPACTS OF HABITAT CONVERSION ON THE ENVIRONMENT

Until recently, the effects on the environment of forest conversion to agriculture, and oil palm in particular, in Southeast Asia have been remarkably understudied (Turner et al., 2008). However, in the last decade, a large expansion of research has occurred in this area, in response to increasing concerns over effects of tropical agricultural expansion on the region's biodiversity and global carbon storage. It is clear that oil palm expansion in Southeast Asia is a major driver of deforestation and has impacts on both carbon storage (Danielsen et al., 2009) and regional biodiversity (Wilcove et al., 2013).

Calculations of the levels of CO_2 released from converting forest to other land uses and the potential value of palm oil as a feedstock for a carbon-neutral biofuel has received considerable research attention. The type of land being converted plays an important role in determining the volume of CO_2 released and the time it would take for palm oil grown on converted land to become carbon-neutral. For example, owing to its high soil carbon stores and faster rates of organic matter decomposition in soil following drainage, conversion of peat swamp forest results in a very large release of CO_2 (Koh et al., 2011; Manik and Halog, 2012). One recent study calculated that Peninsular Malaysia, Sumatra, and Borneo may have net annual emissions of 230–310 million metric tons of carbon dioxide, as a result of the conversion of 20% of their total peatland areas to oil palm (Miettinen et al., 2012). In a comparative study, Danielson et al. (2009) estimated that it would take 75 years before biofuel produced from oil palm grown on converted rain forest areas would result in a net uptake of CO_2 (compared to the net carbon balance for refining and using an energy-equivalent volume of fossil fuel). This estimate increased to 93 years if fire was used as part of the land-conversion process and up to 692 years if the land cleared was on peat soil! In contrast, the establishment of an oil palm plantation on a degraded grassland area could result in a net uptake of CO_2 within only 10 years, as a result of low existing carbon stores in degraded grassland habitats. As well as the volume of carbon dioxide released, concerns about oil palm expansion have also focused on other atmospheric effects, including transboundary haze, as a result of forest fires associated with land conversion (Forsyth, 2014), and changes in the concentrations of volatile organic compounds released from forest and oil palm plantations, which have potentially widespread effects on atmospheric chemistry (Mackenzie et al., 2011).

Conversion of forest to agriculture, particularly oil palm plantation, can have a large negative effect on the biodiversity of an area. Oil palm plantations have been found to contain a lower number of species than forest and most other commercial crops (Fitzherbert et al., 2008; Donald, 2004; Foster et al., 2011), probably owing to their low structural complexity compared to forest and timber plantations and more extreme microclimatic conditions (Luskin and Potts, 2011); oil palm is up to 6.5 °C hotter than primary forest (Hardwick et al., 2015). As well as an altered habitat structure, other factors may also contribute to biodiversity changes in converted landscapes, such as the high density of roads leading to easier access for illegal hunting and road deaths (Azhar et al., 2013a; Luskin et al., 2014). Bird species richness has been found to be lower in degraded habitats, including oil palm, than in forest (Edwards et al., 2010) and lower in oil palm plantations than rubber plantations (Peh et al., 2006). In a meta-analysis of published literature, Danielson et al. (2009) found that the species richness of vertebrates in oil palm was only 38% of that found in natural forests and that only 23% of species occurring in forests were still present in plantations. In contrast, the diversity of invertebrates in oil palm was 89% of the total species richness of forests. This difference was probably as a result of the arrival of open-habitat, nonforest species, because only 31% of invertebrate forest species were still occurring in plantations (Danielsen et al., 2009). Few studies have assessed the impacts of oil palm expansion on below-ground communities, although one recent study found no difference in the alpha diversity of soil bacterial communities between forest and plantation areas, but a significant shift in the community structure (Lee-Cruz et al., 2013). Another recent study on soil termites and ants, found a reduced termite occurrence in oil palm plantations compared to forest, but no change in the occurrence of ants. The same study also recorded a shift in functional groups of termites present in more disturbed habitats, with soil-feeding termites being the most sensitive to disturbance and absent from plantations (Luke et al., 2014).

Losses of forest species on conversion to oil palm plantation are nonrandom with some groups fairing worse than others. For example, one study comparing a wide range of different arthropod groups between forest and oil palm sites found lower overall arthropod abundance and biomass in oil palm, but varying levels of change across different arthropod taxa, with some orders actually increasing in abundance in oil palm plantations, resulting in an altered community structure (Turner and Foster, 2009). Another study comparing bird species in forest, oil palm and rubber plantations in Southern Thailand reported a 60% reduction in species richness between forest and oil palm and that species remaining were more widespread, smaller, and with a lower conservation status (Aratrakorn et al., 2006). A study comparing frogs between secondary forest and oil palm plantations in Peninsular Malaysia, found a similar abundance and diversity of frogs between the two habitats, but a reduction in forest species, with a corresponding increase in widespread and disturbance-tolerant species in the oil palm habitat (Faruk et al., 2013). Another study in Sabah, Malaysia, found that oil palm plantations also contained a lower proportion of endemic frog species (Gillespie et al., 2012). Plantations and agricultural areas also commonly contain a higher proportion of nonnative and invasive species. Several studies on ants have revealed a high number of nonnative species within plantation habitats (Fayle et al., 2010; Pfeiffer et al., 2008), including the invasive yellow crazy ant (*Anoplolepis gracilipes*) that has been reported to cause severe shifts in ecosystem processes elsewhere in the world (Bruhl and Eltz, 2010).

Synergistic relationships between environmental changes may also lead to more species going extinct than would result from forest loss alone (Brook et al., 2008), particularly as a result of interactions between habitat conversion, climate change, and fire (Brodie et al., 2012). For example, it is possible that some of the impacts of fragmentation could have more severe effects in Southeast Asia than other tropical regions, owing to the limited dispersal ability of the dominant forest tree taxa (particularly Fagaceae and Dipterocarpaceae) (Brodie et al., 2012; Corlett, 2009), which may limit tree dispersal between forest patches and recolonization in patches following local extinction. If rates of forest loss and conversion to agriculture continue at the current level, it has been calculated that 7–52% of lowland forest-bird species and 9–36% of lowland forest-mammal species may go extinct by 2100 (Wilcove et al., 2013).

The role of biodiversity within plantations in supporting important ecosystem functions has received little attention, but may prove important for long-term sustainable production and may be threatened by species losses (Foster et al., 2011). For example, a recent meta-analysis found that higher trophic level and larger body-sized, ant, beetle, and bird species were more abundant in forest and declined most severely when forest was converted to oil palm plantation (Senior et al., 2013), potentially reducing levels of predation in the oil palm landscape, with consequences for pest control. Several studies have noted nonrandom loss of different functional groups and a loss of functional diversity in degraded habitats. A study in Sabah found the functional richness of dung beetles to be lower in oil palm than in the forest habitat (Edwards et al., 2014a). Another study comparing peat swamp forest and oil palm plantation in Peninsular Malaysia found a lower diversity of foraging guilds in oil palm plantations and a lower abundance of insectivorous, granivorous, and omnivorous birds, but a higher abundance of raptors and wetland species (Azhar et al., 2013b). It remains to be seen whether such recorded changes have a long-term effect on ecosystem functioning within tropical agricultural landscapes and whether this ultimately compromises crop yield.

12.1.5 CONSERVATION OPTIONS

A number of different options have been put forward to reduce the level of environmental degradation as a result of habitat conversion in Southeast Asia, and to protect species diversity and related ecosystem processes. Central to this is sustained community engagement to protect endangered wildlife and deliver forest conservation programs (e.g., Ancrenaz et al., 2007). The number of people intimately associated with forest areas in Southeast Asia is far from insignificant. For example, roughly one-third of the populations of Cambodia, Indonesia, the Philippines, Thailand, and Vietnam are estimated to be dependent on forests; 140 million people in total. Over the last few decades, many countries in Southeast Asia have moved towards sustainable forestry management policies, which aim to balance social, economic, and environmental components of forestry. This generally involves a move away from centrally controlled industrial forestry paradigms toward a community forest management approach, with associated laws and governance and a decentralized system of management (Poffenberger, 2006). However, appropriate policies, legislation, and institutional arrangements that can achieve sustainable forest management are often difficult to achieve and require the balancing of the different goals of forestry—economic development, the provision of forest products and services, and social benefits (Yasmi et al., 2010).

For biodiversity conservation, proper protection of remaining forest areas is paramount. As well as formally gazetted reserves, designating land for timber production may still reduce the level of forest loss. For example, a study based in Kalimantan, Indonesia, found that levels of forest loss in timber concession areas were not significantly different from protected areas (at 1.5% and 1.2% respectively for 2000–2010) and much lower than those recorded in oil palm concession plots (at 14.1%) (Gaveau et al., 2013). Careful landscape planning of new reserve networks may well also prove vital to the long-term survival of forest species. Such reserve designs must not only take into account projected forest loss, but also interactions with other major drivers of species losses, such as climate change. For example, a recent study on the impacts of projected deforestation and climate change on mammals in Borneo found that 11–36% of Bornean mammal species could lose more than 30% of their habitat by 2080 as a result of a changing climate, but that this proportion could increase to 30–49%, when deforestation also occurs. More optimistically, the study also found that a relatively small increase in protected land in higher-elevation areas could yield significant benefits for conservation, as many species are predicted to shift to higher altitudes in response to climate change (Struebig et al., 2015). Another study investigating the present distribution of protected areas in Borneo and projected climate change has recently flagged up the vulnerability of existing low-altitude protected areas to climate change. They found that the majority of protected areas (87.5–89%) were not topographically diverse enough to provide present-day climate characteristics in the future with predicted changes in temperature. Lack of connectivity in existing protected areas was also flagged up as an issue, with many areas being too isolated for poorly dispersive species to successively migrate into suitable climatic areas following climate change (Scriven et al., 2015).

Careful planning of future agricultural expansion, with social and environmental considerations in mind, can also minimize some of the adverse effects of agricultural expansion and maximize long-term environmental sustainability (Smit et al., 2013; Danielsen et al., 2009). This is particularly important in terms of choosing land suitable for agricultural cultivation and sparing land which may be important for the ecosystem services it supplies, such as flood prevention (Abram et al., 2014), food production, carbon sequestration, and biodiversity protection (Koh and Ghazoul, 2010). To successfully identify important ecosystem service areas and prioritize them for conservation protection, high-resolution mapping of ecosystem services over these regions will be required (Sumarga and Hein, 2014). For example, degraded habitats with lower biodiversity value or reduced carbon stores could be selected for future oil palm expansion, reducing the risks to existing forest habitats of high-conservation importance (Edwards and Laurance, 2012). In Southeast Asia, the extent of such degraded areas may be considerable. One study calculated that 12 Mha of oil palm could be accommodated on existing highly degraded land in the region, including some 8 Mha of anthropogenic grassland in Indonesia (Corley, 2009), although much of this land is likely to be intensively used by local communities and under other management, highlighting the need for social considerations to be taken into account in any decision making. Such an expansion would meet the global vegetable oil demands for palm oil that are calculated to be required by 2050, with potentially only limited impacts on biodiversity (Corley, 2009). Another study has suggested that total demand for oil palm production could be met by restricting expansion to small forest fragments of no bigger than 400 ha for Indonesia and 700 ha for Malaysia, thereby, sparing larger forest patches, which contain a greater diversity of species (Edwards et al., 2012b). However, it is possible that many of these areas may be on steep or flood-prone areas and therefore not suitable for oil palm cultivation. Ultimately the continued protection of forest areas will also require an economic incentive. Reduced emissions from deforestation and forest degradation payments may represent one way to pay for the protection of high-carbon stock areas (Venter et al., 2009) and to relocate oil palm allocation leases from higher- to lower-carbon stock areas (Venter et al., 2013). However, given the high value of commercial logging practices and subsequent conversion to oil palm, some studies have questioned whether revenue from the global carbon market will be sufficient to offset these profits in Southeast Asia (Fisher et al., 2011b; Butler et al., 2009).

As well as prioritizing the protection of remaining forested habitats in the region, the development of more sustainable and high-yielding agricultural areas will also help to lessen forest loss, by reducing the incentives to expand into forest areas still further (so-called "land-sparing," see Phalan et al. (2011)). More environment-friendly oil palm management may also act to facilitate wildlife dispersal through agricultural landscapes and reduce fragmentation effects within forest patches, although it is important that any management options within plantations should not come at a cost to yield, in case this promotes further deforestation elsewhere. Such ideas have led some researchers to suggest that oil palm landscapes should be specifically designed with both biodiversity and productivity in mind (Koh et al., 2009). The type of plantation management and the level of habitat heterogeneity may influence the diversity of species that agricultural areas can support. For example, a study in Peninsular Malaysia found that smallholder oil palm plantations had a higher diversity of bird species than large estates (Azhar et al., 2011), perhaps as a result of a more diverse mix of crops being grown. The presence of some habitat types within an oil palm landscape may also act as a refuge for more sensitive species and reduce the effects of forest conversion on biodiversity at the landscape scale. The value of forest fragments has already been discussed above, but even patches of degraded land within plantations can be important for maintaining biodiversity. A study in Jambi Province, Indonesia, found 38 medium to large mammal species, including the Sumatran Tiger surviving in an oil palm landscape, probably as a result of other land use types surrounding or within the oil palm plantation itself (Maddox et al., 2007). On a smaller scale, epiphytic, bird's-nest ferns have also been found to support similar diversities of ant communities in oil palm plantations compared to forest, whereas canopy and leaf-litter communities show a much reduced diversity (Fayle et al., 2010), indicating that encouraging epiphytes within plantation ecosystems may increase ant diversity.

The greatest chance of making tropical agricultural production more environment-friendly and sustainable in the long-term rests with enforced government guidance or voluntary accreditation schemes, such as registered "tiger-friendly palm oil" (Bateman et al., 2010), which provide a financial incentive in the form of increased market value of the sustainably marketed products produced. The most important accreditation scheme associated with oil palm cultivation is the Round Table on Sustainable Palm Oil (RSPO), which has developed since it was first established in 2004 and has considerable potential to make oil palm more sustainable (Paoli et al., 2010). However, the ability to render the oil palm industry more sustainable has provoked high levels of controversy among the scientific and conservation community

(Anderson, 2008; Laurance et al., 2010), particularly relating to its ability to reduce deforestation. For example, one of the key pieces of guidance within the RSPO related to forest protection focuses on protecting high conservation value forest areas within plantation landscapes. However, the current focus on rare, threatened, or endangered species (RSPO, 2013) could mean that small but biologically important forest patches within oil palm landscapes are cleared, resulting in a loss of biodiversity and important ecosystem services (Edwards et al., 2012b).

Finally, supporting any conservation initiatives in the region should be a greater scope of research to investigate the effects of habitat change and different management options, to inform decisions and hone long-term policy. For example, improved remote sensing and processing techniques are vital for rigorous quantification of the rates of forest loss in the region (Broich et al., 2011) and to source lower-value areas for future expansion. A clearer understanding of the effects of forest fragmentation and habitat change is also required, perhaps, associated with the design of new studies to experimentally investigate the large-scale effects of forest fragmentation and habitat conversion in the region (for example, Ewers et al., 2011). Finally, very little is yet known about viable management options for restoring forests in Southeast Asia, and this too would merit further research (Kettle, 2010).

12.1.6 CONCLUSION

No doubt exists that rates of forest loss in Southeast Asia have been severe, with concurrent reductions in local biodiversity and high levels of predicted extinctions to come. However, an expansion of recent research has demonstrated that the effects of habitat disturbance are far from homogeneous, with even extreme logging events only producing a limited change in species diversity, but conversion to agriculture resulting in much more severe shifts in the species community, both within the agricultural area itself and in remaining forest fragments. Although the impacts of habitat change on biodiversity are now fairly well known, much less is understood about the long-term effects on important ecosystem functions and sustainable crop productivity, making this a research priority.

The high level of biodiversity within forests and threats to forest habitats in Southeast Asia represent a challenge but also an opportunity for local communities, nongovernment organizations, the research and conservation community, and logging and agricultural companies to work together to reduce some of the detrimental impacts of further logging and agricultural expansion. It is possible that modern technology, particularly, satellite-mapping techniques will be key to this by identifying areas of remaining high-value forest that should be protected and possible degraded areas that could be prioritized for further expansion.

REFERENCES

Abood, S.A., et al., 2014. Relative contributions of the logging, fiber, oil palm, and mining industries to forest loss in Indonesia. Conserv. Lett. 8, 58–67.

Abram, N.K., et al., 2014. Synergies for improving oil palm production and forest conservation in floodplain landscapes. PloS One 9 (6), e95388.

Ancrenaz, M., Dabek, L., O'Neil, S., 2007. The costs of exclusion: recognizing a role for local communities in biodiversity conservation. PLoS Biol. 5 (11), 2443–2448.

Anderson, J.M., October 2008. Eco-friendly approaches to sustainable palm oil production. J. Oil Palm Res. 127–142.

Aratrakorn, S., Thunhikorn, S., Donald, P.F., 2006. Changes in bird communities following conversion of lowland forest to oil palm and rubber plantations in southern Thailand. Bird Conserv. Int. 16 (1), 71–82.

Azhar, B., Lindenmayer, D.B., Wood, J., Fischer, J., Manning, A., McElhinny, C., et al., 2013a. Contribution of illegal hunting, culling of pest species, road accidents and feral dogs to biodiversity loss in established oil-palm landscapes. Wildl. Res. 40, 1–9.

Azhar, B., et al., 2014. Ecological impacts of oil palm agriculture on forest mammals in plantation estates and smallholdings. Biodivers. Conserv. 23, 1175–1191.

Azhar, B., et al., 2011. The conservation value of oil palm plantation estates, smallholdings and logged peat swamp forest for birds. For. Ecol. Manag. 262 (12), 2306–2315.

Azhar, B., Lindenmayer, D.B., Wood, J., Fischer, J., Manning, A., Mcelhinny, C., et al., 2013b. The influence of agricultural system, stand structural complexity and landscape context on foraging birds in oil palm landscapes. Ibis 155, 297–312.

Bateman, I.J., et al., 2010. Tigers, markets and palm oil: market potential for conservation. Oryx 44 (2), 230–234.

Bickel, T., et al., 2006. Influence of habitat fragmentation on the genetic variability in leaf litter ant populations in tropical rainforests of Sabah, Borneo. Biodivers. Conserv. 15, 157–175.

Brodie, J., Post, E., Laurance, W.F., 2012. Climate change and tropical biodiversity: a new focus. Trends Ecol. Evol. 27 (3), 145–150.

Broich, M., et al., 2011. Time-series analysis of multi-resolution optical imagery for quantifying forest cover loss in Sumatra and Kalimantan, Indonesia. Int. J. Appl. Earth Obs. Geoinformation 13 (2), 277–291.

Brook, B.W., Sodhi, N.S., Bradshaw, C.J.A., 2008. Synergies among extinction drivers under global change. Trends Ecol. Evol. 23 (8), 453–460.

Brook, B.W., Sodhi, N.S., Ng, P.K.L., 2003. Catastrophic extinctions follow deforestation in Singapore. Nature 424 (6947), 420–426.

Brooks, T.M., et al., 2002. Habitat loss and extinction in the hotspots of biodiversity. Conserv. Biol. 16 (4), 909–923.

Bruhl, C., Eltz, T., 2010. Fuelling the biodiversity crisis: species loss of ground-dwelling forest ants in oil palm plantations in Sabah, Malaysia (Borneo). Biodivers. Conserv. 19, 519–529.

Bryan, J.E., et al., 2013. Extreme differences in forest degradation in Borneo: comparing practices in Sarawak, Sabah, and Brunei. PloS One 8 (7), e69679.

Butler, R.A., Koh, L.P., Ghazoul, J., 2009. LETTER REDD in the red: palm oil could undermine carbon payment schemes. Conserv. Lett. 2, 67–73.

Corlett, R., Primack, R., 2005. Dipterocarps: trees that dominate the Asian rain forest. Arnoldia 63 (3), 3–7.

Corlett, R.T., 2009. Seed dispersal distances and plant migration potential in tropical East Asia. Biotropica 41 (5), 592–598.

Corley, R.H.V., 2009. How much palm oil do we need? Environ. Sci. Policy 12, 134–139.

Danielsen, F., et al., 2009. Biofuel plantations on forested lands : double jeopardy for biodiversity and climate. Conserv. Biol. 23 (2), 348–358.

Didham, R.K., 2011. Life after logging: strategic withdrawal from the Garden of Eden or tactical error for wilderness conservation. Biotropica 43 (4), 393–395.

Donald, P.F., 2004. Biodiversity impacts of some agricultural commodity production systems. Conserv. Biol. 18 (1), 17–37.

Edwards, D.P., et al., 2011. Degraded lands worth protecting: the biological importance of Southeast Asia's repeatedly logged forests. Proc. R. Soc. Lond. Ser. B 278, 82–90.

Edwards, D.P., Woodcock, P., et al., 2012a. Reduced-impact logging and biodiversity conservation: a case study from Borneo. Ecol. Appl. 22 (2), 561–571.

Edwards, D.P., et al., 2010. Wildlife-friendly oil palm plantations fail to protect biodiversity effectively. Conserv. Lett. 3, 236–242.

Edwards, D.P., Fisher, B., Wilcove, D.S., 2012b. High conservation value or high confusion value? Sustainable agriculture and biodiversity conservation in the tropics. Conserv. Lett. 5, 20–27.

Edwards, D.P., Laurance, S.G., 2012. Green labelling, sustainability and the expansion of tropical agriculture: critical issues for certification schemes. Biol. Conserv. 151 (1), 60–64.

Edwards, F.A., et al., 2014a. Does logging and forest conversion to oil palm agriculture alter functional diversity in a biodiversity hotspot? Anim. Conserv. 17, 163–173.

Edwards, F.A., et al., 2014b. Sustainable management in crop monocultures: the impact of retaining forest on oil palm yield. PloS One 9 (3), e91695.

Ewers, R.M., et al., 2011. A large-scale forest fragmentation experiment: the stability of altered Forest ecosystems project A large-scale forest fragmentation experiment: the stability of altered forest ecosystems project. Philos. Trans. R. Soc. B Biol. Sci. 366, 3292–3302.

Ewers, R.M., et al., 2015. Logging cuts the functional importance of invertebrates in tropical rainforest. Nat. Commun. 6, 6836.

Eynck, C., et al., 2013. Sustainable oil crops production. In: Singh, B. (Ed.), Biofuel Crop Sustainability. John Wiley and Sons, pp. 165–204.

FAOSTAT, 2015. Food and Agriculture Organisation of the United Nations. http://faostat3.fao.org/home/E (accessed 07.05.15.).

Faruk, A., et al., 2013. Effects of oil-palm plantations on diversity of tropical anurans. Conserv. Biol. 27 (3), 615–624.

Fayle, T.M., et al., 2010. Oil palm expansion into rain forest greatly reduces ant biodiversity in canopy, epiphytes and leaf-litter. Basic Appl. Ecol. 11, 337–345.

Fisher, B., Edwards, D.P., Larsen, T.H., et al., 2011a. Cost-effective conservation : calculating biodiversity and logging trade-offs in Southeast Asia. Conserv. Lett. 4, 443–450.

Fisher, B., Edwards, D.P., Giam, X., et al., 2011b. The high costs of conserving Southeast Asia's lowland rainforests. Front. Ecol. Environ. 9 (6), 329–334.

Fitzherbert, E., et al., 2008. How will oil palm expansion affect biodiversity? Trends Ecol. Evol. 23, 538–545.

Food and Agricultural Organization of the United Nations, 2012. State of the World's Forests, ISBN 978-92-5-107292-9.

Forsyth, T., 2014. Public concerns about transboundary haze: a comparison of Indonesia, Singapore and Malaysia. Glob. Environ. Change 25, 76–86.

Foster, W.A., et al., 2011. Establishing the evidence base for maintaining biodiversity and ecosystem function in the oil palm landscapes of South East Asia. Philos. Trans. R. Soc. B Biol. Sci. 366, 3277–3291.

Garibaldi, L.A., et al., 2011. Stability of pollination services decreases with isolation from natural areas despite honey bee visits. Ecol. Lett. 14, 1062–1072.

Gaveau, D.L.A., et al., 2014. Four decades of forest persistence, clearance and logging on borneo. PloS One 9 (7), 1–11.

Gaveau, D.L.A., et al., 2013. Reconciling forest conservation and logging in Indonesian borneo. PloS One 8 (8), e69887.

Gibbs, H.K., et al., 2010. Tropical forests were the primary sources of new agricultural land in the 1980s and 1990s. Proc. Natl. Acad. Sci. U.S.A. 107 (38), 16732–16737.

Gillespie, G.R., et al., 2012. Conservation of amphibians in Borneo: relative value of secondary tropical forest and non-forest habitats. Biol. Conserv. 152, 136–144.

Gray, C.L., et al., 2014. Do riparian reserves support dung beetle biodiversity and ecosystem services in oil palm-dominated tropical landscapes? Ecol. Evol. 4 (7), 1049–1060.

Hardwick, S.R., et al., 2015. The relationship between leaf area index and microclimate in tropical forest and oil palm plantation: forest disturbance drives changes in microclimate. Agric. For. Meteorol. 201, 187–195.

Hill, J.K., et al., 2011. Ecological impacts of tropical forest fragmentation: how consistent are patterns in species richness and nestedness? Philos. Trans. R. Soc. Lond. Ser. B Biol. Sci. 366 (1582), 3265–3276.

Kettle, C.J., 2010. Ecological considerations for using dipterocarps for restoration of lowland rainforest in Southeast Asia. Biodivers. Conserv. 19 (4), 1137–1151.

Koh, L.P., 2008. Can oil palm plantations be made more hospitable for forest butterflies and birds? J. Appl. Ecol. 45, 1002–1009.

Koh, L.P., et al., 2011. Remotely sensed evidence of tropical peatland conversion to oil palm. Proc. Natl. Acad. Sci. U.S.A. 108 (12), 5127–5132.

Koh, L.P., Ghazoul, J., 2010. Spatially explicit scenario analysis for reconciling agricultural expansion, forest protection, and carbon conservation in Indonesia. Proc. Natl. Acad. Sci. U.S.A. 107 (24), 11140–11144.

Koh, L.P., Levang, P., Ghazoul, J., 2009. Designer landscapes for sustainable biofuels. Trends Ecol. Evol. 24 (8), 431–438.

Koh, L.P., Wilcove, D.S., 2008. Is oil palm agriculture really destroying tropical biodiversity? Conserv. Lett. 1, 60–64.

Laurance, W.F., et al., 2010. Improving the performance of the roundtable on sustainable palm oil for nature conservation. Conserv. Biol. 24 (2), 377–381.

Laurance, W.F., et al., 2011. The fate of Amazonian forest fragments: a 32-year investigation. Biol. Conserv. 144 (1), 56–67.

Lee-cruz, L., et al., 2013. Impact of logging and forest conversion to oil palm plantations on soil bacterial communities in borneo. Appl. Environ. Microbiol. 79 (23), 7290–7297.

Lucey, J.M., et al., 2014. Tropical forest fragments contribute to species richness in adjacent oil palm plantations. Biol. Conserv. 169, 268–276.

Lucey, J.M., Hill, J.K., 2012. Spillover of insects from rain forest into adjacent oil palm plantations. Biotropica 44 (3), 368–377.

Luke, S.H., et al., 2014. Functional structure of ant and termite assemblages in old growth forest, logged forest and oil palm plantation in Malaysian Borneo. Biodivers. Conserv. 23 (11), 2817–2832.

Luskin, M.S., et al., 2014. Modern hunting practices and wild meat trade in the oil palm plantation-dominated landscapes of Sumatra, Indonesia. Hum. Ecol. 42, 35–45.

Luskin, M.S., Potts, M.D., 2011. Microclimate and habitat heterogeneity through the oil palm lifecycle. Basic Appl. Ecol. 12 (6), 540–551.

Mackenzie, A.R., et al., 2011. The atmospheric chemistry of trace gases and particulate matter emitted by different land uses in Borneo. Philos. Trans. R. Soc. B Biol. Sci. 366, 3177–3195.

Maddox, T., et al., 2007. ZSL Conservation Report No. 7 The Conservation of Tigers and Other Wildlife in Oil Palm Plantations. ZSL Conservation Report No. 7, (7).

Manik, Y., Halog, A., 2012. A meta-analytic review of life cycle assessment and flow analyses studies of palm oil biodiesel. Integr. Environ. Assess. Manag. 9 (1), 134–141.

Matthews, E., Grainger, A., 2002. Evaluation of FAO's global forest resources assessment from the user perspective. Unisylva 53, 42–55.

Miettinen, J., et al., 2012. Extent of industrial plantations on Southeast Asian peatlands in 2010 with analysis of historical expansion and future projections. Bioenergy 4, 908–918.

Miettinen, J., Stibig, H.-J., Achard, F., 2014. Remote sensing of forest degradation in Southeast Asia—Aiming for a regional view through 5–30 m satellite data. Glob. Ecol. Conserv. 2, 24–36.

Myers, N., et al., 2000. Biodiversity hotspots for conservation priorities. Nature 403, 853–858.

Paoli, G.D., Yaap, B., Wells, P.L., 2010. CSR, oil Palm and the RSPO: translating boardroom philosophy into conservation action on the ground. Trop. Conserv. Sci. 3 (4), 438–446.

Peh, K.S., et al., 2006. Conservation value of degraded habitats for forest birds in southern Peninsular Malaysia. Divers. Distrib. 12, 572–581.

Pfeifer, M., et al., 2015. Deadwood biomass: an underestimated carbon stock in degraded tropical forests? Environ. Res. Lett. 10, 044019.

Pfeiffer, M., Tuck, H.C., Lay, T.C., 2008. Exploring arboreal ant community composition and co-occurrence patterns in plantations of oil palm *Elaeis guineensis* in Borneo and Peninsular Malaysia. Ecography 31, 21–32.

Phalan, B., et al., 2011. Reconciling food production and biodiversity conservation: land sharing and land sparing compared. Science 333, 1289–1291.

Poffenberger, M., 2006. People in the forest: community forestry experiences from Southeast Asia. Int. J. Environ. Sustain. Dev. 5 (1), 57–69.

Putz, F.E., et al., 2008. Improved tropical forest management for carbon retention. PLoS Biol. 6 (7), 1368–1369.

Rist, L., Feintrenie, L., Levang, P., 2010. The livelihood impacts of oil palm : smallholders in Indonesia. Biodivers. Conserv. 19, 1009–1024.

Round Table on Sustainable Palm Oil, 2013. Adoption of principles and criteria for the production of sustainable palm oil. In: Submitted by the RSPO Executive Board for the Extraordinary General Assembly, pp. 1–70.

Scriven, S.A., et al., 2015. Protected areas in Borneo may fail to conserve tropical forest biodiversity under climate change. Biol. Conserv. 184, 414–423.

Senior, M.J.M., et al., 2013. Trait-dependent declines of species following conversion of rain forest to oil palm plantations. Biodivers. Conserv. 22, 253–268.

Slade, E.M., Mann, D.J., Lewis, O.T., 2011. Biodiversity and ecosystem function of tropical forest dung beetles under contrasting logging regimes. Biol. Conserv. 144 (1), 166–174.

Smit, H.H., et al., 2013. Breaking the link between environmental degradation and oil palm expansion : a method for enabling sustainable oil palm expansion. PloS One 8 (9), 1–12.

Sodhi, N.S., et al., 2010. Conserving Southeast Asian forest biodiversity in human-modified landscapes. Biol. Conserv. 143 (10), 2375–2384.

Sodhi, N.S., et al., 2004. Southeast Asian biodiversity: an impending disaster. Trends Ecol. Evol. 19 (12), 654–660.

Struebig, M.J., et al., 2008. Conservation value of forest fragments to Palaeotropical bats. Biol. Conserv. 1, 2112–2126.

Struebig, M.J., et al., 2011. Parallel declines in species and genetic diversity in tropical forest fragments. Ecol. Lett. 14, 582–590.

Struebig, M.J., et al., 2015. Targeted conservation to safeguard a biodiversity hotspot from climate and land-cover change. Curr. Biol. 25 (3), 372–378.

Sumarga, E., Hein, L., 2014. Mapping ecosystem services for land use planning, the case of Central Kalimantan. Environ. Manag. 54, 84–97.

Tawatao, N., et al., 2014. Biodiversity of leaf-litter ants in fragmented tropical rainforests of Borneo: the value of publically and privately managed forest fragments. Biodivers. Conserv. 3113–3126.

Turner, E.C., et al., 2008. Oil palm research in context: identifying the need for biodiversity assessment. PloS One 3 (2), e1572.

Turner, E.C., Foster, W.A., 2009. The impact of forest conversion to oil palm on arthropod abundance and biomass in Sabah, Malaysia. J. Trop. Ecol. 25, 23–30.

Venter, O., et al., 2009. Carbon payments as a safeguard for threatened tropical mammals. Conserv. Lett. 2, 123–129.

Venter, O., et al., 2013. Using systematic conservation planning to minimize REDD+ conflict with agriculture and logging in the tropics. Conserv. Lett. 6, 116–124.

Wilcove, D.S., et al., 2013. Navjot's nightmare revisited: logging, agriculture, and biodiversity in Southeast Asia. Trends Ecol. Evol. 28 (9), 531–540.

Woodcock, P., et al., 2011. The conservation value of South East Asia's highly degraded forests: evidence from leaf-litter ants. Philos. Trans. R. Soc. B Biol. Sci. 366, 3256–3264.

Yasmi, Y., et al., 2010. Forestry Policies, Legislation and Institutions in Asia and the Pacific. Trends and Emerging Needs for 2020. Asia-Pacific Forestry Sector Outlook Study II, Working Paper Series, Working Paper No. APFSOS II/WP/2010/34.

Chapter 12.2

Deforestation in Nepal: Causes, Consequences, and Responses

Ram P. Chaudhary[1], **Yadav Uprety**[1] **and Sagar Kumar Rimal**[2]
[1] *Research Centre for Applied Science and Technology, and Central Department of Botany, Tribhuvan University, Kirtipur, Kathmandu, Nepal,* [2] *Ministry of Forests and Soil Conservation, Government of Nepal, Singh Durbar, Kathmandu, Nepal*

ABSTRACT

Nepal is a relatively small mountainous country surrounded by India to the south, east, and west and China to the north. Forest is one of the most important natural resources of Nepal. Some 35 major forest types occur in Nepal that change abruptly, owing to a wide variation of topography, climate, and edaphic conditions. Forests are the source of livelihoods of millions of people, in particular, for rural communities. Deforestation is one of the major environmental issues in Nepal. Deforestation pressures occur throughout Nepal but are most strongly felt in many parts of Tarai and Churia. In general, the drivers of deforestation and degradation are the mixture of direct and indirect causes, such as high dependency on forest resources, unsustainable harvesting practices, illegal harvest of forest products, infrastructure development, forest fire, natural calamities, encroachment, overgrazing, lack of good governance, and ambiguous policy. Deforestation has immediate consequences for the local population in terms of increased fuel scarcity, reduced supply of fodder, and leaf-litter manure. The unpredicted erosion, landslide, and lowland flooding, due to deforestation, are also major concerns in Nepal as well as in downstream countries. Several attempts have been made so far to control the deforestation and mixed success has been achieved. The focus of the government is on good forest governance through its long-term and short-term policy provisions. Community forestry and protected areas systems in Nepal have contributed significantly in forest conservation. A priority has also been given to private forestry. Since the fuelwood is still the major energy source, alternative sources of energy should be provided on a subsidized basis throughout the country to reduce dependency on forests. Public awareness about the importance of forests and the consequences of deforestation are also important to control deforestation.

Biological and Environmental Hazards, Risks, and Disasters. http://dx.doi.org/10.1016/B978-0-12-394847-2.00020-6

12.2.1 INTRODUCTION

Forest is defined as a land spanning more than 0.5 ha with trees higher than 5 m and a canopy cover of more than 10%, or trees able to reach these thresholds in situ (FAO, 2006). Forests in Nepal are the most productive and self-sustaining ecosystem that supports human society ecologically, economically, culturally, and spiritually. Forests perform a number of vital ecosystem goods and services such as provisioning (supply of food, fodder, timber, etc.), regulating (regulation of climate, water, and pollination), supporting (soil formation, nutrient cycling, and primary production), and cultural (recreational, spiritual, religious, and nonmaterial benefits). It is a source of livelihoods for people, in particular for local communities who suffer most when forest resources are lost.

Conversion of forest to another land use or the long-term reduction of the tree canopy cover below the minimum 10% threshold is deforestation (FAO, 2001). In developing countries, basically deforestation is due to population growth and agricultural expansion, aggravated over the long term by wood harvesting for fuel and export (Allen and Barnes, 1985). The main causes of deforestation in Nepal, however, cannot be attributed simply to population pressures accompanied by migration and settlements, but a variety of economic, social, and governance factors that interact to cause deforestation. Deforestation is complex. Three major agents—subsistence agriculture expansion encroaching forestlands; perverse incentives related to illegal timber harvesting and forest fire; and livestock grazing are considered the most immediate causes of deforestation. Other reasons for the destruction of forests in the country include daily fuelwood consumption, development activities, and conflicting policies.

The rate, causes, and consequences of deforestation are diverse in different countries. Rate of deforestation, however, is considered often debatable because of different ways of defining forests, deforestation, and degradation; and political and economic factors that cause countries to hide or exaggerate deforestation (Miller, 2004). In general, deforestation affects a particular geographic area, but widespread deforestation can have global repercussions. Small-scale loss of forest cover can have adverse effects on the supply of fuelwood for household energy, soil and water resources, and the quality of rural life. Large-scale loss of forest area has been implicated in changes in global wood supply, the hydrologic balance, genetic resources, and global cycles of carbon and other elements (Allen and Barnes, 1985). The differentiation on forest quality was officially recognized in Nepal since the first Forest Resources Assessment (FRA) conducted in the early 1960s (Acharya and Dangi, 2009). Since then, the causes and consequences of deforestation have been remained a major concern for Nepal. This chapter describes the physiographic setting of Nepal, major forests types, forest categories from management perspectives, causes and consequences of deforestation, and various efforts of government and private sectors to mitigate the process of deforestation. Based on review, we have also provided a brief conclusion and a few recommendations for better response to deforestation.

12.2.2 NEPAL

Nepal is a relatively small mountainous country surrounded by India to the south, east, and west and China to the north. The country is located between the latitudes 26°22′ and 30°27′N and the longitudes 80°40′ and 88°12′E. It occupies a total area of 147,181 km^2. From east to west the average length is 885 km and north-south width varies from 145 to 241 km with a mean of 193 km. About 86% of its total land area is occupied by high mountains and rolling hills, and the remaining 14% by the flat lands of the Tarai. In altitude, it ranges from 60 m above sea level in the south-eastern Tarai to 8,848 m at the summit of Mount Everest, the highest point of the world. Wide altitudinal variation and diverse climatic conditions within a small area make the physiography of the country unique in the world (Figure 12.2.1). Nepal can be divided into seven physiographic zones in the following order from south to north: (1) **Tarai** belt (60–300 m), a flat land, is a part of alluvial Gangetic plain and rises in the north as **Bhabar**; (2) **Foothills** (up to 1,500 m), also known as *Siwaliks* or *Churia* hills, rise abruptly and reach to an elevation from 700 to 1,500 m comprising gently sloping valleys called **Dun** valley;

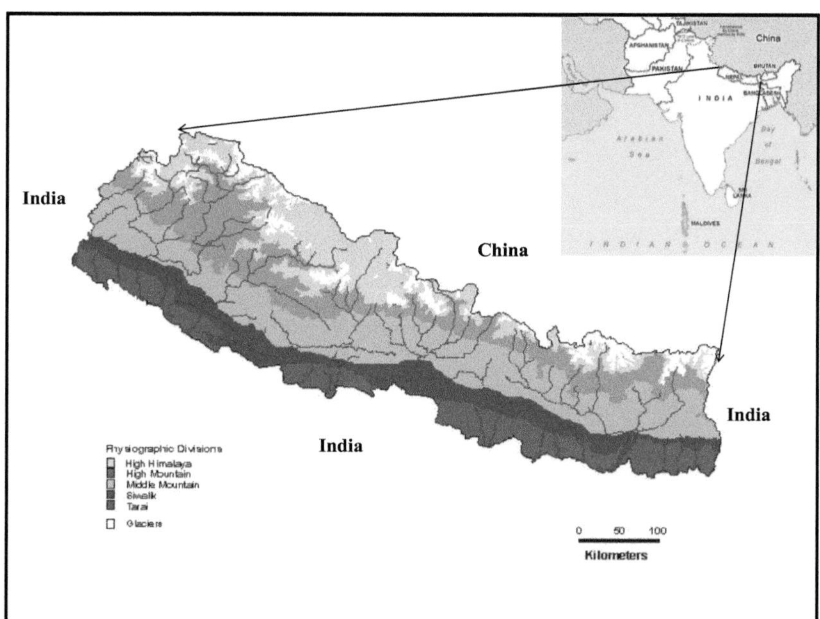

FIGURE 12.2.1 Map of Nepal showing major physiographic regions.

(3) **Mahabharat range** (1,500–2,700 m), composed of hard rocks such as granite or quartzite and limestone; (4) **Midlands** (up to 3,500 m) with an average altitude of 2,000 m comprising high-lying valleys; (5) **Himalayas** (3,000–4,500 m) comprising subalpine and alpine zones; (6) **Inner Himalayas** (above 3,600 m) are dry; and (7) **Arid zone** (above 3,000 m), also known as Tibetan Plateau, lying in the northern part, is almost treeless. Rainfall, winter snowfall, temperature, aspect, and latitude are important factors determining forest and vegetation. Rainfall varies from 250 to 4,500 mm. Some 27 million peoples inhabit the country, with the highest portion (51%) living in the Tarai region (CBS, 2012).

12.2.3 FOREST TYPES IN NEPAL

In Nepal, forest refers to all land having trees with more than 10% crown cover and not used primarily for purposes other than forestry (DFRS, 1999a). Some 35 major forest types occur in Nepal (Stainton, 1972). The forests of Nepal change abruptly, owing to a wide variation of topography, climate, and edaphic conditions (Chaudhary, 1998). The forests of Nepal, for convenience, can be described on the basis of the levels of altitude, and with different types of climate, under the following groups (Table 12.2.1).

In terms of growing stocks, *Abies spectabilis*, *Albizia lebbeck*, *Alnus nepalensis*, *Betula utilis*, *Castanopsis indica*, *Cedrus deodara*, *Cupressus torulosa*, *Dalbergia sissoo*, *Juglans regia*, *Pinus roxburghii*, *Pinus wallichiana*, *Quercus semecarpifolia*, *Rhododendron arboreum*, *Schima wallichii*, *Shorea robusta*, *Terminalia alata*, and *Tsuga dumosa* are the major tree species that are also commonly used for construction and fuelwood (MoFSC, 2013).

12.2.4 FOREST STATUS AND RATE OF DEFORESTATION IN NEPAL

The forests of Nepal are the second largest natural resource after water. The forestry sector in Nepal plays a crucial role in maintaining the economic development and social life of the people (Chaudhary, 2000). The Food and Agriculture Organization of the United Nations (FAO) has estimated that Nepal's forestry sector contributed 3.5% to the GDP of the country in 2000 and 4.4% for the period 1990 to 2000. Nevertheless, the contribution of the forestry sector in the national economy is yet to be critically and comprehensively assessed.

Forest coverage is highest in the midhills (31.95%), followed by High mountains (29.04%) and Siwaliks (25.77%). In terms of coverage, the Tarai harbors the least but highly productive forest of Nepal (Table 12.2.2).

The Department of Forest Research and Survey (DFRS) and Department of Forest (DoF) under the Ministry of Forests and Soil Conservation (MoFSC), Government of Nepal have conducted a number of FRA over the past 40 years

TABLE 12.2.1 Major groups and forest types in Nepal

Major Groups/Altitude	Forest Types
Tropical (up to 1,000 m)	Sal (*Shorea robusta*) forest; Tropical deciduous riverine (*Dalbergia sissoo–Acacia catechu*) forest; Tropical evergreen (*Michelia champaca, Litsea* species, *Persea* species, etc.) forest
Subtropical (up to 2,000 m)	Chilaune–Katush (*Schima–Castanopsis*) forest; Chir pine (*Pinus roxburghii*) forest; Alder (*Alnus nepalensis*) forest
Temperate (2,000–3,000 m)	Lower temperate mixed broadleaved (*Persea* species, *Phoebe* species, *Cinnamomum* species, *Lindera* species, *Neolitsea* species) forest; Temperate mixed evergreen forest (broadleaved oaks—*Quercus lamellosa, Quercus glauca, Quercus semecarpifolia* and conifers—*Picea smithiana, Abies pindrow, Cupressus torulosa, Cedrus deodara*); Upper temperate mixed broadleaved forest (*Aesculus–Juglans–Acer* association in the west, *Magnolia–Acer–Osmanthus* association in the east and central Nepal)
Subalpine (3,000–4,100 m)	Silver fir (*Abies spectabilis*) forest; Birch–Rhododendron (*Betula utilis* and *Rhododendron* species) forest
Alpine (above 4,100 m)	*Juniperus–Rhododendron, Caragana–Lonicera*, and alpine meadows associations

Source: Stainton (1972).

TABLE 12.2.2 Forest coverage in different physiographic zones

Physiographic Zone	Forest Area (ha)	Percentage
High Himalayas	154,300	2.75
High Mountains	1,628,300	29.04
Midhills	1,791,100	31.95
Siwaliks	1,445,000	25.77
Tarai	586,500	10.46
Total	**5,605,200**	**100**

Source: Kleinn (1994).

TABLE 12.2.3 Change in forest and shrub cover (1978–1994)

	Year		
Type	**1978–1979**	**1985–1986**	**1994**
Forest	38.0	37.4	29.0
Shrub	4.7	4.8	10.6
Total	**42.7**	**42.2**	**39.6**

Source: DFRS (1999a).

(Acharya and Dangi, 2009). The last FRA under the name National Forest Inventory (NFI) was carried out in the early 1990s. According to this inventory, about 29% of the area of Nepal is under forest coverage and 10.6% is under shrubs that makes forest and shrub cover about 5.6 million ha which is 39.6% of the total land area of the country (Tables 12.2.2 and 12.2.3). The forest area, which was 45% in 1966 and 37% in 1986, has declined considerably. Conversely, the shrub land area doubled from 5% in the early 1980s to 10.6% in the mid-1990s. The annual deforestation rate was estimated to be 1.7% during the 1980s to mid-1990s (DFRS, 1999a). However, deforestation rates and causes differ according to physiographic regions. For instance, from 1978 to 1994, the rate of estimated annual deforestation varied greatly between the Hills (2.3%) and the Tarai (1.3%) (DFRS, 1999a). The highest rate of deforestation that Nepal witnessed was between 1947 and 1980, where Nepal's forest cover declined at an annual rate of 2.7% (from 57% to 23% of the national territory), and subsequently at an annual rate of 1.8% between 1980 and 2000 (UNEP, 2001). Since the early 1990s no FRA/NFI existed to update data on forest cover change except for Churia and Tarai. The trend shows that the rate of deforestation is slowing down, however, deforestation and forest degradation are still of serious concerns (FAO, 2009).

Very recently, FRA is being conducted by DFRS with the support from the Government of Finland. This project is in the process of completing NFI and only data from Tarai and Churia regions have been completed and published (FRA/DFRS, 2014a,b). The FRA from 20 Tarai districts revealed that the rate of forest cover change was at an annual rate of 0.06% during 1990/1991 to 2010/2011 (Table 12.2.4) whereas the total forest area in Churia decreased by loss of 38,051 ha forest over the period of 15 years (1995–2010), with the annual rate of change about −0.18%. Nevertheless, the macrolevel studies and visual interpretations revealed that Nepal's total forest coverage and condition is significantly improving due to the Community Forestry (CF) intervention (FAO, 2009; see Section 12.2.7).

TABLE 12.2.4 Forest cover change in the Tarai between 1984 and 2010/2011 ('000 ha)

Year of Assessment	Forest Area ('000 ha)
Land Resource Mapping Project (LRMP) 1984	492.1
Department of Forest (DoF) 1991	440.1
Department of Forest (DoF) 2001	424.6
Forest Resources Assessment (FRA) 2010/2011	408.1

Source: FRA/DFRS (2014a).

12.2.5 CAUSES OF DEFORESTATION IN NEPAL

Though the FRA estimates started providing quantitative data on forest cover, deforestation in Nepal has a long history. Intensive deforestation started during the unification phase of Nepal (before 1768), where forests were converted into agriculture land to feed the huge number of militants in different states. Their strength was their army, and military endeavors were typically rewarded with land grants, which often led to further deforestation (Soussan et al., 1995). It was further intensified with initiation of the system of selling timber to foreign contractors during the Rana regime in 1846. Nevertheless, the intensity and consequences of deforestation were noticed only in the 1950s by the early foreign visitors (see also Robbe, 1954; Hagen, 1961).

The extensive Tarai forests were little disturbed until the late 1920s, when the government initiated expansion of cultivated areas by clearing some forests and extracting timber in other forests for export to India to collect revenue (Joshi, 1993). The government hired an experienced British forester (Mr J.V. Collier) who had a long working experience in India from 1925 to 1930 to supervise and improve timber felling in the Tarai. Mr Collier of the Indian Forest Service, whom Rana Prime Minister Chandra Shamsher Jung Bahadur Rana entrusted with the direction of the forest department in Nepal, came in 1923 to Nepal as an adviser who recommended the bulk removal of Sal from the virgin forests of the Tarai, such as Morang district in the east and Kailali–Kanchanpur districts in the west. Mr Collier was entrusted with the task of extracting from Nepal 200,000 railway sleepers offered by the Government of Nepal to British Government as a war gift. This also brought tremendous wealth to Rana rulers but also paved the way to reckless destruction of forests in Nepal, and the situation is being continued with an increasing trend of overexploitation in more accessible parts of Tarai belt, where easy transportation helps removal of timber logs across the border to India (Bhatt, 1977).

FIGURE 12.2.2 Deforestation in high-altitude region of Nepal. *Photo: Y. Shakya.*

Deforestation is one of the major environmental issues in Nepal. Overharvesting of fuelwood and fodder, forest fire, lopping and grazing, slash and burn cultivation, and timber extraction are generally considered to be the major factors responsible for deforestation and forest degradation (Figure 12.2.2). In general, the drivers of deforestation and degradation are the mixture of direct and indirect causes such as high dependency on forest resources, unsustainable harvesting practices, illegal harvest of forest products, infrastructure development, forest fire, natural calamities, encroachment, overgrazing, lack of good governance, and ambiguous policy (Table 12.2.5). Some major causes of deforestation are briefly discussed in this paper.

12.2.5.1 Population Growth

Population growth (annual rate 1.40% in 2011 compared to 2.25% in 2001) appears to be the most important factor behind decreasing forest cover in Nepal. The number of people dependent on agriculture is rising; and as a result, agricultural land has increased, mostly by encroaching upon forest areas (UNEP, 2001; CBS, 2012). Table 12.2.6 illustrates that the annual growth rate of population is higher in Tarai. Until the 1980s, the majority of this growing population had migrated from the hills to the Tarai, which was mostly inhabited by clearing forest (also see Section 12.2.5.4).

12.2.5.2 Fuelwood as a Major Source of Energy

Fuelwood is the major source of energy for majority of the people in Nepal (Table 12.2.7). In 1998, fuelwood derived from the forest constituted the largest proportion of the total fuel consumption (78%), whereas this dependency has been reduced to 64% in 2013 (WECS, 2013). However, in the absence of alternative energy sources, most rural people, which constitute about 80% of total population of Nepal, have to depend on fuelwood for energy generation. Though the collection and/or harvesting of fuelwood is done at the individual level, it has contributed toward deforestation (Figures 12.2.3 and 12.2.4).

TABLE 12.2.5 Drivers of deforestation and forest degradation and their underlying causes

Drivers	Underlying Causes	Nature of Causes	Affected Regions[b]	Area of Issues[a]	In/Out of Forestry[c]
1. High dependency on forests and forest products (timber, firewood, and other nontimber forest products)	1.1 Poverty and lack of livelihood alternatives	Direct	T, H	S, E	O
	1.2 Limited access to alternatives for fuelwood and timber	Direct	T, H, M	E	B
	1.3 Inefficient forest product use	Direct	T, H, M	T, S, E	B
2. Illegal harvest of forest products	2.1 Weak law enforcement and impunity	Direct/Indirect	T, H	G	I
	2.2 Weak governance and governance vacuum	Indirect/Direct	T, H, M	G	I
	2.3 Inefficient distribution mechanisms for timber and firewood	Indirect	T, M	M, G	I
	2.4 Market failure	Indirect	T, H, M	M, E	B
	2.5 Poverty and lack of livelihoods opportunities	Indirect	T, M	S, E	O
	2.6 High cross border demand for forest products	Direct	T, M	E	O
3. Unsustainable harvesting practices	3.1 Weak law enforcement and impunity	Direct	T, H, M	G, political	B
	3.2 Inadequate resources for research and development	Direct	T, H, M	E	B
	3.3 Insecure tenure	Direct	T, M	G	I
	3.4 Insufficient technical inputs	Direct	T, H, M	T	I

Continued

TABLE 12.2.5 Drivers of deforestation and forest degradation and their underlying causes—cont'd

Drivers	Underlying Causes	Nature of Causes	Affected Regions[b]	Area of Issues[a]	In/Out of Forestry[c]
4. Forest fire	4.1 Carelessness	Direct	T, H, M	S	B
	4.2 Intentional	Direct	T, H, M	S, E	B
	4.3 Weak forest fire management practice	Indirect	T, H, M	G, T	I
5. Encroachment	5.1 Expansion of agriculture	Direct	T, H	S, E	B
	5.2 Poverty and landlessness	Indirect	T, H	S, E	O
	5.3 Political motivation	Indirect	T, M	P	O
	5.4 Unclear land tenure, policy, and planning	Indirect	T, H	G, P, T	B
6. Overgrazing	6.1 Governance vacuum	Indirect	T, M, H	G, P	B
	6.2 High number of low productive livestock	Direct	T, H, M	S, E, M	O
	6.3 Limited alternatives for fodder	Direct	T, H, M	E, T	B
	6.4 Poverty and lack of livelihoods opportunities	Indirect	T, H, M	E	O
7. Infrastructure development	7.1 Ad hoc policy process and weak coordination between and within sectors	Indirect	T, H	G, P	B
	7.2 Weak mechanism for planning and compensation including EIA and IEE, and approval and monitoring of development projects	Direct	T, H	G, P, T	B
	7.3 Undervaluation of forest land and forest services	Indirect	T, H	E	B
	7.4 New economic growth prospects (e.g., oil and gas, transmission lines, cement factory, airport, hydropower dam, etc.)	Indirect	T, H	E	O

8. Resettlement	8.1 Undervaluation of forestland and services	Indirect	T, H	E	B
	8.2 Increase demand of land for new settlements	Indirect	T	E, P	O
	8.3 Poorly enforced planning regulations	Indirect	T	G, E, T	B
9. Expansion of invasive species	9.1 Lack of proven eradication practices	Direct	T	T	I
	9.2 Frequent forest fires	Indirect	T, H, M	S, T	O
	9.3 Overgrazing	Direct	T	S, E, T	B
	9.4 Opening of canopy	Direct	T	S, G, T	B

[a] *S—Social, T—Technical, M—Market, G—Governance, E—Economic, P—Policy, and N—Natural.*
[b] *T—Tarai, M—Midhills, and H—High Mountain.*
[c] *I—Internal factors, B—Both internal and external, and O—Factors outside the MoFSC.*
EIA, Environmental Impact Assessment; IEE, Initial Environmental Examination.
Source: MoFSC (2010).

TABLE 12.2.6 Population growth rates by physiographic regions from 1961 to 2011

	Average Annual Growth Rate of Population 1961–2011				
Period	**Mountain**	**Hill**	**Mountain and Hill**	**Tarai**	**Total**
1961–1971	–	–	1.85	2.39	2.05
1971–1981	1.35	1.65	1.61	4.11	2.62
1981–1991	1.02	1.61	1.52	2.75	2.08
1991–2001	1.57	1.97	1.91	2.62	2.25
2001–2011	0.62	1.13	1.06	1.75	1.40

Source: CBS (2011).

TABLE 12.2.7 Ten-year trend of fuelwood consumption in Nepal (Unit in 000 GJ)

	Sector				
Year	**Residential**	**Industrial**	**Commercial**	**Total from Fuelwood**	**Total Energy Consumption**
2008–2009	308,604.3	721.1	1841.9	311,167.3	400,506.4
2007–2008	302,251.5	717.5	1752.1	304,721.2	388,382.1
2006–2007	295,994.3	716.2	1614.9	298,325.4	381,049.9
2005–2006	289,449.0	802.5	2209.0	292,460.4	376,785.8
2004–2005	284,138.4	772.7	2048.9	286,960.0	367,207.4
2003–2004	278,220.0	742.9	1925.5	280,888.3	361,837.0
2002–2003	272,323.0	712.5	1924.7	274,960.2	353,452.5
2001–2002	266,724.4	683.8	1749.6	269,157.7	347,329.3
2000–2001	256,416.3	644.1	1575.3	258,635.6	335,420.9
1999–2000	251,142.3	618.1	1438.6	253,199.0	330,706.0

Source: WECS (2010).

FIGURE 12.2.3 Fuelwood harvesting in midhills of Nepal. *Photo: Y. Uprety.*

FIGURE 12.2.4 Wood stacking on the roof in trans-Himalayan region contributing deforestation (seen in the backdrop). *Photo: R.P. Chaudhary.*

12.2.5.3 Livestock and Grazing

Much forest area in the country is used as open grazing land for livestock. This has led to further degradation of the forest area. The number of livestock (cattle, pigs, goats, and sheep) increased by about 20% between 1985 and 1998, whereas the grazing area remained constant. Raut (1997/1998) indicated that about 10% of the fodder comes from the forest. Consumption of fodder exceeds the sustainable supply in some regions, oak species being the most heavily lopped and exploited (Måren et al., 2014) (Figure 12.2.5).

12.2.5.4 Migration and Settlement in Tarai

The low-lying Tarai region experienced the greatest deforestation since the 1950s due to influx of migration. In general, the major push factors of migration from

FIGURE 12.2.5 Livestock pressure on forest is severe throughout Nepal. *Photo: S.K. Rimal.*

hills to Tarai are population pressure, insufficient land holding, deficiency of food production, indebtedness, deteriorating environmental conditions, natural calamities, lack of employment opportunities, etc. (Regmi, 1994), followed by government policies for resettlement in the Tarai. A total of 103,968 ha of forest was cleared under the planned resettlement program from 1950 to 1986. About an equal area of the forest was occupied illegally by migrants for settlement during the same period. Eradication of malaria, easy access by road connections, availability of land, better availability of food, health facilities, good education, and possibilities of employment in Tarai region motivated people for migration. Comparative statistics show that in 1952, the Tarai had 35% of the total country's population and this increased to almost 51% in 2011 (CBS, 2012). The Tarai received 74% of the total internal migration in three decades from 1961 (Gurung, 1989). Deforestation in the Tarai has been caused mainly by the clearing of forestland for agricultural purposes. From 1963 to 1979, the proportion of cropland in the Tarai increased from 38.5% to 49.8% (Gurung, 1989).

12.2.5.5 Illegal Timber Harvesting and Trade

Commercial logging is banned in Nepal. The domestic consumption of timber is partly fulfilled from the salvage operations at national forests, trees harvested from private lands, and through the management operations from community forests. To a larger extent, timber smuggling along the Nepal–India border is also responsible for deforestation in Nepal. Such timber smuggling is usually done in an organized manner by influential people (Regmi, 1994). Owing to the higher prices of timber in India and China, smugglers are motivated to export timber from Nepal. The activity intensifies when the price of timber is higher in India and China than in Nepal. However, in recent years the smuggling has declined due to improved monitoring and also decline of forest area in Tarai. District forest offices, forest users, and other stakeholders are actively monitoring the forest and also a multistakeholder monitoring mechanism does exist at the district level for monitoring purposes.

FIGURE 12.2.6 Largely undocumented but widely seen timber smuggling in high-altitude regions of Nepal. *Photo: Y. Shakya.*

Timber smuggling from Nepal to Tibetan autonomous region of China, is also noticeable. Though not well documented, this is also a well-established practice from the northern regions of Nepal, particularly from the Solukhumbu, Rasuwa, Gorkha, Dolpa, Humla, and Mugu districts. Such kind of trans-boundary leakage has a tremendous impact in the forests of High mountain region (COMFORTC, 2012) where rate of tree growth is very slow, and once land is devoid of vegetation, it triggers erosion of top soil (Figure 12.2.6).

12.2.5.6 Developmental Activities

As in many developing countries, deforestation in Nepal has been directly linked with the process of development (Regmi, 1994). Deforestation has proliferated in recent years by the development activities related to road construction, schools, hospitals, graveyards, irrigation canals, dam building, and expansion of settlements. Unplanned and unregulated developmental activities, generally without environmental impact assessments, are causing forest loss and degradation in the country. For example, in the construction of east-west highway in the beginning of the twentieth century, hundreds of hectares of forest area was cleared along the road side, and huge extractions of timber were made for use of construction of bridges, culverts, heating charcoal for pitching, etc. (Regmi, 1994). Recent initiatives of the government to build large-scale hydropower and transmission lines will certainly pose serious threats to the forests in many parts of the country. Between 1986 and 1999, about 0.155 million ha of forest area were consumed by settlement, agriculture, institutional buildings, and roads (DFRS, 1999a). This trend continues to increase in recent years owing to growing demand of forest area for hydropower projects, road, irrigation canals, and other infrastructure.

12.2.5.7 Tourism and Trekking

Mountaineering and trekking are the popular tourism activities in Nepal. Although tourism has very positive impact in local peoples' livelihood, their impact on forests is serious. For example, a series of accounts and documents for national park planning in the 1970s and early 1980s portrayed a deforestation crisis in Khumbu region and identified tourism as a key cause (Stevens, 2003). This is a widespread problem in all major trekking routes. The hotels and lodges on the trekking routes need large amounts of timber for construction and firewood for cooking and heating.

12.2.5.8 Forest Policy

Gaps that exist in the government forest policy are themselves additional factors that have contributed to deforestation in Nepal (UNEP, 2001). One of the major driving factors was the enactment of Private Forest Nationalization Act in 1957. The act was implemented on the assumption that it could consolidate the protection and management of the forests, but conversely, it rather led to degradation of the national forests by providing uncontrolled local access to them. It also completely ignored the traditional forest management practices of the people (UNEP, 2001). After nationalization of the forests the villagers reacted negatively, believing that their traditional rights of access and use of forests had been limited (Shrestha, 1999). As there were no land records, villagers had a strong incentive to destroy the forest so that the land could be claimed as private property after it was cleared and cultivated (Wallace, 1983). Likewise, the Land Tax Act 1977 defined lands with forests as government lands that encouraged local inhabitants to cut down trees standing around their farms.

Conflicting cross-sector policies and inconsistencies in the statutory frameworks are other challenges (Belbase and Thapa, 2007). Forest acts and bylaws are contradictory to other sectoral acts and regulations, the major ones being the Mines and Minerals Act of 1985, and the Local Self Governance Act (LGSA) of 1999. For example, LSGA empowers village development committees (VDCs) to sell specified natural resources and products, and stipulates that the proceeds from such sales are to be deposited in the VDC fund, whereas the Forest act empowers community forest user groups to sell the same products (Belbase and Thapa, 2007).

12.2.5.9 Land Tenure Systems

In Nepal, various types of land tenure systems notably the *Raiker*,[1] *Birta*,[2] and *Kipat*[3] have historically contributed to deforestation and forest degradation.

1. *Raiker* is a state landlordism in which the state retains full rights of its use and alienation.
2. *Birta* was the land given out by the state to individuals.
3. *Kipat* is a communal form of land tenure practiced among *Rais* and *Limbus* communities who inhabit in the eastern part of Nepal.

After the unification of Nepal (1768–1769), the monarchy started granting large areas of *Raiker* cultivated and cultivable lands. The emergence of the Rana autocratic regime in 1846 began a new era in the history of the *Birta* system in Nepal. In both cases, the land was granted to their relatives and friends to assure support from them. The *Birta* system had a great influence on deforestation processes in the Tarai. Most members of the families of Ranas were granted *Birta* in Tarai because of the existence of most fertile land and forest resources. The *Kipat* also contributed toward deforestation as the communities claim autonomous rights for forest utilization. The process of timber export to India by the *Birta* holders continued until 1969, when it was formally abolished (Regmi, 1994). According to one estimate, almost one-third of the total forests and cultivated lands were under *Birta* tenure by 1950, and 75% of that belonged to the *Rana* family (Joshi, 1993).

12.2.5.10 Political and Nonpolitical Use and Misuse of Forest

It is very important to consider the political context of deforestation in Nepal. Political instability in the country since 1950s remained as one of the major drivers of deforestation. The *Panchayat* system that was introduced in 1960 accelerated deforestation in the Tarai. In this regime, various political and nonpolitical persons were able to acquire land through personal decrees of the king. These peoples, who were considered the key to the political system by the king, were able to acquire property rights over uncultivated forest or wasteland at no cost to them. The opportunistic exploitation of timber and forest encroachment was further intensified in the Tarai after the introduction of multiparty system in 1990. During the transition phase the forest encroachment was so intensified that the interim government was compelled to constitute a Forest Protection Task Force. On the other hand, some other groups were actively involved in deforestation in the name of *Sukumbasis* (landless peoples) in the Tarai. The landless people also endeavored to have access to a patch of land through forest clearance during this transitional phase. The timber smugglers, with the help of the political supporters, had also then increased the timber exports from various districts during this political transition (Regmi, 1994).

12.2.5.11 Cultural Use of Wood

A considerable amount of wood is also used for cultural purposes in Nepal; most notably the wood used for cremating dead human bodies, and that used for flag poles in the mountainous regions. Though the official data are not available, significant volumes of fuelwood are considered to have been used for cremation. In rural areas, the wood for cultural purposes, including cremating human bodies, is provided free of cost mostly from the community forests but it is an issue in urban areas. In urban areas, the wood for such purposes should be purchased and that is provided on a subsidized basis by the government bodies. In mountainous regions, it is a cultural practice that every household fixes flag poles in foreyard. Hundreds of immature poles of conifers with clear boles are cut for this purpose. In many instances a large number of immature trees are felled while selecting the clear ones (COMFORTC, 2012).

12.2.5.12 Impact of Earthquake

On Saturday, April 25, 2015, a 7.6 magnitude of earthquake which was followed by another major aftershock of 6.8 magnitude, as recorded by Nepal's National Seismological Centre (NSC), struck Nepal almost after 80 years. Besides big casualties and injuries in 31 (of 75) districts of the country, 14 were declared "crisis hit." In addition to damage and loss of settlements and agricultural land, the devasted earthquake has destroyed large areas of forests compromising the capacity of natural forests ecosystems to deliver important services and benefits to the people (NPC, 2015a). Losses of ecosystem services due to earthquake have been estimated NPR 34,021.3 million (NPC, 2015a,b).

The 31 earthquake-affected districts contain around 41% (2,393,535 ha) of all forest area. Around 48% of this area is community forest managed by 11,554 community forest user groups and involving more than 55% (over 2.28 million) of the households in these districts. Another 48% forest area is managed by the Ministry of Forests and Soil Conservation (MoFSC), Government of Nepal (GoN) including protected areas (PAs). The rest includes leasehold forests (0.73%), private forests, and other categories (NPC, 2015b).

Earthquake damages in the forest resources of 31 districts. FAO, Rome has estimated forest loss rate of 2.2% for only six earthquake districts (NPC, 2015b). Mainly two types of forests are damaged, pine forest (30%) and subtemperate forest (70%). The total damages and losses in the forestry sector are estimated to be NPR 32,960.3 million and NPR 1061 million respectively, excluding loss of ecosystem services value.

Damage and loss of forest and biodiversity have also been observed in 7 PAs (out of total 20 PAs) that have received hard earthquake hit covering 15,988 km^2, i.e., 46.8% of total PAs coverage in Nepal. Among them, severely affected are Langtang NP, Makalu Barun NP, Sagarmatha NP, Manaslu NP, Shivapuri-Nagarjun NP, Gaurishankar CA, and Annapurna CA, and two Ramsar sites, namely Gosainkunda and associated lakes, and Gokyo and associated lakes. The PAs provide refuge to several endangered keystone fauna of global significance as well as nationally protected species protected under National Parks and Wildlife Conservation Act (NPWC) 1973, such as red panda, musk deer, Himalayan tahr. Important habitats of mammals such as Blue pipe forest, Temperate Oak forest, subalpine Fir, and Birch forest are expected to be severally damaged.

A number of landslides and cracks are reported, particularly in the catchment areas (Figures 12.2.8 and 12.2.9.). More landslides and erosion on slopes are expected in monsoon increasing further loss of forest and wildlife habitat.

12.2.6 CONSEQUENCES OF DEFORESTATION IN NEPAL

Deforestation has immediate consequences for the local population in terms of increased fuel scarcity, reduced supply of fodder, and leaf-litter manure. The unpredicted erosion, landslide, and lowland flooding due to deforestation are also major concerns in Nepal as well as in downstream countries (Metz, 1991; but also see Ives, 2006). The following are the major consequences of deforestation in Nepal.

12.2.6.1 Impacts on Forest Structure and Ecosystem Services

One major impact of deforestation is on forest structure of Nepal. The inventory shows that the growing stock of the forests has decreased in all physiographic regions. In 1985/1986, the total growing stock was 522 million m^3 over bark up to 10-cm top diameter (HMGN/ADB/FINIDA, 1988). DFRS estimated that the stock decreased to 387.5 million m^3 in 1999 (DFRS, 1999b).

Another conspicuous impact of deforestation is on the floral and faunal diversity. Continuous loss of forest is a major threat to biodiversity. Various plant and animal species are considered threatened as a result of deforestation and increasing pressure on their uses.

The occurrence of landslides, soil erosion, and floods has become normal phenomena during the monsoon season (Figure 12.2.7). Deforestation may

FIGURE 12.2.7 Impact of deforestation—commonly seen landslides in hills of Nepal. *Photo: C.K. Subedi.*

FIGURE 12.2.8 Landslide causing loss and damage of temperate forest in Rasuwa district, Nepal.

FIGURE 12.2.9 Land crack after earthquake damaging montane Sal forest in Nuwakot district.

be leading to an increase in some of these natural disasters. A thin overstorey canopy of trees, with virtually no regeneration, severe erosion, and low organic matter content of soils, characterize most of the degraded forest. Over the sloping areas of the middle hills, the farmers have cleared forests for cultivation to meet their food requirements. This has resulted in environmental degradation in the form of accelerated soil erosion leading to land degradation, declining productivity, and sedimentation in downstream areas. The occurrence of floods and landslides as a result of deforestation has affected not only the degradation of land but also human lives and property (UNEP, 2001). The accelerated soil erosion and consequent loss of top soil due to deforestation has adversely affected yields on agriculture land, and increased the downstream sedimentation of dams, reservoirs, and irrigation systems. A high dependency on forests for fuelwood and resulting deforestation have increased the CO_2 concentration. It is estimated that annual deforestation of 26,602 ha in 1999 has emitted 7.77 million tonnes of carbon into the atmosphere (SEAMCAP, 2000).

Forests are the basis of many rural livelihoods in Nepal; they provide ecosystem services, including fuelwood, timber, fodder, and nontimber forest products and regulation of water quality and flow, carbon sequestration, erosion, and regional climate (MEA, 2005). Deforestation has direct impact on these ecosystem services. According to one estimate from 22 VDCs of eastern Nepal provisioning, forest ecosystem services accounts for $93 million (Pant et al., 2012).

12.2.6.2 Increase in Distance to Forest Access

As a result of deforestation, the distance traveled by rural people to reach the forests has increased considerably. In 1985/1986, the total accessible forestland in the country was 5.8 million ha. In 1992/1993, it declined to 4.6 million ha (CBS, 1998). This means that rural people must travel additional distances to collect forest products, thereby reducing the time available for other productive activities.

12.2.6.3 Export of Forest Products

Export of forest products, including timber and nontimber, to India declined sharply from 1975 to 1985. Total exports, including those to India, have remained more or less stable since 1990. Timber export was banned in 1984. According to the statistics of Ministry of Finance (1999), in 1975, timber exports reached a maximum value of about US$28 million and then declined sharply to a minimum of US$0.1 million in 1981. In 1980, the value of nontimber exports was US$5.6 million and this declined to US$2.7 million in 1998.

12.2.7 RESPONSES TO DEFORESTATION

12.2.7.1 Forest Policy and Governance

As governance reform is considered to be one of the main approaches to address the causes of deforestation, the Government of Nepal has made several attempts to respond to this by formulating effective policies with an aim to delivering good governance. The MoFSC is the apex body to create an enabling environment for the conservation and sustainable management of forests. Five departments and regional directorate offices exist under the MoFSC that are coordinating forest conservation and management programs and forest-related activities. Most importantly, the district forest offices in each district under the DoF are responsible for the forestry programs. Enforcing legislations through the district forest offices is one of the priority programs of DoF to control forest encroachment and conserve forest resources under its jurisdiction.

The forestry sector policy in Nepal can be divided into three broad groups, viz., privatization (pre-1950), nationalization (1957 and up to the mid-1970s), and the community orientation that began in the late 1970s with the introduction of the community forestry concept. Currently, the Master Plan for the Forestry Sector (MPFS) (1988); Forest Act (1993); Forest Regulations (1995); different periodic plans; fiscal policies; forest and forestry laws and regulations are the policy guidelines and legal instruments facilitating sustainable forest management in Nepal (Chaudhary, 2000).

The MPFS was prepared in 1988 for a comprehensive long-term plan to meet the basic needs of the people by sustainably managing the forest resources in Nepal. The Master Plan, endorsed by the government in 1989, presents a comprehensive strategy for 21 years of management of forestry in Nepal. The approach focuses on developing procedures to enable the handing over of the forest to the forest user's group (FUG) and the private sector, based on a partnership between the MoFSC and the local forest users that encourages management of all aspects of forests, including trees, shrubs, grasses, and medicinal plants.

The main long-term management strategies for the forestry sector are: (1) to meet the people's basic needs for fuel, timber, fodder, and other forest products on a sustainable basis; (2) to contribute to food production through an effective interaction between forestry and farming practices; (3) to protect the land against degradation by soil erosion, floods, landslides, desertification, and other effects of ecological imbalance; (4) to conserve ecosystems and genetic resources; and (5) to contribute to the growth of local and national economies by managing the forest resources and forest-based industries, and creating opportunities for income generation and employment.

The other medium-term objectives are to: (1) promote people's participation in forestry resource development, management, and conservation; (2) develop the legal framework needed to enhance the contribution of individuals, communities, and the institutions; and (3) strengthen the organizational framework and develop the institutions of the forestry sector to enable them to carry out their allotted tasks. The primary programs formulated by MPFS are: (1) community and private forestry; (2) national and leasehold forestry; (3) soil conservation and watershed management; (4) conservation of ecosystems and genetic resources; (5) medicinal and aromatic plants and other minor forest products; (6) wood-based industries; with supportive programs such as: (7) policy and legal reform; (8) monitoring and evaluation; (9) human resources development; (10) research and extension; (11) institutional reform; and (12) resource information and planning assistance.

After the launch of MPFS (1988), many issues and perspectives have emerged into the arena of national and international policies such as Millennium Development Goals, decentralized, community-based, forest management systems, to: landscape approach to biodiversity conservation, climate change adaptation and mitigation, the green economy, promotion of concept of ecosystem services, forest certification and payments for environmental services, REDD+ and local communities, indigenous peoples' rights to access and benefit sharing, market opportunities and private sector services, and so on. Against this backdrop, the MoFSC has initiated a review of the Master Plan of 1988 and prepared Nepal's Forest Sector Strategy in 2013.

The Ninth Five Year Plan (1997–2002) of Government of Nepal includes "Forest" as a separate chapter It includes, among other things, material to: (1) ensure regular supplies of day-to-day needs of timber, firewood, fodder plants, and other forest products through community forestry development; (2) develop policy for encouragement of private sectors in forest management; (3) create directives for development of leasehold forestry by eradicating the issues; (4) create opportunities for popularization, management, job opportunity, and earnings through large-scale cultivation of medicinal plants; (5) promote the cultivation of fodder, fruit trees, and other agricultural cash crops in the afforestation programs related to forest development; (6) regulate the monitoring program effectively; (7) emphasize proper conservation and management of threatened medicinal plants; (8) implement the programs by developing institutional infrastructure and spelling out the role and responsibilities of government, private sector, and nongovernmental organizations.

The Forest Act (1993) and Forest Regulations (1995) recognize the importance of forests in maintaining a healthy environment. Section 23 empowers the government to delineate any part of a national forest that has a "special environmental, scientific, or cultural importance" as a protected forest. Such delineation should be seriously taken up so as to designate and include national "hotspot" areas also. The Environment Protection Act (1996) and Environmental Protection Act Regulations (1997) have made environmental impact assessments mandatory also in the forest sector.

The Forest Fire Management Strategy (2010) has considered the following as four pillars in forest fire management: (1) policy, legal, and institutional development and improvement; (2) education, awareness raising, capacity building, and technology development; (3) participatory fire management and research; and (4) coordination and collaboration, networking and infrastructure development, and international cooperation.

To address the severe degradation of Churia resulting from overexploitation of the resources, the Churia Area Program Strategy (CAPS) 2008 was framed to provide holistic guidance for all stakeholders interested in supporting development initiatives in the Churia area in the future. Though not yet approved by the MoFSC, the strategy aims to provide a common vision for all stakeholders interested to work in the Churia area. The goal of CAPS is to create an enabling environment for all stakeholders so that they can contribute to the conservation as well as to the livelihood of the resource-dependent people in an equitable manner (MoFSC, 2008).

Forest Encroachment Control Strategy (2012) was framed to rehabilitate and restore the encroached forest area, aiming to achieve the national policy of maintaining 40% of the total area under forests. It envisions a National Forest Encroachment Control Coordination Committee chaired by MoFSC at the central level. The Central Forest Encroachment Control Unit is chaired by the secretary at the MoFSC and District Forest Encroachment and Management Task Force chaired by chief district officer at the district level. The Task Force at the district is responsible for carrying out a collaborative protection activity, whereas the committee at the central level is responsible for annual planning, coordination with political parties and other stakeholders, and monitoring of the field activities. Apart from all these policies and governance-related arrangements, armed forest guards are mobilized by the DoF for controlling forest area encroachment, wildlife protection, controlling illegal felling of the trees and smuggling of the timber.

Forestry policy 2015 provides guidance to all other policies of forestry sector. It aims to conserve, promote, and utilize forest, vegetation, wildlife, protected areas, watershed, and biodiversity for employment generation, livelihood improvement, and maintaining ecosystem balance.

12.2.7.1.1 Forest Category in Nepal

For effective management of forests, national forests and private forests are recognized in Nepal as the broad categories of forest on the basis of land ownership. The national forest includes all forests, excluding private forests, whether marked or unmarked with forest boundary markers. The term includes waste or uncultivated lands or unregistered lands surrounded by or adjoining forests, as well as paths, ponds, lakes, rivers or streams, and riverine lands within forests.

Private forest is the forest which is planted, nurtured, or conserved in any private land owned by an individual under current law. But no data are available about the extent of private forest.

On the basis of management objectives and rights, government forests have been further categorized under the Forest Act (1993).

1. **Government-managed forest**: The national forest that is managed by Government with production for internal consumption [as well as biodiversity conservation] as the main objective.
2. **Protection forest**: Under this category, national forest that is considered to be of special environmental, scientific, and cultural significance is declared by Government as a protection forest.
3. **Community forest**: National forest is handed over to a user group for its development, conservation, and utilization for collective benefit.
4. **Leasehold forest**: Leasehold forest means a National Forest handed over as a Leasehold Forest to any institution established under prevailing laws, industry based on Forest Products or community for afforestation, ecotourism, wildlife farming, production of raw materials for forest-based industries. Under this provision Government may also allocate portions of national forests to the group of people who are below the poverty line for livelihood enhancement and environment conservation.
5. **Religious forest**: National forest is handed over to any religious body, group, or community for its development, conservation, and utilization.

12.2.7.1.2 Community Forestry

Community-based forestry is the second largest forest management regime after the government-managed forest. Since its adoption in 1978, this participatory forest management has become a successful model for forming the capital (natural, human, financial, and physical) and reforming forest governance (FAO, 2009). The community forests are national forests handed over to an FUG for development, conservation, and utilization for the collective benefit of the community. Approximately, 3.56 million ha of forests have been estimated potential for community forest in Nepal (Tamrakar and Nelson, 1991). The latest figure shows that approximately 1.798 million ha of forests (51.4% of the potential community forest area) are handed over to 18,960 community forest user groups (CFUGs) involving 2.392 million households under the community forestry program by the end of September 2015 (http://dof.gov.np/image/data/Community_Forestry/Summary.pdf. Accessed on September 28, 2015). These CFUGs have reversed past trends of deforestation significantly. A total of 34,359 ha forests were handed over to the communities before 1992. The area increased to 1.02 million ha between 1992 and 2002. The trend to hand over the national forests was at a high rate (2882%) in one decade during 1992–2002, whereas the trend has been rather slow (about 20%) after

FIGURE 12.2.10 A community-managed forest in Tarai. *Photo: S.K. Rimal.*

2002. One of the reasons for the slow handing-over process can be assumed partly due to over a decade of conflict in the country. The trend of community forest handing over is higher in the hills than the Tarai (MoFSC, 2009) and also the program is much more successful in the hills. The success of the community forestry program in the hills can partly be attributed to many successful indigenous systems of forest management that were in existence before the forests were nationalized in 1957 (Gautam et al., 2004) (Figure 12.2.10).

12.2.7.1.3 Protection Forests

Eight forests covering a total area of 133,754.8 ha have been declared as protected forests since 2002 (MoFSC, 2014). These forests are important wildlife corridors and rich in biodiversity (Table 12.2.8). Eight other forests, covering a total area of 223,107 ha, are in the process of being declared as protected forests in the near future. Enhancing biodiversity through rehabilitation of habitats of rare and important species, biological corridors, and wetlands, and enhancing local livelihoods through implementation of income-generating activities are the main objectives of protected forest management. Promotion of alternative energy, such as solar energy, improved cooking stoves, and biobriquettes, has been initiated in some protected forest sites to reduce the pressure on these forests for fuelwood.[4,5]

---

4. Government of Nepal may declare any part of a national forest (as per the provisions of Forest Act 1993, aimed at managing national forest outside protected areas network) as protected forest that has special environmental, scientific, or cultural importance or of any other special importance. This type of forest falls outside the jurisdiction of protected area network.
5. Government can also (as per the provision of National Parks and Wildlife Conservation Act 1973) declare national parks, wildlife reserve, hunting reserve, conservation area, and buffer zone for the conservation of wild flora, fauna, natural landscape, and wildlife habitat of unique ecosystems and its sustainable management. These are commonly known as PAs. In essence, protected forest and protected areas are legally, as well as technically, different and are managed for different purposes.

TABLE 12.2.8 Protected forests in Nepal

Forest and Location	Year of Establishment	Size (ha)	Conservation Significance
Kankre Bihar (Surkhet)	2002	175.5	Historical, archeological, and biodiversity
Madhane (Gulmi)	2010	13,761	Biodiversity, ecotourism
Barandabhar (Chitwan)	2011	10,466	Biological corridor, wetland, habitat for several endangered species
Panchase (Kaski, Parbat, Syangja)	2011	5,775.7	Biodiversity, ecotourism, religious
Laljhadi-Mohana (Kailali, Kanchanpur)	2011	29,641.7	Biological corridor, wetland
Basanta (Kailali)	2011	69,001.2	Wildlife habitat and corridor
Khata (Bardia)	2011	4,503.7	Wildlife habitat and corridor
Dhanushadham (Dhanusha)	2012	430	Historical, religious, biodiversity

Source: MoFSC (2014).

12.2.7.1.4 Leasehold Forestry

Leasehold forestry is one of the pro-poor innovative approaches adopted by the Government in the forestry sector. Under this program, patches of government-owned forest are handed over to groups of people that are below the poverty line, on the basis of a 40-year lease agreement. They adopt agroforestry practices to grow tree and food crops on the leased land. So far 7,419 such groups have been formed and more than 42,335 ha of forestland has been handed over to them. Nearly 75,021 households are engaged in this program by the end of September, 2015 (DoF, Unpublished data).

12.2.7.1.5 Collaborative Forest

Apart from the forest categories according to the Forest Act (1993), collaborative forest management is also a practice that is an innovative approach adopted by the Government of Nepal to provide forest access to traditional forest users of the Tarai who are currently living far away from the forest. Under this management regime, selected forests of the Tarai are managed through collaborative efforts of the DoF, forest users, local government, and civil society. A multistakeholder governance mechanism has been established at the district level commonly known as the District Forest Sector Coordination Committee (DFSCC). Started in 2001 in selected forests of Nepal as a pilot scheme, this management regime now covers more than 57,000 ha in 10 districts of the country, benefitting 5 million of the population (DoF unpublished record).

12.2.7.1.6 REDD+ Program

The Government of Nepal has been promoting REDD+ since 2008 as a mechanism to control forest loss and degradation. The Readiness Preparation Proposal has identified five activities for payment under REDD+ schemes. These include (1) reducing deforestation, (2) reducing forest degradation, (3) sustainable forest management, (4) conservation of forest carbon, and (5) enhancement of forest carbon stock (MoFSC, 2010). Currently, the National REDD+ Strategy development process has been in progress and is expected to be completed by 2015.

12.2.7.1.7 Afforestation and Reforestation Programs

Regular afforestation and reforestation activities of the district forest offices and CFUGs are helping to restore degraded forests in Nepal. The MoFSC through district forest offices provide technical support as well as seedlings for afforestation and reforestation in degraded lands including forests. Since seedlings are provided free of cost, the CFUGs and the owners of the private land are encouraged for plantation. Below is the table showing seedling production and plantation area in last 5 years. It shows that ≈40% of the seedlings are planted on private land illumining the importance of private forestry in recent years (Table 12.2.9).

TABLE 12.2.9 Seedling production and plantation in Nepal

Year	Seedling Production (in '000)	Equivalent Plantation Area (Assuming 2,000 Seedlings per ha)	Plantation on Private Land
2009	8,000	4,000	1,600
2010	17,950	8,975	3,590
2011	12,352	6,176	2,470.4
2012	13,878	6,939	2,775.6
2013	10,625	5,312.5	2,125
Total	**62,805**	**31,402.5**	**12,561**

Source: Hamro Ban (2009–2013).

12.2.7.1.8 Agroforestry

The agroforestry system in Nepal is diversified and integrated with livestock, trees, and crops. The dependency of local people on forests has decreased considerably due to agroforestry as plantation of multipurpose trees is the main practice. The species are either deliberately planted or are managed to improve agriculture productivity. These trees provide fodder, fuelwood, and timber for household consumption and also for sell at local and regional markets. But, existing government policy undermines the value of protection and marketing of on-farm agroforestry species and products (Regmi, 2003) (Figure 12.2.11).

12.2.7.1.9 Protected Areas System

Protected areas have significantly contributed to forest conservation in Nepal. Currently, protected areas cover a total of 34,185 km^2 or 23.23% of the country's total area including 10 national parks, 3 wildlife reserves, 1 hunting reserve, 6 conservation areas, and 5602.67 km^2 of buffer zone areas (Figure 12.2.12, Tables 12.2.10 and 12.2.11). These protected areas are managed under four types of management modalities. The national parks, wildlife reserves, and the hunting reserve are exclusively managed by the Department of National Parks and Wildlife Conservation (DNPWC). Among the six conservation areas, two, Api Nampa Conservation Area and Blackbuck Conservation Area, are directly managed by the DNPWC. The Annapurna, Manaslu, and Gaurishankar Conservation Areas are managed by the National Trust for Nature Conservation under a policy of multiple use. Kangchenjunga Conservation Area has been managed by a local management council since 2006 with the support from DNPWC and WWF Nepal. Buffer zones (Table 12.2.11) for the protected areas are managed by local buffer zone councils.

FIGURE 12.2.11 Logging from private forest/agroforestry from eastern Nepal. *Photo: R.P. Chaudhary.*

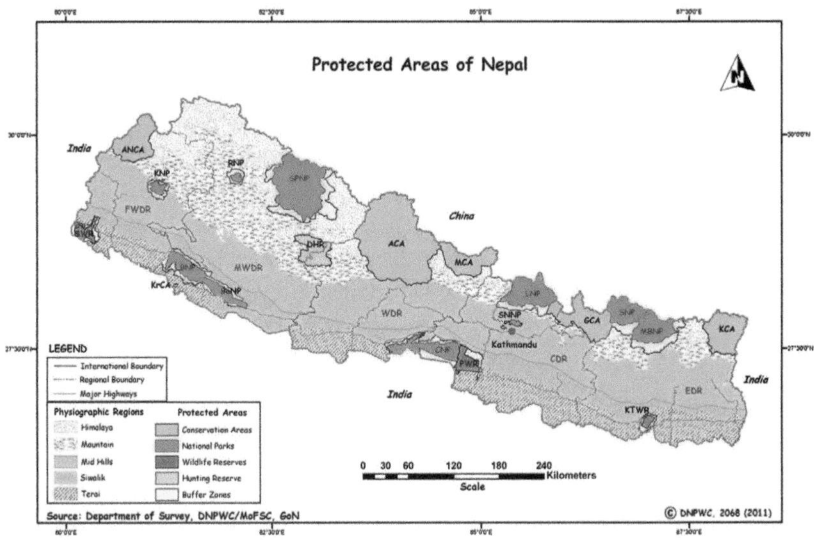

FIGURE 12.2.12 Protected areas of Nepal.

12.2.7.1.10 Extension and Outreach in Forestry Sector

Nepal has successfully demonstrated that people's participation in natural-resource management in general and forest management in particular can bring fundamental positive changes in restoring the resource base, reversing the rate of deforestation and forest degradation, access to resources, decision-making processes, and benefit sharing. Extension, sensitization, capacity building, and institutional-strengthening activities are part of regular program of the MoFSC. In other words, forestry extension and outreach activities are implemented on a regular basis throughout the country. This has helped not only in establishing a communication system with forest users living in rural areas of the country but also in developing a sense of ownership. Community members are actively engaged in forest protection, controlling illegal harvest activities, biodiversity conservation, containing forest fire, and engaging in sustainable management in close coordination with forest offices across the country. Extension and outreach activities have accommodated contemporary issues related to the forestry sector, such as climate change mitigation and adaptation, development of forest product-based enterprise, green economy, and payment for environmental services, to name a few. Some of the programs related to extension and outreach are conservation education, production of audio visual materials; poster; pamphlets; brochures; etc., street drama, training, workshops, seminars on forest management, governance, office management, as well as biodiversity conservation, soil conservation, and watershed management.

TABLE 12.2.10 National parks, wildlife and hunting reserves, and conservation areas of Nepal

Protected Areas	Area (km^2)/ Altitude (m)	Year of Notification	IUCN Management Category	Notable Biodiversity: Forest Type/ Vegetation	Notable Biodiversity: Fauna	Major Problems
National Parks						
1. Chitwan (included in World Heritage Site)	932/150–815	1973 1984	II X	Sal, sal-pine, riverine forest, grassland	Rhinoceros, tiger, leopard, wild dog, sloth bear, crocodile, gharial, king cobra, Bengal florican	Collection of firewood, grazing, crop raiding by wild animals, rhino and tiger poaching, environmental pressure from tourism, factory effluent pollution
2. Langtang	1,710/792–7,245	1976	II	Sal, *Schima–Castanopsis*, oak, blue pine, oak and hemlock, fir and *Rhododendron*, fir and birch, alpine meadows	Red panda, snow leopard, black bear, musk deer, tahr, goral, serow, rosefinch, ibisbill, scarlet finch, smoky warbler, rock lizard	Poaching for musk deer, crop raiding by wild boars, refuse and garbage, collection of medicinal plants, illegal transboundary timber trade
3. Sagarmatha (included in World Heritage Site)	1,148/2,945–8,848	1976 1979	II X	Blue pine, fir, juniper scrub, alpine meadows	Red panda, snow leopard, goral serow, Himalayan musk deer, Himalayan black bear, Indian muntjac, pheasant, robin accentor	Environmental pressure from tourism, waste disposal, tree felling, heavy grazing by yak and sheep
4. Rara	106/1,800–4,048	1976	II	Blue pine, fir, birch–*Rhododendron*, alpine meadow	Himalayan musk deer, black bear, leopard, red panda, coots, snow cock, impeyan pheasant	Grazing, felling of trees, collection of firewood and medicinal plants
5. Shey-Phoksundo	3,555/2,000–6,885	1984	II	Oak, blue pine, spruce, fir, birch, junipers, alpine meadows	Snow leopard, blue sheep, musk deer, red fox, red panda, Tibetan partridge, brown accentor, cheer pheasant, impeyan pheasant	Grazing, poaching for musk deer, hunting for blue sheep, collection of medicinal plants
6. Khaptad	225/1,000–3,276	1984	II	Montane sal, pine, alder, oak, conifer, mixed broadleaved, fir and oak, birch and *Rhododendron*, meadows	Leopard, musk deer, tahr, common langur, Himalayan black beer, pheasants, monal, red and blue magpie, flycatcher, thrushes, Himalayan griffin, finches, bulbuls	Grazing, crop depredation by wild boars, firewood collection, fires in the chir pine forest
7. Bardiya	968/152–1,441	1988[a]	II	Sal, sal-pine, khair-sissoo, riverine evergreen, grassland	Wild elephant, tiger, leopard, sloth bear, spotted deer, wild boar, sambhar, striped hyena, Bengal florican, Changeable hawk-eagle	Poaching, hunting, grazing, fishing using explosive and poison, hydropower plant construction

Continued

TABLE 12.2.10 National parks, wildlife and hunting reserves, and conservation areas of Nepal—cont'd

				Notable Biodiversity		
Protected Areas	Area (km^2)/ Altitude (m)	Year of Notification	IUCN Management Category	Forest Type/ Vegetation	Fauna	Major Problems
8. Makalu Barun	1,500/435–8,463	1991	I, II, VI	Sal, pine, *Schima–Castanopsis, Macaranga, Castanopsis,* oak-laurel, *Rhododendron,* oak, birch, fir, junipers, Berberis	Snow leopard, red panda, musk deer, weasel, Himalayan marten, marmot, woolly hare, thrush, tesia, monal, Darjeeling pied woodpecker	Excessive human encroachment, slash and burn agriculture, poaching for bears, collection of medicinal plants, illegal transboundary timber trade
9. Shivapuri-Nagarjun	159/1,366–2,732	CA 1984 NP in 2002	IV II	*Schima–Castanopsis*, pine, oak, *Rhododendron*	Common langur, leopard, yellow-throated marten, wild boar, yellow-bellied bush warbler, gray-sided laughing thrush	Collection of firewood and fodder, grazing, deforestation
10. Banke	550/360–480	2010	II	Sal, sal-pine, khair-sissoo, riverine evergreen, grassland	Antelope, wild elephant, rhesus macaque, langur, cobra, krait, python	Poaching, hunting, grazing
Wildlife Reserves						
1. Shuklaphanta	305/90–270	1976	IV	Sal, khair-sissoo, grassland, marsh	Swamp deer, wild elephant, tiger, hispid here, blue bull, leopard, hog deer, Bengal florican, swamp francolin, grass owl, cobra, python	Collection of wood, grazing, crop raiding by wild animals, illegal transboundary timber trade
2. Koshi Tappu	175/80–100	1976	IV	Khair-sissoo, tropical mixed deciduous riverine, grassland, wetland	Wild water buffalo, leopard, fishing cat, Gangetic dolphin, otter, deer, wild boar, swamp francolin, gharial, python	Grazing, genetic erosion of wild buffalo population, overfishing, high-tension electrical transmission, flooding situation
3. Parsa	499/150–815	1984	IV	Mixed deciduous riverine, mixed deciduous hardwood, sal, sal-pine, pine, acacia	Wild elephant, tiger, sambhar deer, leopard, leopard cat, rhesus macaque, langur, giant hornbill, cobra, king cobra, krait, python	Collection of wood, poaching, grazing, illegal transboundary timber trade

Hunting Reserve						
1. Dhorpatan	1,325/2,850–7,000	1987	VIII	Upper temperate mixed forest, fir, hemlock, spruce and birch, *Rhododendron*, juniper, grass/sedge	Blue sheep, goral, tahr, serow, musk deer, wolf, red panda, leopard, lynx, pheasants, Himalayan pied woodpecker, satyr tragopan	Overgrazing, grass burning, firewood cutting
Conservation Areas						
1. Annapurna	7,629/1,000–8,092	1986 1992	VI	Hill sal forest, *Schima–Castanopsis*, alder, oak, mixed broadleaved, *Rhododendron*, birch, blue pine, juniper	Langur, yellow-throated marten, jungle cat, Indian muntjac, goral, serow, tahr, bharal, snow leopard, red panda, patridge, pheasants	Environmental deterioration, cultural deterioration, tourism pressure, collection of wood, hunting, waste disposal
2. Kangchenjunga	2,035/1,200–8,598	1997	VI	Larch, juniper, oak, *Magnolia*, fir, and hemlock	Snow leopard, red panda, musk deer, blue sheep	Grazing, poaching for musk deer, hunting, collection of medicinal plants, illegal transboundary trade
3. Manaslu	1,663/1,360–8,163	1998	VI	Oak, blue pine, larch, birch,	Snow leopard, musk deer, blue sheep, red panda, Himalayan tahr	Grazing, poaching for musk deer, hunting, collection of medicinal plants
4. Gaurishankar	2,179/1,000–7,200	2010	VI	Riverine, *Schima–Castanopsis*, pine, alder, oak, temperate mixed–broadleaved, *Rhododendron*, birch	Snow leopard, musk deer, blue sheep, red panda, Himalayan tahr, ibisbill	Grazing, poaching for musk deer, hunting, collection of medicinal plants
5. Api Nampa	1,903/539–7,132	2010	VI	Hill sal, chir pine, alder, oak, mixed broadleaved, blue pine, junipers, birch, alpine scrub	Snow leopard, musk deer, clouded leopard, ghoral, Himalayan black bear, Himalayan tahr, cheer pheasant, Himalayan monal, satyr tragopan	Grazing, poaching for musk deer, hunting, collection of medicinal plants, illegal transboundary timber trade
6. Khairapur	16/120–230	2009	VI	Sal, sal-pine, khair-sissoo, riverine evergreen, grassland	Blackbuck, spotted deer, wild boar, sambhar, striped hyaena, Bengal florican, changeable hawk-eagle	Collection of fuelwood, poaching of blackbuck

[a]*Initially established as Royal Karnali Wildlife Reserve in 1976 but later expanded and gazetted as a national park in 1988.*
Source: Modified after Chaudhary (1998) and MoFSC (2002).

TABLE 12.2.11 List of protected areas' buffer zones

Buffer Zone	Area	Year of Establishment
1. Chitwan NP	750.00	1996
2. Bardiya NP	327.00	1996
3. Langtang NP	420.00	1998
4. Shey-Phoksundo NP	1,349.00	1998
5. Makalu Barun NP	830.00	1999
6. Sagarmatha NP	275.00	2000
7. Shuklaphanta WR	243.50	2004
8. Koshi Tappu WR	173.00	2004
9. Parsa WR	298.17	2005
10. Rara NP	198.00	2006
11. Khaptad NP	216.00	2006
12. Banke NP	343.00	2010

12.2.7.1.11 Public Awareness Programs

Extension and outreach programs have contributed to public awareness toward the importance of forests. Involvement of governmental and nongovernmental organizations in forums such as mass media (radio, television), workshops, seminars, publications, and exhibitions has contributed to create awareness among the people. Furthermore, the community forests have played an important role in increasing awareness toward the importance of forests to the local people. In the areas where the community forestry programs have been implemented, the level of awareness can be considered high; people are motivated for conservation of the forests since they are getting many ecosystem services from their forests.

12.2.8 CONCLUSIONS AND RECOMMENDATIONS

Deforestation is a serious environmental problem in Nepal. Deforestation pressures occur throughout Nepal but are most strongly felt in many parts of Tarai and Churia. It has many driving factors ranging from socioeconomic to governance. Past efforts to control deforestation and degradation have largely failed because many drivers of deforestation operate outside the forest sector, such as the transboundary economy across the open borders, political unrest, and the lack of incentives to the marginal poor who have depended on local resources for their livelihoods. Though many of these causes still persist, in recent years, several successful efforts have occurred to reverse the trend with some remarkable progress, notably the one achieved by the community-based forest management approach. The decline in forestry resources in Nepal took place in the past due to the lack of appropriate policy to guide the legal, institutional, and operational development for the forestry sector. Forestry policy in Nepal in the past has been shaped by political and economic motives, rather than by ecological considerations. Therefore, research and development on the forestry sector should focus on basic issues of forest/biodiversity conservation, green economy, and climate change adaptation and mitigation at the local level so that more realistic measures and policy programs can be devised.

The recent declaration of the Tarai—Madesh—Churia conservation area would have a positive contribution in controlling deforestation in these very vulnerable physiographic regions of Nepal. The community-based forest management program would require more support and active participation of all the stakeholders and the dispute over forests should be resolved in a participatory manner. Extensive plantation of trees should be undertaken wherever there are vacant public lands such as village wastelands, in and around farms, rural and urban parks, temple and school premises, and along roads and rivers. Alternative sources of energy should be provided on a subsidized basis throughout the country to reduce dependency in fuelwood. The contribution of private forestry in mitigating deforestation should be recognized and promoted. The agroforestry practice should be supported by a conducive policy environment. The private sector should be encouraged to participate in the use and management of forests as far as possible. The use of wood in cremating human bodies can be significantly reduced by using an electric crematorium throughout the urban centers. Some religious groups are against the use of the electric crematorium but that can be overcome by awareness programs. Public awareness about the importance of forests and the consequences of deforestation is very important. A need exists need to establish long-term forest monitoring program to assess the impact of earthquake in forest degradation. Landscape conservation and development initiatives taken by the Government of Nepal, in collaboration with different partners, would require both national as well as transboundary commitment, coordination, cooperation, and collaboration. Moreover, alternative livelihood options should be provided to the people who are engaged in illegal smuggling of timber both in the Tarai and high-altitude regions.

REFERENCES

Acharya, K.P., Dangi, R.B., 2009. Case studies on measuring and assessing forest degradation. In: Forest Resources Assessment Working Paper 163, FAO, Rome, Italy.

Allen, J.C., Barnes, D.F., 1985. The causes of deforestation in developing countries. Ann. Assoc. Am. Geogr. 75 (2), 163—184.

Belbase, N., Thapa, L.B., 2007. Environmental justice and rural communities, Nepal. In: Moore, P., Pastakia, F. (Eds.), Environmental Justice and Rural Communities, Studies from India and Nepal. IUCN Bangkok, Thailand and Gland, Switzerland.

Bhatt, D.D., 1977. Natural History and Economic Botany of Nepal. Orient Longman Limited, New Delhi.

CBS, 1998. A Compendium on Environmental Statistics 1998 Nepal. Central Bureau of Statistics, His Majesty's Government of Nepal, Kathmandu.

CBS, 2011. Environment Statistics of Nepal. Central Bureau of Statistics, Government of Nepal, Kathmandu.

CBS, 2012. National Population and Housing Census 2011 (National Report). Central Bureau of Statistics, Government of Nepal, Kathmandu.

Chaudhary, R.P., 1998. Biodiversity in Nepal: Status and Conservation. S. Devi, Saharanpur (U.P.). India and Teckpress Books, Bangkok, Thailand.

Chaudhary, R.P., 2000. Forest conservation and environmental management in Nepal: a review. Biodivers. Conserv. 9, 1235–1260.

COMFORTC, 2012. Study on Drivers of Deforestation and Degradation of Forests in High Mountain Regions of Nepal. Community Forestry Research and Training Centre (COMFORTC), Kathmandu.

DFRS, 1999a. Forest and Shrub Cover of Nepal 1994 (1989–1996). Department of Forest Research and Survey, Kathmandu, Nepal.

DFRS, 1999b. Forest resources of Nepal (1987–1998). Department of Forest Research and Survey, Publication No 74, Kathmandu, Nepal.

FAO, 2001. Global Forest Resources Assessment 2000. Food and Agriculture Organization of the United Nations, Rome, Italy.

FAO, 2006. Global Forest Resources Assessment 2005. Food and Agriculture Organization of the United Nations, Rome, Italy.

FAO, 2009. Nepal Forestry Outlook Study. FAO Working Paper No. APFSOS II/WP/2009/05. Food and Agriculture Organization of the United Nations, Thailand.

FRA/DFRS, 2014a. Terai Forests of Nepal (2010–2012). Forest Resource Assessment Nepal Project/Department of Forest Research and Survey, Babarmahal, Kathmandu.

FRA/DFRS, 2014b. Churia Forests of Nepal (2011–2013). Forest Resource Assessment Nepal Project/Department of Forest Research and Survey, Babarmahal, Kathmandu.

Gautam, A.P., Shivakoti, G.P., Webb, E.L., 2004. A review of forest policies, institutions, and changes in the resource condition in Nepal. Int. For. Rev. 6 (2), 136–148.

Gurung, H., 1989. Regional Patterns of Migration in Nepal. Papers of the East-West Population Institute, No. 113. E-W Institute, Hawaii.

Hagen, T., 1961. Nepal: The Kingdom in the Himalayas. Geographical Publisher, Berne.

Hamro Ban, 2009–2013. Annual Progress Reports of Department of Forest. Government of Nepal, Ministry of Forests and Soil Conservation, Department of Forest, Kathmandu.

HMGN/ADB/FINIDA, 1988. Master Plan for Forestry Sector Nepal (Main Report). His Majesty's Government of Nepal, Ministry of Forests and Soil Conservation, Kathmandu.

Ives, J.D., 2006. Himalayan Perceptions: Environmental Change and the Well-being of Mountain Peoples. Himalayan Association for the Advancement of Science, Lalitpur, Nepal.

Joshi, A.L., 1993. Effects on administration of changed forest policies in Nepal. In: Policy and Legislation in Community Forestry. Proceedings of a Workshop Held in Bangkok, January 27–29. Regional Community Forestry Training Centre, Bangkok.

Kleinn, C., 1994. Forest resources Inventories in Nepal: Status Quo, Needs, Recommendations. Forest Resource Information System Project Paper No. 1, Kathmandu, Nepal.

Måren, I.E., Bhattarai, K.R., Chaudhary, R.P., 2014. Forest ecosystem services and biodiversity in contrasting Himalayan forest management systems. Environ. Conserv. 41 (01), 73–83.

MEA, 2005. Ecosystems and Human Well-being: Current State and Trends, vol. 1. Millennium Ecosystem Assessment. Island Press, Washington, DC.

Metz, J.J., 1991. A reassessment of the causes and severity of Nepal's environmental crisis. World Dev. 19 (7), 805–820.

Miller Jr., G.T., 2004. Living in the Environment – Principles, Connections, and Solutions. Thomson Learning Inc., CA, USA.

MoFSC, 2002. Nepal Biodiversity Strategy. His Majesty's Government of Nepal, Ministry of Forests and Soil Conservation, Kathmandu.

MoFSC, 2008. Churia Area Programme Strategy. Government of Nepal, Ministry of Forests and Soil Conservation, Kathmandu.

MoFSC, 2009. Nepal Fourth National Report to the Convention on Biological Diversity. Government of Nepal, Ministry of Forests and Soil Conservation, Kathmandu.

MoFSC, 2010. Nepal's Readiness Preparation Proposal: REDD 2010–2013. Government of Nepal, Ministry of Forests and Soil Conservation, Kathmandu.

MoFSC, 2013. Country Report on the State of Forest Genetic Resources, Nepal. Government of Nepal, Ministry of Forests and Soil Conservation, Kathmandu.

MoFSC, 2014. National Biodiversity Strategy and Action Plan. Government of Nepal, Ministry of Forests and Soil Conservation, Kathmandu.

NPC, 2015a. Nepal Earthquake 2015 Post Disaster Need Assessment. In: Key Findings, vol. A. National Planning Commission, Government of Nepal, Singha Durbar, Kathmandu.

NPC, 2015b. Nepal Earthquake 2015 Post Disaster Need Assessment. In: Sector Reports, vol. B. National Planning Commission, Government of Nepal, Singha Durbar, Kathmandu.

Pant, K.P., Rasul, G., Chettri, N., Rai, K.R., Sharma, E., 2012. Value of Forest Ecosystem Services: A Quantitative Estimation from the Kangchenjunga Landscape in Eastern Nepal. ICIMOD Working Paper 2012/5. ICIMOD, Kathmandu.

Raut, Y., 1997/98. A Handbook of Animal Husbandry (Part 1: Pasture Production). Ministry of Agriculture, Department of Livestock Services, Kathmandu.

Regmi, R.R., 1994. Deforestation and Rural Society in the Nepalese Terai. In: Occasional Papers in Sociology and Anthropology, vol. 4, 72–89.

Regmi, B.N., 2003. Contribution of agroforestry for rural livelihoods: a case of Dhading district, Nepal. In: Paper Presented at the International Conference on Rural Livelihoods, Forests and Biodiversity 19–23 May 2003, Bonn, Germany.

Robbe, E., 1954. Report to the Government of Nepal on Forestry. ETAP Report No. 209. FAO, Italy.

SEAMCAP, 2000. State of Environment Dataset Report: Strengthening Environment Assessment and Monitoring Capabilities in Nepal. UNEP and ICIMOD, Kathmandu.

Shrestha, V.P., 1999. Forest resources of Nepal: destruction and environmental implications. Contrib. Nepal. Stud. 26 (2), 295–307.

Soussan, J., Shrestha, B.K., Uprety, L.P., 1995. The Social Dynamics of Deforestation: A Case Study from Nepal. The Parthenon Publishing Group, London and New York.

Stainton, J.D.A., 1972. Forests of Nepal. John Murray, London.

Stevens, S., 2003. Tourism and deforestation in the Mt Everest region of Nepal. Geogr. J. 169 (3), 255–277.

Tamrakar, S.M., Nelson, D., 1991. Potential Community Forest Land in Nepal. Part 2, Field Documentation No. 16, NEP/85/017. Government of Nepal, Ministry of Forests and Soil Conservation, Kathmandu.

UNEP, 2001. Nepal: State of the Environment 2001. United Nations Environment Programme, Thailand.

Wallace, M., 1983. Managing resources that are common property: from Kathmandu to Capitol Hill. J. Pub. Anal. Manag. 2 (2), 220–237.

WECS, 2010. Energy Sector Synopsis Report Nepal-2010. Water and Energy Commission Secretariat, Kathmandu.

WECS, 2013. Energy Sector Synopsis Report Nepal-2013. Water and Energy Commission Secretariat, Kathmandu.

Chapter 12.3

Deforestation in the Brazilian Amazon

Tim Boekhout van Solinge
Utrecht University, Utrecht, Netherlands

ABSTRACT
This essay takes a (green) criminological and multidisciplinary perspective on deforestation in the Brazilian Amazon, by focusing on the crimes and damages that are associated with Amazonian deforestation. The analysis and results are partly based on longer ethnographic stays in North Brazil (Amazon region). If focuses on the human victimization of deforestation such as violence against forest inhabitants, which is usually committed by large landholders (e.g., cattle and soy farmers, timber traders) or their henchmen. Ultimately, deforestation also leads to the disappearance of communities and traditional lifestyles. This essay takes a more sociological and political science perspective on the question of why (illegal) Amazonia is accompanied by so much deforestation-related crime and violence, and on the question as to how and why Amazonian deforestation has arrived on the political agenda. The prolonged drought of 2014 and 2015 in populous southern Brazil seems to change the Brazilian debate and discourse with regard to deforestation and development.

At first I thought I was fighting to save rubber trees, then I thought I was fighting to save the Amazon rainforest. Now I realise I am fighting for humanity.
Chico Mendes

12.3.1 INTRODUCTION

On December 22, 1988, rubber tapper and defender of the Amazon, Chico Mendes, was murdered deep in the Amazon near the Brazilian–Bolivian border. Global news headlines made Chico Mendes an internationally recognized icon and martyr for the protection of the Amazon rain forest—also called Amazonia—which is by far the planet's largest tropical rain forest that covers parts of nine countries.

Biological and Environmental Hazards, Risks, and Disasters. http://dx.doi.org/10.1016/B978-0-12-394847-2.00021-8

Chico Mendes' murder, committed by the son of a rancher, led to a wave of international concern and protests about Amazonian deforestation. Deforestation in Brazil—where two-thirds of Amazonia lie—received much attention from scientists, journalists, NGOs, policy-makers, writers, celebrities, and the general public (Shoumatoff, 1990). A few years after the unfortunate incident, the 1992 Earth Summit in Rio de Janeiro took place with over 100 heads of state or government present. One of the three major outcomes was the Statement of Forest Principles, aimed at attaining sustainable forest management worldwide.

Despite Amazonian deforestation being the focus of widespread attention since the late 1980s, the process has not stopped. The speed of Amazonian deforestation has since decreased considerably, but deforestation through logging, burning, and land conversion nonetheless continues. One human generation after Mendes' murder, Amazonian rain forest still disappears with an average speed of more than one football pitch per minute.[1] A substantial part—according to different estimates for the timber, cattle, and soy between 60% and 90%—of this deforestation in the Brazilian Amazon has been illegal; it thus concerns crime. However, Brazil's new Forest Code of 2012 granted amnesty to some illegal deforestation prior to 2008, which makes it difficult to determine how much deforestation can now be considered illegal. (Boekhout van Solinge, 2010a,b, Lawson, 2014, 24−36).

The violence against people trying to protect the Amazon rain forest has continued during the 27 years that have passed since Mendes' murder. As Amazonia is home to millions of people, deforestation meets resistance, first and foremost from the people living in or near the forest. Chico Mendes still is the most recognized victim, but many more people have been killed for their efforts to protect and preserve the Amazon rain forest. The report *Deadly Environment* by Global Witness (2014) showed that worldwide, 900 environmental protector people were killed during the last decade; 448 of them occurred in Brazil and many in the Brazilian Amazon, with two Brazilian states (Pará and Mato Grosso) standing out as being particularly violent. Hence, much Amazonian deforestation is illegal and strongly associated with violent crimes.

This essay gives an overview of deforestation in the Brazilian Amazon. What are the trends, geographical patterns, and driving forces behind Amazonian deforestation? Why is this deforestation so strongly associated with violence and

1. One football field of Amazon rain forest disappearing per minute means that 60 disappear every hour and 1440 every day (of 24 h). On a yearly basis this comes down to 525,600 football fields. Considering that a standard football pitch of many professional teams measures 7140 m^2 (105 by 68 m) 525,600 pitches equals 3753 km^2. Over the last few years, deforestation in the Brazil Amazon was larger: respectively 4571 km^2 in 2012 (1.2 pitch per minute), 5891 km^2 in 2013 (1.6 pitch per minute), and 4.848 km^2 in 2014 (1.3 pitch per minute). In previous decades, however, the destruction of Amazonian rain forest occurred several times faster (see further). It should be noted that some football pitches are smaller. Nobre (2014) 23, for example, uses a smaller pitch of 4136 m^2 as a basis for calculating Amazonian deforestation in number of football fields, which logically leads to a higher number of "deforested (football) fields."

who are the (main) victims and perpetrators? Also addressed is the perception or framing of the Amazonian deforestation question. What is being done—or not—to limit Amazonian deforestation and its associated crimes and damages, and which arguments are used? In this essay, the Amazonian deforestation of the late 1980s, when it came on the political agenda, will be juxtaposed with today's situation. What has changed and what has remained unchanged?

12.3.2 THEORETICAL PERSPECTIVES AND METHODS

Deforestation can be studied from several academic perspectives. An uncommon perspective is from criminology, although one could argue that this is logical, considering that most tropical deforestation is not only illegal, but is also related to other deforestation-related crimes such as violence and corruption (Boekhout van Solinge, 2014a). Still, one can count on one hand the number of criminologists studying illegal deforestation.

Criminology can be simply defined as the study of crime, but a more common approach among practitioners—mostly social scientists—is that it looks at crime as a social phenomena. According to a much used definition by Edwin Sutherland (1883–1950), it includes the process of making law, of breaking laws, and the social reaction toward the breaking of law (Sutherland et al., 1992, 3).

Criminology's young and small branch, green criminology, which has existed as such for a decade or so, has expanded the study area in two ways. First, green criminology distances itself from the traditional anthropocentric demarcation of the domain of conventional criminology. Whereas most criminologists are preoccupied with acts against humans, green criminologists have included acts of harm and violence against animals and ecosystems as well. A second way in which green criminology has expanded the terrain of criminology is by not only considering acts that are criminalized—as conventional criminology does—but also by taking harm as a point of departure (Beirne and South, 2007). Although (expected or perceived) harm is generally the basis for criminalizing certain behavior, no direct relationship exists between the extent of harm and the extent to which something is criminalized (in terms of punishment or other sanctions). Also morals, power, and the level of knowledge or ignorance about certain activities influence whether they are considered harmful and consequently declared illegal. Put differently, there is not always a congruent relationship between the extent of harm and the extent of criminalization. Some acts that are (severely) criminalized[2] are in reality not so harmful, whereas some harmful acts are not criminalized—or they are criminalized but not enforced.

2. Here one could think, for example, of certain psychoactive substances. Several medical studies (published, for example, in *The Lancet* or *British Medical Journal*) have shown that there is no direct or logical relationship between the criminalization of certain illicit drugs and the harm or danger they pose to consumers. The way societies and the "international community" (such as the UN) deals with psychoactive substances is the author's former area of specialization (see for example Boekhout van Solinge (2004)).

Tropical deforestation is a good example of something that is clearly harmful, but which is not correspondingly criminalized and enforced. The different human and nonhuman harms that are related to tropical deforestation, such as in the planet's largest rain forest, Amazonia, are the subject of discussion in this essay. These damages and risks, including crimes, are addressed by using data from various disciplines.

Different research methods were used for this essay. Scientific literature was consulted, as well as reports from NGOs and news sources. Some Web sites were also used to gather information about forest cover in the Amazon, in particular, the recent and valuable Web site of Global Forest Watch.[3]

Anthropological methods were also used, as the author lived halftime in the Brazilian Amazon between 2009 and 2012; first in Amazonas state, later in Pará state (Boekhout van Solinge, 2014b).[4] Amazonas and Pará are Brazil's largest states—each measuring more than twice continental (metropolitan) France.[5] They are located in northern Brazil, around the Amazon River. Pará is located in the lower Amazon, where the Amazon River flows into the Atlantic Ocean, and Amazonas is located upstream, west of Pará. The difference in forest cover between these two states is enormous. Although Amazonas is still mostly forested, Pará is known for its high deforestation rates. The author's stays in the two states, and traveling back and forth, mostly by boat and sometimes by plane,[6] helped to understand the dynamics of Amazonian deforestation and the forest's vulnerability to deforestation.[7]

One explanation for the contrasting forest cover between Amazonas and Pará is economical. Amazonas' economy is more industrial, with the capital Manaus being Brazil's industrial center in the north; an urban and industrial enclave in Amazonia, mostly accessed by rivers. The economy of downstream Pará is dominated by natural resource exploitation: forestry, mining, and agriculture. These sectors directly imply deforestation.

---

3. See www.globalforestwatch.org
4. These stays were partly facilitated by a project that he coordinates: Lands and Rights in Troubled Water (LAR), funded by the Dutch Organisation of Scientific Research (NWO). See the blog of this project, landsandrights.blog.com. This project is part of an NWO programme on Conflicts and Cooperation over Natural Resources (CoCooN), (see Bavinck et al., 2014).
5. The size of continental or metropolitan France is 551,695 km^2. The size of Pará state (capital Belém) is 1,248,000 km^2, which is more than twice France. The size of Amazonas (capital Manaus) is 1,559,000 km^2, which is almost three times France's size.
6. The most common public transport in the Amazon is by riverboat where the travelers sit or lie in their hammock. From the river, one can notice that Amazonas state has more trees along the riverbanks than Pará state. An aerial view, such as from the airplane from Manaus to Belém, makes the difference in forest cover more obvious. Web sites such as from Global Forest Watch also show the difference.
7. The author lived for some time in the small town of Presidente Figueiredo (Amazonas state) that is surrounded by rain forest. Some of the forest there grows on sand or solid rock. When trees grow on rocks, the tree roots form a horizontal patchwork, or tree root tapestry. If such an area is deforested, it will take many human generations before a forest can come back, if it comes back.

Besides this economic explanation, a criminological or criminal justice explanation is also possible. Whereas the rule of law in Amazonas state is relatively adequate, Pará state is known as the "conflict state" where the rule of law does not function properly (Greenpeace International, 2003).[8] In Pará, a powerful agricultural lobby is present and large landholders traditionally have had much power in the rural and forested areas, which may also mean that in some parts they do, in fact, monopolize the violence.

As Max Weber formulated a century ago, the monopoly on the use of violence is a key characteristic of a (functioning) state. When this monopoly is not in the hand of the state, this undermines the state's power. The factual power relations in rural and forested Pará are indeed such—more than in Amazonas—that large landholders have much power. State institutions and state actors are not very common, which provides opportunities for other actors, such as large landholders and their henchmen. This also explains why (illegal) deforestation is much more prevalent in Pará than in Amazonas.

The author's stays in the Brazilian Amazon and work with NGOs influenced the perspective that is taken in this essay.[9] When deforestation is discussed in the western world, the—threatened—nonhuman inhabitants of Amazonia are often emphasized: the flora and especially the fauna. International NGOs commonly emphasize the nonhuman inhabitants of tropical rain forests such as Amazonia. Their public campaigns aimed at raising awareness and money often emphasize the natural beauty. The media have played their part too, familiarizing the public fairly well with Amazonia and its unparalleled biodiversity (Goulding et al. 2003:229).

Staff members of local NGOs in the Brazilian Amazon regularly express being bothered by the fact that international NGOs tend to emphasize the nonhuman victimization of deforestation (Boekhout van Solinge, 2014b). By contrast, local NGOs usually emphasize the human victimization of (illegal) deforestation and the deforestation-related violence.

---

8. This difference in the rule of law is something that is widely known among locals, especially among those who are familiar with both states (Pará and Amazonas), such as people from Pará who have moved to Amazonas, for example, to find work in Manaus. Traveling by riverboat between the two states, when one spends one, two, or three days and nights in a hammock, the difference in rule of law between the two states was a subject that the author discussed several times with fellow travelers. Also at other occasions this issue was discussed. In 2012, when the author lived in Pará and visited Amazonas State University in Manaus, he had a conversation with three guards at the reception, one of whom was from Pará. This guard spontaneously started telling his two colleagues from Amazonas about the lawlessness in Pará. "In Pará, there is no law," he told them. Although his statement was clear, it is more accurate to say that the problem in Pará is the lack of law enforcement.
9. A requirement of the before-mentioned NWO-funded project that the author coordinates is that academic researchers not only work with other academics, but also with NGOs. In the case of Brazil, academics work with two NGOs in the Brazilian Amazon: the Pastoral Land Commission (CPT) and *Terra de Direito*, an NGO with mostly lawyers specializing in land right issues.

Living in Amazonia—following the news, reading, meeting people, and visiting communities—makes one realize that millions of people are living there, in the cities and towns and along rivers and lakes. In line with the dominant, local perspective, the emphasis in this essay is put on the human victimization.

12.3.3 THE CONTEXT OF CHICO MENDES' MURDER

As explained at the start of this essay, the international uproar about Amazonian deforestation started after the murder of Chico Mendes in 1988. Who was he and why was he killed?

Chico Mendes was a third-generation rubber tapper. His grandfather had migrated in the early twentieth century from Brazil's northeast to work in the rubber industry (Shoumatoff, 1990). Since the mid-nineteenth century, rubber extraction had been the mainstay of the Amazonian economy. Until the beginning of the twentieth century, the Brazilian Amazon supplied almost all of the world's rubber—in increasing demand since the industrial age—and Amazonian latex made up 40% of Brazil's total exports (Grandin, 2010, 26). This rubber boom attracted national and international migrants; many poor from Brazil's northeast—still the country's poorest region—and merchants from other continents, such as Arab and Jewish merchants from North Africa and the Middle East (Grandin, 2010).[10] Chico Mendes' grandfather was among the poor national migrants who found work in Acre state, in the southwestern Amazon near Bolivia.

As a child during the 1950s, Chico Mendes (1944–1988) started tapping rubber. As Mendes (1989) explained in a book published after his death (based on interviews), ranchers had arrived in Acre during the 1960s and 1970s, turning rain forest into pastureland. This brought them in almost immediate conflict with the already present and often-impoverished rubber tappers, for whom the forest is the main source of livelihood. This land conversion, transforming Amazon rain forest into pastureland, substantially increased much during the 1980s and the conflicts of interest between rubber tappers and ranchers grew sharply.

These conflicts also became more violent, especially from the side of the ranchers. Shoumatoff (1990) described that there were at the time 130 ranchers in Acre. In a period of 20 years they had, supported by their henchmen, *pistoleiros,* driven tens of thousands of rubber tappers out of the forest. Shoumatoff further mentioned that many dozens of union leaders had been killed and that, during the 1980s, over one thousand people had been killed in the struggle over land use in the Brazilian Amazon, but this had only led to two

10. The rubber boom also attracted people attempting to take seeds from the precious rubber tree, in order to commercialize them elsewhere. It led to the historical case of eco-piracy when in 1876 the Briton, Henry Wickham, smuggled some 70,000 rubber seeds out of the Brazilian Amazon to England. They were later shipped to British tropical Asia to help start rubber plantations. Eventually this would lead to the demise of the Amazonian economy until far into the twentieth century (see Jackson, 2008).

condemnations. Also, local policemen had either been paid off or they were afraid, as most murders did not lead to court cases.

The charismatic and articulate rubber tapper Chico Mendes started organizing rubber tappers in unions. Mendes also increasingly gave speeches, in and outside Brazil, promoting preservation of the Amazon rain forest. In 1987, he went to the United States, where he was invited for the annual meeting of the Inter-American Development Bank. He also met members of the U.S. Senate in Washington. In 1987, he was awarded international prizes from the UN's Environmental Programme and from the Better World Society.

The ranchers increasingly considered him as a troublemaker and threat, which made him a target. As Shoumatoff (1990) described in detail, Mendes had been in danger since the early 1980s, but in 1988, when he received the *anunciado,* the announcing death threat, he knew his last days were counted. He also knew which rancher family was going to kill him, as they had openly announced it—a sign of the climate of impunity that was and still is prevalent in many places in the Brazilian Amazon.

12.3.4 AMAZONIAN DEFORESTATION ON THE INTERNATIONAL AGENDA

An interesting question is why the murder of a rubber tapper, committed in a remote, southwestern corner of the Amazon rain forest, far from any city, made international headlines. Why did this murder lead to so much international uproar?

One reason is that there already was a certain level of international concern about deforestation in the Amazon since the 1970s, related to the construction of Trans-Amazon Highway of the early 1970s. *Time* and *Newsweek* articles pointed at the possible negative effect of this highway on the Amazon's flora and fauna, which were still largely unknown to the scientific community (Shoumatoff, 1990).

Second, deforestation in the Amazon was increasing rapidly. Fearnside (1982) noted that deforestation in the Brazilian Amazon grew exponentially in the late 1970s and early 1980s. By the late 1980s, annual forest loss in the Amazon increased to several tens of thousands of square kilometers. This forest loss was mainly caused through burning. Usually it is only at the end of the dry season, in September, that the rain forest is dry enough to successfully burn it.

In 1987, a considerably dry year, at least 80,000 km^2 of Amazon rain forest were destroyed (Stoddard, 1992, 527), an area almost twice the Netherlands or Switzerland. Some sources even put the forest loss at 200,000 km^2 (Shoumatoff, 1990; Setzer and Pereira, 1991, 19). Satellites detected over 350,000 fires during the dry season, from June to October (Simons, 1988). The consequent giant smoke and haze clouds extended over millions of square kilometers. The fires produced so much different emissions—of carbon monoxide, methane, ozone, nitrogen oxides, and others—that it was comparable to the outburst of a very large volcano. These emissions caused severe atmospheric pollution effects, with possible global implications.[11] Scientists and journalists consequently reported about it in a worrisome manner (Simons, 1988; Setzer and Pereira, 1991, 19).

The year 1988 was another dry one in the Amazon, and large areas of Amazonia went up in smoke again. Smoke clouds rose to 12,000 feet (3.6 km), and as gases and particles, including methane and nitrogen oxides, were lifted up into the jet streams and blown south toward Antarctica, scientists believed that this could affect the stratosphere by directly or indirectly depleting ozone (Simons, 1988). That Amazonian deforestation could negatively affect the ozone layer made Amazonian deforestation a pressing global issue.[12]

Third, next to these possible stratospheric consequences and risks, what was really gaining attention in 1988 were the potential atmospheric consequences of Amazonian deforestation, particularly, global warming. Scientists' warnings started appearing in the media: the releasing into the air of large quantities of carbon, such as from the rain forest's biomass, might be warming up the atmosphere (London and Kelly, 2007, 40).

An important reason as to why this came on the agenda was that 1988 was globally the hottest year on record. Not only the Amazon was hot and dry; North America experienced a heat wave, the so-called North American Drought of 1988. Thus, while the Amazon rain forest in South America suffered from fires, North America experienced a heat wave. The two were easily combined and causally related: the South American Amazon was burning and North Americans were feeling the heat (see London and Kelly, 2007, 40).

Amidst of all these global concerns, Chico Mendes, a rain forest protector from the interior of the Amazon rain forest was killed, which gave the Amazon deforestation disaster a human face. "For the first time, an issue emerged from the jungle with a human element. Not only were trees dying, but people were dying, too. (…) His death reminded the world that there were millions of people living in the Amazon (London and Kelly, 2007, 44)."

12.3.5 DEFORESTATION DRIVERS AND MECHANISMS

Large-scale deforestation in the Amazon is relatively recent. In 1970, only 2% of the Amazon had been deforested (Loureiro, 2011, 102). Deforestation on a larger scale started in the 1970s, first driven and facilitated by the construction of two roads that would cut through the rain forest.

---

11. In a later article, Setzer and Pereira (1991: 19) made the following estimates of the emissions of the 1987 fires in the Brazilian Amazon, in millions of tonnes: 1700 for CO_2, 94 for CO, 6 for particulates, 9 for ozone (secondary reactions), 10 for CH_4, 1 for NO_x, and 0.1 for CH_3Cl.
12. At the time, the depleting ozone layer was high on the political agenda. In 1985, the Antarctic ozone hole had been discovered and in 1987 an agreement was reached over an international treaty, the Montreal Protocol, which took effect in 1989.

PICTURE 12.3.1 Fires in the Brazilian Amazon (Mato Grosso state, August 12, 2007) http://earthobservatory.nasa.gov/NaturalHazards/view.php?id=18862&eocn=image&eoci=related_image Text: Deforestation and fires in Mato Grosso state are marked in red (August 12, 2007).

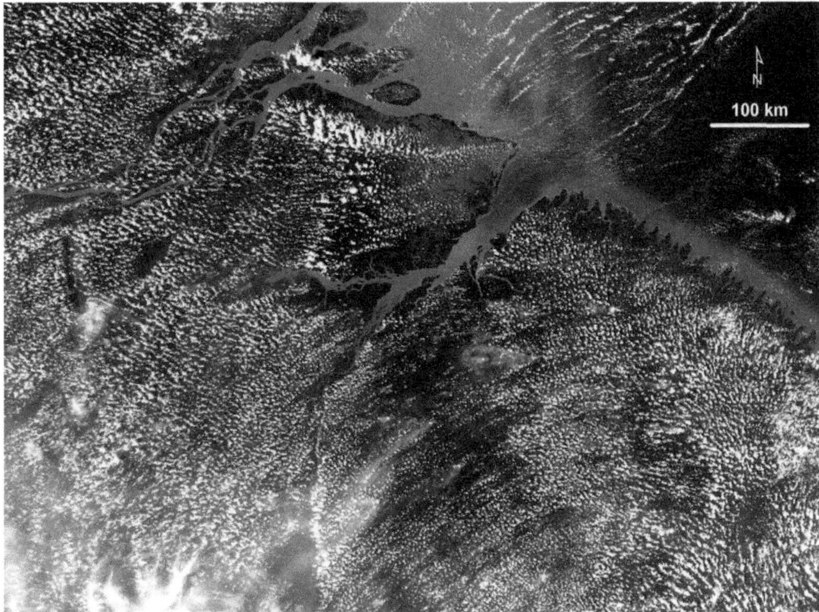

PICTURE 12.3.2 Fires at the Amazon River mouth (Pará state, November 22, 2004) http://earthobservatory.nasa.gov/NaturalHazards/view.php?id=14225&eocn=image&eoci=related_image Text: Numerous fires burning (in red) near the Amazon River mouth in Brazil's Pará state.

In 1970, the Brazilian military government made a commitment to integrate the relatively "disconnected" Amazon basin with the rest of the economy (Goulding et al., 2003, 56). Roads were a key first step. Military and geopolitical reasons also existed for them, to be able to better control the Brazilian Amazon, which represents 40% of Brazil's territory and which is immensely rich in natural resources. Not only does this refer to well-known resources such as gold and iron—for example, the world's largest iron mine is located in the lower Amazon—but water as well. After all, the Amazon basin is by far the world's largest river basin. And Brazil is heavily dependent on water for its national energy supplies; today 80% of all energy comes from hydroelectric sources.

Trans-Amazonian roads would also enable settlers to colonize the Amazon. In order to reduce poverty, Brazil started a colonization program, aimed at transferring Brazilians from the poor northeast to the Amazon. Initially, most Brazilian colonists migrated to the southwestern part of the Brazilian Amazon (Vosti et al., 2002, 18). Using the road (BR-364) built in 1968, they could easily access the states of Rondônia and Acre, respectively near Bolivia and Peru.

Loggers and ranchers were the first who started clearing forests on a large scale. They would often work together; before a rancher lights a forest, he would commonly allow a timber trader to extract valuable trees. The presence of a road obviously facilitates log transports. In this economic context, the tensions grew between the already present rubber tappers and the later arrival of ranchers, which eventually also led to Mendes' murder in Acre.

During the late 1980s and 1990s, most large-scale deforestation took place on the other side of the Brazilian Amazon, on the eastern and southeastern side, in the states of Maranhão, Mato Grosso, and Pará. More than half of the forest cleared between 1988 and 1994 occurred in Mato Grosso and Pará (Faminow and Vosti, n.d.). The principal driver during this period was pastureland expansion for livestock grazing.

Since the early twentieth century, most deforestation still occurred in the east and southeastern Amazon, but now a second additional agricultural activity has been expanding and driving Amazonian deforestation: the mechanized cultivation of soybeans that are mostly destined for exports to Asia and Europe, where they are mainly used as animal feed (Boekhout van Solinge and Kuijpers, 2013, 209). Brazil's soy production has grown impressively and Brazil has become, after the US, the second largest soy producer and exporter.[13]

Large-scale soybean cultivation started in southern Brazil in the 1960s and its cultivation has gradually moved northward toward Central Brazil, such as to today's Brazil's "soy heartland," Mato Grosso state. The southern half of Mato Grosso has a dry savannah climate whereas the northern humid half is part of the Amazon biome. After Brazilian scientists developed a new soybean variety that could grow in humid climates, the soybean frontier moved farther

13. See e.g., the Web site soystats.com, or see www.fao.org.

north into the Amazon. This started in the northern part of Mato Grosso, and later moved to Pará state as well, stimulated by the Pará state government which saw much economic potential in soy cultivation (Steward, 2007).

Since 2004, soy has been cultivated near the Amazon River and in the lower or eastern Amazon, particularly near the town Santarém, situated at the confluence of the clearwater Tapajós River and the Amazon River. Local businessmen from the town Santarém, together with Pará state officials had traveled to Mato Grosso in order to stimulate soy producers to invest in the Santarém region (Steward, 2007, 113). But what really prompted soy farmers from central and southern Brazil to go there, was the—initially illegal (Santana 2010)—Cargill-led construction of a soy export harbor in Santarém (Boekhout van Solinge and Kuijpers, 2013).[14] This soy export harbor in the west of Pará state, at some 700 km from the river mouth, was built for logistical reasons: to export soy from Central Brazil. An indirect effect of the harbor, however, was that it attracted soy farmers from south and central Brazil to the lower Amazon, which led to much (illegal) deforestation in the Santarém region (see also Boekhout van Solinge, 2014b).

In the process of Amazonian deforestation, logging, and land conversion often go hand-in-hand. The conversion of rain forest to agriculture is the main driver of Amazonian deforestation. As mentioned before, much of this happens illegally, as many farmers deforest (much) more than is allowed (Lawson, 2014, 35–36). Both ranchers and soy farmers often work together with loggers, who generally take out commercially attractive trees before the forest is burned and cleared in the dry season. Logging thus has remained an economic driver for deforestation as well and the Brazilian Amazon continues to export much illegal timber, especially from notorious "conflict state" Pará.

Environmental inspectors (e.g., IBAMA) do exist, but their presence and enforcement is quite weak in Pará state. Considering that environmental inspectors in the Amazon may face threats from farmers and loggers, they may prefer to be bribed (see also London and Kelly, 2007). Brazil's Forest Code states that if a farmer controls a piece of forestland, a maximum of 20% can be cleared. However, landholders commonly remove more, as this law is not well enforced, especially in "conflict state" Pará. It also regularly happens that inspectors are corrupt. In 2014, the Federal Police shut down the IBAMA office in Santarém, Pará. This closure was related to a huge quantity of illegal timber (500,000 m^3) that was "legalized" by hackers who obtained access to the governmental timber monitoring system (Boekhout van Solinge, 2014b).

When the state intervenes and gives fines to timber traders or loggers, it is generally known that most of the fines are not paid (Lawson and MacFaul, 2010, 15). In practice, a general lack of law enforcement and monitoring occurs.

14. The harbor construction was allegedly illegal because the legally required Environmental Impact Assessment was not executed.

The problem is acknowledged at the Federal level in Brazil, where corruption is lower than the state or local level, and improvements are being made, but there still is a long way to go.[15] The state and its institutions need to perform better in order to limit illegal deforestation and deforestation-related crimes.

As Acemogul and Robinson (2012) argued in their noted book *Why Nations Fail*, well-functioning institutions make the difference whether countries perform well economically or not. The problem of Amazonian deforestation and deforestation-related crimes, however, is not only the weak functioning of the state and some of its institutions. Responsible behavior of private actors and the monitoring or governance of private actors' promises—such as codes of conduct and corporate social responsibility—also need improvements. For example, the soy moratorium, a promise of private actors to no longer buy soy that is grown on recently deforested land, has resulted in less soy-led deforestation (see Gibbs et al., 2015). However, in the region of Santarém in western Pará, deforestation for soy continues. Some of the claims, such as by soy-giant Cargill, that deforestation has stopped since 2006 in the Santarém area, do not correspond with what locals say and what satellites show (Boekhout van Solinge, 2014b).

Beef and soy, in combination with timber extraction, are the chief proximate agricultural drivers of Amazonian deforestation. However, as a substantial part of the beef, leather, soy and timber are for export markets, global demand for meat and timber indirectly influences deforestation in Brazil. This shows that it is thus too simple to solely blame "the Brazilians" for this deforestation, as this would only highlight the supply side of the problem, and ignore the demand side or destination markets of timber, cattle, and soy.

12.3.6 AMAZONIAN (DEFORESTATION) SCALES AND TRENDS

At the time of Chico Mendes's murder in the late 1980s, around 10% of the Amazon had been deforested. Today this has doubled to around 19%; 80% of which occurred in Brazil. The area of the Amazon that has been deforested to date is generally estimated to be around 762,979 km^2 (Nobre, 2014). This is an area larger than the size of two Germanys or two Japans. It should be noted, however, that an even larger part of Amazonia is considered degraded: an estimated 1,225,100 km^2 (Nobre, 2014: 24). The latter figure is commonly not taken into account when Amazonian deforestation is discussed. What is also often underestimated and not observed by most satellite systems that only detect large-scale deforestation is the effect of selective

15. An example of a recent improvement is a special anti-logging security force that was created early 2015. However, although enforcement may improve with regard to illegal logging, illegal deforestation for land conversion (for agriculture) seems to get less attention. It is known that at state level (such as in Pará and Mato Grosso) as well as Federal level the agricultural lobby is powerful.

logging on deforestation. When selective logging is factored in, deforestation in the Amazon might actually be twice as high as the conventional (governmental) satellite surveys reveal (Asner et al., 2005). Ecologists think that if 30–40% of the forest cover were to be removed, the Amazon would be pushed into a permanently drier climate (Verweij et al., 2009: 7, 32, Nobre, 2014).

A relative positive trend is that the general level of deforestation today is much lower than in the late 1980s. Between 1988 and 2006, the annual deforestation rate in the Brazilian Amazon averaged 18,100 km^2 (Malhi et al., 2008, 169). This is comparable to forest areas being destroyed almost half the size of the Netherlands or Switzerland. A deforestation peak occurred in the mid-1990s (such as 1995 with 29,000 km^2) and also in 2003 and 2004 (respectively 25,000 and 27,000 km^2), but since 2004 a clear downward deforestation trend has occurred.

Since 2009, annual deforestation in the Brazilian Amazon has been under 10,000 km^2 and since 2012 it has been around 5000 km^2 per year.[16] This is the equivalent of 80 football pitches per hour, or 1.3 pitch per minute of Amazon rain forest that is destroyed in Brazil.[17] The general downward trend since 2004 was broken in 2013, when deforestation suddenly increased by 29% as compared to 2012 (from 4571 km^2 in 2012 to 5891 km^2 in 2013), but in 2014 deforestation decreased again to 4848 km^2, which is still higher than in 2012.

On an aggregate level, cattle ranching has been responsible for some 70% of deforestation of the Brazilian Amazon (Malhi et al., 2008, 169). Brazil's cattle herd has grown rapidly and especially in the Amazon. Brazil now has, after India, the largest cattle herd, and has become a main producer and exporter of beef and leather. The numbers of cows and steers are now comparable to the number of humans: both are around 200 million. In the Brazilian Amazon, the cattle number is three times higher than the number of people (of 25 million people). The continued expansion of cattle ranches in the Brazilian Amazon has led to a new term, "cattelization" (Verweij et al., 2009), next to the earlier expression "Africanization of the Amazon," as African grasses for grazing animals replace the rain forest and its rich arboreal fauna (Goulding et al., 2003, 56).

Cattle numbers have grown much particularly in the states Mato Grosso and Pará (Greenpeace Brazil, 2009). Over the last few years, these two states have been responsible for over half of all deforestation in the Brazilian

16. See the Web site of Brazil's Federal Ministry of the Environment that monitors deforestation with satellites: http://www.obt.inpe.br/prodes/index/. Sources such as Wikipedia also use this database. See http://en.wikipedia.org/wiki/Deforestation_of_the_Amazon_Rainforest.
17. As explained in the first footnote of this contribution, this number is based on a football field size of 7140 m^2 (105 by 68 m). Nobre (2014), 23 takes a smaller football field as a basis for his calculations, which would increase the number of pitches of Amazon rain forest that has disappeared.

Amazon. In 2013 and 2014, 59% of all deforestation occurred in these two states.[18] Cattle ranching has been driving most deforestation in these two states, but some deforestation is attributed to soy cultivation. Soy is, after cattle, the second agricultural driver of Amazonian deforestation.

The dynamics of deforestation in the Amazon are complex. A debate occurs as to whether the more recent mechanized soy cultivation has become a more important driver of Amazonian deforestation than cattle. Barona et al. (2010) studied this question for the period 2000–2006 in Mato Grosso state. Their study showed that deforestation was mainly the result of pasture expansion. However, the study also found evidence for the hypothesis that the growth of the soy industry was indirectly contributing to deforestation, as soybean expansion sometimes replaced pastures, and displaced them into forest areas. The term is used in this context is the soy–beef complex (Verweij et al., 2009, 40). Although common, this does not happen everywhere. In the Santarém area in western Parà, rain forest areas have been transformed directly into soy fields, without the intermediary step of cattle farms (Boekhout van Solinge and Kuijpers, 2013).

A clear geographical trend can be observed with regard to deforestation in the Brazilian Amazon. In the 1970s, it started in the west and southwest of the Brazilian Amazon. In the 1990s, it jumped to the east and southeast of the Brazilian Amazon. Since the early twentieth century the deforestation frontier is moving northward, from Mato Grosso state into Pará State. Hence, besides the major shift from deforestation in the western Amazon to the eastern Amazon, this more recent deforestation frontier in the east is moving from the southeastern Amazon to the northeastern Amazon.

This geographical dimension of deforestation is important in order to understand some of the risks this may entail (Monbiot, 1991). It was already mentioned that the Amazon basin is by far the planet's largest river basin. It occupies more than 6.8 million km^2 and is nearly twice the size of the second largest river basin of the world, the Congo (Goulding et al., 2003, 18). The Amazon River, the largest and main river in the Amazon basin, is by far the planet's largest river in terms of water discharge,[19] discharging 15–16% of all freshwater that is delivered to oceans (Goulding et al., 2003, 26–27).

18. In Brazil, nine states occur in the Amazon region. See (again) the following Web site where Brazilian deforestation data are located for each state: http://www.obt.inpe.br/prodes/index.php. For example, they show that in 2014 in total 4848 km^2 was deforested. 1829 km^2 of it occurred in Pará and 1048 km^2 in Mato Grosso, which together accounts for 59% of the total. In 2013, these two states also accounted for 59% of the total: 2346 km^2 in Pará and 1139 km^2 in Mato Grosso, of a total of 5891 km^2. If one looks back a few more years, more than half of all deforestation in the Brazilian Amazon occurred in these two states.

19. The Amazon basin contains 14 large rivers, including the largest and best known, the Amazon River. The names of some of the other large rivers in the Amazon River are generally not very known, except among locals. For example, it is not much known that two of the Amazon's transnational tributaries—the Madeira and the Negro—are among the 10 largest rivers of the world (Goulding et al., 2003: 10).

Interestingly, of all the rain that falls in the Amazon basin, only 41% makes it to the sea. Some 59% of the rainwater is recycled back into the atmosphere through evapotranspiration (Goulding et al., 2003, 24). The original source of all this rainwater is the Atlantic Ocean. In South America, eastern winds dominate, bringing evaporated Atlantic Ocean water formed into clouds. Much rain from these clouds first falls in the eastern Amazon and is then recycled back through evapotranspiration several times before reaching the western Amazon and other parts of the continent. In order words, the eastern Amazon can be considered as the first water recycling area; it represents the first phase of a series of water cycles of the Amazonian water pump (Nobre, 2014).[20]

This process is referred to as the theory of the biotic pump of Makarieva and Gorshkov (2007). They argued that land with extended natural forests attracts much more moisture from the sea than land without these forest areas. As natural forests with high leaf indexes maintain high evaporation fluxes, this water evaporation supports the ascending air motion over the forest, which attracts or "sucks in" moist air from the ocean. According to this theory, when a substantial part of the (eastern) Amazon is deforested, less water can be contained in the eastern Amazon that can later be pumped to the western Amazon and elsewhere on the South American continent. Several studies indicate that the Amazon is drier than it used to be as a result of less rain (see Hilker et al., 2014).

12.3.6.1 Perpetrators and Victims

That Chico Mendes was murdered by the son of a rancher was not coincidental. The roads of the 1960s and 1970s that cut the Amazon rain forest open brought settlers and farmers, but not all was planned. Unplanned colonization through squatting, the traditional means of settlement, "engendered many bloody fights throughout Amazonia between squatters and either landowners holding documents for legal ownership or the more feared *grileiros*,

20. See Chapter 1 of Monbiot (1991). That the Amazon rain forest contains so much water can be experienced when living there (but not in the large urban enclave Manaus, which as Monbiot already pointed out, has a climate on its own). The author lived for some time in Presidente Figueiredo (Amazonas state), a small town that is surrounded by rain forest. In the wet season, the humidity is so high that it can take days before clothes are dry, if the days are cloudy. Often, and not only in the rainy season, (heavy) rain falls during the night. During the morning hours the forest (floor) stays wet. During the day, it heats up and water evaporates. Near Santarém in Pará state, the Tapajós National Park has a 45-m high tower with monitoring instruments. In 2014, the author stood on top of the tower at midday when it started to rain. Immediately after the rain stopped, the humidity (clouds) could be observed rising up from the forest. Rainfall is abundant near the Amazon's river mouth, such as in Pará's capital Belém, where it rains every early afternoon. The local expression "after the rain," is much used, such as for setting appointments.

speculators making their living by contracting thugs (*jagunços* or *pistoleiros*) to drive small farmers off the land they occupy (Fearnside, 1984, 50)."

All along the deforestation frontier or deforestation arc in Amazonia, there are land conflicts between recent arrivals, mainly farmers who are often large landholders or land-grabbers (*grileiros*), and the people who already lived there. This includes rubber tappers (Mendes, 1989) and traditional communities: indigenous (who today form a small minority in the Amazon), *quilombos* (descendants of runaway slaves), and *caboclos* (people of mixed Indian and white or Indian and black ancestry). Many traditional forest communities have existed for generations, but the inhabitants never registered officially at a governmental institution. As a result, they do not always have legal papers that show or confirm that they are formally living where they live. Considering that land grabbing is a "way of life" in the Amazon (London and Kelly, 2007, 151), a farmer, large landholder or speculator may turn up one day claiming the land in or around where these communities live. This claim may be supported by documents, that are maybe authentic, and the claim may be strengthened by the presence of some armed men.

In South America and its history of latifundia, the Weberian ideal type of nation-states having the monopoly on violence, does not exist in all territories of states (Koonings and Kruijt, 2004). In order words, the state and its institutions that are supposed to be responsible for public order, control of (non-state) violence, and justice, and not always around. In many parts of Amazonia, large landowners traditionally have had much power; not only political power, but sometimes also in control of the monopoly on violence, through henchmen. It still is a quite common practice that large landholders such as ranchers and soy farmers, in collusion with loggers, use *pistoleiros* to threaten or kill opposition (Monbiot, 1991; Brooks, 2011; Boekhout van Solinge, 2010b, 2014a). Opposition more commonly comes from forest inhabitants or from environmental activists.

Timber traders and farmers can log and deforest illegally and threaten and kill opposition, because in most cases they get away with it. Much of this violence goes unpunished, as there is a climate of impunity. As Amazonian deforestation often goes hand-in-hand with violence and corruption of public officials, it actually fits within the common definitions of organized crime (Boekhout van Solinge, 2014a,b). Boekhout van Solinge (2014b) coined the term violent business subcultures, as agribusinesses and timber traders are often involved as perpetrators.[21]

21. In Europe, farmers and timber traders are generally considered normal professions without any negative stigma attached to them. In the Brazilian Amazon, this is very different. Large landowners and timber traders often have a bad name among a large part of the population. They have the reputation for being violent, to easily give orders to violence, and for being involved in other illegal activities as well, deforesting or illegal logging, or falsifying paper work. In the Amazon, most parents are not happy if their daughter would date a timber trader.

Violence used against people trying to protect and preserve Amazonia generally does not reach international media. It only seems to reach those media when westerners with access to western media witness the violence (Boekhout van Solinge, 2010a), when the victim is western, or when the victim is an internationally known local. This was the case in 2005 with the murder, in Pará, of Dorothy Stang, a 73-old woman, originally from the USA, defending the Amazon and its inhabitants. It also was the case with the murder of Chico Mendes in 1988, as well as in 2011, when the Da Silvas, a well-known Amazonian activist couple were killed in Pará (Boekhout van Solinge, 2010b).

Since the murder of Chico Mendes, over 1150 people have been killed in the Brazilian Amazon in the context of these types of land disputes over deforestation (Brooks, 2011). The Brazilian NGOs, CPT, and CIMI monitor the violence over land conflicts, but their annual reports rarely reach international media.[22]

The direct human victims of deforestation-related land conflicts are forest inhabitants. Most of them are not indigenous.[23] Indigenous people now form a clear minority in the region of around half a million,[24] out of a total of around 25 million Brazilians who live in the Amazon. However, indigenous populations are clearly overrepresented among the human victims of deforestation, which is partly caused by their lower socioeconomic status.[25] They do have rights, but these rights are often not respected in practice. One subcategory of indigenous tribes is particularly vulnerable: the strongly decreasing (estimated) numbers of tribal groups that (deliberately) live without contact with

22. CPT (Comissão Pastoral da Terra, Pastoral Land Commission) is a well-known and well respected (by both rural and forest communities, as well as by governmental institutions) NGO that is affiliated to the Catholic Church. CIMI is the Brazilian Missionary Indigenous Council, which also publishes annual reports detailing violence against indigenous people in Brazil.
23. The population that is dominant in quantitative numbers in Amazonia are the so-called *caboclos* (literally copper colored), people who are of mixed Indian/white or Indian/black ancestry. Waves of migration brought different groups: escaped African slaves fleeing into the forest, Europeans, traders from the Middle East, Japanese, Confederates from the United States, and adventurers from different corners.
24. Estimates are that in pre-Colombian times, hence before the Europeans arrived, the human population—obviously exclusively indigenous—in the Amazon region counted several million.
25. In a book by pop artists Sting and Jean-Pierre Dutilleux (1989), the latter writes that a cattle farmer told him that he does not make any difference between Indians and wild pigs. If they come on my land, I shoot them without hesitation, the cattle farmer explains. A similar picture emerges from the book by Dutch journalist and former Brazil correspondent Ineke Holtwijk (2006). For her book about uncontacted tribes, she also met large landholders who basically explain her to not see any problem in killing Indians. Actually, this sometimes is a strategy. The law says that when uncontacted Indians are found to be living in a part of a forest, that piece of forestland will get protected status. This same law can also endanger the Indians. When a farmer interested in that piece of land knows there are Indians living, he might decide to have them killed, as this will mean there is no-one living, which makes exploitation (starting with deforestation) easier. From Holtwijk's book the picture emerges that this is a quite common attitude of ranchers and large landholders.

modern society. The numbers of these sometimes very small tribal groups are dwindling and are now estimated at several dozens. These groups of often seminomadic hunters-gatherers represent the oldest form of human societies and the Amazon is the world's main location where they occur (Boekhout van Solinge, 2010b).

Forest inhabitants are, however, not the only (potential) human victims of Amazonian deforestation. A substantial portion of modern medicines is derived from traditional botanical medicine, in particular, from species-rich tropical rain forests. This botanical knowledge of medicinal usages of plants is especially prevalent among forest inhabitants. Further destroying (Amazonian) rain forest could, in the long run, have repercussions for humanity, as it might mean that potential medicinal plants are being destroyed. Hence, besides the direct human victims, in the long term if future generations are included, indirect victims occur as well. From this perspective, the human victimization of Amazonian deforestation is much larger.

The disappearance of species as a result of Amazon rain forest destruction is especially plausible when one takes into account that biodiversity has geographical dimensions, as Darwin's friend and colleague Alfred R. Wallace already discovered during his mid-nineteenth century research in Amazonia and later Southeast Asia (Boekhout van Solinge, 2008). Wallace developed the river or riverine barrier hypothesis. Amazonia's many waterways, which really form a water labyrinth, seem to have formed barriers that have, in evolutionary terms, led to a differentiation of species.

Tropical rain forests, in general, are the "headquarters of biodiversity," covering 6% of the land surface and containing more than half of the known species (Wilson, 2003, 59). Amazonia represents 40% of the remaining tropical rain forest and is considered the most biodiverse region on the planet, even though large parts are still unexplored by scientists and many specimens that were collected have not been studied yet. Maybe over 50,000 plants still remain to be discovered by science (Verweij et al., 2009, 12). Ten square kilometers of Amazon rain forest contain more plant and animal species than in all of Europe (Wilson, 2003, 62). Destroying (large) parts of Amazonia logically implies that animal and plants species are being lost.

Considering their high biodiversity, tropical rain forests today are in general the "leading abattoir" of species extinction (Wilson, 2003, 59).[26] The destruction of species-rich rain forests such as Amazonia, therefore, contributes to the already ongoing extinction crisis (Leakey and Lewin, 1996). Besides the human victimization, one could argue that many nonhuman victims

26. One beautiful book that deserves being mentioned here is *The Tree* by Tudge (2006). One of the things that Tudge explains well (pp. 323 and 343–344) is how animals (including insects) are vital for the dispersal of seeds and thus the reproduction of plants and trees. Whereas in temperate climates trees reproduce by using the wind and gravity, in tropical rain forests many trees are dependant on animals for their successful reproduction.

of Amazonian destruction also occur. As much of this destruction is illegal, one could label this as flora and fauna crime.

12.3.6.2 Discussion: Deforestation on the Agenda

Deforestation in Brazil is a sensitive political issue to discuss as a foreigner. Ranked fifth in size and population, in combination with its emerging economy, Brazil is increasingly self-confident. Brazil's (educated) middle and upper classes easily perceive criticism on Amazon deforestation as an intermingling into internal Brazilian affairs. Why should Brazil not be allowed to develop itself and use its natural resources for that purpose? A Brazilian viewpoint is that in order to compete with the subsidized agricultural sectors in the Europe and North America, Brazil has to use its relatively unused land in the Amazon.

What seems a quite widespread viewpoint in urban Brazil, such as Brazil's south, is that Brazil is developing and has many issues with which to deal. Amazonian deforestation is a problem, but it does not have priority. What seems a quite common viewpoint in forested Brazil, the Amazon, is that there still remains so much forest and that it is hard to imagine that it will disappear.

The year 1988 spurred international interest in global warming and brought Amazonian deforestation to the forefront with Chico Mendes' murder. Deforestation has been back on the international agenda for a few years. The main reason is global warming. Since 1988, the global average temperatures have increased. The year 1988 was warm at the time, but today 1988 does not even feature in the top 20 warmest years.[27] What has increased as well since 1988, is scientific knowledge about global warming, including knowledge about the share that deforestation causes, which the IPCC (2007) estimates at 18%. Although an influential corporate lobby has attempted to raise doubt about global warming, as Oreskes and Conway (2010) showed in their important book, international agreements aim to limit global warming to 2 °C by 2100.

In September 2014, during a UN Climate Summit in New York, the New York Declaration on Forests was announced. Brazil was not among the signatory countries because it had not been consulted beforehand.[28] Also in September 2014, news reports started appearing in Brazilian media about a drought in central and especially southern Brazil. These reports only increased during the

---

27. As remarked James Renwick, of the School of Geography, Environment and Earth Sciences, Victoria University in Wellington, New Zealand. Taken from http://www.sciencemediacentre.co.nz/2015/01/20/2014-hottest-year-on-record-experts-respond/. Consulted on March 23, 2015.
28. Brazil's position about climate policies is not entirely clear. Although President Rousseff made statements that suggest she takes global warming seriously, in January 2015 she appointed a new government minister of Science, Technology and Innovation, Aldo Rebelo, who doubts the existence of global warming. See: http://blogs.edf.org/climatetalks/2015/01/05/climate-change-denier-named-brazils-science-minister/#more-4299.

following months. Particularly São Paulo, Brazil's economic center and South America's largest city with a population of 20 million, suffered a severe drought and water crisis. As so many Brazilians and also (hydroelectric) energy sources were effected, Brazil's water crisis led to discussions about the cause.

One scientist was getting much attention in Brazil's (social) media: Antonio Nobre, who argued that deforestation in Brazil was a major contributor of the drought. Nobre, a scientist at the Brazil's Centre for Earth Systems Science, had been pointing at the relationship between deforestation and droughts before, but due to Brazil's serious drought, media and public opinion seemed more receptive to his arguments. Nobre (2014) has argued that Amazonia acts like a giant water pump, delivering precipitation across a considerable portion of South America.

Brazil's strong agricultural lobby denied the connection between deforestation and the drought, but it seems that Brazil's drought of 2014/2105 was a wake-up call for the Brazilians, making clear that it is in the general interest and in Brazil's interest to preserve Amazonia. Interestingly, the risk of drought for Brazil's south is one of the potential risks that Mendes (1989) explicitly mentioned in interviews not long before being murdered. This intelligent man from the rain forest, from poor descent, and mostly self-educated, explained that southern Brazil receives rain clouds from the north, the Amazon. The humid Amazon guarantees a constant supply of rain clouds for the south. Amazonian deforestation, he explained, could lead to a drier climate in the south.

REFERENCES

Acemogul, D., Robinson, J.A., 2012. Why Nations Fail. The Origins of Power, Prosperity and Poverty. Profile Books, London.

Asner, G.P., Knap, D.E., Broadbent, E.N., Oliveira, P.J.C., Keller, M., Silva, J.N., 2005. Selective logging in the Brazilian Amazon. Science 310, 480–482.

Barona, E., Ramankutty, N., Hyman, G., Coomes, O.T., 2010. The role of pasture and soybean in deforestation of the Brazilian Amazon. Environ. Res. Lett. 5, 024002. http://dx.doi.org/10.1088/1748-9326/5/2/024002 (pp. 9).

Bavink, M., Pellegrini, L., Mostert, E., 2014. Conflicts over Natural Resources in the Global South. Conceptual Approaches. Taylor and Francis, London.

Beirne, P., South, N. (Eds.), 2007. Issues in Green Criminology. Confronting Harms against Environments, Humanity and Other Animals. Devon, Willan.

Boekhout van Solinge, T., 2004. Dealing with Drugs in Europe. An Investigation of European Drug Control Experiences: France, the Netherlands and Sweden. Boom Legal Publishers, The Hague.

Boekhout van Solinge, T., 2008. The land of the Orangutan and Bird of Paradise under threat. In: Sollund, R. (Ed.), Global Harms. Ecological Crime and Speciesism. Nova Science, New York, pp. 51–70.

Boekhout van Solinge, T., 2010a. Equatorial deforestation as a harmful practice and criminological issue. In: White, R. (Ed.), Global Environmental Harm. Criminological Perspectives. Willan, Devon, pp. 20–36.

Boekhout van Solinge, T., 2010b. Deforestation crimes and conflicts in the Amazon. Crit. Criminol. 18 (4), 263–277.

Boekhout van Solinge, T., 2014a. Natural resources and organized crime. In: Paoli, L. (Ed.), Oxford Handbook on Organized Crime. Oxford University Press, New York, pp. 500–528.

Boekhout van Solinge, T., 2014b. Researching illegal logging and deforestation. Int. J. Crime, Justice Soc. Democr. 3 (2), 35–48.

Boekhout van Solinge, T., Kuijpers, K., 2013. The amazon rainforest: a Green criminological perspective. In: South, N., Brisman, A. (Eds.), Routledge International Handbook on Green Criminology. Routledge, New York, pp. 199–213.

Brooks, B., 2011. Like Many before, Amazon Activists Silenced by Gun. The Boston Globe, 28 May 2011. http://articles.boston.com/2011-05-28/news/29601132_1_rain-forest-amazon-activists-amazon-state.

Faminow, M.D., Vosti S.A., n.d. Livestock-Deforestation links: Policy Issues in the Western Brazilian Amazon, FAO (Agriculture and Consumer Protection). On-line: http://www.fao.org/wairdocs/lead/x6139e/x6139e00.htm.

Fearnside, P.M., 1982. Deforestation in the Brazilian Amazon: how fast is it occurring? Interciencia 7 (2), 82–85.

Fearnside, P.M., 1984. Brazil's settlement amazon schemes. Habitat Intl. 8 (1), 45–61.

Gibbs, H.K., Rausch, L., Munger, J., Schelly, I., Morton, D.C., Noojipady, P., Soares-Filho, B., Barreto, P., Micol, L., Walker, N.F., 2015. Brazil's soy moratorium. Supply-chain governance is needed to avoid deforestation. Science 347 (6220), 377, published January 23, 2015. on-line: http://www.sage.wisc.edu/pubs/articles/Gibbs/GibbsetalScience2015.pdf.

Global Witness, 2014. Deadly Environment. The Dramatic Increase in Killings of Environmental and Land Defenders. Global Witness, London.

Goulding, M., Barthem, R., Ferreira, E., 2003. The Smithsonian Atlas of the Amazon. Smithsonian, Washington, London.

Greenpeace Brazil, 2009. Amazon Cattle Footprint. Mato Grosso: State of Destruction. Manaus/São Paulo, Greenpeace, Brasil.

Grandin, G., 2010. Fordlandia. The Rise and Fall of Henry Ford's Forgotten Jungle City. Icon Books, London.

Greenpeace International, 2003. State of Conflict. An Investigation into the Landgrabbers, Loggers and Lawless Frontiers in Pará State, Amazon. Greenpeace International, Amsterdam.

Hilker, T., Lyapustin, A.I., Tucker, C.J., Hall, F.G., Myneni, R.B., Wang, Y., Bi, J., Moura, Y.M.de, Sellers, P.J., 2014. Vegetation dynamics and rainfall sensitivity of the Amazon. Proc. Natl. Acad. Sci. U.S.A. 111 (45), 16041–16046. http://dx.doi.org/10.1073/pnas.1404870111.

Holtwijk, I., 2006. Rooksignalen. Op zoek naar de laatste verborgen indianen in Brazilië (Translation: Smoke Signals. In search of the last hidden Indians in Brazil). Atlas, Amsterdam.

IPCC (Intergovernmental Panel on Climate Change), 2007. Summary for policy makers. In: Solomon, S., Qin, D., Manning, M., Chen, Z., Marquis, M., Averyt, K.B., Tignor, M., Miller, H.L. (Eds.), Climate Change 2007: The Physical Science Basis. Contribution of Working Group I to the Fourth Assessment Report of the Intergovernmental Panel on Climate Change. Cambridge University Press, Cambridge, New York.

Jackson, J., 2008. The Thief at the End of the World: Rubber, Power, and the Seeds of Empire. Viking, New York.

Koonings, K., Kruijt, D., 2004. Armed Actors. Organized Violence and State Failure in Latin America. Zed Books, London/New York.

Lawson, S., MacFaul, L., 2010. Illegal Logging and Related Trade. Indicators of the Global Response. Chatham House, London.

Lawson, S., 2014. Consumer Goods and Deforestation: An Analysis of the Extent and Nature of Illegality in Forest Conversion for Agriculture and Timber Plantations. Forest Trends, Washington.

Leakey, R.E., Lewin, R., 1996. The Sixth Extinction. Patterns of Life and the Future of Humankind. Weidenfeld and Nicolson, London.

London, M., Kelly, B., 2007. The Last Forest. The Amazon in the Age of Globalisation. Random House, New York.

Loureiro, V.R., 2011. Amazônia. História e Análise de Problemas. Editora Cejup, Belém.

Makarieva, A.M., Gorshkov, V.G., 2007. Biotic pump of atmospheric moisture as driver of the hydrological cycle on land. Hydrol. Earth Syst. Sci. 11, 1013–1033.

Malhi, Y., Roberts, J.T., Betts, R.A., Killeen, T.J., Li, W., Nobre, C.A., 2008. Climate change, deforestation, and the fate of the Amazon. Science 319 (5860), 169–192. January 11, 2008.

Mendes, C., 1989. Fight for the Forest. Chico Mendes in His Own Words. Latin America Bureau, London.

Monbiot, G., 1991. Amazon Watershed. The New Environmental Investigation. Michael Joseph, London.

Nobre, A.D., 2014. The Future Climate of Amazonia. Scientific Assessment Report. Articulación Regional Amazonica (ARA), São José dos Campos. Online at: http://www.ccst.inpe.br/wp-content/uploads/2014/11/The_Future_Climate_of_Amazonia_Report.pdf.

Oreskes, N., Conway, E.M., 2010. Merchants of Doubt: How a Handful of Scientists Obscured the Truth on Issues from Tobacco Smoke to Global Warming. Bloomsbury Press, New York.

Santana, R.R., 2010. Justiça Ambiental na Amazônia. Análise de Casos Emblemáticos. Juruá Editora, Curitiba.

Setzer, A.W., Pereira, M.C., 1991. Amazonia biomass burnings in 1987 and an estimate of their tropospheric emissions. Ambio 20 (1), 19–22. February 1991.

Shoumatoff, A., 1990. The World Is Burning. Murder in the Rainforest. Little Brown and Co, Boston.

Simons, M., 1988. Vast Amazon Fires, Man-made, Linked to Global Warming. New York Times. August 12, 1988. Online: http://www.nytimes.com/1988/08/12/world/vast-amazon-fires-man-made-linked-to-global-warming.html (Consulted March 23, 2015).

Steward, C., 2007. From colonization to "environmental soy": a case study of environmental and socio-economic valuation in the Amazon soy frontier. Agric. Hum. Values 24, 107–122.

Sting, Dutilleux, J.-P., 1989. Jungle Stories. The Fight for the Amazon. Barrie and Jenkins, London.

Stoddard, R.H., 1992. The disaster of deforestation in the Brazilian rainforest. In: Majumdar, S.K., Forbes, G.S., Miller, E.W., Schmalz, R.F. (Eds.), Natural and Technological Disasters: Causes, Effects and Preventive Measures. The Pennsylvania Academy of Science, Pittsburgh, pp. 527–535.

Sutherland, E.H., Cressey, D.R., Luckenbill, D.F., 1992. Principles of Criminology. General Hall, Lanham.

Tudge, C., 2006. The Tree. A Natural History of what Trees Are, How They Live and Why They Matter. Crown Publishers, New York.

Verweij, P., Schouten, M., Beukering van, P., Triana, J., van der Leeuw, K., Hess, S., 2009. Keeping the Amazon Forests Standing: A Matter of Values. WWF Netherlands, Zeist.

Vosti, S.A., Witcover, J., Carpentier, C.L., 2002. Agricultural Intensification by Smallholders in the Brazilian Amazon. From Deforestation to Sustainable Land Use. International Food Policy Research Institute, Washington, DC.

Wilson, E.O., 2003. The Future of Life. Vintage Books, New York.

Chapter 13

Ecological Impacts of Climate Change

George P. Malanson and Kathryn J. Alftine
Department of Geographical & Sustainability Sciences, University of Iowa, Iowa City, IA, USA

ABSTRACT
The effects of climate change on species and ecosystems are reviewed in the context of hazards and disasters. Other recent assessments, from other perspectives, are introduced. Examples of past on-ongoing effects of climate change in marine and terrestrial systems, for individual species and for biomes, are presented. Marine examples include sea turtles (affected by rising sea level), corals (directly affected by temperature), shorebirds (affected through synchrony of migration with prey), and the organisms associated with sea ice such as penguins and polar bears (through reductions in available food). Terrestrial examples include tropical forests (through drought and pathogens), Mediterranean-type shrublands (though fire frequency), coniferous forests (affected by drought, fire, and insect outbreaks), and the tundra-treeline ecotone (possible advancing into tundra). How ecosystem services are affected by climate change is examined. The provisioning of goods (such as in agroecosystems), the regulating of functions (such as carbon regulation), the support of these processes (e.g., pollinators for agriculture), and cultural services (such as aesthetic experiences) are all likely to be reduced in some places by climate change. Within the context of climate change in the past, the possibility of mitigation (in the hazards and disasters perspective) is considered through the notion of "assisted migration." Assisted migration, however, may create as many problems as it solves—but the problems are being evaluated. Ex situ mitigation—through species/seed/gene banks—is advocated.

13.1 INTRODUCTION

The ecological impacts of climate change are not often addressed in the literature of natural hazards and disasters. By ecological impacts we mean the changes in the abundance, location, and functions of organisms, with functions including their interactions and their effects on the abiotic environment. These impacts can be hazardous to people, reaching disaster proportions, and if one takes the perspective of the organisms themselves then most of the effects might be hazardous. The effects of climate change on people, including the direct effects on people as animals, are addressed elsewhere in this series.

Biological and Environmental Hazards, Risks, and Disasters. http://dx.doi.org/10.1016/B978-0-12-394847-2.00022-X

Ecological responses to climate change were recognized long before global warming was a concern. The relationship of organisms and climate was recognized early in the study of the environment (von Humboldt and Bonpland, 1807), and consequences of climate change were discussed directly following eighteenth-century recognition of past climatic change. Following the initial use of glacial geomorphology to infer past climates, ecological evidence was used to refine understanding of climatic change (Auer, 1927), and this work continues (Webb and Bartlein, 1992; Caffrey and Doerner, 2012; Hupy, 2012).

The most salient response of organisms to large-scale climatic change is a shift in their geographic ranges. Species move so that they can exist in the climates to which they are adapted. Thus, we have seen poleward shifts in the broad ranges of species during the past ~20,000 years following the Last Glacial Maximum (LGM). At shorter timescales within the period of moving ranges, species adjust through physiological, behavioral, and life-history changes that can be part of natural selection, possibly changing the genome. The most significant of these responses is change in phenology, e.g., spring greening in North America (Schwartz, 1993; Schwartz et al., 2006). Several studies have demonstrated that the length of the vegetative growing season has increased with changing climate. This trend has been more pronounced at higher latitudes where regional temperature rise has been the greatest. In a Europe-wide, controlled study of growing season in gardens, spring events (e.g., leaf-unfolding) were found to occur 6 days earlier, and autumn events (e.g., leaf coloring) were delayed by 4.8 days. In this study, the average annual growing season lengthened by 10.8 days between 1959 and 1993 (Menzel and Fabian, 1999). A 61-year record of phenophases, including arrival dates for migratory birds and dates of first bloom of spring flowers, in southern Wisconsin (43°N) showed an increase in earliness of −0.12 day per year, or 7.3 days over the length of the study (Bradley et al., 1999).

From the perspective of the organisms, the responses to climate change are hazards for some because they can increase the probability of extinction by reducing the size of populations and/or spatially fragmenting populations. One of the broad considerations raised by Botkin et al. (2007) is that as the ranges of species shift poleward with climatic warming, the area in any climate must shrink; combined with the most reliable relationship in all of biogeography—that the number of species is directly related to area, the predicted consequence is a loss of species. Malanson (2008) noted that such a pattern could be part of an extinction debt—a time lag in when the eventual, inevitable extinctions would take place. The species we are discussing do not anticipate extinctions, however, and their responses are not directed or chosen except in the case of some animal migrations.

From the perspective of humans, two aspects should be considered. First, as animals, *Homo sapiens* can experience hazard secondarily as other organisms respond to climate change. Changes in living disease vectors are discussed in other chapters in this volume, and changes in other environmental conditions directly affecting people, such as agriculture, are discussed in other volumes of this series. Here, we focus on changes to ecosystem services that can result from climatic change.

We organize the following discussion to first report effects of climate change in selected terrestrial and marine ecosystems. Within terrestrial ecosystems we review biomes (tropical forest and savanna, Mediterranean-type shrublands, and boreal forest). In marine ecosystems we review impacts on coral, pelagic birds, penguins, and marine mammals (especially polar bears) and on ecosystem level effects of changing sea ice. Second, we discuss ecosystem services per se, and how they can be affected by climate change and consequent organism-level responses. Last, we discuss possible mitigation and post-problem responses.

13.2 OBSERVED EFFECTS OF CLIMATE CHANGE

A symposium in 1987 addressed key questions on the effects of a warming climate on ecological patterns and processes. Most of what we knew and still need to know was compiled in an edited volume (Peters and Lovejoy, 1992). Although it remains the best reference, we have added considerable detail and refinement in both observations and concepts to that base.

13.2.1 Broad Assessments

Root et al. (2003) analyzed 143 studies that addressed twentieth-century response of species to global warming over at least 10 years (averaging 34.5 years). The studies covered a wide range of species, and each study found a strong association between a trait of the species and site-specific temperature. The types of change in species' traits included changes in density, range shifts either poleward or up in elevation, changes in phenology, morphology. or genetic frequency. The authors concluded that more than 80% of the species in the studies that showed changes were shifting in the direction expected on the basis of understood physiological constraints of the species in response to climate change.

Menzel et al. (2006) analyzed a phenological data set that included 542 plant species through 21 European countries from the years 1971–2000. Results indicated that 78% of species were showing advancing dates of leafing, flowering, or fruiting, and that the average advance of spring was 2.5 days per decade.

In an assessment of 250 observational and experimental studies that addressed climate and either a biological or physical process, and met other criteria, Gitay and Raudsepp-Hearne (2005) determined that 61% of the species in the studies showed a response in areas such as phenology, range shifts to higher elevations or latitudes, changes in densities, development, genetics, or morphology.

Staudinger et al. (2013) summarized global responses of species. This most recent review noted the observations that document the expected changes in species ranges and phenology, and noted the species-specific (individualistic) nature of these responses. They noted the continuing concern that the rate of climate change could be faster than the capacities of species to adapt or respond.

Studies and reviews continue to assess the link between ongoing climatological and ecological change (e.g., Malanson et al., 2004; Diffenbaugh and Field, 2013; Meyer and Rannow, 2013; Jaeschke et al., 2014).

13.2.2 Marine Ecosystems

The effects of climate change are included in the changes that can be hazards and disasters in marine settings in other papers in this volume and other volumes in this series. Here, we briefly touch on examples that illustrate ecological principles: direct effects of temperature (sea turtles), trophic cascades (corals), phenology (shorebirds), and connectivity (sea ice and polar bears and penguins).

13.2.2.1 Sea Turtles

Sea turtles illustrate a direct dependence on temperature. Warming may lead to maximum sand temperatures that exceed the survivable limit for turtle eggs (34 °C), or change the length of the incubation period. Sand temperatures also determine the sex of the hatchling, and rising temperatures may bias the sex ratio toward females (LeBlanc et al., 2012). Female sea turtles commonly migrate long distances from foraging habitat to suitable nesting beaches. Sea turtles are affected by climate change as low-level nesting beaches erode due to rising sea levels and increased storm intensity. Adult foraging sites are also affected by climate change. Sea grass beds are in decline (Orth et al., 2006), the supply of horseshoe crab eggs has decreased, and coral reefs, which serve as feeding grounds for green turtles, have been affected by climate-related bleaching. All sea turtles (including green, hawksbill, loggerhead, Kemps' ridley, olive ridley, and leatherback) occurring in US waters are currently listed under the Endangered Species Act (ESA).

13.2.2.2 Corals

Reef-building corals have a direct dependence on temperature and are the basis of a trophic cascade. Corals normally form in warm, shallow sea waters and obtain a large portion of their energy needs through a symbiotic relationship with zooxanthellae, which are photosynthetic microalgae. The pigment contained in the zooxanthellae contributes to the color of the coral. Expulsion of zooxanthellae or loss of algal pigmentation leaves coral appearing white or bleached. Without zooxanthellae, most corals will starve unless the stress is quickly removed and symbionts recover. Bleaching events may occur following changes in water chemistry including acidification, increased solar irradiance, changing sea levels, bacterial infection, and other triggers. Commonly located near their upper thermal limits, coral-bleaching events occur when ocean temperatures exceed average summer temperatures by more than 1 °C for several weeks or longer (Hoegh-Guldberg, 1999). The first documented coral-bleaching event occurred in Bird Key Reef in Florida in 1911. Unusually hot and calm conditions at that time also killed many fish and mollusks (Mayer, 1914). In 1929, a bleaching event in the Great Barrier Reef was recorded during hot, calm conditions (Young and Nicholls, 1931). Further bleaching was not documented until 1961 and occurred in Key Largo, Florida (Shinn, 1966). Over the last 35 years, the number, intensity, and geographic extent of bleaching events has grown, in large part, due to climate change (Hoegh-Guldberg, 1999). Short periods of high temperature stress the corals and result in predictable patterns of bleaching as documented in the Great Barrier Reef during extreme bleaching events in 1998 and 2002 (Berkelmans et al., 2004).

Bleaching events may lead to reduced disease resistance, reproduction, growth, and survival of affected corals (Douglas, 2003). If bleaching events become more frequent and severe, and corals cannot adapt to changing conditions, many species of coral may suffer irreversible declines and potential extinction. Reef-building corals support one of the most productive and diverse ecosystems known. Impacts resulting from the functional degradation of coral reefs will range from loss of coastal protection, economic impacts on fisheries, to potential loss of global biodiversity.

13.2.2.3 Shorebirds

Phenology is the timing of naturally recurring seasonal activities of plants and animals. Substantial data have been collected on the phenology of migratory birds including timing of arrival and the condition of migrant birds, timing of breeding, nesting, hatching, and departure. Several studies have demonstrated an earlier onset of spring events that coincides with climate change (Sparks and Menzel, 2002; Walther et al., 2002; Root et al., 2003).

Many populations of migrant shorebirds are in serious decline across the globe. At Delaware Bay, on the northeastern seaboard of the United States, shorebirds, such as the Atlantic red knot (*Calidris canutus rufa*) (currently at critical population levels) and Ruddy turnstones (*Arenaria interpres*), arrive at spring foraging sites at their lean migrating weight (Baker et al., 2004), and it is imperative that food is immediately available at these refueling sites. The eggs of the horseshoe crab (*Limulus polyphemus*) provide an almost exclusive food source for shorebirds and other species. The crabs lay their eggs largely during high tides in May and June. Climate change could impact the timing of spawning or the arrival of the migratory birds, disrupting the synchrony, and affecting populations that are already at critical levels. Egg availability is also influenced by commercial harvesting of crabs and decreasing habitat due to development and beach erosion. Erosion is exacerbated by sea-level rise, specifically in areas where the coastline is unable to move inland due to topography or seawalls. In addition, climate change may lead to delayed spring bird migrations, which may lead to increased competition for arctic nest sites (Both and Visser, 2001).

13.2.2.4 Sea Ice and Related Organisms

In the Arctic, 80% of low-elevation tundra lies within 100 km of winter sea ice (Post et al., 2009) and the extent and duration of ice cover significantly influences both marine and terrestrial populations. Multiyear ice is the thick component of the ice cover that is less vulnerable to seasonal melting than thinner perennial ice and first-year ice. Satellite data collected during the winters of 1979–2011 demonstrated that the multiyear ice extent declined at a rate of −15.1% per decade (Comiso, 2012). Declines in sea-ice extent since the mid-1970s have reduced abundances of ice algae, leading to declines in krill populations from 38% to 75% per decade (Atkinson et al., 2004).

Sea ice provides habitat for microorganisms, including ice algae, which is a primary food source for zooplankton such as krill (*Euphausia superba*), as well as vertebrate communities. Krill, which are crustaceans, are a major food source for a variety of species, including penguins, seals, whales, and pelagic seabirds. Krill are dependent on sea ice as the larvae need to feed on phytoplankton beneath the ice to survive winter. As sea ice diminishes, krill habitat is shrinking as well. Sites along the western Antarctic Peninsula have experienced an 80% decrease in krill abundance over the last 30 years as a result of loss of sea ice (Atkinson et al., 2004). Sea ice decline will be detrimental to krill but may favor other species such as filter-feeding salps, macroalgae, and possibly crabs and other predators (McClintock et al., 2008).

Sea ice loss is not only an indicator of climate change, but is also a major component in driving the ice-albedo positive feedback. An ice-albedo positive feedback is created when highly reflective sea ice melts and the darker ocean surface is exposed, which absorbs more insolation and amplifies warming. Much of the recent warming in the Arctic is attributable to declining sea ice (Screen et al., 2012). The Arctic sea ice cover has been identified as a tipping element: A critical threshold at which small changes can qualitatively alter the state of a system, often abruptly and with global implications (Lenton et al., 2008). As the area of open water in the fall increases, the ice cover in the following spring is increasingly comprised of thin, new ice, which is more vulnerable to summer melting. Increased area of open water also brings higher levels of evaporation and humidity, which has impacted precipitation in areas considered polar deserts.

Alexander (1992) identified the crucial role that sea ice will play in the ecological consequences of climate change in the Arctic. Here, we highlight the effects on archetypes of sea ice ecosystems: polar bears in the Arctic and penguins in the Antarctic.

Polar bears (*Ursus maritimus*) occur throughout the ice-covered waters of the Arctic, concentrated in areas of seasonal ice cover over the continental shelf where biological productivity is highest (Derocher et al., 2004). In 2008, the US Fish and Wildlife Service listed the polar bear as a threatened species under the ESA. Polar bears are affected by climate change most obviously through the reduction of sea ice cover, and also by a decrease in the abundance of food sources such as the ringed seal (*Phoca hispida*). Polar bears use sea ice for dens, hunting, resting, reproducing, and traveling. The bears seek prey at holes in the sea ice and in ice caves. As rising temperatures increase the area of open water, and arctic ice coverage diminishes, the amount of energy required for sustaining activities will increase as bears must swim or walk longer distances to reach preferred habitat.

Southern populations of polar bears, such as those around Hudson Bay, are believed to fast for months during summer ice-free period (Stirling and Derocher, 1993). As warming occurs, bears may not be able to store enough fat to survive the ice-free months. Bears in this region have exhibited declines in population abundance and body condition as a result of longer ice-free seasons (Stirling and Derocher, 2012); recent observations of terrestrial hunting with less ice indicate that it can cost energy (Iles et al., 2013). Reduced survival in cubs and increased bear–human interactions may also result from increased nutritional stress among bears. In 2007, the USGS estimated that the polar bear population of 22,000 would decrease by two-thirds by 2050 (Amstrup et al., 2007). The Arctic Climate Impact Assessment Report of 2005 stated that polar bears are unlikely to survive the complete loss of summer sea ice cover. As bear populations decrease, harvest quotas for native humans may be reduced or eliminated and the economic impacts of tourism based on polar bear viewing may disappear. As a top predator, a decline in polar bear population will impact natural selection and the entire arctic marine food web.

Penguins have been dramatically affected by recent climate change, primarily through reduction in sea ice coverage (Croxall et al., 2002) and associated decrease in krill populations (Atkinson et al., 2004), but also as a result of increasing snowfall including spring blizzards during the breeding season. With a research history spanning 70 years in some areas (longer where carbon-14 dating is appropriate), penguins are a valuable indicator of climate change as data on population, range, reproductive success, and diet allow researchers to identify trends and response to changing environmental conditions (McClintock et al., 2008). These top predators are relatively visible, easily studied, and abundant. Adélie penguins (*Pygoscelis adeliae*) and emperor penguins (*Aptenodytes forsteri*) both consume krill as a primary food source and are dependent on sea ice for protection from predators, resting, breeding, and traveling to preferred feeding sites. These penguins occur only in areas where sea ice is present throughout the winter. During years with heavy sea ice, the ice covers regions of upwellings of the Antarctic Circumpolar Current and the penguins have access to productive foraging sites where the upwellings bring nutrients to the surface. As ice cover decreases, the penguins will lose access to the most productive winter-feeding regions (McClintock et al., 2008). Emperor penguins have declined from 300 breeding pairs in 1970 to 9 in one study area on the western Antarctic Peninsula (Gross, 2005). On the northern Anvers Island, the Adélie penguin population has lost 10,000 breeding pairs (a 70% decrease) since 1975 (Emslie et al., 1998; Fraser et al., 1992) but populations are not declining on the more southerly Ross Island (Wilson et al., 2001), which is effectively shifting the range of this species to higher latitudes. Conversely, open-ocean feeding penguins, the chinstrap and the gentoo, are growing more abundant in areas previously inhabited solely by the Adélie. Along with sea-level rise and ocean acidification, climate change will impact the range and population density of penguin colonies as some species are able to survive ice-free conditions and others must find alternate suitable habitat, or face extinction.

These effects are only the tip of the iceberg. Post et al. (2013) recently described the ecological connections affected by sea ice in the Arctic. Notable in their thinking was the effect on bottom-dwelling organisms that then had effects up through the food web (these effects had been described by Alexander, 1992), including terrestrial organisms such as Arctic fox (*Vulpes lagopus*).

13.2.3 Terrestrial

The response of plant species to climate change since the LGM is the most thoroughly studied phenomena in all of ecology. Still, we are left with the Quaternary conundrum: Why do we see so little extinction during this period of climate fluctuation? Moreover, the response of many plant species, especially for trees, has been projected for various future climate scenarios. Here, we review the general lessons from the past, discuss some of the projections for the future, and examine one species, whitebark pine (*Pinus albicaulis*).

13.2.3.1 Tropical Forest

The diverse tropical regions of the globe will experience the impacts of climate change through range shifts and changes in disturbance regimes and growth and mortality rates. Studies of the effects of past climate changes have found that the tropics were not as immune as once thought, largely in response to changing precipitation regimes (Mayle et al., 2004), but that stability was the norm (Colinvaux et al., 2000). In the shorter term, changes in the seasonal distribution of rainfall may be more important than changes in temperature (Hartshorn, 1992). Zhou et al. (2014) studied trends in species composition, size, and density in 13 plots during the period from 1978 to 2012. The study spanned an area between 21°N and 31°N latitude, and included stable communities that had not been subject to significant artificial management or natural disturbance. The plots were surveyed at less than every 5-year intervals for species, size, health, and location of each tree. Soil moisture and meteorological data were also compiled for each plot. The meteorological data demonstrated that the number of days with little or no rainfall significantly increased, whereas annual precipitation remained unchanged, implying intensification of storm activity in this region. The data showed rapid, synchronous, and directional changes in tree size toward denser and smaller individuals during the study. The authors concluded that the evidence points to regional warming (0.2 °C $decade^{-1}$) and associated drying of soils as the driver of the change observed in size and density of trees.

Reductions in size can have negative impacts on carbon storage in trees. In a global study of over 400 tree species, Stephenson et al. (2014) demonstrated that mass growth rate increases continuously with tree size. For most species, large, old trees actively store more carbon than smaller trees. A positive feedback is then formed when climate change creates conditions leading to decreased tree size, and smaller trees are able to fix less carbon, leaving more atmospheric CO_2 which drives more warming.

During the twenty-first century, an approximate 3 °C increase in temperatures and a 20% reduction in precipitation have been predicted for the Amazon Basin (Houghton et al., 2001). Mayle et al. (2004) used previously published data sets, current pollen studies, and the simulation results from a dynamic vegetation model to make predictions about possible ecosystem responses to significant changes in temperature, precipitation, and atmospheric CO_2 in the Amazon. Proxy data from the Early-Mid-Holocene (6000 cal yr BP) when aridity and possibly temperatures were higher than present, suggest that future warmer and more arid conditions will result in increased fires through southern and eastern Amazonia. Frequent, widespread fires would lead to the replacement of lowland evergreen rainforest species by drought and fire tolerant semi-deciduous forest taxa. The combination of deforestation due to land-use conversion, species composition changes brought about by warming and the potential increase of fire would have a major detrimental effect on the ecosystems of the Amazon. Much of the precipitation received in the western Amazon results from evapotranspiration from the eastern rainforest. Deforestation in eastern Amazonia would reduce precipitation in other parts of the region, increasing the length of the dry season, and having serious implications for the reestablishment of rainforests following deforestation (Nobre et al., 1991).

Climate change may impact tropical birds through increased spread of pathogens. Hawaiian forest birds have experienced extreme rates of extinction since the arrival of humans on the islands. The anthropogenic factors responsible for the losses include habitat destruction and degradation due to logging, ranching, and agriculture, the introduction of predators and competitors, and climate change. A growing concern exists that Hawaiian bird populations are most threatened by introduced mosquito-borne avian malaria (Atkinson et al., 1995). As mosquitoes are not native to the Hawaiian Islands, native forest birds evolved in an environment free of this parasite. The southern house mosquito (*Culex quinquefasciatus*) was introduced in the early 1800s and spread rapidly throughout the islands. Transmission of avian malaria is strongly affected by temperature. Temperatures lower than 16 °C inhibit parasite development and transmission does not occur where the mean ambient temperature is 13 °C or cooler (LaPointe, 2000). A 2 °C increase in temperature could greatly reduce the extent of disease-free habitat on the islands, particularly if upward expansion is limited by anthropogenic deforestation or the summit of the island.

Recent warming trends in the tropics have been associated with the movements of species into more temperate areas. Banding and sight reports have revealed a remarkable shift in the winter range of the Rufous hummingbird (*Selasphorus rufus*) over the last 30 years. Prior to 1990, winter sightings along the Gulf Coast never numbered more than 30. In the six winters from 1991 to 1996, 1,643 sightings were reported in a 5-state study (Hill et al., 1998). As part of a discovery of a number of Odonata species apparently extending their ranges to higher latitudes, two species of a tropical dragonfly previously found only in Cuba, were discovered to have established in Florida in 2000 (Paulson, 2001). The author proposed that climate change would allow these species to expand the northern limits of their range northward in part because adult flight seasons have become longer.

13.2.3.2 Mediterranean-Type Shrublands

The Mediterranean biome occurs found on five continents, supports a high plant species diversity and density, and provides many regulatory, cultural, and provisioning ecosystem services (Millennium Ecosystem Assessment, 2005). The Mediterranean climate coincides with conditions optimal for many agricultural uses, tourism, and other human development, resulting in a disproportionately high conversion of land use and a low level of land protection. As a result, this biome is projected to lose the largest proportion of biodiversity among all terrestrial biomes by the year 2100 (Sala et al., 2000).

Field studies have shown that experimental warming decreases growth and reproduction in some Mediterranean plant species. Peñuelas et al. (2004) found that 2 years of 1 °C experimental warming decreased soil moisture at the peak of the growing season in a Spanish site where temperatures were already close to the optimum for photosynthesis and water was the limiting factor for growth. The result was a 14% decrease in total aboveground plant biomass growth and a 40% decrease in flowering relative to control.

Fire has been a dominant ecological and evolutionary factor throughout the Mediterranean biome, and its response to climate change can have greater consequences than the direct effects of changing temperature or precipitation (Westman and Malanson, 1992). Climate change can increase the frequency of fire to a point where productivity is decreased (Diaz-Delgado et al., 2002) and the organic content of the soil is reduced leading to a decrease in the water infiltration capacity and a general deterioration of the soils (Garcia et al., 2002). Fires may also become more intense in areas where an increased concentration of CO_2 has increased biomass productivity and subsequent fuel load.

Modeling studies have demonstrated that fire frequency and the potential for fire spread are highly sensitive to climate (Moriondo et al., 2006; Malanson and Westman, 1991). Mouillot et al. (2002) combined different scenarios of climate change with a multispecies functional model for vegetation dynamics of maquis shrubland in central Corsica, France, to simulate the impacts of increasing temperatures and changing rainfall patterns. The model results indicate that climate change will decrease the time interval between successive fires from 20 to 16 years in this region, leading to more shrub-dominated landscapes. The ability of plants to regenerate following fire will influence species composition. Plant species that are able to resprout following moderate fire can quickly reoccupy the space, often before plants that regenerate through reseeding (Riera et al., 2007).

The ecological impacts of climate change in the Mediterranean biome will likely result from the complex interactions between increased water stress and changes to the fire regime that may alter plant species composition, abundance, and biomass. Such changes may have implications for ecosystems as well as associated human activities.

13.2.3.3 Western Coniferous Forests

Evidence exists that climate change has and will continue to affect the abundance and geographical distribution of western coniferous forests through the effects of climate on processes such as growth, mortality, and disturbance. Western forests may be sensitive to climate change due to the importance of dry summers (Franklin et al., 1992). Climate change will likely alter snowpack, cause earlier snowmelt and reduced summer stream flow, increase summer evapotranspiration, and increase drought stress (Chmura et al., 2011). Changes in temperature and precipitation are generally expected to increase the severity and frequency of disturbance, such as insect outbreaks, disease, and wildfire (Gedalof et al., 2005; Littell et al., 2010).

Climate variability in coniferous forests has influenced species growth in the recent past. Peterson and Peterson (2001) established tree-ring chronologies from 31 sites across the latitudinal and elevational ranges of mountain hemlock (*Tsuga mertensiana*) in the Pacific Northwest. The authors found that growth appeared to be limited by late snowmelt, short growing seasons, and cool summer temperatures throughout much of the study area. Increasing temperatures would likely increase growth in the northern sites as earlier snowmelt would reduce spring snowpack. However, earlier snowmelt, higher summer temperatures, and lower summer precipitation would increase summer drought stress and would likely reduce growth in southern sites and near the lower elevation limit of the species.

Using tree-ring and ecological plot analysis, Millar et al. (2004) demonstrated that annual branch growth and dates of vertical branch emergence, as well as invasion by whitebark pine and lodgepole pine (*Pinus contorta*) into formerly treeless meadows in upper-montane and treeline ecosystems, is correlated with regional climate in the central Sierra Nevada, California. Branch emergence and meadow invasion occurred during distinct pulses between 1945 and 1976, and all ecological responses correlated with climatic variables. Future climate change may continue to facilitate the advance of trees into previously treeless meadows and tundra can alter the hydrology of the area. Shading from trees delays snow melt, increasing snowpack retention later into the year, may alter soil moisture, and modify nutrient transport by overland flow.

In an analysis of data from 76 plots in unmanaged old forests in the western United States, van Mantgem et al. (2009) found that across elevation zones, dominant genera, tree size, and fire history, noncatastrophic mortality rates have risen rapidly in recent decades. Density and basal area declined during this period, which suggests that the increase in mortality was not due to competition. Since 1955, both temperature and water deficit has increased, predominantly in the higher elevation sites occupied by these forests. The authors suggested that regional warming and resulting drought stress changes are major contributors to the increase in tree mortality. Experimentally induced warmer temperatures (~4 °C) decreased time to drought-induced mortality trees by nearly one-third in piñon pine (*Pinus edulis*) (Adams et al., 2009).

Climatic variability is a significant factor affecting wildfire activity in the western United States. During the twentieth century, the extent of area burned by western wildfires was greater in years with low precipitation and high temperature (Heyerdahl et al., 2008). Warmer springtime temperatures cause snowpack to melt earlier, which causes forests to dry out sooner and lengthens the overall potential fire season. Increased pest and pathogen activity, drought stress, and mortality can increase the dead fuel load.

During the period between 1970 and 2003, forest fires in the western US increased in area of wildfire, duration of fire, and length of the wildfire seasons. Despite the influence of fire suppression, the correlation between spring and summer temperatures, low precipitation and area burned indicate that wildfire activity in the western US is substantially controlled by climate (Littell et al., 2009). Increasing temperatures and changing precipitation and soil moisture will likely extend the fire seasons and increase the total area burned. McKenzie et al. (2004) used the historical (twentieth century) statistical relationship between observed climate and fire extent to infer the sign and magnitude of future wildfire activity under several climate projections. The authors found that using a predicted mean temperature increase of 2 °C, their model indicated that the area burned would roughly double by the end of the twenty-first century in most Western states.

Wildfire has been linked to changes in climate through studies of Holocene charcoal sediments, fire-scar reconstructions, and other methods (Littell et al., 2009). Increasing extent and intensity of wildfire resulting directly through climate change, and indirectly through changes in pest and pathogen activity, which may increase mortality are of particular concern. In 2006, the Tripod Complex Fire burned over 80,000 ha in north-central Washington, and was one of the largest fire events in the state over the last 50 years. The fire burned more intensely than had been anticipated based on historical fires. Over 60% of the area burned was classified as moderate to high severity (US Forest Service, 2008). The lightning-caused fire was preceded by unusually hot and dry weather, and an extensive outbreak of mountain pine beetles.

The effects of climate change may be magnified in the mountains, as increasing temperatures reduce the depth of snowpacks and the timing of snowmelt (Knowles et al., 2006). Drought stress is often a limiting factor for tree growth at higher elevations (Peterson and Peterson, 2001).

Climate change has been and will continue to be an important driver of forest dynamics in the western US, affecting processes such as growth, mortality, and disturbance. Regional warming and water deficits will impact the range and community composition of forest species as trees are impacted by drought stress, changing fire regime, and increased insect and pathogen activity.

13.2.3.4 Tundra—Treeline

The treeline ecotone is the transition zone between a closed canopy, upright forest and low growing, primarily herbaceous tundra. We emphasize the ecotone as a zone, rather than a line, because any line one could identify is more scale dependent and has less ecological meaning. The ecotone occurs on mountain slopes as the alpine treeline and at high latitudes as the Arctic treeline (even in Tierra del Fuego the treeline seems to be determined by altitude (Cuevas, 2002)). At global scales, the latitudinal and elevational limits of trees are controlled by temperature. The limitation is either direct damage by frost, a lack of energy that does not allow individual plants to accumulate enough carbon via photosynthesis to form a tree (e.g., Cairns and Malanson, 1998), or, as more precisely demonstrated, at lower temperatures plants are not able to reallocate the energy they can gain in photosynthesis sufficient to form a tree (e.g., Körner, 1998). At finer spatial scales other factors, particularly geomorphology and available water for photosynthesis, could also be limiting factors (Malanson et al., 2011); however, at the scale where the ecological response may be considered a disaster, the focus should be on temperature (e.g., Billings and Peterson (1992) noted the importance of melting and eroding thermokarst for effects in the Arctic). The resulting hypothesis is that as climate warms, trees will move upslope and to higher latitude.

We have good evidence that such geographic responses have occurred in the past (Webb, 1992; Lloyd et al., 2002). Although the retreat of continental glaciers following the Late Glacial Maximum allowed trees to expand poleward, the details of the connection to climate are only seen at millennial or finer temporal scale. More recently, observations of current treelines using tree rings and of somewhat older treelines using dead trees indicate that the ecotone can fluctuate with changes in global climate (e.g., Lloyd and Graumlich, 1997; MacDonald et al., 1998).

The advance of tree cover upslope or to higher latitude also has some implications for climate change itself. Trees will absorb and store more carbon than do tundra plants. Thus, this response could have a negative feedback on the driving force of climate change. However, the strength of this feedback is not well quantified. The rise of alpine treelines to higher elevations would amount to a minor effect, given that the area is limited. A shift in latitudinal treelines could be more significant in the northern hemisphere. Northern forests are an important store of carbon, and through increase in biomass and soil organic matter have absorbed a significant amount of the carbon released by the burning of fossil fuels over the past two centuries. A northward expansion could increase this effect. Another feedback, a positive one, is that forests have a lower albedo than tundra. Absorbing more radiation over a larger area, they can contribute at least locally to warming—and thus to their own expansion.

Although the transition zone between trees and tundra has been the focus of research on possible ecological impacts of climate change, recently the role of shrubs has received more attention (Naito and Cairns, 2011). Although some tundra is clearly recognized as dominated by shrubs, the potential for the expansion of shrubs to replace herbaceous tundra could be an outcome of climate change. This expansion could affect carbon storage and albedo, but with lower canopy depth probably less than trees, but it also has effects on snow cover (Myers-Smith and Hik, 2013).

To further illustrate the effects of climate change on the treeline ecotone, we examine a single species, whitebark pine (*Pinus albicaulis*), as an example of the multiple factors of ecological response to climatic change (Tomback et al., 2001). Whitebark pine is a keystone species and plays an important role in maintaining ecosystem functions in subalpine environments. The large, highly nutritious seeds of the whitebark pine provide a vital food source for many species, including the threatened grizzly bear (*Ursus arctos horribilis*), Clark's nutcracker (*Nucifraga columbiana*), and other birds. The tree occupies high-elevation sites, often in steep, rocky areas, and serves to increase slope stability and soil formation in those sites. Individuals and small groups of whitebark pine provide rare shelter in those sites, blocking wind, and shading underlying snow which impacts the hydrology and succession. At higher elevations in undisturbed sites, whitebark pine communities may persist for more than 1,000 years. Among the threats to whitebark pine are mountain pine beetle (*Dendroctonus ponderosae*), white pine blister rust (*Cronartium ribicola*), and replacement by successional species as a result of fire suppression, all of which are impacted by climate change.

The mountain pine beetle spends the majority of its lifecycle as a larva feeding in the phloem tissue of pine trees. The host trees are eventually girdled and killed (Amman and Cole, 1983). Beetle survival and growth is sensitive to temperature, and outbreaks have been correlated with shifts in temperature (Powell and Logan, 2005). Climate change may affect beetle infestations through increasing drought stress, which inhibits the ability of trees to defend against the beetles, and warmer winters that may increase winter survival for the beetles. Predictions from population models suggest that range expansion will occur as beetles will be able to survive at higher latitudes and elevations in the coming century (Bentz et al., 2010). Range expansion into higher elevation forests will allow beetles to infect more whitebark pine, a species that has not evolved any defenses against the beetles (Raffa et al., 2008). Changes in forests will have secondary consequences (Saab et al., 2014).

Whitebark pine is unique in that has a mutually beneficial, even dependent, relationship with the Clark's nutcracker. The bird is dependent on the whitebark pine for food, has profoundly influenced the ecology and evolution of the pine as the tree is dependent on the nutcracker for seed dispersal, and hence the birds are responsible for locating many of these trees (Hutchins and Lanner, 1982). Clark's nutcrackers can store over 30,000 whitebark pine seeds in a mast season (Tomback, 1982): a number exceeding its nutritional requirements. The bird caches seeds at a depth of approximately 2 cm, often in the shelter of rocks, in sites compatible with germination requirements (Tomback, 1982).

White pine blister rust (WPBR) is a stem rust introduced from Europe early in the nineteenth century, and has spread throughout the ranges of the five-needled pines (including white, sugar, limber, and southwestern white pine). The fungus enters white pines through needle stomata and erupts as sport-producing cankers commonly on upper, cone-bearing branches, and tree mortality occurs through girdling or following the loss of branches from multiple cankers (Hoff and Hagle, 1989). Mortality can take many years in a mature tree; death can be hastened by pine beetle infections, root diseases, and other pathogens (Krebill and Hoff, 1995).

WPBR has the potential for causing local, if not global, extinction of whitebark pine (Kendall and Keane, 2001; Tomback and Achuff, 2010). The spread of blister rust and infection in whitebark pine is an intricate process, though small climatic changes, including an increase in the frequency of extreme precipitation events, could accelerate the spread of WPBR through whitebark pine habitat (Koteen, 2002). Changes in temperature, precipitation, relative humidity, and soil moisture influence sporulation and colonization of fungal diseases (Lonsdale and Gibbs, 1996; Smith-McKenna et al., 2013). Mortality from WPBR, warming and associated increased evapotranspiration, related changes in the fire regime, and competition from lower elevation species would push the whitebark pine to higher elevations (e.g., Millar et al., 2004). As with all species occurring at high-elevation, mountain peaks serve as a hard limit to the species' ability to move uphill to find habitable sites, and even at treeline WPBR is a threat (Tomback and Resler, 2007).

13.3 ECOSYSTEM SERVICES

Ecosystems across the globe affect services that are crucial not only to human survival, but also the functioning of all of the Earth's systems. From a human perspective, ecosystem services are those that are provided by nature when a product (e.g., timber) or a function (e.g., purification of water) benefits human well-being (Daniel et al., 2012). The United Nations Millennium Ecosystem Assessment (MA) grouped the ecosystem services that benefit humankind into four categories: provisioning, regulating, supporting, and cultural services (MA, 2005). Provisioning services are the products that we obtain from ecosystems for basic human needs such as food, water, minerals, shelter, and fuel. Regulating services provided by ecosystems include climate regulation, carbon feedbacks, the detoxification, and decomposition of waste, and associated purification of air and water. Supporting services are those that are required for the functioning of all other ecosystem services such as nutrient cycling, pollination, and seed dispersal. Cultural services are those that are nonmaterial but are directly experienced in nature and are essential to the human pursuit of meaning, such as spiritual and aesthetic experiences, cultural heritage, recreation, scientific exploration, and discovery.

Grimm et al. (2013a) noted that the impact of climate change on ecosystem services (as detailed by Nelson et al. (2013)) depended on impacts on ecosystem structure and function (as detailed by Grimm et al. (2013b)). Grimm et al. (2013b) noted that climate change would affect ecosystem functions such as productivity. Specifics included aquatic system responses to increased temperature and organic inputs and forest responses to changes in fire regimes and insect outbreaks. Nelson et al. (2013) noted the ecosystem services that would be affected included crop, seafood, and freshwater provisioning as well as the regulating value of coastal protection.

13.3.1 Provisioning

The provisioning services that would be directly affected by the ecological impacts of climatic change would be those wherein agroecosystems and silvicultural systems suffer reduced productivity. Climate change and agriculture is discussed elsewhere in this series, but we will discuss agroecosystems below in the context of supporting services. Silvicultural services are important components of economies globally, and changes in the productivity of trees are what underlie projections of change in commercially valuable tree species.

13.3.2 Regulating

The biosphere plays a major role in the global carbon cycle at temporal scales ranging from annual to geological. The annual flux in atmospheric CO_2 is the uneven balance of photosynthesis and respiration between summer and winter in the northern hemisphere; the existence of fossil fuels is in part a result of a longer nonequilibrium of photosynthesis and respiration in biological communities in the distant past. In between, the biosphere has absorbed a significant fraction of the carbon released by burning fossil fuels over the last two centuries. Ecological responses to climate change could affect photosynthesis and respiration and this regulating service; notably, photosynthesis can be limited by higher temperatures while respiration increases. Relative to the systems discussed above, we can see changes that could be significant in tropical, boreal, and oceanic ecosystems.

In boreal ecosystems, the advance of the Arctic treeline ecotone into tundra could allow for increased storage of carbon, given that boreal forests, particularly their soils, store relatively large amounts of carbon. However, this increase in storage in the ecotone could be balanced or overwhelmed by having less-continued storage of carbon with higher respiration rates and the release of currently stored carbon (having accumulated in recent decades) in higher temperatures. Given the extensive wetlands in the boreal and Arctic regions, some of this released carbon could be in the form of methane, which has a stronger radiative trapping effect per atom of C than does CO_2. Decreasing water level and increased temperature may decrease carbon sequestration in subarctic and boreal peatlands, especially in more southern latitudes (Gorham, 1991). The process may also be repeated in the tropics (Page et al., 2004). Mitsch and Gosselink (2007) provided a useful summary for wetlands.

13.3.3 Supporting

The supporting services that are of concern in the context of climate change are mostly those relevant to agroecosystems. Pollinators are the organisms of primary interest. Although crops and their pollinators share climatic niche space, it is not certain that their fundamental niches closely align, and so they could become decoupled during a period of climate change.

13.3.4 Cultural

The spiritual and aesthetic experiences, cultural heritage, recreation, scientific exploration, and discovery opportunities that make up cultural services can be based on individual species or entire ecosystems. Some of the individual species and even biomes that provide such services have been discussed above, and here we will focus on the broadest of the cultural services, biodiversity (cf. Mace et al., 2012). Although we discuss biodiversity in these terms, it is also a secondary (indirectly supporting) component of other ecosystem services in providing systems level redundancy for provisioning and regulating services (e.g., Tscharntke et al., 2005; Worm et al., 2006; Fisher et al., 2011).

13.3.4.1 Biodiversity

Biological diversity is the sum of the species, ecosystems, and genetic diversity in a given area. Species diversity is the species richness or the number of species in an area. Ecosystem diversity refers to the diversity of species assemblages and their environments over a specified scale, and genetic diversity is the variation of genes with and between species populations. Biodiversity is impacted by humans in many ways, including habitat degradation and loss, fragmentation, introduction of invasive species and pollution. Biodiversity was the focus of the 1987 symposium on ecological effects of climate change (Peters, 1992).

Climate change has the potential to reduce species diversity globally through extinction. Given that many species have limits in their tolerances to the ranges of some climate variables, climate change can kill individuals and result in extinction. Local biodiversity can also be changed in the same way. Climate change can reduce ecosystem diversity by eliminating the environment in which an entire group of species coexist. This impact is more likely to be local than global. One can imagine that in a drying climate the wetlands in an area can disappear. Relative to the ecosystems discussed above, the potential exists for a rising treeline to locally eliminate alpine tundra. Malanson et al. (2007) and Gret-Regamey et al. (2008) have discussed the aesthetic loss that would accompany this scenario.

13.4 THE PAST AND THE FUTURE

From studies of ecological responses to past climate change, we have learned that geographic range shifts are important responses to global climate change (Webb, 1992). But the most important problem that we have identified is the potential impact of the cultural ecosystem service of biodiversity. However, climate change in the past has not had a significant impact on biodiversity. Botkin et al. (2007) coined the term the "Quaternary conundrum." They noted that although our theoretical and model expectations would lead us to conclude that climate change would lead to extinctions, few extinctions observed during the glacial–interglacial cycles of the past 2 million years can be attributed to climate change. Instead, we can more directly attribute extinctions to the appearance of *Homo sapiens* and/or developments in our technology. One should note that extinctions, and related speciation, in the more distant past have been linked to climate change (e.g., Alroy et al., 2000).

Consider alpine tundra in the Rocky Mountains. At the LGM, tundra was at lower elevations and more extensive, possibly with donut holes of ice and rock that fell at higher elevations. During Holocene warming, tundra species migrated upslope and perhaps northward, but in both cases may have at least started in a more continuous band, except in the south where the alpine zone was already separated on mountain top islands (Lomolino et al., 2010; Harris, 2007). Now, however, alpine tundra is everywhere separated on such islands. Moreover, the area at highest elevations on any mountain range is smaller; even if the rock and ice could be colonized at the same rate as treeline migrated upward, and so the elevation band of alpine tundra remained constant, its areal extent would necessarily shrink. From the LGM to the peak altithermal the tundra responded to a similar amount of shrinkage.

An additional consideration is the high endemicity in alpine tundra. For data that we have examined for 56 mountain ranges in the western USA and southwestern Canada (by no means all ranges, especially in the Great Basin), approximately 40% of the species are found on a single range. Many endemics are, however, closely related to other endemics.

While the Quaternary conundrum exists, we recognize that current and future responses will be in a world with a different geography (Pitelka and The Plant Migration Workshop Group, 1997). Where we have the best detail, for the past 20,000 years in North America, we can see that species shifted their ranges northward; they did so at varying rates rather than synchronously (i.e., "individualistically"), and only major geographic barriers made a difference (i.e., the Great Lakes for the northwestern movement of American beech (*Fagus grandifolia*)) (Davis, 1989). However, more geographic barriers exist today (e.g., those created by land uses such as agriculture and urbanization that fragment habitat) and they are more extensive or even ubiquitous (Dyer, 1994; Malanson and Armstrong, 1996; Hellmann et al., 2010). Thus not only do barriers need to be crossed fewer target areas exist into which to move, and fewer sources occur (and thus fewer potential migrants or less "source strength") from which to start.

13.4.1 Mitigation

Given our observations and what we know, what we know we do not know, and the unknowns we do not yet know exist, what are possible responses, if any, to the ecological impacts of climate change in terms of mitigation? Given the geographical responses of the past and the geography of the present, the discussion has been on "assisted colonization," also known as "assisted migration." That is, can, and should people assist species to successfully change their range?

McLachlan et al. (2007) opened the debate on assisted migration. They noted that efforts at such assistance had started, but no science-based policy was in place; indeed, no directly applicable science was to be found. They developed options that could be debated, and outlined a framework under which research could be initiated. To date, this journal article has been cited 394 times (according to Google Scholar, 8/29/2014). These articles range from specific plans for individual species (e.g., for whitebark pine, discussed above; McLane and Aitken, 2012) to ethical considerations (Schwartz et al., 2012) to analyses of the rhetoric (Klenk and Larson, 2013). Among the notable follow-ons, Hoegh-Guldberg et al. (2008) developed a rationale and decision framework for undertaking assisted colonization. The decision to assist in colonization would be reached only if the following conditions were met: (1) a high risk of decline or extinction under climate change; (2) translocation and establishment are technically possible; and (3) the benefits of translocation outweigh the biological and socioeconomic costs and constraints.

Ricciardi and Simberloff (2009) criticized this approach. Basing their argument on the evidence from biological invasions, a topic covered elsewhere in this volume, they noted that translocations can decrease biodiversity and disrupt ecosystem function, with the latter including changes in fire regime and landscape pathology. Their arguments were countered, however, by Vitt et al. (2009), who noted that much of their argument was based on intercontinental transfers or those for systems with historically low connectivity. Vitt et al. (2009) argued that the problems envisioned were outweighed by the benefits in systems that historically were extensively connected but are now fragmented, such as the prairies of central North America (cf. Vitt et al., 2010). Schwartz et al. (2009) specifically recommended that the precautionary principle advocated by Ricciardi and Simberloff (2009) be balanced with a consideration of the costs of inaction.

People can move species. We could, as we observe climatic warming, move species into new areas where the climate is becoming more like that of their current range. However, the unknowns make this problematic and risky. The unknowns include not knowing the actual climatic limits of most species. We can observe the climate limits in which they do live, but these will seldom be the full climatic limits of where they can live (i.e., the difference between their realized and fundamental niche; Hutchinson, 1959; cf. Malanson et al., 1992). This means we may not know when we need to move a species. The difference between the realized and fundamental niche of species may be more important: the difference can depend on interactions with other species—thus any assisted migration must be in the context of multispecies relations. As noted above, mutualists are codependent. We also are uncertain about how to select individuals, or seeds, to move. Ecotypes within species vary across ranges, and those in ecotones may be more successful in advancing into new territory given adaptations to an atypical environment (Davis et al., 2005).

A related factor is that species are probably able to exist in climates that are outside their currently observed (realized) range in the absence of competitors. Evidence for such inertia comes from sites to which dispersal is limited (e.g., Cole, 1985). Current analogies to assisted migration exist where people have planted individuals in new places. For example, boreal forest conifers grace yards, parks, and cemeteries throughout the region of mid-latitude deciduous forest in the eastern US. These species are not able to establish in any numbers in those surrounding forests. If some of those deciduous forest species were to be assisted to move northwards as climate warms, would they be able to expand into the now-boreal forest area and hasten the demise of the current occupants? And thus would their move require subsequent assistance to move those occupants farther north?

Thinking on assisted migration can benefit from two considerations. First, the criteria laid out by Hoegh-Guldberg et al. (2008) need more precise explanation. In particular, their third criterion, that "the benefits of translocation outweigh the biological and socioeconomic costs and constraints," needs to be clearly understood. In many cases, it seems that the biological benefits and costs are not defined—a problem that has plagued ecological cost–benefit analysis. Second, as the overall decision framework of Hoegh-Guldberg et al. (2008) indicates, assisted migration must be part of a broader strategy.

Millar et al. (2007) laid out such a strategy, which they defined in three parts: resistance, resilience, and response. Resistance does not deny change, but allows for uncertainty. They noted several resistance tactics: modifying fire regimes through fuel breaks; active removal of invasive species; and defending against invasive pests or pathogens. Resilience, the ability to recover from disturbance, also buys time in the face of climate change. In addition to the tactics of resistance, assisting in reestablishment following disturbance is recommended. Response includes assisted migration; other tactics on their list include related actions such as establishing "neo-native" (considering paleo-ranges) distributions and promoting connectivity. Heller and Zavaleta (2009) followed up with a call for more specific and practical examples of planning and the planning process, especially some that integrated more rigorous social science.

Stein et al. (2013) emphasized the flexibility needed in addressing the ecological impacts of climate change. Not only do the adaptations of species and ecosystems need to be considered, but the adaptations of people and institutions, including our underlying goals for species and ecosystem conservation, may need to be constantly reviewed and adjusted as climate change continues.

To summarize, moving one or more species northward has been opposed largely on grounds of unknown consequences, based in part on evidence from the problems of invasive species. However, key points can be extracted from ecological theory.

- Points against assisted migration:
  - Quaternary conundrum
  - Species ability to migrate at fast rates
  - Ecotypes, possible in ecotones, better represent abilities to adapt
  - Inertia—species at southern end of their range could be affected negatively more by competition from new immigrants than by temperature (leaving water aside).

- Extinction debt—given an approximate equilibrium number of species in a given area, the addition of new species could lead to an increased rate of extinction with no certainty of which species would be lost. Extinction debt theory suggests that losses would depend on the degree of isolation of the place and the dispersal ability of the species.
- Points for assisted migration:
  - Increased fragmentation of landscapes leaves more barriers, to which species migration rates are not evolved; this fragmentation also means lower source strength.
  - Endemism means that local extinctions can be global extinctions.
- Points even:
  - Although the rate of climate change in the next century will be faster than the average rate of any millennium since the LGM, high rates did exist in the past. Although we do not know the spatial extent of these episodes, we do not know they were necessarily isolated—and in any case species will be responding locally, not globally.

For alpine tundra, the increasing geographic isolation, the high endemicity, and the aesthetic value of biodiversity argue for active mitigation.

13.4.1.1 Conservative Mitigation

For plant species, particularly endemics, an active, conservative mitigation strategy would be to create species/seed/gene banks, i.e., gardens (Havens et al., 2006; Hunter, 2007). Such gardens would need to maintain genetic diversity and thus would require isolation and replication. The existing model for this approach is the zoo. Conservationists have argued the merits of zoos as a source for reintroduction at some later time (e.g., Millar et al., 2004), but a definitive answer is not yet available (Would zoo animals be maladapted for the wild?). For plants, some problems are lessened. Combined with other strategies, to avoid the *Silent Running* endgame, adding ex situ conservation for plants, is a mitigation action that keeps future options open.

REFERENCES

Adams, H.D., Guardiola-Claramonte, M., Barron-Gafford, G.A., Villegas, J.C., Breshears, D.D., 2009. Temperature sensitivity of drought-induced tree mortality portends increased regional die-off under global-change type drought. Proc. Natl. Acad. Sci. U.S.A. 106 (17), 7063–7066.

Alexander, V., 1992. Arctic marine ecosystems. In: Peters, R.L., Lovejoy, T.E. (Eds.), Global Warming and Biological Diversity. Yale University Press, New Haven, CT, pp. 221–232.

Alroy, J., Koch, P.L., Zachos, J.C., 2000. Global climate change and North American mammalian evolution. Paleobiology 26, 259–288.

Amman, G.D., Cole, W.E., 1983. Mountain pine beetle dynamics in lodgepole pine forests. Part II: population dynamics. General Tech. Rep. Intermt. For. Range Exp. Stn. USDA Forest Service No INT-145, 63 pp.

Amstrup, S.C., Marcot, B.G., Douglas, D.C., 2007. Forecasting the Range-wide Status of Polar Bears at Selected Times in the 21st Century. US Geological Survey, Washington, DC.

Atkinson, A., Siegel, V., Pakhomov, E., Rothery, P., 2004. Long-term decline in krill stock and increase in salps within the Southern Ocean. Nature 432, 100–103.

Atkinson, C.T., Woods, K.L., Dusek, R.J., Sileo, L.S., Iko, W.M., 1995. Wildlife disease and conservation in Hawaii: pathogenicity of avian malaria in experimentally infected iiwi. Parasitology 111, S59–S69.

Auer, V., 1927. Stratigraphical and morphological investigation of peat bogs in southeastern Canada. Quaetonum Foresta, Finlandiae, Edita 12.

Baker, A.J., González, P.M., Piersma, T., Niles, L.J., do Nascimento, I.D.L.S., Atkinson, P.W., Clark, N.A., Minton, C.D.T., Peck, M.K., Aarts, G., 2004. Rapid population decline in red knots: fitness consequences of decreased refuelling rates and late arrival in Delaware Bay. Proc. R. Soc. B 271, 875–882.

Bentz, B.J., Régnière, J., Fettig, C.J., Hansen, E.M., Hayes, J.L., Hicke, J.A., Kelsey, R.G., Negrón, J.F., Seybold, S.J., 2010. Climate change and bark beetles of the western United States and Canada: direct and indirect effects. BioScience 60, 602–613.

Berkelmans, R., De'ath, G., Kininmonth, S., Skirving, W.J., 2004. A comparison of the 1998 and 2002 coral bleaching events on the Great Barrier Reef: spatial correlation, patterns, and predictions. Coral Reefs 23, 74–83.

Billings, W.D., Peterson, K.M., 1992. Some possible effects of climatic warming on arctic tundra ecosystems of the Alaskan North Slope. In: Peters, R.L., Lovejoy, T.E. (Eds.), Global Warming and Biological Diversity. Yale University Press, New Haven, CT, pp. 233–243.

Both, C., Visser, M.E., 2001. Adjustment to climate change is constrained by arrival date in a long-distance migrant bird. Nature 411, 296–298.

Botkin, D.B., Saxe, H., Araujo, M.B., Betts, R., Bradshaw, R.H.W., Cedhagen, T., Chesson, P., Dawson, T.P., Etterson, J.R., Faith, D.P., Ferrier, S., Guisan, A., Hansen, A.S., Hilbert, D.W., Loehle, C., Margules, C., New, M., Sobel, M.J., Stockwell, D.R.B., 2007. Forecasting the effects of global warming on biodiversity. BioScience 57, 227–236.

Bradley, N.L., Leopold, A.C., Ross, J., Huffaker, W., 1999. Phenological changes reflect climate change in Wisconsin. Proc. Natl. Acad. Sci. U.S.A. 96, 9701–9704.

Caffrey, M.A., Doerner, J.P., 2012. A 7000-year record of environmental change, Bear Lake, rocky mountain National Park, USA. Phys. Geogr. 33, 438–456.

Cairns, D.M., Malanson, G.P., 1998. Environmental variables influencing the carbon balance at the alpine treeline: a modeling approach. J. Veg. Sci. 9 (5), 679–692.

Chmura, D.J., Anderson, P.D., Howe, G.T., Harrington, C.A., Halofsky, J.E., 2011. Forest responses to climate change in the northwestern United States: ecophysiological foundations for adaptive management. For. Ecol. Manag. 261, 1121–1142.

Colinvaux, P.A., De Oliveira, P.E., Bush, M.B., 2000. Amazonian and neotropical plant communities on glacial time-scales: the failure of the aridity and refuge hypotheses. Quat. Sci. Rev. 19, 141–169.

Cole, K., 1985. Past rates of change, species richness, and a model of vegetational inertia in the Grand Canyon, Arizona. Am. Nat. 125, 289–303.

Comiso, J.C., 2012. Large decadal decline of the Arctic multiyear ice cover. J. Clim. 25, 1176–1193.

Croxall, J.P., Trathan, P.N., Murphy, E.J., 2002. Environmental change and Antarctic seabird populations. Science 297, 1510–1514.

Cuevas, J.G., 2002. Episodic regeneration at the *Nothofagus pumilio* alpine timberline in Tierra del Fuego Chile. J. Ecol. 90, 52–60.

Daniel, T.C., Muhar, A., Arnberger, A., Aznar, O., Boyd, J.W., Chan, K.M., Costanza, R., Elmqvist, T., Flint, C.G., Gobster, P.H., Gret-Regamey, A., Lave, R., Muhar, S., Penker, M., Ribe, R.G., Schauppenlehner, T., Sikor, T., Soloviy, I., Spierenburg, M., Taczanowska, K., Tam, J., von der Dun, C., 2012. Contributions of cultural services to the ecosystem services agenda. Proc. Natl. Acad. Sci. U.S.A. 109, 8812–8819.

Davis, M.B., 1989. Lags in vegetation response to greenhouse warming. Clim. Change 15, 75–82.

Davis, M.B., Shaw, R.G., Etterson, J.R., 2005. Evolutionary responses to changing climate. Ecology 86, 1704–1714.

Derocher, A.E., Lunn, N.J., Stirling, I., 2004. Polar bears in a warming climate. Integr. Comp. Biol. 44, 163–176.

Diaz-Delgado, R., Lloret, F., Pons, X., Terradas, J., 2002. Satellite evidence of decreasing resilience in Mediterranean plant communities after recurrent wildfires. Ecology 83, 2293–2303.

Diffenbaugh, N.S., Field, C.B., 2013. Changes in ecologically critical terrestrial climate conditions. Science 341, 486–492.

Douglas, A.E., 2003. Coral bleaching– how and why? Mar. Pollut. Bull. 46, 385–392.

Dyer, J.M., 1994. Implications of habitat fragmentation on climate change-induced forest migration. Prof. Geogr. 46, 449–459.

Emslie, S.D., Fraser, W., Smith, R.C., Walker, W., 1998. Abandoned penguin colonies and environmental change in the Palmer Station area, Anves Island, Antarctic Peninsula. Antarct. Sci. 10, 257–268.

Fisher, B., Turner, R.K., Burgess, N.D., Swetnam, R.D., Green, J., Green, R.E., Balmford, A., 2011. Measuring, modeling and mapping ecosystem services in the Eastern Arc Mountains of Tanzania. Prog. Phys. Geogr. 35, 595–611.

Franklin, J.F., Swanson, F.J., Harmon, M.E., Perry, D.A., Spies, T.A., Dale, V.H., McKee, A., Ferrell, W.K., Means, J.E., Gregory, S.V., Lattin, J.D., Schowalter, T.D., Larsen, D., 1992. Effects of global climate change in forests of northwestern North America. In: Peters, R.L., Lovejoy, T.E. (Eds.), Global Warming and Biological Diversity. Yale University Press, New Haven, CT, pp. 244–257.

Fraser, W.R., Trivelpiece, W.Z., Ainley, D.G., Trivelpiece, S.G., 1992. Increases in Antarctic penguin populations: reduced competition with whales or a loss of sea ice due to environmental warming? Polar Biol. 11, 525–531.

Garcia, C.T., Hernandez, T., Roldan, A., Martin, A., 2002. Effect of plant cover decline on chemical and microbiological parameters under Mediterranean climate. Soil Biol. Biochem. 34, 635–642.

Gedalof, Z., Peterson, D.L., Mantua, N.J., 2005. Atmospheric, climatic, and ecological controls on extreme wildfire years in the Northwest United States. Ecol. Appl. 15, 154–174.

Gitay, H., Raudsepp-Hearne, C., 2005. Assessment process. In: Assessments, M.-S. (Ed.), Findings of the Sub-global Assessment Working Group (Millennium Ecosystem Assessment). Island Press, Washington.

Gorham, E., 1991. Northern peatlands: role in the carbon cycle and probable responses to climatic warming. Ecol. Appl. 1, 182–195.

Gret-Regamey, A., Walz, A., Bebi, P., 2008. Valuing ecosystem services for sustainable landscape planning in Alpine regions. Mt. Res. Dev. 28, 156–165.

Grimm, N.B., Chapin, F.S., Bierwagen, B., Gonzalez, P., Groffman, P.M., Luo, Y.Q., Melton, F., Nadelhoffer, K., Pairis, A., Raymond, P.A., Schimel, J., Williamson, C.E., 2013a. The impacts of climate change on ecosystem structure and function. Front. Ecol. Environ. 11, 474–482.

Grimm, N.B., Staudinger, M.D., Staudt, A., Carter, S.L., Chapin, F.S., Kareiva, P., Ruckelshaus, M., Stein, B.A., 2013b. Climate-change impacts on ecological systems: introduction to a US assessment. Front. Ecol. Environ. 11, 456–464.

Gross, L., 2005. As the antarctic ice pack recedes, a fragile ecosystem hangs in the balance. PLoS Biol. 3, e127.

Harris, S.A., 2007. Biodiversity of the alpine vascular flora of the NW North American Cordillera: the evidence from phyto-geography. Erdkunde 61, 344–357.

Hartshorn, G.S., 1992. Possible effects of global warming on the biological diversity of tropical forests. In: Peters, R.L., Lovejoy, T.E. (Eds.), Global Warming and Biological Diversity. Yale University Press, New Haven, CT, pp. 137–146.

Havens, K., Vitt, P., Maunder, M., Guerrant Jr., E.O., Dixon, K., 2006. Ex situ plant conservation and beyond. BioScience 56, 525–531.

Heller, N.E., Zavaleta, E.S., 2009. Biodiversity management in the face of climate change: a review of 22 years of recommendations. Biol. Conserv. 142, 14–32.

Hellmann, J.J., Nadelhoffer, K.J., Iverson, L.R., Ziska, L.H., Matthews, S.N., Myers, P., Prasad, A.M., Peters, M.P., 2010. Climate change impacts on terrestrial ecosystems in metropolitan Chicago and its surrounding, multi-state region. J. Gt. Lakes. Res. 36, 74–85.

Heyerdahl, E.K., McKenzie, D., Daniels, L.D., Hessl, A.E., Littell, J.S., Mantua, N.J., 2008. Climate drivers of regionally synchronous fires in the inland Northwest (1651–1900). Int. J. Wildland Fire 17, 40–49.

Hill, G.E., Sargent, R.R., Sargent, M.B., 1998. Recent change in the winter distribution of rufous hummingbirds. Auk 1151, 240–245.

Hoegh-Guldberg, O., 1999. Climate change, coral bleaching and the future of the world's coral reefs. Mar. Freshw. Res. 50, 839–866.

Hoegh-Guldberg, O., Hughes, L., McIntyre, S., Lindenmayer, D.B., Parmesan, C., Possingham, H., Thomas, C.D., 2008. Assisted colonization and rapid climate change. Science 321, 345–346.

Hoff, R.J., Hagle, S., 1989. Disease of whitebark pine with special emphasis on white pine blister rust. In: Schmidt, W.C., MacDonald, K.J., (Compilers) (Eds.). Proceedings–Symposium on Whitebark Pine Ecosystems: Ecology and Management of a High-mountain Resource, USDA Forest Service General Technical Report INT-270, pp. 179–190.

Houghton, R.A., Lawrence, K.T., Hackler, J.L., Brown, S., 2001. The spatial distribution of forest biomass in the Brazilian Amazon: a comparison of estimates. Glob. Change Biol. 7, 731–746.

Hunter Jr., M.L., 2007. Climate change and moving species: furthering the debate on assisted colonization. Conserv. Biol. 21, 1356–1358.

Hupy, C.M., 2012. Mapping ecotone movements: holocene dynamics of the forest tension zone in central Lower Michigan, USA. Phys. Geogr. 33, 473–490.

Hutchins, H.E., Lanner, R.M., 1982. The central role of Clark's nutcracker in the dispersal and establishment of whitebark pine. Oecologia 55, 192–201.

Hutchinson, G.E., 1959. Homage to Santa Rosalia or why are there so many kinds of animals? Am. Nat. 93, 145–159.

von Humboldt, A., Bonpland, A., 1807. Essai sur la géographie des plantes. Fr. Schoell, Bookseller, Paris. Essay on the Geography of Plants, University of Chicago Press, 2013.

Iles, D.T., Peterson, S.L., Gormezano, L.J., Koons, D.N., Rockwell, R.F., 2013. Terrestrial predation by polar bears: not just a wild goose chase. Polar Biol. 36, 1373–1379.

Jaeschke, A., Bittner, T., Jentsch, A., Beierkuhnlein, C., 2014. The last decade in ecological climate change impact research: where are we now? Naturwissenschaften 101, 1–9.

Kendall, K.C., Keane, R.E., 2001. Whitebark pine decline: infection, mortality, and population trends. In: Tomback, D.F., Arno, S.F., Keane, R.E. (Eds.), Whitebark Pine Communities: Ecology and Restoration. Island Press, Washington, DC, pp. 221–242.

Klenk, N.L., Larson, B.M., 2013. A rhetorical analysis of the scientific debate over assisted colonization. Environ. Sci. Policy 33, 9–18.

Knowles, N., Dettinger, M.D., Cayan, D.R., 2006. Trends in snowfall versus rainfall in the western United States. J. Clim. 19, 4545–4559.

Körner, C., 1998. A re-assessment of high elevation treeline positions and their explanation. Oecologia 115, 445–459.

Koteen, L., 2002. Climate change, whitebark pine, and grizzly bears in the Greater Yellowstone Ecosystem. In: Schneider, S.H., Root, T.L. (Eds.), Wildlife Responses to Climate Change: North American Case Studies. Island Press, Washington, pp. 343–414.

Krebill, R.G., Hoff, R.J., 1995. Update on *Cronartium ribicola* in *Pinus albicaulis* in the Rocky Mountains, USA. In: Proceedings 4th IUFRO Rusts of Pines Working Party Conference, Tsukuba, Japan, pp. 119–126.

LaPointe, D.A., 2000. Avian Malaria in Hawaii: The Distribution, Ecology and Vector Potential of Forest-dwelling Mosquitoes (Ph.D. thesis). University of Hawaii, Honolulu.

LeBlanc, A.M., Drake, K.K., Williams, K.L., Frick, M.G., Rostal, D.C., 2012. Nest temperatures and hatchling sex ratios from loggerhead turtle nests incubated under natural field conditions in Georgia, U.S.A. Chelonian Conserv. Biol. 11, 108–116.

Lenton, T.M., Held, H., Kriegler, E., Hall, J.W., Lucht, W., Rahmstorf, S., Schellnhuber, H.J., 2008. Tipping elements in the Earth's climate system. Proc. Natl. Acad. Sci. U.S.A. 105, 1786–1793.

Littell, J.S., McKenzie, D., Peterson, D.L., Westerling, A.L., 2009. Climate and wildfire area burned in western U.S. ecoprovinces, 1916–2003. Ecol. Appl. 19, 1003–1021.

Littell, J.S., Oneil, E.E., McKenzie, D., Hicke, J.A., Lutz, J.A., Norheim, R.A., Elsner, M.M., 2010. Forest ecosystems, disturbance, and climatic change in Washington State, USA. Clim. Change 102, 129–158.

Lloyd, A.H., Graumlich, L.J., 1997. Holocene dynamics of treeline forests in the Sierra Nevada. Ecology 78, 1199–1210.

Lloyd, A.H., Rupp, T.S., Fastie, C.L., Starfield, A.M., 2002. Patterns and dynamics of treeline advance on the Seward Peninsula, Alaska. J. Geophys. Res. Atmos. 108 (D2) art. no. 8161.

Lomolino, M.V., Riddle, B.R., Whittaker, R.J., Brown, J.H., 2010. Biogeography. Sinauer, Sunderland, MA.

Lonsdale, G., Gibbs, J.N., 1996. Effects of climate change on fungal diseases of trees. In: Frankland, J.C., Magan, N., Gadd, G.M. (Eds.), Fungi and Environmental Change. Cambridge University Press, Cambridge, pp. 1–19.

MacDonald, G.M., Szeicz, J.M., Claricoates, J., Dale, K.A., 1998. Responses of the central Canadian treeline to recent climatic changes. Ann. Assoc. Am. Geogr. 88, 183–208.

Mace, G.M., Norris, K., Fitter, A.H., 2012. Biodiversity and ecosystem services: a multilayered relationship. Trends Ecol. Evol. 27, 19–26.

Malanson, G.P., Armstrong, M.P., 1996. Dispersal probability and forest diversity in a fragmented landscape. Ecol. Model. 87, 91–102.

Malanson, G.P., Westman, W.E., 1991. Climatic change and the modeling of fire effects in coastal sage scrub and chaparral. In: Fire and the Environment: Ecological and Cultural Perspectives: Proceedings of an International Symposium. USDA Forest Service, Southeastern Forest Experiment Station, General Technical Report SE-69, pp. 91–96.

Malanson, G.P., 2008. Extinction debt: origins, developments, and applications of a biogeographic trope. Prog. Phys. Geogr. 32, 277–291.

Malanson, G.P., Butler, D.R., Fagre, D.B., 2007. Alpine ecosystem dynamics and change: a view from the heights. In: Prato, T., Fagre, D. (Eds.), Sustaining Rocky Mountain Landscapes: Science, Policy and Management of the Crown of the Continent Ecosystem. RFF Press, Washington, pp. 85–101.

Malanson, G.P., Butler, D.R., Walsh, S.J., 2004. Ecological response to global climatic change. In: Janelle, D.G., Warf, B., Hansen, K. (Eds.), WorldMinds. Kluwer, Dordrecht, pp. 469–473.

Malanson, G.P., Resler, L.M., Bader, M.Y., Holtmeier, F.-K., Weiss, D.J., Butler, D.R., Fagre, D.B., Daniels, L.D., 2011. Mountain treelines: a roadmap for research orientation. Arct. Antarct. Alp. Res. 43, 167–177.

Malanson, G.P., Westman, W.E., Yan, Y.-L., 1992. Realized versus fundamental niche functions in a model of chaparral response to climatic change. Ecol. Model. 64, 261–277.

Mayer, A.G., 1914. The Effects of Temperature upon Marine Animals. Published Papers, Marine Lab Tortugas (6). Cam Institute, Washington, pp. 3–24.

Mayle, F.E., Beerling, D.J., Gosling, W.D., Bush, M.B., 2004. Responses of Amazonian ecosystems to climatic and atmospheric carbon dioxide changes since the last glacial maximum. Philos. Trans. R. Soc. B 359, 499–514.

van Mantgem, P.J., Stephenson, N.L., Byrne, J.C., Daniels, L.D., Franklin, J.F., Fulé, P.Z., Harmon, M.E., Larson, A.J., Smith, J.M., Taylor, A.H., Veblen, T.T., 2009. Widespread increase of tree mortality rates in the western United States. Science 323, 521–524.

McClintock, J., Ducklow, H., Fraser, W., 2008. Ecological responses to climate change on the Antarctic Peninsula. Am. Sci. 96, 302–310.

McKenzie, D., Ze'ev, G., Peterson, D.L., Mote, P., 2004. Climatic change, wildfire, and conservation. Conserv. Biol. 18, 890–902.

McLachlan, J.S., Hellmann, J.J., Schwartz, M.W., 2007. A framework for debate of assisted migration in an era of climate change. Conserv. Biol. 21, 297–302.

McLane, S.C., Aitken, S.N., 2012. Whitebark pine (*Pinus albicaulis*) assisted migration potential: testing establishment north of the species range. Ecol. Appl. 22, 142–153.

Menzel, A., Sparks, T.H., Estrella, N., Koch, E., 2006. European phenological response to climate change matches the warming pattern. Glob. Change Biol. 12, 1969–1976.

Menzel, A., Fabian, P., 1999. Growing season extended in Europe. Nature 397, 659.

Meyer, B.C., Rannow, S., 2013. Landscape ecology and climate change adaptation: new perspectives in managing the change. Reg. Environ. Change 13, 739–741.

Myers-Smith, I.H., Hik, D.S., 2013. Shrub canopies influence soil temperatures but not nutrient dynamics: an experimental test of tundra snow–shrub interactions. Ecol. Evol. 3 (11), 3683–3700.

Millar, C.I., Westfall, R.D., Delany, D.L., King, J.C., 2004. Response of subalpine conifers in the Sierra Nevada, California, USA, to 20th century warming and decadal climate variability. Arct. Antarct. Alp. Res. 36, 181–200.

Millar, C.I., Stephenson, N.L., Stephens, S.L., 2007. Climate change and forests of the future: managing in the face of uncertainty. Ecol. Appl. 17, 2145–2151.

Millennium Ecosystem Assessment, 2005. Island Press, Washington, DC.

Mitsch, W.J., Gosselink, J.G., 2007. Wetlands. Wiley, Hoboken, NJ.

Moriondo, M., Good, P., Durao, R., Bindi, M., 2006. Potential impact of climate change on fire risk in the Mediterranean area. Clim. Res. 31, 85–95.

Mouillot, F., Rambal, S., Joffre, R., 2002. Simulating climate change impacts on fire frequency and vegetation dynamics in a Mediterranean-type ecosystem. Glob. Change Biol. 8, 423–437.

Naito, A.T., Cairns, D.M., 2011. Patterns and processes of global shrub expansion. Prog. Phys. Geogr. 35 (4), 423–442.

Nelson, E.J., Kareiva, P., Ruckelshaus, M., Arkema, K., Geller, G., Girvetz, E., Goodrich, D., Matzek, V., Pinsky, M., Reid, W., Saunders, M., Semmens, D., Tallis, H., 2013. Climate change's impact on key ecosystem services and the human well-being they support in the US. Front. Ecol. Environ. 11, 483–493.

Nobre, C.A., Sellers, P.J., Shulka, J., 1991. Amazonian deforestation and regional climate change. J. Clim. 4, 957–988.

Orth, R.J., Carruthers, T.J.B., Dennison, W.C., Duarte, C.M., Fourqurean, J.W., 2006. A global crisis for seagrass ecosystems. BioScience 56, 987–996.

Page, S.E., Wust, R.A.J., Weiss, D., Rieley, J.O., Shotyk, W., Limin, S.H., 2004. A record of Late Pleistocene and Holocene carbon accumulation and climate change from an equatorial peat bog (Kalimantan, Indonesia): implications for past, present and future carbon dynamics. J. Quat. Sci. 19, 625–635.

Paulson, D.R., 2001. Recent Odonata records from southern Florida-effects of global warming? Int. J. Odonatology 4, 57–69.

Peñuelas, J., Gordon, C., Llorens, L., Nielsen, T., Tietema, A., 2004. Nonintrusive field experiments show different plant responses to warming and drought among sites, seasons, and species in a north-south European gradient. Ecosystems 7, 598–612.

Peters, R.L., Lovejoy, T.E., 1992. Global Warming and Biological Diversity. Yale University Press, New Haven, CT.

Peters, R.L., 1992. Conservation of biological diversity in the face of climate change. In: Peters, R.L., Lovejoy, T.E. (Eds.), Global Warming and Biological Diversity. Yale University Press, New Haven, CT, pp. 15–30.

Peterson, D.W., Peterson, D.L., 2001. Mountain hemlock growth responds to climatic variability at annual and decadal time scales. Ecology 82, 3330.

Pitelka, L.F., The Plant Migration Workshop Group, 1997. Plant migration and climate change. Am. Sci. 85, 464–473.

Post, E., Forchhammer, M.C., Bret-Harte, M.S., Callaghan, T.V., Christensen, T.R., Elberling, B., Fox, A.D., Gilg, O., Hik, D.S., Høye, T.T., Ims, R.A., Jeppesen, E., Klein, D.R., Madsen, J., McGuire, A.D., Rysgaard, S., Schindler, D.E., Stirling, I., Tamstorf, M.P., Tyler, N.J.C., van der Wal, R., Welker, J., Wookey, P.A., Schmidt, N.M., Aastrup, P., 2009. Ecological dynamics across the Arctic associated with recent climate change. Science 325, 1355–1358.

Post, E., Bhatt, U.S., Bitz, C.M., Brodie, J.F., Fulton, T.L., Hebblewhite, M., Kerby, J., Kutz, S.J., Stirling, I., Walker, D.A., 2013. Ecological consequences of sea-ice decline. Science 341 (6145), 519–524.

Powell, J.A., Logan, J.A., 2005. Insect seasonality: circle map analysis of temperature- driven life cycles. Theor. Popul. Biol. 67, 161–179.

Raffa, K.F., Aukema, B.H., Bentz, B.J., Carroll, A.L., Hicke, J.A., Turner, M.G., 2008. Cross-scale drivers of natural disturbance prone to anthropogenic amplification: the dynamics of bark beetle eruptions. BioScience 58, 501–517.

Ricciardi, A., Simberloff, D., 2009. Assisted colonization in not a viable conservation strategy. Trends Ecol. Evol. 24, 248–253.

Riera, P., Peñuelas, J., Farreras, J., Estiarte, V., 2007. Valuation of climate-change effects on Mediterranean shrublands. Ecol. Appl. 17, 91–100.

Root, T.L., Price, J.T., Hall, K.R., Schneider, S.H., Rosenzweig, C., Pounds, J.A., 2003. Fingerprints of global warming on wild animals and plants. Nature 421, 57–60.

Saab, V.A., Latif, Q.S., Rowland, M.M., Johnson, T.N., Chalfoun, A.D., Buskirk, S.W., Heyward, J.E., Dresser, M.A., 2014. Ecological consequences of mountain pine beetle outbreaks for wildlife in western North American forests. For. Sci. 60, 539–559.

Sala, O.E., Chapin, F.S., Armesto, J.J., Berlow, E., Bloomfield, J., 2000. Biodiversity- global biodiversity scenarios for the year 2100. Science 287, 1770–1774.

Schwartz, M.D., 1993. Assessing the onset of spring: a climatological perspective. Phys. Geogr. 14, 536–550.

Schwartz, M.D., Ahas, R., Aasa, A., 2006. Onset of spring starting earlier across the Northern Hemisphere. Glob. Change Biol. 12, 343–351.

Schwartz, M.W., Hellmann, J.J., McLachlan, J.S., 2009. The precautionary principle in managed relocation is misguided advice. Trends Ecol. Evol. 24, 474.

Schwartz, M.W., et al., 2012. Managed relocation: integrating the scientific, regulatory, and ethical challenges. Bioscience 62, 732–743.

Screen, J.A., Deser, C., Simmonds, I., 2012. Local and remote controls on observed Arctic warming. Geophys. Res. Lett. 39, L10709.

Shinn, E.A., 1966. Coral growth rate, an environmental indicator. J. Paleontol. 40, 233–240.

Smith-McKenna, E., Resler, L.M., Tomback, D.F., Zhang, H., Malanson, G.P., 2013. Topographic influences on the distribution of white pine blister rust in *Pinus albicaulis* treeline communities. Ecoscience 20, 215–229.

Sparks, T., Menzel, A., 2002. Observed changes in seasons: an overview. Int. J. Climatol. 22, 1715–1725.

Staudinger, M.D., Carter, S.L., Cross, M.S., Dubois, N.S., Duffy, J.E., Enquist, C., Griffis, R., Hellmann, J.J., Lawler, J.J., O'Leary, J., Morrison, S.A., Sneddon, L., Stein, B.A., Thompson, L.M., Turner, W., 2013. Biodiversity in a changing climate: a synthesis of current and projected trends in the US. Front. Ecol. Environ. 11, 465–473.

Stein, B.A., Staudt, A., Cross, M.S., Dubois, N.S., Enquist, C., Griffis, R., Hansen, L.J., Hellmann, J.J., Lawler, J.J., Nelson, E.J., Pairis, A., 2013. Preparing for and managing change: climate adaptation for biodiversity and ecosystems. Front. Ecol. Environ. 11, 502–510.

Stephenson, N.L., Das, A.J., Condit, R., Russo, S.E., Baker, P.J., Beckman, N.G., Coomes, D.A., Lines, E.R., Morris, W.K., Rüger, N., Álvarez, E., Blundo, C., Bunyavejchewin, S., Chuyong, G., Davies, S.J., Duque, Á., Ewango, C.N., Flores, O., Franklin, J.F., Grau, H.R., Hao, Z., Harmon, M.E., Hubbell, S.P., Kenfack, D., Lin, Y., Zavala, M.A., 2014. Rate of tree carbon accumulation increases continuously with tree size. Nature. Advanced online publication January 15, 2014. Nature 507, 90–93.

Stirling, I., Derocher, A.E., 1993. Possible impacts of climatic warming on polar bears. Arctic 46, 240–245.

Stirling, I., Derocher, A.E., 2012. Effects of climate warming on polar bears: a review of the evidence. Glob. Change Biol. 18, 2694–2706.

Tomback, D.F., 1982. Dispersal of whitebark pine seeds by Clark's nutcracker: a mutualism hypothesis. J. Animal Ecol. 51, 451–467.

Tomback, D.F., Resler, L.M., 2007. Invasive pathogens at alpine treeline: consequences for treeline dynamics. Phys. Geogr. 28, 397–418.

Tomback, D.F., Achuff, P., 2010. Blister rust and western forest biodiversity: ecology, values and outlook for white pines. For. Pathol. 40, 186–225.

Tomback, D.F., Arno, S.F., Keane, R.E., 2001. Whitebark Pine Communities: Ecology and Restoration. Island Press, Washington, DC.

Tscharntke, T., Klein, A.M., Kruess, A., Steffan-Dewenter, I., Thies, C., 2005. Landscape perspectives on agricultural intensification and biodiversity—ecosystem service management. Ecol. Lett. 8, 857–874.

U.S. Forest Service, 2008. Final Environmental Impact Statement. Methow Valley Ranger District. Winthrop, Washington.

Vitt, P., Havens, K., Hoegh-Gulberg, O., 2009. Assisted migration: part of an integrated conservation strategy. Trends Ecol. Evol. 24, 473–474.

Vitt, P., Havens, K., Kramer, A.T., Sollenberger, D., Yates, E., 2010. Assisted migration of plants: changes in latitudes, changes in attitudes. Biol. Conserv. 143, 18–27.

Walther, G.R., Post, E., Convey, P., Menzel, A., Parmesan, C., Beebee, T.J., Bairlein, F., 2002. Ecological responses to recent climate change. Nature 416, 389–395.

Webb, T., Bartlein, P.J., 1992. Global changes during the last 3 million years- climatic controls and biotic responses. Annu. Rev. Ecol. Syst. 23, 141–173.

Webb III, T., 1992. Past changes in vegetation and climate: lessons for the future. In: Peters, R.L., Lovejoy, T.E. (Eds.), Global Warming and Biological Diversity. Yale University Press, New Haven, CT, pp. 59–75.

Westman, W.E., Malanson, G.P., 1992. Effects of climate change in Mediterranean-type ecosystems in California and Baja California. In: Peters, R.L., Lovejoy, T.E. (Eds.), Global Warming and Biological Diversity. Yale University Press, New Haven, CT, pp. 258–276.

Wilson, P.R., Ainley, D.G., Nur, N., Jacobs, S.S., Barton, K.J., Ballard, G., 2001. Adélie penguin population change in the Pacific sector of Antarctica: relation to sea ice extent and the Antarctic Circumpolar Current. Mar. Ecol. Prog. Ser. 213, 301–309.

Worm, B., Barbier, E.B., Beaumont, N., Duffy, J.E., Folke, C., Halpern, B.S., Watson, R., 2006. Impacts of biodiversity loss on ocean ecosystem services. Science 314, 787–790.

Young, C.M., Nicholls, A.G., 1931. Studies on the physiology of corals. IV. The structure, distribution and physiology of the zooxanthellae. Sci. Rep. Gt. Barrier Reef Exped. 1928–1929, 135–176.

Zhou, G., Houlton, B., Zhang, D., 2014. Substantial reorganization of China's tropical and subtropical forests: based on the permanent plots. Glob. Change Biol. 20, 240–250.

Chapter 14

Meteor Impact Hazard

René A. De Hon
Department of Geography, Texas State University, San Marcos, TX, USA

ABSTRACT
The Earth sits in a cosmic shooting gallery. In its orbit around the Sun, Earth intercepts meteoroidal and asteroidal debris left over from the earliest days of the solar system. Meteor impact threat ranges from small impacts—providing local disruption of human activities similar to large storms or earthquakes—to catastrophic extinction events similar to the one that led to the extinction of the dinosaurs. It is simply a matter of time before a big impact threatens a large portion of the population. International cooperation is required to detect and counter potential threats from large impacts.

14.1 INTRODUCTION

At 9:22 AM (UTC/GMT + 6 h) on Febuary 15, 2013, a 17- to 20-m-diameter meteor entered the atmosphere above the Southern Ural region of Russia near the town of Chelyabinsk at a speed of 18.6 m/s. As it traversed the atmosphere at a shallow angle to the surface, the 10,000-ton meteor was slowed by friction and heated to brilliant radiance (Figure 14.1). A bright light flared in the early morning grayness as a bolide sped across the sky (Kuzmin, 2013). At a height of 23.3 km, the atmosphere in front of the meteor had been compressed to such an extent that the meteor exploded with the energy equivalent of 500 kilotons of TNT (Brown et al., 2013) and sent a shock wave that struck the surface more than 2 min later with the equivalent force of a 2.7 earthquake—shattering windows and toppling unsupported walls. Because the meteor was traveling at a low angle to the surface, the shock wave was spread out along its ground path. Authorities estimate (Popova et al., 2013) nearly 1500 people were injured—mostly by shattered glass, and over 7200 buildings were damaged at a cost estimated at a billion rupees ($33 million) (Zhang, 2013). Fragments of the asteroid, an ordinary chondrite, were scattered across the surface west of Chelyabinsk and on the frozen surface of Lake Chebarkul. A 6-m-wide hole in the lake ice prompted a magnetic survey, and in October, 2013, a 654-kg fragment of the meteor was recovered from the lake. The Chelyabinsk

Biological and Environmental Hazards, Risks, and Disasters. http://dx.doi.org/10.1016/B978-0-12-394847-2.00023-1

FIGURE 14.1 Chelyabinsk bolide as imaged by dashboard camera on the morning of February 15, 2013.

meteorite was a timely reminder that the Earth is a target in the cosmic shooting gallery (Brown et al., 2013).

14.2 PLANETARY EVIDENCE

The early solar system was awash in a chaotic jumble of asteroidal bodies, many of which were perturbed into highly elliptical orbits, that crossed the orbits of newly formed planets. A reign of asteroidal bombardment peppered the face of the planets in the early days of the solar system. Without the shield of an atmosphere, the Moon and Mercury stand as witness to unrelenting bombardment by asteroidal impact. Because Earth and Venus have atmospheres and active surface processes, they have a much lower number of preserved impacts. Although the rates of impacts are greatly reduced from the early days of the solar system, recent events give evidence that the frequency of impacts is not zero.

The surface of the Moon provides a record of heavy bombardment very early in its history, followed by a rapid decline in impacts until the present. The early impact rate produced a saturated cratered surface (Figure 14.2) in which any new crater would destroy existing craters to maintain a constant crater count. Formation of new surfaces by volcanic eruptions provides a glimpse at lower cratering rates as time passed. It is, therefore, possible to establish an impact-crater-age versus time for the Moon and—by extension—for the Earth. As asteroidal debris was swept up by planetary gravity fields, the impact rate declined drastically from their initial high rates. Currently, asteroidal impact rate by bodies of appreciable size is low, but not negligible. The NASA monitoring system (Suggs et al., 2008) reports approximately 300 observable impacts recorded on the Moon between 2005 and 2013. The largest recorded strike was thought to be by a meteor approximately 40 kg or about 0.3- to 0.4-m diameter.

FIGURE 14.2 Lunar surface illustrating saturated cratering in which the crater count for the surface becomes constant as any new crater will destroy older craters. Saturated cratering occurred early in the history of the solar system as the larger bodies swept out smaller bodies in their path. Lunar Orbiter IV frame 4106-h2.

14.3 HISTORICAL RECORD

The Earth's atmosphere acts as a shield against small meteors. A casual watch of the night sky reveals a flash of meteorites at a rate of 6–10/h. Although these are mere specs of dust, they are part of the population of asteroidal debris intersected by the Earth's orbit in its journey through space. Infrequently a meteor large enough to brighten the daytime sky—or to be seen to flare and break apart—gives evidence of larger fragments. Occasionally pieces of debris large enough to survive the fiery passage through the atmosphere make it to the surface. Those that do are most often traveling at relatively low velocities. It is the larger meteors traveling at high velocities that become a concern.

Throughout history, reports have occurred of stones falling from the sky, but the scientific community did not recognize the extraterrestrial origin of meteorites until the 1700s. Within historic times (D'Orazio, 2007) Aristotle recounted "the stone at Aegospotami fell out of the air" in 465 BC. A more complete account was described by Pliny the Elder. "The Greeks tell the story that Anaxagoras of Clazomenae, in the 2nd year of the 78th Olympiad was enabled by his knowledge of astronomical literature to prophecy that in a certain number of days a rock would fall from the sun. And that this occurred in the daytime in the Goat's River district of Thrace (the stone is still shown—it is of the size of a wagon-load and brown in color), a comet also blazing in the nights at the time." Thomas Jefferson, when president of the American Philosophical Society, is reported to have said, "That it was easier to believe that two Yankee Professors could lie than to admit that stones could fall from heaven" (Silliman, 1874). The veracity of the quote has been questioned, but it does convey the skepticism of that time.

Records of deaths by meteor impact are rare and so deeply buried in the past as to be unverifiable (Spratt, 1991). The Ch'ing-yang event in March or April of 1490 (Yau et al., 1994) is the only record of a large number of fatalities by what is presumed to be the result of an atmospheric burst of a meteorite that resulted in a rain of stones from the sky. The Ensisheim fall of 1492 appears to be the first well-documented fall in which fragments are still preserved (Marvin, 1992). Between 1790 and 1990, 61 buildings have been struck by meteorites (Spratt, 1991).

In June 30, 1908, a mysterious explosion in remote Siberia, felled trees over an area of 2000 km^2 and put enough dust in the atmosphere to decrease temperatures in Europe for several months. Various exotic hypotheses, ranging from a chance encounter with antimatter, to a crashed alien spaceship, abound (Jackson and Ryan, 1973). The remoteness of the impact site assured that no casualties occurred, and that the site was not visited until 1929, over two decades after the event. Although evidence of a very large explosive event was observed in the flattened forest with felled trees pointing away from the center of the explosion (Figure 14.3), neither crater nor meteorite debris was found. It is now assumed that the Tunguska event was similar to, but much larger than, the more recent Chelyabinsk event (Chyba et al., 1993; Farinella et al., 2001; Kvasnytsya et al., 2013).

More recent records of less striking falls serve as clear examples of the continuing rain of meteoric debris. In 1913, a ship was reported to have been struck and damaged by a meteorite while sailing between Sydney and South America (Mercury, 1913). An air blast and fragmentation of a stony meteor is reported to have occurred over Chicora, PA, in June 1938 that may have killed a cow (Preston et al., 1941). The Sikhote-Alin meteor shower in 1947 was an atmospheric airburst of an iron meteorite that spread fragments over an area of about 1.3 km^2. The shower included one fragment of 1.75 tons in a 4-m-deep crater as well as many other small craters up to 26.5-m diameter (Gallant, 1996). The Sylacauga (Alabama) meteorite fell on November 30, 1954. The brick-sized meteorite punched through the roof of a house and struck Ann Elizabeth Hodges while napping on the couch (Spratt, 1991). She received large bruising on her thigh but no lasting damage. The Allende meteor shower (Mexico) of February 8, 1969, produced a bright fireball and thousands of stones over a 50 × 20-km ellipse (Clarke et al., 1970). Over 2 tons of material

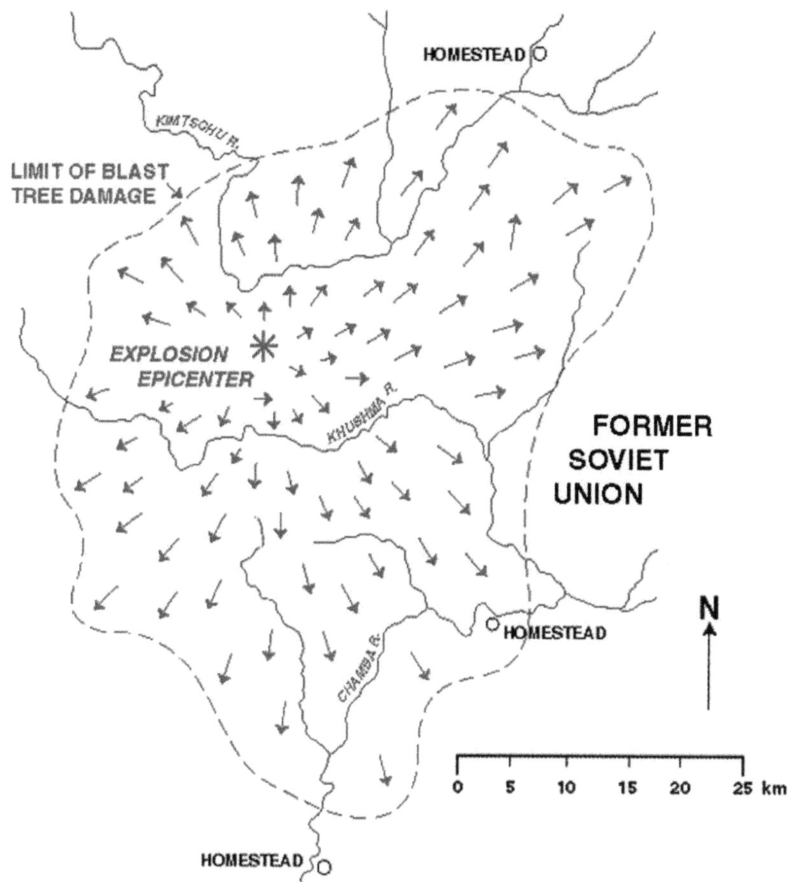

FIGURE 14.3 The radial pattern of flattens trees indicated the center of the Tunguska airburst, but early expeditions were unable to find a crater (*after Kridec, 1963*).

have been recovered. The largest fragment is 110 kg. In 1992, a very small fragment (3 g) of Mbale meteorite (Jenniskens, 1994) hit a young Ugandan boy but it had been slowed down by a tree and did not cause any injury. Also in 1992, the Peekskill fireball of October 9 in New York (Beech et al., 1995), struck and damaged a parked car (Figure 14.4). In September of 2007, a meteorite struck near Carancas, Peru (Jackson et al., 2008), causing a 20-m-diameter crater and causing arsenic poisoning to a large number of people—apparently by interaction of the hot meteor and groundwater.

14.4 GEOLOGIC RECORD

The preceding accounts of encounters with meteors or asteroids involve relatively small bodies slowed by atmospheric drag. Larger bodies survive

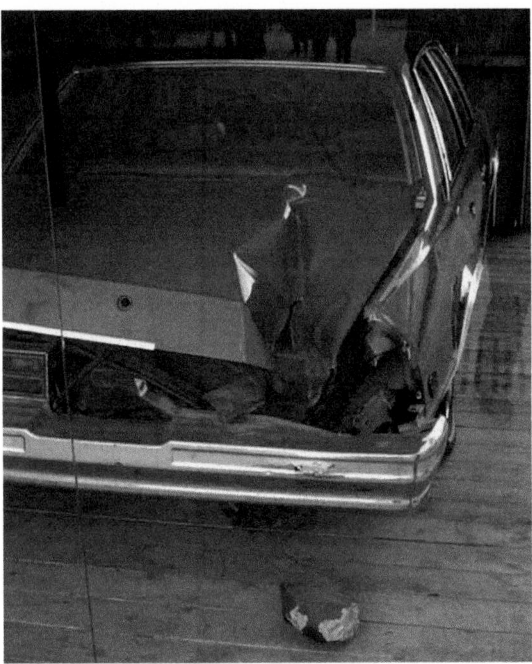

FIGURE 14.4 The Peekskill Meteorite penetrated the roof of a garage and damaged a car.

passage through the atmosphere with a significant portion of their original mass and velocity. The resulting impact of a fast-moving body with a solid, rock surface is a nearly instantaneous conversion of energy from the moving body to the target material. Craters produced by meteor impact are essentially explosion craters. The expanding shock wave in the surface rocks excavates a crater and heaves rocks onto the surface to form a raised rim. On airless bodies, such as the Moon, craters persist for longer periods of time until buried by subsequent lava flows or degraded by later impacts, On Earth, active surface processes wear away at the crater until their surface expression is removed, subcrater structures and petrographic evidence of the impact may last for extremely long times, especially in tectonically stable regions (Figure 14.5). Only the geologically more recent impact sites are identified by their characteristic craterform morphology. Most of the nearly 175 confirmed terrestrial impact sites (Hodges, 1994) are recognized by the characteristic deformation of impacted rocks subjacent to the crater, shatter cones (Dietz, 1959), and the high-pressure polymorph of silica–coesite (Chao et al., 1960).

Small meteorites traveling at subsonic velocities do little more than imbed themselves into the soil. As meteoroid velocities reach hypervelocity, speeds greater than the speed of sound in the target, the energy of motion is converted into explosive energy (heat and shock wave) that shatters the impactor and blasts a crater in the surface. Small craters of a few kilometers in diameter

FIGURE 14.5 Clearwater Lakes of Quebec, Canada, are twin craters formed in crystalline Precambrian rocks of the Canadian Shield. The inner ring of the larger lake is about 10 km in diameter.

primarily produce local effects that are limited to an area, a few times larger than the crater rim-crest diameter. Large impacts produce atmospheric effects that may trigger long-term changes in climate. Early life would have had a difficult time becoming established during the period of high impact rates.

Quite a few ancient crater remnants are recognized in stable Precambrian terrains as those of the Canadian Shield of North America (Figure 14.6; Robertson and Grieve, 1975). Stable continental interiors, such as the North American mid-continent region, chiefly contain almost flat-lying strata of shallow marine strata overlying Precambrian basement rocks. Craters formed in these rocks have been eroded to the point that the surface expression of an impact crater is lost, but the deformation patterns in rock layers beneath the crater persist. Identified as cryptovolcanic structures by Bucher (1936), these areas of circular, concentric, and radial structural patterns have been verified as impact structures (French, 1998) by the presence of shatter cones (Figure 14.7) and coesite. Sixty structures in North America are currently verified meteor impact sites and another 11 are suspected (Rajmon, 2009).

Major extinctions may have been triggered by disruptions associated with very large impacts. Generally two to five taxonomic families of marine invertebrates and vertebrates become extinct every million years. Mass extinctions are recognized when rates of extinction exceed rates of speciation. Spikes in the rate of extinction are marked by a sharp change in the diversity and abundance of macroscopic life. Depending on what level of change is

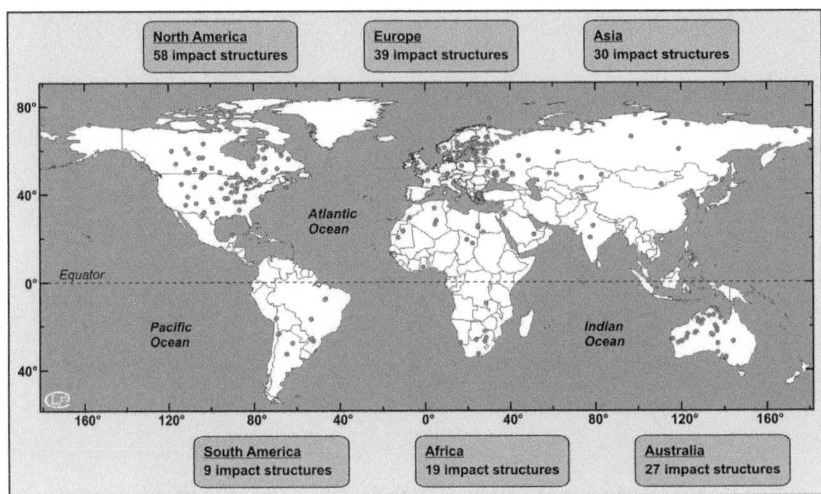

FIGURE 14.6 World map showing the location of most of the known meteor impact sites (after L. Ferriere). Most sites have been modified by erosion so that their surface expression is no longer a crater. The pattern of deformation and the presence of coesite and shatter cones provide evidence of ancient impact.

FIGURE 14.7 Shatter cones east of Santa Fe, New Mexico.

taken as the threshold, 5–10 mass extinction events are recognized in the fossil record. The causes for these crises may be related to major changes in habitat caused by changes in climate or changes in ocean chemistry or temperature. These changes may have been precipitated by abrupt variations caused by extensive volcanism (Wignall, 2005); asteroidal impact; changes in sea level; depletion of oxygen in the oceans; widespread glaciation; or the introduction of a new efficient predator (Sole and Newmann, 2002; Alroy, 2008; Jenniskens et al., 2013). Geologists currently debate which of the mass extinctions might be attributed to asteroidal impact, but the acceptance of an impact at the end of Cretaceous has gained strong support.

The five largest extinction events (Raup and Sepkoski, 1982) include the end of the Ordovician Period, 450 million years before the present (MYBP); the late Devonian Period, 375 MYBP; the Permian–Triassic transition, 251 MYBP; the Triassic–Jurassic event, 200 MYBP; and the end of Cretaceous event, 65 MYBP (Figure 14.8).

The Permian–Triassic extinction event—at 251 MYBP, the largest mass extinction and known as the "Great Dying"—killed 57% of all families; 83% of all genera; and more than 90% of all marine species (Retallack, 1995; Erwin, 1990). This is the only extinction event to have greatly affected insects. Ocean temperature and anoxic conditions reached record levels. This event cleared the field for the rise of the dinosaurs. Recovery time was one of the longest, requiring 10 million years. Current evidence (Benton and Twitchett, 2003) suggests that massive volcanic eruptions in Siberia are most likely responsible for this event.

The Triassic–Jurassic extinction event—at 200 MYBP during the Age of the Dinosaurs—lost 23% of the families, 48% of the genera, and more than 70% of all species were eliminated. Most of the large amphibians were lost, as well as the early mammals, providing even less competition for the rising dinosaur populations. No event has been unequivocally identified as responsible for this mass extinction. No crater of sufficient size has been found.

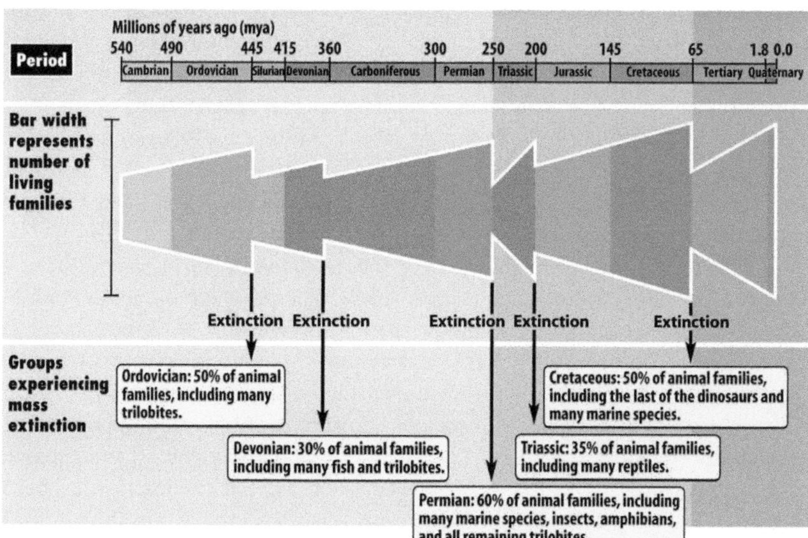

FIGURE 14.8 Major disruptions in the ecology of the planet are marked massive extinction events (Singh and Cain, 2011). The loss of the dinosaurs at the end of the Cretaceous Period is thought to have been caused by the impact that formed the Chicxulub crater in the southern Gulf of Mexico.

Possible volcanic activity in the Mid-Atlantic Volcanic Province appears to be the most likely cause for this extinction event (Blackburn et al., 2013).

The end of Cretaceous extinction event (K–T boundary)—at 65 MYBP—brought the dinosaurs to an abrupt end. The Cretaceous ended with 17% of all families, 50% of genera, and 75% of all species extinct. Dinosaurs, which ruled for nearly 200 MY, were virtually eliminated, and mammals and birds began their rise to dominance. An impact event and/or volcanism have been suggested for the demise of the dinosaurs. The proposal by Alvarez et al. (1980) that a large iridium spike at the K–T boundary was the smoking gun for impact is generally accepted. The 150-km-diameter Chicxulub crater in the Gulf of Mexico is regarded as the probable crater responsible for the extinction event (Schulte et al., 2010; Pope et al., 1994). The impact in a shallow marine shelf is credited with producing mega-conglomerates in coastal sediments observed in Texas, Cuba, Yucatan, and Belize.

14.5 SURVIVING METEOR CRATERS

Several young craters (Pleistocene or younger), less than 2 million years old, retain their craterform morphologies. Within the United States, Barringer crater in Arizona is the most notable. The Odessa Craters in West Texas are probably the same age, but they are smaller and less well preserved. The Impact Crater Database lists several moderately young craters in Australia, Namibia, Chad, Guyana, Egypt, and India.

Barringer Crater, also known simply as Meteor Crater, Arizona, is a well-preserved impact crater approximately 1.2 km in diameter surrounded by a raised rim of overturned strata and fractured materials. Although it was originally declared to be of volcanic origin by Gilbert (1896), Barringer (1906) and others (Fairchild, 1907; Pickering, 1909; Nininger, 1954) argued for its meteoritic origin. Barringer went so far as to try to find and mine for profit the large mass of iron meteor he thought would remain beneath the crater floor. Its impact origin was verified by Shoemaker unpublished: Chao et al. (1960). The crater extends 170 m below the level of the surrounding surface, and the rim extends 45 m above the surface (Figure 14.9). The presence of iron meteorite fragments and coesite—a high-pressure polymorph of quartz—points convincingly to an impact origin. A minimum of 20 tons of iron meteorite fragments have been collected from the surface outside the crater. The crater was produced by the impact of a meteor approximately 50 m in diameter that released the equivalent of a 10 MT TNT explosion upon impact. It is estimated that the rim has been lowered by 30 m by erosion and that the interior has been filled with approximately 200 m of material washed from the rim (Shoemaker and Kieffer, 1979; Roddy and Shoemaker, 1995). The age of the crater is estimated as 50,000 years.

The Odessa Crater field consists of five small craters, the largest of which is about 160 m in diameter and 10 m deep. The small craters are extensively

FIGURE 14.9 The well-preserved Meteor Crater (Barringer Crater) in Arizona provides a natural laboratory for the study of impact structures. Although privately owned, the Astrogeology Branch of the U.S. Geological Survey has provided extensive information for the museum at the crater and the crater served as a training ground for the lunar astronauts.

reduced in relief by erosion. Trenching by the Civilian Conservation Corp in the 1930s and a drill hole near the middle of the largest crater reveals the structure of the crater and an original depth of 30 m (Holliday et al., 2005). Meteorite fragments found in the area resemble those of the Barringer Crater, raising the speculation that these craters may have been produced by a fragmented meteor at the same time.

Australia has 30 verified impact sites and 20 more that are suspected as being impact sites. Wolfe Creek Crater is 880-m diameter and 60-m deep is the best preserved, although nearly 300,000 years old (Reeves and Chalmers, 1949). It is associated with iron meteor fragments. The Henbury crater field contains 13 or 14 small craters ranging from 7 to 180 m in diameter and up to 15-m deep, and replace with-the craters are estimated to be approximately 4200 years old (Alderman, 1931).

14.6 ORIGINS AND PROPERTIES OF ASTEROIDS, METEOROIDS, AND COMETS

Planetesimals are bodies of solid materials formed in the earliest planetary nebulae that have grown large enough to begin to draw other materials to them by gravitational attraction. The larger planetesimals are differentiated bodies consisting of an iron-rich core, a rocky mantle, and an impacted, brecciated regolith. A few planetesimals grew to near planetary dimensions by capturing nearby smaller planetesimals. Others were shattered into smaller bodies by repeated impacts in the chaotic jumble of the early solar nebula. Over the millennia, planets swept up asteroidal debris from their orbits, but a very large concentration remains in the asteroid belt between Mars and Jupiter.

The term "asteroid" lacks a formal definition, but generally applies to those solid bodies of the inner solar system that lack sufficient mass to self-gravitate into a spherical body. Asteroids are small planetesimals or fragments of planetismals shattered by random impacts in the chaotic early days of the solar system. Some asteroids are little more than loose collections of debris reassembled by local gravitational forces. Others are solid fragments identified by their spectra as iron (M-type); silicate-rich rock (S-type); or carbon-rich (C-type). Asteroidal densities varying from 7 (M-type) to 2 (C-type); sizes range from 1000-km to 1-m diameter; and orbital velocities range from 10 to 30 km/s relative to the moving Earth (Figure 14.10).

Most asteroids are in orbits between Mars and Jupiter, but some exist within the orbit of Jupiter. Perturbations of asteroidal orbits by the gravitational attraction of the planets and large asteroids result in unstable and ever-changing orbits. Perturbations cause many asteroids to be in highly elliptical orbits that intersect the orbit of the inner planets. Earth-crossing asteroids with potential to intersect with the Earth are in elliptical orbits that originate in the asteroid belt and have perihelion inside the Earth's orbit (Figure 14.11).

FIGURE 14.10 A collage of asteroids showing varying shapes and sizes. The irregular surfaces are pitted from repeated collisions with other asteroids. Some bodies may be little more than a loose gravitation pile of rubble.

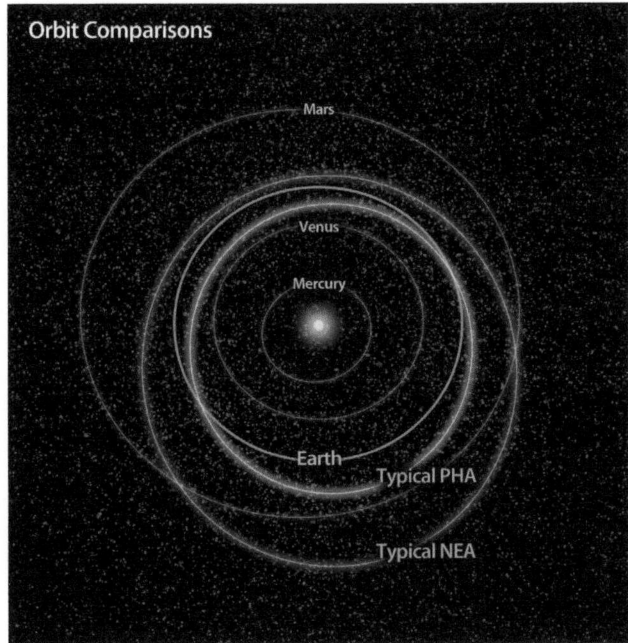

FIGURE 14.11 Example of a potentially hazard asteroid PHA with and Earth crossing orbit and a Near Near Asteroid (NEA) that does not intersect the Earth's orbit.

A meteoroid is a solid object smaller than an asteroid moving in space (Rubin and Grossman 2010). Meteoroids are small pieces—often described as less than 10 m—of shattered asteroids or stony remnants of burned out comets. A meteor is the visible streak seen in the sky as a micrometeoroid or meteoroid flares by friction with the atmosphere. A meteorite is a portion of a meteoroid or an asteroid that survived passage through the atmosphere to impact on the surface. Meteorites may arrive at the surface greatly slowed from their original orbital velocity, or if sufficiently large, they may impact the surface with a significant portion of their original velocity. If the impacting velocity is greater than the velocity of sound in the target rocks, it is known as a hypervelocity impact. The result of a hypervelocity impact into rock is essentially an explosion in which the kinetic energy of the moving mass is converted to shock and heat energy in the target rocks and the meteorite.

Meteorites are classified as irons, stony-irons, and stones. The irons are largely iron–nickel alloys with minor sulfide and silicate components. The stony-irons are iron meteorites with a significant portion of silicate minerals. The stony meteorites are subdivided into carbonaceous chondrites, ordinary chondrites, and achondrites. Chondritic meteorites are recognized by the presence of small, bb-sized, silicate spherical bodies. The silicate components of chondrites are chiefly olivine and pyroxene. A subclass of the achondrites, known as the basaltic achondrites, contains feldspars.

Most comets are icy bodies that originate in the Oort cloud beyond the orbit of Neptune. Perturbations bring these long-period comets into the inner solar system with very high velocities relative to the planets. Short-period comets originate within the orbit of Neptune and have orbits similar to those of the asteroids. Comets are identified in the inner solar system by the development of an elongated tail developed as the solar wind drives gases from the comets' surfaces. Orbital velocities of short-term comets, when approaching the Earth, are similar to those of the asteroids and meteoroids. The long-period comets are quite variable, ranging from 30 to 150 mps owing to the widely variable places of origin beyond Neptune. As orbits decay and perihelion approaches the Sun, comets may ablate entirely, breaking into smaller bodies, or plunging into the Sun. Comet Shoemaker–Levy 9 provided comet watchers with a rare view of a fragmented comet and the rarer spectacle of multiple impacts onto the planet Jupiter (Figure 14.12). After many passages into the inner solar system, ice sublimation may reduce the comet to rocky bodies that would be largely unobserved as they continue to circle the Sun with very long periods. The short-period comets that lose their icy component may continue to exist as meteoroids.

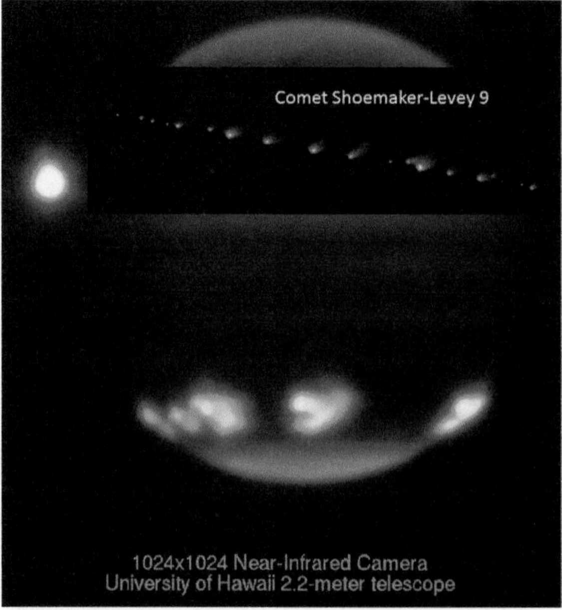

FIGURE 14.12 The shoemaker–Levy 9 comet was a unique string of 21 comet fragments trapped in an orbit around Jupiter. The string impacted Jupiter in July, 1994.

14.7 HYPERVELOCITY IMPACT

Meteors and asteroids encounter the Earth with relative velocities of 15–70 km/s. Owing to their origins much further out in the Solar System, comet velocities are on the order of 70–300 km/s. The energy released by a meteorite encounter depends upon its diameter, density, velocity, and angle of approach. The kinetic energy of the impact (KE) = 1/2 mv^2; where m is the mass of the meteor and v is velocity. Upon entering the Earth's atmosphere, friction will begin to slow the body as it plunges toward the surface. Friction with the atmosphere will heat the outer layer, and ablation will strip melted material from the meteor. Small meteors flare to incandescence and ablate completely in the upper atmosphere. Meteor size determines how much it is slowed. Drag in the atmosphere slows small bodies (<1 ton) to terminal velocity of approximately 90 m/s or 320 km/h. Larger bodies maintain a portion of their original velocity. Meteors larger than 10–20 tons strike the surface with velocities greater than the velocity of sound waves in the target material (hypervelocity impact greater than 2–2.5 km/s). The larger the body, the greater the proportion of original velocity upon impact. A 1 kiloton meteorite will retain almost 75% of its original velocity.

Stony meteorites up to 100-m diameter are prone to atmospheric explosions in which air in front of the meteor becomes compressed to the point that the meteor is shattered in an airburst that spreads high-speed projectiles over an extended area. Such an airburst was responsible for the Tunguska event of 1908 and the Chelyabinsk event of 2013. Airbursts distribute their meteoritic fragments over an oval-shaped area of many square kilometers known as a strewn field (Figure 14.13). Impact velocities of the fragments are generally slowed to less than hypervelocity, and these impacts do not create significant craters. Another recent airburst was the Allende meteorite (carbonaceous chondrite) of February 1969, which created an 8 by 50 km strewn field from which more than 3 tons of specimens have been recovered. Other past airbursts are evidenced by strewn fields in which large numbers of mostly stony meteorite fragments occur.

Meteors and asteroids greater than 100 m are likely to survive passage through the atmosphere. Upon striking the surface, the meteorites' energy is converted to heat and hypervelocity shock in both the target rocks and the meteorite (Melosh, 1989). The meteorite is stopped in a distance less than its own diameter. The expanding shock wave in the target shatters the rock and ejects it from the growing crater as an expanding cone of high-speed ballistic ejecta (Figure 14.14). The shock wave in the meteorite melts and vaporizes part of the impactor. The remaining solid portion is shattered and ejected from the crater with target material. As the shock wave slows in the target, bulk ejecta is heaved at low velocity to form a raised rim and some materials fall back to the crater floor. The crater floor consists of a layer of fall-back ejecta overlying shattered target rock. In very large craters, the floor may contain a central rise of material compressed by the impact and then rebounded back toward the surface.

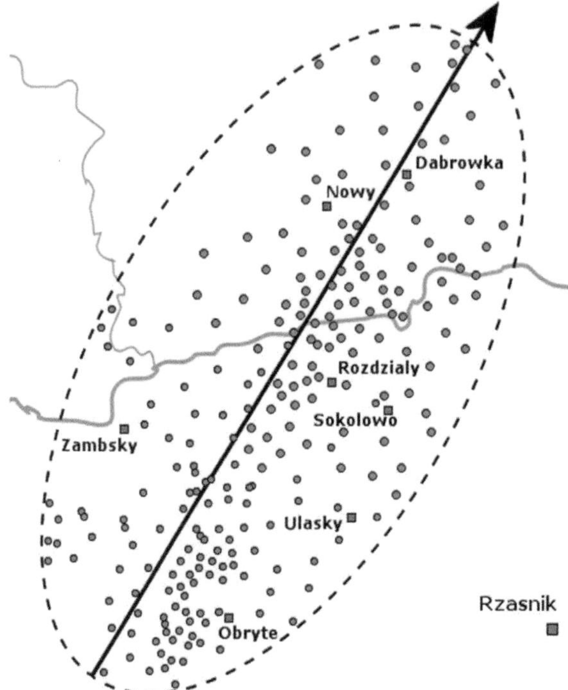

FIGURE 14.13 Most asteroids are in orbits between Mars and Jupiter, but collisions and perturbations by the gravitational attraction of Jupiter forces some asteroids into elliptical, planet-crossing orbits.

FIGURE 14.14 Impact of a hypervelocity projectile with the Earth produces a shock wave in both the target and the projectile. The results are essentially an explosive ejection of material from the growing crater along at angles.

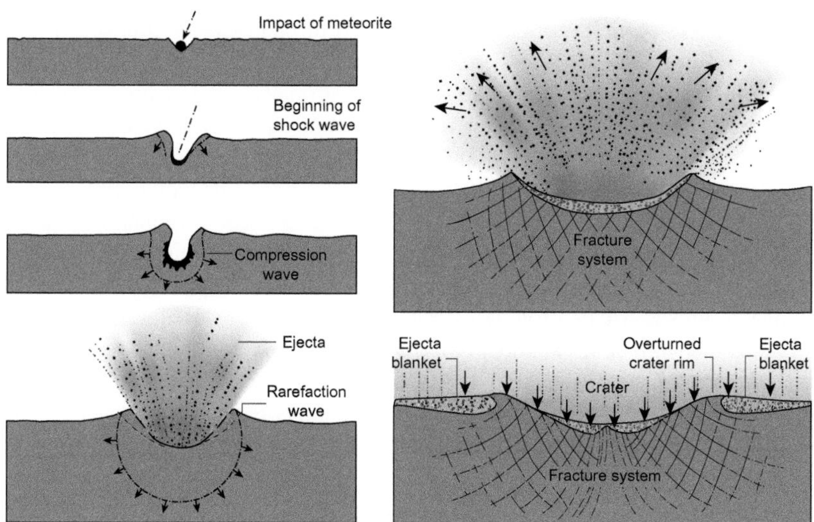

FIGURE 14.15 The young crater Kepler on the Moon exhibits typical fresh-crater morphology. The bright radial streaks radiating from the crater are bright rays formed by high-speed ejecta falling back to the surface creating overlapping, small, secondary craters and churning bright materials to the surface.

A large amount of fine-grained target material entrained in heated air and gas generated by the explosion flows along the ground as a fast-moving dense cloud of suspended particles, which form bedded, base-surge deposits. The outer rim and down range for several crater diameters is subjected to the impact of high-speed ballistic ejecta that is ejected from the crater along high-angle ballistic paths (Figure 14.15).

In addition to mechanical effects of shock waves, excavation of a crater, and fall-back of ejecta, tornado-velocity winds radiate as much as five times greater than the crater diameter. High-speed, dense asteroids (>3.5) create a fireball that can be expected to radiate from the impact site as a firestorm. Changes in weather patterns may be similar to those postulated by large-scale nuclear war (Turco et al., 1983). Dust raised by large impact explosion and smoke from the firestorm will rise into the upper atmosphere and promote a climatic cooling effect by blocking incoming sunlight. It is the possibility of a long-lasting, global climate change that leads to catastrophic consequences of mass extinction by cessation of photosynthesis leading to the loss of plant life for herbivores and the loss of feed stock for carnivores. Even smaller impacts that reduce sunlight and temperatures would result in loss of crops and widespread famine in human populations (Engvild, 2003).

In studies, reported by Covey et al. (1994) and Toon et al. (1997), it was found that a meteorite about 10 km (6.2 mi) in diameter had the explosive force of about 108 MT, and it could send upward of about 2.5 × 1015 kg of 1-μm-sized particles into the upper atmosphere. Anything larger would fall quickly back to the surface. These particles would then be spread throughout the atmosphere and absorb or refract the sunlight before it is able to reach the surface, cooling the planet in the same fashion as sulfuric acid droplets rising from a volcanic eruption. These particles would remain in the atmosphere for about a year before precipitating out, but even then only about 85% of the Sun's radiation would reach the surface. After the first 20 days, the land temperature would drop quickly by about 13° K. After about a year, the temperature would rebound by about 6° K, but by this time about one-third of the Northern Hemisphere would be covered in ice. Atmospheric blocking would be further enhanced by smoke rising from wildfires, perhaps prolonging the period of cooling. Reduced photosynthesis would stunt or kill a large portion of the plant community accompanied by the starvation of herbivores and the carnivores that feed on them. With approximate 30% loss of the feed stock and food holdings required for a year, widespread famine and disruption of civilization may be expected.

14.7.1 High-Velocity, Low-Density Impact

A 2500 kg/m^3 projectile at 15 mps and 45° to surface into sedimentary rock (2500 kg/m^3), the following effects are determined from Collins et al. (2005):

A 100-m stony asteroid is expected to release approximately 40 MT of energy and form a crater slightly over 1 km in diameter. The area of total destruction would extend out another crater diameter at a minimum. At 2-km distance, the shaking is equivalent of IV–V earthquake damage; the wind velocity is 365 mps (900 mph); the overpressure is 41 psi and results in almost total destruction by air blast. At 5-km distance, the seismic energy is slight; ejecta is scarce; the air blast is 90 mps (200 mph); and the overpressure greater than 6 psi causes collapse of multistory, wall-bearing buildings and wood buildings. At 10 km from the impact the air blast 30 mps (less than 70 mph) and overpressure less than 2 psi can cause windows to shatter. The strike would be catastrophic if it occurred in a populated region.

A 1 km meteorite would release nearly 47,000 MT of energy and create a crater 12 km in diameter. Rim ejecta would radiate outward for another 10–15 km, causing total devastation over an area greater than 1000 km^2. At 50 km distance, the seismic effects are equivalent to a 7.6 earthquake causing considerable damage. The air blast of 350 m/s is enough to do nearly complete destruction of man-made structures. The air blast would topple buildings beyond 100 km.

A 5-km meteorite would release 4.4×10^6 MT TNT equivalent and create a 52-km crater with ejecta extending another 50 km. Local seismic magnitude of 9 would be equal to that of the largest occurring earthquake. At 100 km, seismic damage would approach 6 or 7 causing major seismic damage to structures. Near the outer edge of the ejecta blanket, an air blast of 30 psi and wind velocity greater than 1000 mps would cause widespread, near total destruction. At a 200-km distance, seismic damage is slight to moderate, but air blast and wind damage would still be major (most buildings would collapse). At 500 km, the air blast is still major. At a 1000-km distance, no seismic damage is expected and only slight air blast effects (shattered windows) are the major concern. No thermal radiation effects are expected. An area greater than 17,000 km^2 is within the area of influence.

A 10-km impact releases 3.5×10^7 MT TNT equivalent energy and produces a 96.5-km crater. At 500-km distance, the seismic effects are slight to moderate, but widespread air blast effects are significant. At 1000 km there are no seismic effects, but air blast effects remain a problem. At 2000 km, air blast is reduced to shattering windows. The area of total destruction exceeds 70,000 km^2. Except for areas of very low population density, an impact of this size can be expected to approach catastrophic conditions with a large number of fatalities and damage to infrastructure.

14.7.2 High-Velocity and High-Density Impact

By comparison, a high-density meteorite (>3.5 density) at 30 km/s into sediment undergoes less fragmentation; has higher impact energy which increases the area of destruction; and adds the generation of a fireball. The higher impact energy translates into a larger crater; wider range of ejecta; greater area of seismic damage; wider area of air blast effects; the possibility of widespread fire; and greater concentration of dust in the upper atmosphere.

A 100-m meteorite would excavate a 1.7-km-diameter crater. The energy at the impact surface is 48 MT TNT. At 10-km distance from ground zero the seismic disturbance is 5.7, sufficient to do moderate damage to structures. The air blast peak overpressure at almost 12 psi will do as much damage or more than the seismic shaking. At 50 km, neither significant seismic nor air blast effects occur.

A 500-m meteor would release 2.4×10^4 MT and excavate an 11-km crater. The continuous ejecta blank will extend over 20 km from impact center, and scattered missile ejecta may reach beyond 25 km. At 25 km, the seismic damage is nearly complete to most structures, and air blast effects are devastating to almost all structures. Trees and wooden buildings would ignited by a fast-moving fireball. At 50-km distance from impact, seismic energy affects masonry and wood-frame buildings. The air blast with peak overpressure in excess of 9 psi and wind velocity of 124 mps, similar to a tornado, would cause damage to many structures.

A 1-km impactor releases 3.5×10^4 MT TNT at the surface and excavates a 20-km crater. At 25 km from the crater rim, damage from impact seismic and air blast effects would cause nearly total destruction. At 50 km, the fireball would ignite deciduous trees. Seismic effects are equivalent to those of 1908 San Francisco earthquake. The air blast effect would be damaging out to 300 km from the impact site.

A 10-km impactor would release 2×10^8 MT TNT and create a 155-km-diameter crater at a scale similar to the impact that is responsible for the end of the Cretaceous Period. At a 500-km distance, seismic shaking is moderate, but the air blast is still destructive. Fire and major wind damage extends beyond 1000 km. The developing firestorm and dust in the atmosphere would lead to total global cooling for several years and although mammals may have more adaptability than the dinosaur, the probability of a major mass extinction event is very high.

14.8 OCEAN IMPACT

Impact into the ocean is a much more likely scenario as oceans cover 71% of the globe. Ocean impact can be expected to produce major tsunamis (Chesley and Ward, 2005). Their model indicates that waves radiating from the impact of a 300-m-wide asteroid would carry 300 times more energy than the 2004 Asian tsunami. The most common asteroids, between 100 and 400 m, would yield tsunami waves up to 10 m when they arrived at the coast. A total of about 50 million coastal residents are vulnerable to such waves, though no single impact would affect them all. The models predict a tsunami-generating impact should occur about once every 6000 years. It would affect more than one million people and cause $110 billion in property damage. Their analysis suggests that tsunamis are the biggest risk posed by asteroid impacts. The risks of big impacts on climate by atmospheric dust and smoke are about two-thirds that of tsunamis, whereas those of land impacts are about one-third of the tsunami risk.

In contrast, Melosh (2003), based on work by Van Dorn et al. (1968), only recently declassified, reports that impacts by asteroids in the 100–1000 m range do not pose as great a threat as originally presumed. The disruption is primarily in the column of water, producing a large ocean wave. Results of the study indicate that wave height can never exceed 0.39 of the ocean depth; the amplitude decreases with distance traveled; and the waves break at the edge of the continental shelf. Melosh agrees that impacts greater than 1 km are a global threat. It is assumed that impact into a shallow shelf area around the continents could produce large tsunamis. Impacts that punch through shallow water to form craters in the sea floor may produce effects similar to impacts on land. Such impacts maybe catastrophic, at least locally.

14.9 NEW EARTH ASTEROIDS

Collision with some of the debris in Earth-crossing orbits is inevitable. In fact, the Earth receives 90 tons of meteoritic debris each day. Most of this debris is small dust, grain-sized, and basketball-sized material that is abated and slowed in the upper atmosphere. Larger bodies that retain much of their original energy constitute a risk of eventual impact with the Earth.

Recognizing the potential hazard of asteroids and comets striking the Earth has led to a concerted effort to detect and monitor near-Earth objects (NEO). In 1969, while at California Institute of Technology, Eugene Shoemaker began a search for Earth orbit-crossing asteroids using the 18″ Schmidt camera at Palomar Observatory. He is credited with the discovery of many asteroids, and he shares recognition for discovery of the Shoemaker–Levy 9 comet (Marsden, 1993) that impacted Jupiter.

The Spacewatch program was established in 1983 at Kitt Peak, Arizona using the refurbished 0.9-m telescope (Gehrels and Jedicke, 1995, 1996). A 1.8-m telescope was added to the sky survey program in 2001. In 2009, NASA launched the Wide-field Infrared Survey Explorer (WISE), a 0.4-m space telescope to perform an all-sky-survey. In 2011, the survey mission was reconfigured into the NEOWISE extended mission to search for NEOs (asteroids and comets), and in August 2013, the telescope was recommissioned to search for Potentially Hazard Asteroids (PHAs). By December 2012, NASA listed 1360 PHAs. An object is considered a PHA if its minimal orbital insertion distance with respect to Earth is less than 0.05 AU (7,500,000 km—approximately 19.5 lunar distances) and its diameter is at least 100–150 m. NASA estimates that there are 4700 ± 1500 PHAs with a diameter greater than 100 m.

Close encounters include asteroids that come within the orbit of the Moon (384,400 km average distance). The International Astronomical Union lists 24 asteroids smaller than 50 m and 4 greater than 50 m to have passed within one lunar radius since 1914, with most detections in the late twentieth century and early twenty-first century (Figure 14.16). This list is by no means complete.

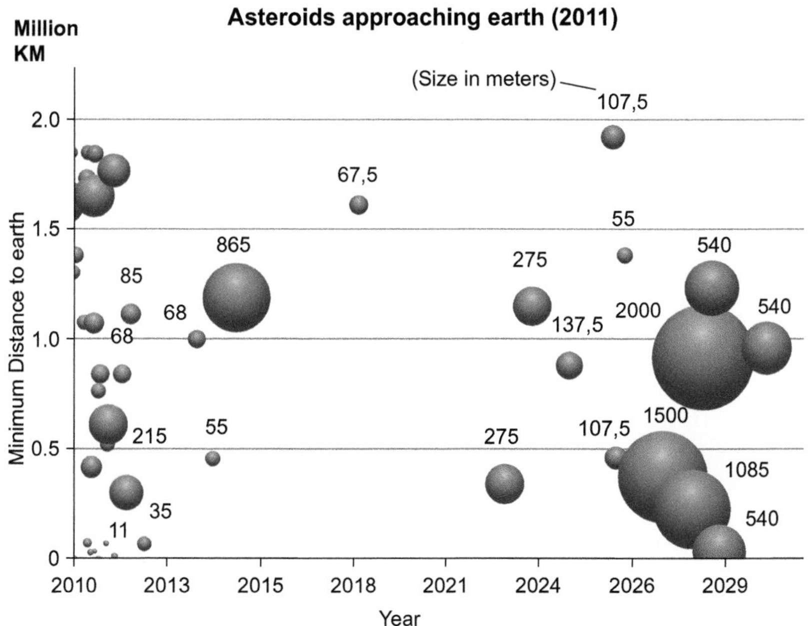

FIGURE 14.16 Pictured are Near-Earth Asteroids that have been identified and their nearest approach to Earth. A possible impact is noted for a 540-m meteor in 2029. Continued observation will refine the orbit. *Source: Nasa.*

14.10 EVALUATION OF IMPACT HAZARD AND MITIGATION

Meteor impact hazards range from purely local to global effects. Local effects are a function of frequency of impacts, size of impactor, and distribution of population. The probability of a significant impact in a populated area is small, considering that 71% of the surface is ocean and more than 50% of the continental land mass is made inhospitable by mountains, deserts, jungles, and glaciers. Thus, only about 15% of the Earth is subject to a potential impact hazard of small to moderate impacts. On the other hand, although a very large event equivalent to the K/T boundary may have a very large reoccurrence time, its effects are global. A large asteroidal impact is one of the very few hazards facing civilization, other than all-out nuclear war which could lead to global catastrophe.

Recent experience and the geologic record stand as evidence that the potential for impact is very real. It is just a matter of time until a large meteor threatens some portion of the population—or civilization itself. The options available for response vary according to the size of the impacting body; the location of the impact; and the lead time from discovery to impact.

A continuation of the Spacewatch program is necessary to identify possible threats with as much lead time as possible; however, we will never be able identify meteors that will cause small impacts. Even large meteorites can arrive undetected. It is assumed that for very large PHAs, a lead time of several years is possible.

14.10.1 Hazard Assessment

How often should we expect an impact to occur? An impact frequency versus time has been prepared for the Moon. After an initial high rate of impacts, the number of impacts decreases as debris is cleared by interception with the planets. A large number of meteoroids enter the Earth's atmosphere every day. Most are small grains that burn up by friction caused by their high-speed passage through the atmosphere. The number of incoming objects is inversely proportional to size. The larger the object, the smaller the frequency. It is estimated that a 1-mm-diameter grain enters the atmosphere every 30 s; a 1-m meteoroid each year; and a 100-m meteoroid every 10,000 years (Figure 14.17). At 15- to 20-m diameter, the Chelyabinsk meteoroid is on the order of a 100-year event. Chapman (2004) came very close to describing the Chelyabinsk event in his description of a hypothetical "one-in-a-century mini-Tunguska atmospheric explosion." Chapman and Morrison (1994) define the threshold of a major catastrophe as one that would kill 25% of the Earth's population, primarily from agricultural losses of an "impact winter." Toon et al. (1997) concluded that the energy range between 10^5 and 10^6 MT is transitional between regional and global effects. A meteoroid capable of causing a worldwide catastrophic event is estimated to be in the range of a

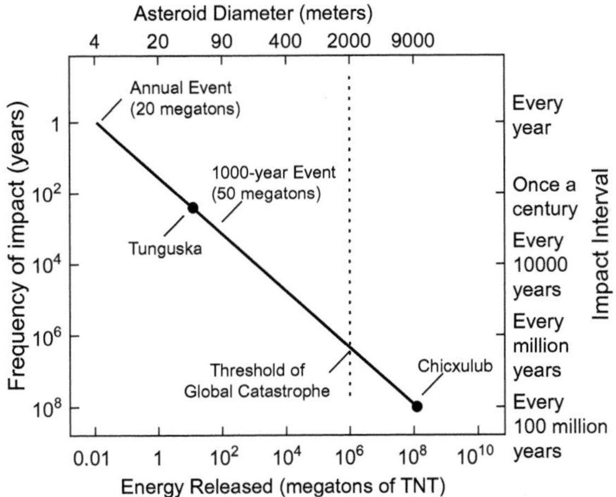

FIGURE 14.17 The frequency of impacts as a function of meteor size.

2-km object with 1 million MT energy. The reoccurrence interval is estimated to be 500,000 years. Morrison et al. (2002) concluded that the greatest risk is associated with large impacts that create a global ecological catastrophe. Despite the large interval between catastrophic events, the large number of casualties brings the chance of death-by-impact into a range comparable to other risks (Table 14.1).

TABLE 14.1 Odds of dying in the United States from selected causes

Cause of Death	Odds of Happening
Motor vehicle accident	1 in 100
Murder	1 in 300
Fire	1 in 800
Firearms accident	1 in 2500
Electrocution	1 in 5000
Asteroid or comet impact	1 in 20,000
Airplane crash	1 in 20,000
Flood	1 in 30,000
Tornado	1 in 60,000
Venomous bite or sting	1 in 100,000
Food poisoning by botulism	1 in 3,000,000

14.10.2 Mitigation

Mitigation for possible impacts depends chiefly on the size of the impact, the length of time of advanced warning, the population of the impact zone, and level of cooperation between responding agencies. The probability of detection of small asteroids similar to those that caused the Tunguska event or craters up to 5 km is extremely low. These events will cause damage to property and infrastructure, similar to that of a large storm or earthquake. An impact in the ocean or in sparsely inhabited terrain would produce minimal effects. Impact in a more densely settled region will cause greater disruptions, but other than a bull's-eye strike on a densely populated center of commerce or government, the effects can largely be dealt with by existing local or regional relief efforts.

Relief for small impacts falls into the general responsibility of local governments and relief agencies. Relief efforts are much the same as required for other local disasters such as earthquakes, tsunamis, and very large storms. Impacts responsible for craters similar to Meteor Crater Arizona may have very little effect over much of the globe, and become important only when occurring in high-density, populated regions.

As the potential impactor size increases, the possibility of prior detection and need for large-scale response increases. Moderately large impacts which create firestorms and short-term climate modifications will require much more massive relief response. Widespread infrastructure and agriculture disruption would require multinational response and coordination. Collateral mortality by famine and disease could reach millions. Very large impacts that might produce widespread fire or dust-laden atmosphere may bring about crop failure over a wide region for a matter of years (Pope et al., 1994).

Preparation for an impact of any size is essentially nonexistent at this time. Surveys to detect and evaluate large bodies are in place, but no plans currently exist for a response to a threat. Presumably once a large asteroid is determined to be on a collision course, the appropriate relief agencies will have several years to respond. With sufficient warning, local or regional governments can decide on a plan of action to mitigate the effects of the impact.

Political and relief bodies must decide whether to attempt preventive measures or to take a more passive attitude. The choices are between preparing a response to take place after an impact; evacuation and resettling populations of the impact zone prior to impact; or attempts to intercept and redirect or destroy the approaching body. No doubt a risk-cost analysis will be considered between evacuation and interception, depending upon impact size and location.

If evacuation is considered as the appropriate response to an impact threat, the resulting displacement of a large part of the population and loss of infrastructure is a major concern. It may be assumed that a small portion of the population will refuse to evacuate. Those that do evacuate will require permanent relocation and reemployment. Examples of small-scale disruptions by evacuation in response to hurricanes or to permanent refugee problems by war provide insight into the potential chaos. International cooperation may be required to stockpile and supply basic needs to a displaced population. Rebuilding infrastructure will place an enormous burden on commercial organizations and governmental agencies.

For very large impacts with planet-wide, near extinction-level possibilities, options are reduced to ones of prevention. If there is a reasonable lead time between discovery and impact, the responsible governing bodies can choose to attempt deflection or destruction of the approaching asteroid. Deflection involves applying external forces to redirect the asteroid to miss the Earth. Total destruction is impossible. Destruction simply reduces the size of the asteroid by fragmentation, allowing smaller pieces to enter the atmosphere and be slowed to below hypervelocity. The surface will then experience a multitude of smaller impacts. For very large potential impacts, all governments must become involved, but the major effort falls on the technologically advanced nations in a cooperative effort to avoid extinction.

Just because the reoccurrence interval for catastrophic impacts is large, does not mean that one cannot occur tomorrow. We have seen "hundred-year floods" happen within a few years of each other. The fate of New Orleans immediately following the 2005 hurricane Katrina notwithstanding, the response to disasters usually brings an immediate response by governmental agencies as well as an outpouring of private volunteer help. Governments and the populace are less responsive to unseen, future risks. We have built large commercial centers in active earthquake zones; towns continue to grow on coasts subject to hurricanes and tsunamis; and people still build homes on the floodplains. Governments are not noted for long-term planning. We can only hope that when a PHA is found and verified that the hazard is not politicized in the manner of global warming.

REFERENCES

Alderman, A.R., 1931. The meteorite craters at Henbury, Central Australia with an addendum by L.J. Spencer. Mineral. Mag. 23, 19–32.

Alroy, J., 2008. Dynamics of origination and extinction in the marine fossil record. Proc. Natl. Acad. Sci. USA 105 (Suppl. 1), 11536–11542.

Alvarez, L.W., Alvarez, W., Asaro, F., Michel, H.V., 1980. Extraterrestrial cause for the Cretaceous–Tertiary extinction. Science 208, 1095–1108.

Barringer, D.M., 1906. Coon mountain and its crater. Proc. Acad. Nat. Sci. Philadelphia 57, 861–886.

Beech, M., Brown, R., Hawkes, R.L., Ceplecha, Z., Mossman, K., Wetherill, G., 1995. The fall of the Peekskill meteorite: video observation, atmospheric path, fragmentation record, and orbit. Earth Moon Planets 68, 189–197.

Benton, M.J., Twitchett, R.J., 2003. How to kill (almost) all life: the end-Permian extinction event. Trends in Ecol. Evol. 18, 358–365.

Blackburn, T.J., Olsen, P.E., Bowring, S.A., McLean, N.M., Kent, D.V., Puffer, J., McHone, G., Rasbury, T., 2013. Zircon U-Pb Geochronology links the end-Triassic extinction with the Central Atlantic Magmatic Province. Science 340, 941–945.

Brown, P.G., 32 co-authors, 2013. A 500-kiloton airburst over Chelyabinsk and an enhanced hazard from small impactors. Nature 503, 238–241.

Bucher, W.H., 1936. Cryptovolcanic structures in the United States (with discussion). In: 16th International Geological Congress, Report, 2, pp. 1055–1084.

Chapman, C.R., 2004. The hazard of near-Earth asteroid impacts on Earth. Earth Planet. Sci. Lett. 222, 1–15.

Chapman, C.R., Morrison, D., 1994. Impacts on the earth by asteroids and comets – assessing the hazard. Nature 367, 33–40.

Chao, E.C.T., Shoemaker, E.M., Madsen, B.M., 1960. First natural occurrence of coesite. Science 132, 220–222.

Chesley, S.R., Ward, S.N., 2005. Impact-generated Tsunami: a quantitative assessment of human and economic hazard. J. Nat. Hazards 38, 355.

Chyba, C.F., Thomas, P.J., Marcus, R.A., 1993. Tunguska explosion: atmospheric disruption of a stony asteroid. Nature 361, 40–44.

Clarke Jr., R.S., Jarosewich, E., Mason, B., Nelen, J., Gomez, M., Hyde, J.R., 1970. The allende, Mexico, meteorite shower. Smithson. Contrib. Earth Sci. 5, 1–53.

Collins, G.S., Melosh, H.J., Marcus, R.A., 2005. Earth impact effects program: a web-based computer program for calculating the regional environmental consequences of a meteoroid impact on Earth. Meteorit. Planet. Sci. 40 (6), 817–840.

Covey, C., Thompson, S.L., Weissman, P.R., MacCracken, M.C., 1994. Global climatic effects of atmospheric dust from an asteroid or comet impact on Earth. Glob. and Planet. Change 9, 263.

Dietz, R.S., 1959. Shatter cones in cryptoexplosion structures (meteorite impact?). J. Geol. 67 (5), 496–505.

D'Orazio, M., 2007. Meteorite Records in the Ancient Greek and Latin Literature: Between History and Myth, 273. Geological Society Special Publications, London, 215–225.

Engvild, K.C., 2003. A review of the risks of sudden global cooling and its effects on agriculture. Agric. and For. Meteorol. 115, 127–137.

Erwin, D.H., 1990. The end-Permian mass extinction. Annu. Rev. Ecol. Syst. 21, 69–91.

Fairchild, H.L., 1907. The origin of Meteor Crater (Coon Butte), Arizona. Bull. Geol. Soc. Amer. 613, 493–504.

Farinella, P., Foschini, L., Froeschl, C., Gonczi, R., Jopek, T.J., Longo, G., Michel, P., 2001. Probable asteroidal origin of the Tunguska cosmic Body. Astron. Astrophys. 377, 1081–1097.

French, B.M., 1998. Traces of Catastrophe: A Handbook of Shock-Metamorphic Effects in Terrestrial Meteorite Impact Structures. LPI Contribution No. 954, Lunar and Planetary Institute, Houston, 120 pp.

Gallant, R., February 1996. Sikhote-alin Revisited. Meteorite magazine. Pallasite Press.

Gehrels, T., Jedicke, R., 1996. Detection of near earth asteroids based upon their rates of motion. Astron. J. 111, 970–982.

Gehrels, T., Jedicke, R., 1995. The population of near-earth objects discovered by spacewatch. Earth Moon Planets 72, 233–242.

Gilbert, G.K., 1896. The origin of hypotheses, illustrated by the discussion of a topographic problem. Science 53, 1–13.

Hodge, P.W., 1994. Meteor craters and impact structures of the Earth. Cambridge University Press, 136p.

Holliday, V.T., Kring, D.A., Mayer, J.H., Goble, R.J., 2005. Age and effects of the Odessa meteorite impact, western Texas, USA. Geology 33, 945–948.

Jackson IV, A.A., Ryan Jr., M.P., 1973. Was the Tungus event due to a Black hole? Nature 245, 88–89.

Jackson, L.E., Brown, P., Melosh, J., Hill, D., 2008. Meteorite strikes Peru. Geotimes 6, 11.

Jenniskens, P., Betlem, H., Betlem, J., Barifaijo, E., Schlüter, T., Keller, C.G., Kerr, A.C., MacLeod, N., 2013. Exploring the causes of mass extinction events. Eos Trans. AGU 94, 200.

Jenniskens, P., Betlem, H., Betlem, J., Barifaijo, E., Schluter, T., Hampton, C., Laubenstein, M., Kunz, J., Heusser, G., 1994. The Mbale meteorite shower. Meteoritics 29, 246–254.

Kridec, E.L., 1963. The Tunguska and Sikhote-Alin meteorites. In: Middlehurst, B., Kuiper, G.P. (Eds.), The Solar System, Vol. IV: Moon, Meteorites and Comets. University of Chicago Press, Chicago, pp. 208–217.

Kvasnytsya, V., Wirth, R., Dobrzhinetskaya, L., Matzel, J., Jacobsen, B., Hutcheon, I., Tappero, R., Kovalyukh, M., 2013. New evidence of meteoritic origin of the Tunguska cosmic body. Planet. Space Sci. 84, 131–140.

Kuzmin, A., 2013. Meteorite explodes over Russia, more than 1000 injured. Reuters. http://www.reuters.com/article/2013/02/15/us-russia-meteorite-idUSBRE91E05Z20130215.

Marsden, B.G., 1993. Comet Shoemaker-Levy (1993e). In: International Astronomical Union Circular 5725.

Marvin, U.B., 1992. The meteorite of Ensisheim: 1492 to 1992. Meteoritics 27, 28–72.

Mercury (news), 1913. Struck by Meteorite; Barque Belfast Damages. Mercury, Hobart, Tasmania. Wednesday, January 8, pp.5.

Melosh, H.J., 2003. Impact-generated tsunamis: an over-rated hazard. In: Lunar and Planetary Institute Science Conference Abstracts, 34, pp. 2013–2014.

Melosh, H.J., 1989. Impact Cratering: A Geologic Process. Oxford University Press, NY, pp. 253.

Morrison, D., Harris, A.W., Sommer, G., Chapman, C.R., Carusi, A., 2002. Dealing with the impact hazard. Asteroids III, 739–756.

Nininger, H., 1954. Impact slag at Barringer crater. Am. J. Sci. 252, 277–290.

Pickering, W.H., 1909. The chance of collision with comets, iron meteorites, and Coon Butte. Pop. Astron. 17, 329–339.

Pope, K.O., Baines, K.H., Ocampo, A.C., Ivanov, B.A., 1994. Impact Winter and the cretaceous-tertiary extinctions—results of a Chicxulub asteroid impact model. Earth Planet. Sci. Lett. 128, 719–725.

Preston, F.W., Henderson, E.P., Randolph, J.R., 1941. The chicora (Butler County, Pa.) meteorite. Proc. US Nat. Mus. 90, 387.

Popova, O.P., 58 co-authors, 2013. Chelyabinsk airburst, damage assessment, meteorite recovery, and characterization. Science 342, 1069–1073.

Rajmon, D., 2009. Impact Database, 2010.1. Online: http://impacts.rajmon.cz.

Raup, D.M., Sepkoski Jr., J.J., 1982. Mass extinctions in the marine fossil record. Science 215, 1501–1503.

Reeves, F., Chalmers, R.O., 1949. The wolf creek crater. Australian J. Sci. 11, 145–156.

Retallack, G.J., 1995. Permian–Triassic life crisis on land. Science 267, 77–80.

Robertson, R.B., Grieve, R.A.F., 1975. Impact structures in Canada: their recognition and characteristics. Royal Astron. Soc. Canada 69, 1–21.

Roddy, D.J., Shoemaker, E.M., 1995. Meteor crater (Barringer meteorite crater), Arizona: summary of impact conditions. Meteoritics 30, 567.

Rubin, A.E., Grossman, J.N., 2010. Meteorite and meteoroid: new comprehensive definitions. Meteorit. Planet. Sci. 45, 114–122.

Schulte, P., 29 co-authors, 2010. The Chicxulub asteroid impact and mass extinction at the cretaceous-Paleogene boundary. Science 327, 1214–1218.

Shoemaker, E.M., Kieffer, S.W., 1979. Guidebook to the Geology of Meteor Crater, Arizona. Center for Meteorite Studies, Arizona State University, Tempe, AZ, pp. 45.

Shoemaker, E.M. Impact Mechanics at Meteor Crater Arizona. Princeton University, pp. 55 (Unpublished PhD Thesis).

Silliman, B., 1874. American Contributions to Chemistry: An Address Delivered on the Occasion of the Celebration of the Centennial of Chemistry, at Northumberland, PA. August 1, 1874. Yale College (reprinted from the American Chemist for August-September and December 1874.).

Singh, A., Cain, M.L., 2011. Discover Biology. W.W. Norton and Company, pp. 795.

Sole, R.V., Newman, M., 2002. Extinctions and Biodiversity in the Fossil Record – Volume Two, The Earth System: Biological and Ecological Dimensions of Global Environment Change. Encyclopedia of Global Environmental Change. John Wilely and Sons, pp. 297–391.

Spratt, C.E., 1991. Possible hazards of meteorite falls. J. Royal Astron. Soc 85, 262–280.

Suggs, R.M., Cooke, W.J., Suggs, R.J., Swift, W.R., Hollon, N., 2008. The NASA lunar impact monitoring program. Earth Moon Planets 102, 293–298.

Toon, O.B., Zahnle, K., Morrison, D., Turco, R.P., Covey, C., 1997. Environmental perturbations caused by the impacts of asteroids and comets. Rev. Geophys. 35, 41–78.

Turco, R.P., Toon, O.B., Ackerman, T.P., Pollack, J.B., Sagan, C., 1983. Nuclear, winter: global consequences of multiple nuclear explosions. Science 222, 1283–1292.

Van Dorn, W.G., LeMehaute, B., Hwang, Li-san, 1968. Handbook of Explosion Generated Water Waves. Vol. 1 State of the Art. Tetra Tech Report TC-130. Office of Naval Research, pp. 166.

Wignall, P.B., 2005. Volcanism and mass extinctions. In: Marti, J., Ernst, G. (Eds.), Volcanoes and the Environment, Cambridge. Cambridge University Press, pp. 207–226.

Yau, K., Weissman, P., Yeomans, D., 1994. Meteorite falls in China and some related human casualty events. Meteoritics 29, 864–871.

Zhang, M., 2013. Russia Meteor 2013; Damage to Top $33 Million; Rescue, Cleanup Team Heads to Meteorite-hit Urals. International Business Times. http://www.ibtimes.com/russia-meteor-2013-damage-top-33-million-rescue-cleanup-team-heads-meteorite-hit-urals-1090104.

Index

Note: Page numbers followed by "f", "t" and "b" indicates figures, tables and boxes, respectively.

A

B

E

F

G

H

I

K

L

M

N

O

P

Q

R

S

T